Body Sensations: The Conscious Aspects of Interoception

Editor: Ivy Fuller

AMERICAN
MEDICAL PUBLISHERS
www.americanmedicalpublishers.com

AMERICAN
MEDICAL PUBLISHERS
www.americanmedicalpublishers.com

Cataloging-in-Publication Data

Body sensations : the conscious aspects of interoception / edited by Ivy Fuller.
p. cm.
Includes bibliographical references and index.
ISBN 978-1-63927-938-8
1. Somesthesia. 2. Senses and sensation. 3. Interoception. 4. Perception. 5. Neuropsychology.
6. Psychophysiology. 7. Psychiatry. I. Fuller, Ivy.
QP448 .B63 2023
612.8--dc23

American Medical Publishers,
41 Flatbush Avenue,
1st Floor, New York,
NY 11217, USA

ISBN 978-1-63927-938-8 (Hardback)

Contents

Preface

The world is advancing at a fast pace like never before. Therefore, the need is to keep up with the latest developments. This book was an idea that came to fruition when the specialists in the area realized the need to coordinate together and document essential themes in the subject. That's when I was requested to be the editor. Editing this book has been an honour as it brings together diverse authors researching on different streams of the field. The book collates essential materials contributed by veterans in the area which can be utilized by students and researchers alike.

Interoception is the sense of the internal state of the body. It is the process of receiving, accessing, and evaluating internal bodily signals, which can be conscious and non-conscious. It is critical for maintaining homeostatic conditions in the body. Interoception is presented in relation to well-being, since it is assumed to be intimately linked to self-regulation. It is developed to assist organisms in maintaining homeostasis. Emotionally balanced body signals play a role in larger mood states, which supports emotional balance in modern lifecycle. The level to which a person is sensitive towards the interoceptive signals, affects the decision making capability of that person. Although, high sensitivity comes at a cost, increased interoceptive sensation can contribute to feelings of anxiety. Therefore, interoceptive sensitivity can either contribute to or detract from happiness, which implies the need for direction in modulating significant visceral signals. This book aims to shed light on some of the unexplored conscious aspects of interoception. Those in search of information to further their knowledge will be greatly assisted by it.

Each chapter is a sole-standing publication that reflects each author´s interpretation. Thus, the book displays a multi-facetted picture of our current understanding of application, resources and aspects of the field. I would like to thank the contributors of this book and my family for their endless support.

Editor

1

Heartfelt Empathy? No Association between Interoceptive Awareness, Questionnaire Measures of Empathy, Reading the Mind in the Eyes Task or the Director Task

Vivien Ainley, Lara Maister and Manos Tsakiris*

Lab of Action and Body, Department of Psychology, Royal Holloway University of London, Egham, UK

***Correspondence:**
*Vivien Ainley,
Lab of Action and Body, Department of Psychology, Royal Holloway University of London, Egham, Surrey TW20 0EX, UK
vivien.ainley.2008@live.rhul.ac.uk*

Interoception, defined as afferent information arising from within the body, is the basis of all emotional experience and underpins the 'self.' However, people vary in the extent to which interoceptive signals reach awareness. This trait modulates both their experience of emotion and their ability to distinguish 'self' from 'other' in multisensory contexts. The experience of emotion and the degree of self/other distinction or overlap are similarly fundamental to empathy, which is an umbrella term comprising affect sharing, empathic concern and perspective-taking (PT). A link has therefore often been assumed between interoceptive awareness (IA) and empathy despite a lack of clear evidence. To test the hypothesis that individual differences in both traits should correlate, we measured IA in four experiments, using a well-validated heartbeat perception task, and compared this with scores on several tests that relate to various aspects of empathy. We firstly measured scores on the Index of Interpersonal Reactivity and secondly on the Questionnaire of Cognitive and Affective Empathy. Thirdly, because the 'simulationist' account assumes that affect sharing is involved in recognizing emotion, we employed the 'Reading the Mind in the Eyes Task' for the recognition of facial expressions. Contrary to expectation, we found no significant relationships between IA and any aspect of these measures. This striking lack of direct links has important consequences for hypotheses about the extent to which empathy is necessarily embodied. Finally, to assess cognitive PT ability, which specifically requires self/other distinction, we used the 'Director Task' but found no relationship. We conclude that the abilities that make up empathy are potentially related to IA in a variety of conflicting ways, such that a direct association between IA and various components of empathy has yet to be established.

Keywords: empathy, perspective-taking, emotion recognition, interoception, heartbeat perception

Introduction

Empathy is an essential aspect of human emotional experience and social interaction (Panksepp, 1998; Preston and de Waal, 2002; Gonzalez-Liencres et al., 2013). It is, however, a notoriously difficult concept to define and operationalise (Singer and Lamm, 2009; Decety and Cowell, 2014).

We adopt in this paper a definition of empathy as an umbrella term comprising 'affect sharing,' otherwise known as 'Emotion Contagion' [which can lead to 'personal distress (PD)'], 'empathic concern (EC)' (defined as the motivation to care for others) and 'perspective-taking (PT),' which is putting oneself in the other's emotional shoes (Bernhardt and Singer, 2012; Decety and Cowell, 2014).

Of these three components, affect sharing is assumed to be an automatic process, whereby perceiving or imagining another person in a particular emotional state activates the same state in the observer, producing similar autonomic and somatic responses (Preston and Hofelich, 2012), such that the emphasiser 'feels with' the person in distress (Singer and Lamm, 2009).

'Empathic concern,' which is also called 'sympathy' or 'compassion,' involves 'feeling for' the other person (Singer and Lamm, 2009) and is associated with motivation to alleviate their suffering (Singer and Lamm, 2009; Decety and Cowell, 2014). This construct is frequently equated with empathy, although it does not necessarily involve any affect sharing (Bernhardt and Singer, 2012).

During 'PT' the observer consciously puts herself in the shoes of the person observed (Jackson et al., 2006; Lamm et al., 2007). This is closely related to theory of mind – the distinction being that during the PT involved in empathy the observer must understand the feelings of another person, whereas theory of mind is defined as understanding the other's beliefs (Lawrence et al., 2004; Preston and Hofelich, 2012).

All three components of empathy involve a process whereby the empathiser represents both 'self' and 'other' (the person observed). This process may be regarded as a continuum. Overlapping representations of self and other are involved in affect sharing. By contrast, accurate self/other distinction (between one's own and the other's viewpoint) is required for PT (Bird et al., 2010). It is unclear how self/other overlap or self/other distinction impacts on EC. On the one hand EC may be a function of previous affect sharing and therefore involve self/other overlap. However, it might equally depend on good self/other distinction if, for example, affect sharing leads to PD which may interfere with the motivation to help (Singer and Klimecki, 2013).

A link between empathy and 'interoceptive awareness (IA)' can be predicted, on the basis of both affect sharing and self/other distinction (or overlap), where IA is defined as the extent to which internal bodily cues reach awareness and influence behavior, feelings and cognition. An extensive literature shows that IA influences the intensity of an individual's own emotional experience (Barrett et al., 2004; Pollatos et al., 2005; Wiens, 2005). For example, people with high IA report more emotional arousal for identical changes in objective physiological indices (Wiens, 2005; Pollatos and Schandry, 2008; Dunn et al., 2010). This implies that high trait IA could be associated with greater affect sharing, because the shared emotion is potentially more intense than it would be in an individual with low IA. Moreover, self/other distinction (or overlap), which is also crucial to empathy, is modulated by IA, specifically in embodied contexts. Individuals with good IA are less susceptible to body ownership illusions (Tsakiris et al., 2011; Tajadura-Jiménez and Tsakiris, 2013) and when interoceptive cues are projected onto a virtual image of their own body, or hand, people have a greater sense of self-identification with, and self-location toward, that image (Aspell et al., 2013; Suzuki et al., 2013). This suggests that IA is associated with better self/other distinction and will therefore correlate positively with PT, although a relationship with EC cannot be unambiguously predicted.

The small number of studies that have attempted to link interoception with empathy provide inconclusive evidence because of differences in the way that both empathy and IA have been operationalised and measured. For example, Fukushima et al. (2011) reported that the amplitude of heartbeat-evoked potentials (HEPs; a potential index of interoceptive cortical processing) was higher when participants judged emotion in a pictured face, compared with judgments of facial symmetry. Ernst et al. (2013) found that when participants were asked to empathize with emotional facial expressions after a period of explicit interoceptive attention, neural activity was enhanced in a number of brain regions involved in interoception and self-processing, compared with a control task of counting exteroceptive tones. These two experiments, however, used indirect tests of *both* IA and empathy. Other studies that compared the two traits using conventional heartbeat perception measures of IA reported conflicting results. Tajadura-Jiménez and Tsakiris (2013) reported correlations between IA and both the PT and Fantasy (FS) subscales of the Interpersonal Reactivity Index (IRI; Davis, 1983). Both of these measures suggest better self/other distinction in people with high IA but the relationships were significant only at the 10% level (Tajadura-Jiménez and Tsakiris, 2013). More recently, Grynberg and Pollatos (2015) found that participants with good IA, who viewed pictures of people in painful situations, rated the pain as greater and felt more compassion for the sufferer, although they reported no greater PD than controls. This result is potentially attributable to the greater emotional arousal also reported by the participants with high IA, which is typical of good heartbeat perceivers and could reflect affect sharing.

Given these conflicting results, we devised a set of experiments in an attempt to elucidate various strands of the potential relationship between IA and the components of empathy, measuring the former by the Mental Tracking Method of heartbeat perception (Schandry, 1981). Empathy has been assessed in many different ways. The measures chosen in the four studies described in this paper were selected as widely used measures of affect sharing, EC and PT and for being relevant to previous empirical research purporting to show links between interoception and empathy.

We first chose the most prominent self-report empathy scale, which is the IRI (Davis, 1983). Its subscales include EC, PT and PD, all of which refer to aspects of empathy that have been described in theoretical accounts. The EC subscale of the IRI has been particularly widely used as a general empathy measure (for a review see Bernhardt and Singer, 2012). The IRI, however, possess no specific measure of affect sharing. We therefore selected as our second measure the recently developed Questionnaire of Cognitive and Affective Empathy (QCAE; Reniers et al., 2011) which includes a subscale for

'Emotion Contagion' (affect sharing) and has the additional merit of effectively summarizing several common self-report empathy measures.

Our third measure was also designed to assess affect sharing, reflecting the proposition that people understand and interpret emotional facial expressions by automatically simulating (i.e., by empathizing with) the observed expression (Preston and de Waal, 2002; Kaplan and Iacoboni, 2006). Although this 'simulation account' has been disputed (Gallese and Sinigaglia, 2011), it implies that people with high IA, who experience their own emotions more strongly, are likely to perform well when recognizing facial expressions in the 'Reading the Mind in the Eyes' test (Baron-Cohen et al., 2001). A further reason for choosing this measure was that the two prominent studies, described above, that report significant relationships between empathy and interoception have used similar facial stimuli (Fukushima et al., 2011; Ernst et al., 2013). Furthermore, it has been reported that people with high IA have an advantage in recognizing the presence of emotion in faces, which might imply that they are sharing affect by mirroring the observed emotion (Terasawa et al., 2014).

Finally, as a direct test of whether people with high IA are better at separating self from other in domains other than multisensory integration, we employed the 'Director Task,' which measures cognitive PT ability (Santiesteban et al., 2012).

The four experiments were approved by the Ethics Committee of the psychology department at Royal Holloway University of London. All participants gave written informed consent and were free to withdraw at will.

Experiment 1

Method

Participants

Ninety students at Royal Holloway University of London took part for course credit. The data for four participants was excluded because movement artifacts made it impossible to count the number of recorded heartbeats. Of the remaining 86 (14 men), mean age was 20.4 years (SD = 6.5).

Procedure

Participants completed the IRI after the heartbeat perception task.

Interoceptive Awareness (IA)

Interoceptive awareness was measured using the Mental Tracking Method of heartbeat perception (Schandry, 1981). Heartbeat perception methods correlate with awareness of gastric cues (Whitehead and Drescher, 1980; Herbert et al., 2012). The Mental Tracking Method has been extensively used in research on emotion (Dunn et al., 2007; Herbert et al., 2010; Werner et al., 2010; Pollatos et al., 2012). It is well-validated, with good test–retest reliability and discriminates well between individuals (Mussgay et al., 1999; Werner et al., 2013). Throughout this paper 'IA' refers to the accuracy with which participants were able to count their own heartbeats on the Mental Tracking task (Cuenen et al., 2012; Garfinkel and Critchley, 2013).

Participants were seated in a comfortable chair and given several minutes to relax, in preparation for the task. All instructions were delivered and behavioral responses were recorded using Presentation software (Neurobehavioral Systems, Albany, CA, USA) on a standard desktop PC. Instructions were presented over noise-attenuating headphones. The onset and offset of each heartbeat counting trial were cued by the words *'go'* and *'stop,'* presented audiovisually. Results are sensitive to the instructions given (Ehlers et al., 1995) so a standard instruction was used, whereby participants were asked to concentrate hard and try to silently count their own heartbeats, simply by *'listening'* to their bodies, without taking their pulse. Heartbeat signals were acquired with a piezo-electric pulse transducer, fitted to the participant's left index finger and connected to a physiological data unit (26T PowerLab, AD Instruments), sampling at 1 kHz, which recorded the derived electrical signal onto a second PC running LabChart6 software (AD Instruments). Three trials (25, 35, 45 s) were presented in random order, after one training interval of 15 s. No feedback was given. IA was calculated as [1/3Σ (1-(|recorded heartbeats – counted heartbeats| /recorded heartbeats); Schandry, 1981)]. Higher scores indicate higher IA. Heart rate is known to correlate with IA and this was also recorded during the heartbeat perception task (Cameron, 2001; Knapp-Kline and Kline, 2005; Ainley et al., 2012).

The Interpersonal Reactivity Index

The IRI (Davis, 1983) is a prominent, well-validated and reliable measure of trait empathy, containing four 7-item subscales. These are EC, PT, PD, and Fantasy. Typical items are: for EC *'I often have tender, concerned feelings for people less fortunate than me'*; for PT *'I believe that there are two sides to every question and try to look at them both'*; for PD *'When I see someone who badly needs help in an emergency, I go to pieces'*; and for Fantasy *'I really get involved with the feelings of the characters in a novel.'* Each item is scored on a five-point Likert scale from 0 for *'does not describe me well'* to 4 *'describes me very well.'*

Results

The mean values and SD obtained were close to published norms (**Table 1**).

TABLE 1 | Descriptive statistics for the Interpersonal Reactivity Index (IRI) and its subscales – empathic concern (EC), personal distress (PD), perspective-taking (PT) and fantasy (FS).

	IRI	EC	PD	PT	FS
Mean (SD)	65.8 (15.0)	19.9 (4.6)	11.7 (5.6)	17.2 (4.9)	16.8 (4.9)
Norms (a)		20.36 (4.02)	10.87 (4.78)	17.37 (4.79)	

(a) Norms value provided by Davis as personal communication, reported in Belllini et al. (2002).

Correlations between IA and the IRI scores are shown in **Table 2**. Bonferroni corrections for multiple comparisons were applied with an alpha value of 0.01. There were no significant correlations.

Mean IA was 0.64 (SD = 0.18). IA was negatively correlated with the participants' heart rate, $r = -0.47, p < 0.001$. Heart rate was not significantly correlated with any IRI measure.

In this sample there were significant differences in mean IRI scores between genders, $F(1,84) = 11.7, p = 0.001$. However, the small number of men amongst the participants makes this comparison unreliable. Correlations for the 72 women participants only were therefore also calculated. An apparent negative correlation between IA and Fantasy did not survive Bonferroni correction, with an alpha level of 0.01. No other correlation was significant.

Discussion of Experiment 1

We investigated the relationship between a standard measure of IA and the most commonly used self-report measure of empathy. There were no significant correlations between IA and the IRI as a whole or with any of its subscales. The only relationship that approached significance was a negative correlation with Fantasy, suggesting that people with high IA tend not put themselves in the position of characters in books and films, which might perhaps reflect good self/other distinction. However, this did not survive Bonferroni correction.

Although the IRI has the advantage of distinct, separable subscales that refer to several theoretically based components of empathy, direct mapping of each subscale onto a particular aspect of empathy has been criticized (Singer and Lamm, 2009). The EC score is a very widely used measure of empathy but did not correlate with IA in this study. In order to verify the null results found in Experiment 1, we selected a further self-report measure – the QCAE (Reniers et al., 2011), which, unlike the IRI, has a specific subscale designed to capture affect sharing.

Experiment 2

Method

Participants

Thirty-four students (five men), mean age 20.2 years (SD = 2.9), at Royal Holloway University of London, took part for payment. There were no exclusions.

Procedure

Interoceptive awareness was measured as in Experiment 1. Participants completed the QCAE after the heartbeat perception task.

Questionnaire of Cognitive and Affective Empathy

The QCAE (Reniers et al., 2011) is an attempt to capitalize on the strengths of several commonly used self-report measures. It is comprised of items from the Empathy Quotient (Lawrence et al., 2004), the Hogan Empathy Scale (Hogan, 1969), the Empathy subscale of the Impulsiveness-Venturesomeness-Empathy Inventory (Esysenck and Eysenck, 1978), and the IRI (Davis, 1983). The scale consists of 31 items to which participants respond on a four-point Likert scale from *'strongly disagree'* to *'strongly agree.'* 'Cognitive Empathy' is made up of two subscales which are (i) 'PT' (defined as 'Intuitively putting oneself in another person's shoes'), assessed by 10 items, e.g., *'I can easily work out what another person might want to talk about'*; and (ii) 'Online Simulation' (defined as 'An effortful attempt to put oneself in another person's position by imagining what that person is feeling'), assessed by nine items, e.g., *'I try to look at everyone's side of a disagreement before I make a decision.'* By contrast, 'Affective Empathy' is made up of three subscales, which are (iii) 'Emotion Contagion' (defined as 'The automatic mirroring of the feelings of others'), assessed by four items, e.g., *'People I am with have a strong influence on my mood'*; (iv) 'Proximal Responsivity' (defined as 'Affective responses when witnessing the mood of others in a close social context'), assessed by four items, e.g., *'I am very unhappy when I see someone cry'*; and (v) Peripheral Responsivity (defined as 'Similar to Proximal Responsivity but in a detached context'), assessed by four items, e.g., *'I usually stay detached when watching a film.'*

Results

The mean values for the QCAE scales were close to published norms (**Table 3**).

Correlations between IA and the QCAE scores are shown in **Table 4** for all participants and for the 29 women participants alone. IA was not significantly correlated with the whole QCAE scale nor with any of the subscale measures.

Mean IA was 0.61 (SD = 0.16). IA was negatively correlated with average resting heart rate, for all participants, $r = -0.37$, $p = 0.03$, and for the women participants alone, $r = -0.45$, $p = 0.02$, as in Experiment 1. Correlations between heart rate and scores on the QCAE and its subscales were non-significant (Bonferroni corrections with an alpha value of 0.006 were applied).

Discussion of Experiment 2

No links were found between IA and any scale of the QCAE. In particular, Emotion Contagion (which measure affect sharing) was not related to IA. The results of Experiment 2 thus confirm those of Experiment 1 and might be explained by the general weakness of self-report measures. Questionnaires

TABLE 2 | Correlations between interoceptive awareness (IA), the Interpersonal Reactivity Index (IRI) and its subscales.

	IRI	EC	PD	PT	FS
Correlations with IA (all participants) *(n = 86)*	$r = -0.10, p = 0.36$	$r = 0.002, p = 0.98$	$r = -0.12, p = 0.28$	$r = 0.10, p = 0.34$	$r = -0.23, p = 0.03$*
Correlations with IA (women only) *(n = 72)*	$r = -0.14, p = 0.23$	$r = 0.04, p = 0.75$	$r = -0.08, p = 0.50$	$r = 0.05, p = 0.66$	$r = -0.27, p = 0.02$*

**non-significant after Bonferroni correction.*

TABLE 3 | Descriptive statistics for the Questionnaire of Cognitive and Affective Empathy (QCEA) and its subscales – cognitive empathy (Cog Emp) perspective-taking (PT), online simulation (Sim), affective empathy (Affect Emp), emotion contagion (Emot Con), proximal responsivity (Prox Res) and peripheral responsivity (Periph Res).

	QCAE	Cog Emp			Affect Emp			
			PT	Sim		Emot Con	Prox Res	Periph Res
Women Mean (SD)	95.6 (7.5)	59.6 (5.4)	31.9 (3.2)	27.7 (3.0)	35.9 (4.7)	11.9 (2.0)	12.5 (2.0)	11.6 (2.8)
Men Mean (SD)	84.6 (7.6)	55.4 (6.1)	28.8 (4.1)	26.6 (2.4)	29.2 (4.7)	9.0 (3.3)	10.2 (2.4)	10.0 (1.0)
Norms for Women (a)		59.4 (6.3)			36.7 (4.3)			
Norms for Men (a)		56.1 (10.5)			32.3 (6.3)			

(a) Reported by Reniers et al. (2011) for University students, aged 20–30, n = 925.

TABLE 4 | Correlations between interoceptive awareness (IA) and the Questionnaire of Affective and Cognitive Empathy (QCAE) and its subscales.

	QCAE	Cog Emp			Affect Emp			
			PT	Sim		Emot Con	Prox Resp	Periph Resp
Correl. with IA $n = 34$	$r = 0.09$ $p = 0.63$	$r = 0.07$ $p = 0.70$	$r = 0.18$ $p = 0.32$	$r = -0.08$ $p = 0.69$	$r = 0.06$ $p = 0.73$	$r = 0.13$ $p = 0.47$	$r = 0.01$ $p = 0.98$	$r < 0.01$ $p = 0.99$
Correl. with IA (women) $n = 29$	$r = -0.02$ $p = 0.91$	$r = -0.05$ $p = 0.79$	$r = 0.09$ $p = 0.64$	$r = -0.19$ $p = 0.32$	$r = 0.02$ $p = 0.91$	$r = 0.16$ $p = 0.42$	$r = -0.01$ $p = 0.96$	$r = -0.07$ $p = 0.73$

can only reflect what people think they feel, rather than what they would truly feel in any given context. Such measures are open to the additional confound that they may index what is socially desirable, rather than recording the respondent's true feelings (Singer and Lamm, 2009). The two further tests we used require judgments that avoid these two confounds. Given that several of the studies that have linked empathy and interoception have used tasks involving the appraisal of facial expressions which, it has been argued, involves affect sharing (Gallese and Sinigaglia, 2011), we next investigated whether scores on the 'Reading the Mind in the Eyes' test are modulated by IA.

Experiment 3

Method

Participants

One hundred and thirteen adult members of the public volunteered to participate in this study, as part of the Live Science installation at the Science Museum London. Sixteen participants were excluded before data analysis, due to incomplete data, movement artifacts or insufficient attention to the tasks, which left 97 participants (40 men) in the final data set. The mean age of these participants was 31.0 years (SD = 12.0).

Procedure

Interoceptive awareness was measured as in Experiment 1. Participants completed the Reading the Mind in the Eyes test after the heartbeat perception task.

'Reading the Mind in the Eyes' Task

The Reading the Mind in the Eyes Task is a commonly used test that is the product of research into empathy deficits in autism. The procedure was developed by Baron-Cohen et al. (2001). Participants are seated at a standard PC and shown a series of 36 images of faces cropped around the eyes. For each image, participants are provided with four single words to describe the possible emotions that the eyes could be displaying (e.g., serious, ashamed, alarmed, or bewildered). They are required to choose which description best matches the emotion displayed in the image. The experiment is self-paced and takes about 7 min.

Results

Interoceptive awareness was calculated as in Experiment 1. The mean IA score of this sample was 0.67 (SD = 0.17). As often reported, average heart rate was significantly negatively correlated with IA, $r = -0.43, p < 0.001$.

Performance on the Reading the Mind in the Eyes Task was assessed by calculating a simple accuracy score that indicated the proportion of the total trials in which the participant identified the correct emotion. Mean accuracy score was 0.74 (SD = 0.09). There were no gender differences in accuracy, $t(95) = -0.36$, $p = 0.72$. To identify links between IA and emotion recognition, a Pearson's correlation coefficient was calculated between IA scores and Reading the Mind in the Eyes scores but this was non-significant, $r = -0.10, p = 0.36$.

Discussion of Experiment 3

Although evidence suggests that individuals with high IA experience their own emotions as more intense than those with lower IA (Wiens et al., 2000), our findings replicate those of Hanford and colleagues who reported that this does not confer any benefits when people are required to recognize *others'* emotions in the Reading the Mind in the Eyes test (Handford et al., 2013). Their experiment, however, used a heartbeat discrimination task based on the Method of Constant Stimuli in which the majority of people score at chance (Lenggenhager et al., 2013; Eshkevari et al., 2014). We had expected that IA measured by the Mental

Tracking Method (Schandry, 1981) might be a more successful index because this test produces a distribution of scores that is approximately Gaussian (Ainley et al., 2012). Terasawa's task, using full faces and morphs containing variable amounts of emotion, also implies that there is a link between IA and recognition of emotion in faces but their results were significant only when comparing the performance of the 10 best and 10 worst heartbeat perceivers (Terasawa et al., 2014).

If the simulation account is correct in claiming that the identification of emotion in another person's face relies on resonating with the observed emotion (Gallese, 2007; Gallese and Sinigaglia, 2011), then the findings of this experiment casts doubt on the hypothesis that interoception is linked to empathy. Alternatively the Reading the Mind in the Eyes test itself may be performed by theory of mind and therefore not involve the observer resonating with the observed emotion.

For our fourth and final experiment we tested IA against the 'Director Task,' which assesses cognitive PT (which requires self/other distinction). Research into bodily illusions has suggested that individuals with high IA may more clearly distinguish between their own and other's perspectives in multisensory contexts (Tsakiris et al., 2011; Aspell et al., 2013). We wished to determine whether this advantage would translate to the Director Task, given that putting oneself in another's *emotional* shoes is a facet of empathy.

Experiment 4

Method

Participants

Sixty-six adults volunteered for this study, as part of the Live Science installation at the Science Museum, London. Six participants were excluded before data analysis due to incomplete data and movement artifacts, which left 60 participants (34 men), mean age 30.8 years (SD = 13.3).

Procedure

Interoceptive awareness was measured as in Experiment 1. Participants completed the Director Task after the heartbeat perception task.

The 'Director Task'

The Director Task requires participants to move objects on a computer screen according to verbal instructions from a 'Director,' delivered via headphones. Crucially, on critical trials the Director's visual perspective on the objects that have to be moved differs from that of participant. Thus participants must understand the Director's point of view in order to perform the task correctly.

The visual stimulus consisted of a 4 × 4 grid containing eight common objects. The image of a character, introduced as 'the Director,' appears behind the grid. On each trial, the Director would name one of the objects and instruct the participant to move it, using the computer mouse, either one space up, down, left, or right. Importantly, five of the spaces in the grid were occluded, so that the participant could see the objects situated in these spaces but the Director, who was standing behind the grid, could not. In experimental trials ('E' trials), there was a conflict between the participant's and the Director's perspective. An example of such a situation would be if the Director asked the participant to 'move the small candle right,' but the smallest candle that the participant could see was occluded from the Director. In this case, the participant should instead move the second smallest candle, which *can* be seen by the Director. In addition to these critical 'E' trials, there were also two types of control trial ('C1' and 'C2' trials). In these trials there was no conflict between the participant's and Director's perspectives. Thus participants were either asked to move an object which was the only item of that type and was visible to both participant and Director (C1 trials); or the participant was asked to move the smallest or largest of a type of object and again that target object was visible to both (C2 trials).

The procedure and stimuli were adapted from Santiesteban et al. (2012). For the participants in the current study, the task was shortened in make it more suitable for members the general public visiting the Science Museum. Participants completed 12 E trials, six C1 trials, and six C2 trials in total. The task took about 5 min. Participants' accuracy and reaction times (RTs) were recorded.

Results

Accuracy and RT data were analyzed with a repeated-measures ANOVA with trial type (E vs. C1 vs. C2) as the within-subjects factor. The data violated the sphericity assumption and therefore all results reported are with Greenhouse-Geisser corrections. For accuracy scores, there was a main effect of trial type, $F(1.31,77.47) = 28.72$, $p < 0.001$. *Post hoc* paired samples *t*-tests revealed that participants were significantly more accurate on C1 trials, $M = 0.91$, $SD = 0.22$, than they were on C2 trials, $M = 0.61$, $SD = 0.17$, $t(59) = 12.19$, $p < 0.001$, or on E trials, $M = 0.61$, $SD = 0.44$, $t(59) = 5.39$, $p < 0.001$. Performance did not differ significantly between C2 and E trials, $t(59) = 0.02$, $p = 0.98$. The same analysis was repeated with gender inserted as a between-groups factor. The main effect of trial type remained, and there was no effect of gender or any interaction.

For RT scores, there was also a main effect of trial type, $F(1.31,95.76) = 40.03$, $p < 0.001$. Paired samples *t*-tests revealed that responses on E trials, $M = 4.54$ s, $SD = 1.20$, were significantly slower than on both C1 trials, $M = 3.79$ s, $SD = 1.04$, $t(59) = 7.30$, $p < 0.001$, and on C2 trials, $M = 3.75$ s, $SD = 0.88$, $t(59) = 6.74$, $p < 0.001$. Performance on C1 and C2 trials did not differ significantly, $t(59) = 0.38$, $p = 0.70$. The same analysis was repeated with gender inserted as a between-groups factor. The main effect of trial type remained, and there was no main effect of gender or any significant interaction.

Interoceptive awareness was calculated as in Experiment 1. Mean IA was 0.71 (SD = 0.16). IA was negatively correlated with the participants' heart rate, $r = -0.33$, $p < 0.001$. Heart rate was not significantly correlated with any aspect of performance on the Director Task.

To assess the relationship between performance on the Director Task and participants' IA, we calculated Pearson's correlation coefficients between participants' performance on the E trials and their IA scores. We focussed on E trials, as it is only on these trials that the participant is required to take a perspective that conflicts with their own, which is the ability relevant to cognitive empathy.

No significant correlations were found between IA and performance on E trials, either as measured by RTs, $r = -0.02, p = 0.91$, or by accuracy, $r = -0.08$, $p = 0.55$. There was also no correlation between IA and performance on C1 or C2 trials (all p-values >0.47).

Discussion of Experiment 4

Research into body ownership illusions has implied that people with high IA make better self/other distinctions in embodied contexts (Tsakiris et al., 2011; Aspell et al., 2013; Suzuki et al., 2013). Given that the ability to distinguish between one's own and another's perspective is crucial for successful performance on the Director Task (Santiesteban et al., 2012), we hypothesized that we would find a positive association between IA and performance on this task.

However, this association was not found. It is likely that the type of sensory and embodied self/other distinction modulated in bodily illusions does not carry over to the cognitive self/other distinction measured by the Director Task, which may be solved using theory of mind.

General Discussion

We investigated the relationship between individual differences in IA and components of empathy, using a well-validated heartbeat perception task, together with four measures. In our first study we found no significant relationship with the IRI or any of its subscales, including EC and PT. In a further group of participants, no links were found between IA and scores on the QCAE or its subscales, including Emotion Contagion (which measures affect sharing) and PT. A further study found no significant relationship between IA and judgment of facial emotion in the Reading the Mind in the Eyes test (Baron-Cohen et al., 2001), which potentially depends on affect sharing, as a result of mirroring the observed emotion (Gallese and Sinigaglia, 2011). In our final experiment, scores on the Director Task which assesses cognitive PT (Santiesteban et al., 2012) were not associated with IA. Our only significant relationships were negative correlations between heart rate and IA, as frequently reported in other studies (Cameron, 2001; Knapp-Kline and Kline, 2005; Fairclough and Goodwin, 2007; Stevens et al., 2011), which potentially reflect physiological variables such as stroke volume of the heart.

These null findings imply that a simple relationship between IA and any of the three components of empathy cannot be predicted. The theoretical basis for supposed links rest on evidence that IA is related to the intensity of one's own emotional experience. If empathy is embodied and involves the vicarious experience of emotion, then it was predicted that people with high IA would have a greater tendency to resonate with the emotion that they observe in others. They should hence demonstrate greater affect sharing and hence higher scores on the Emotion Contagion subscale of the QCAE as well as better ability in identifying emotional expressions in the Reading the Mind in the Eyes test, assuming that the task involves shared affect by mirroring of the observed emotion.

However, a previously overlooked issue, which may explain our results, is that it is by no means obvious how the better self/other distinction that is associated with high IA impacts on the components of empathy. If high IA is accompanied by less self/other overlap, this might involve *less* shared affect and consequently *less* experience of vicarious emotion. Thus if the effects of greater emotional arousal and greater self/other distinction work in opposite directions, as seems probable, then the predicted link between IA and affect sharing becomes a matter of which effect is stronger. We believe that this type of complexity explains our null results for the Emotion Contagion subscale of the QCAE and the Reading the Mind in the Eyes test. Support for his idea is provided by Grynberg and Pollatos' (2015) study in which good heartbeat perceivers reported more compassion when viewing pictures of people in pain. Their results suggest that the higher arousal these participants also reported led them to judge the observed pain as more severe. However, they did not report greater PD, which may have been the result of good self/other distinction which allowed these individuals to avoid sharing the suffering. In support of this we found no relationship between IA and the PD subscale of the IRI.

Empathic concern (the motivation to help others) was not linked to IA in either questionnaire. EC is commonly used as a general measure of empathy. However, a link between the two would be anticipated only if EC is influenced by previous experience of affect sharing or, alternatively, by better self/other distinction (both of which are potentially greater in good heartbeat perceivers). Potentially, EC relies, at least in part, on theory of mind rather than direct affect sharing (Bernhardt and Singer, 2012), which would negate any obvious direct relationship with IA.

With regard to PT, we expected a relationship between IA and the *affective* PT subscales of the two questionnaires (which assess putting oneself in another's emotional shoes), given that people with high IA make better self/other distinction on embodied contexts. That no relationship was found suggests that the type of PT measured by putting oneself in the other person's emotional shoes may depend on theory of mind, rather than on an embodied response. By contrast, the Director Task is a good measure of *cognitive* PT (Decety and Cowell, 2014) but the test may also be solved by theory of mind (Santiesteban et al., 2014), in which case it too is not necessarily comparable with the type of embodied self/other distinction that is involved in body ownership.

The difference between the results presented here and experiments that have previously reported links between empathy and IA can be attributed to differences of definition and measurement. In Ernst et al.'s (2013) study, for example, a heartbeat-counting task primed emotion circuits in the insula such that the effects carry over into a subsequent emotion appraisal task. This significant finding shows that attention to interoception enhances subsequent emotion processing. However, the authors did not assess IA as generally defined (Garfinkel and Critchley, 2013) and did not establish whether IA, as a trait, is linked to empathy. A similar difficulty of definition arises with Fukushima et al.'s (2011) study, which likewise employed an emotional appraisal task. The amplitude of HEPs increased during this empathy manipulation. However, while the amplitude of the HEP is a useful index of interoceptive processing, it has not previously been taken as a measure of trait IA. Fukushima et al.'s (2011) experiment therefore also provides, at best, indirect evidence of a possible link between IA and empathy.

In addition to the criticism of our four measures that are addressed above, further limitations of our study are that we did not screen for personality variables, such as psychopathy, narcissism, altruism, agreeableness, or trait anxiety, which potentially moderate links between IA and empathy.

Our results support the characterisation of empathy as an umbrella term for a distributed set of mechanisms (Singer, 2006; Decety, 2010; Zahavi, 2012). Foremost of these is affect sharing, which may be linked to IA through greater intensity of emotional experience, although good self/other distinction potentially confounds the relationship. An important focus for future research is further direct tests of affect sharing, such as empathy for pain (Singer et al., 2004) and facial EMG (Lamm et al., 2008). EC, by contrast, is perhaps a product of affect sharing, self/other distinction and theory of mind, so that a clear relationship with IA cannot be predicted. Finally, PT is closely related to theory of mind and may not be influenced by the type of embodied self/other distinction that is modulated by IA in body ownership illusions. While it is therefore probable that people with high IA have the ability to use their bodies to simulate the emotions of other people, we conclude that they may do so only in certain contexts, so that reliable associations between empathy measures and IA cannot be predicted.

Acknowledgments

This work was supported by the European Platform for Life Sciences, Mind Sciences and Humanities, Volkswagen Foundation (II/85064), and the European Research Council (ERC-2010-StG-262853) under the FP7 to MT. We would like to thank Idalmis Santiesteban for the materials for the Director Task and the Science Museum, London for hosting our experiments.

References

Ainley, V., Tajadura-Jiménez, A., Fotopoulou, A., and Tsakiris, M. (2012). Looking into myself: changes in interoceptive sensitivity during mirror self-observation. *Psychophysiology* 49, 1672–1676. doi: 10.1111/j.1469-8986.2012.01468.x

Aspell, J. E., Heydrich, L., Marillier, G., Lavanchy, T., Herbelin, B., and Blanke, O. (2013). Turning body and self inside out: visualized heartbeats alter bodily self-consciousness and tactile perception. *Psychol. Sci.* 24, 2445–2453. doi: 10.1177/0956797613498395

Baron-Cohen, S., Wheelwright, S., Hill, J., Raste, Y., and Plumb, I. (2001). The "reading the mind in the eyes" Test revised version: a study with normal adults, and adults with Asperger syndrome or high-functioning autism. *J. Child Psychol. Psychiatry* 42, 241–251. doi: 10.1111/1469-7610.00715

Barrett, L. F., Quigley, K. S., Bliss-Moreau, E., and Aronson, K. R. (2004). Interoceptive sensitivity and self-reports of emotional experience. *J. Pers. Soc. Psychol.* 87, 684–697. doi: 10.1037/0022-3514.87.5.684

Belllini, L. M., Baime, M., and Shea, J. (2002). Variation of mood and empathy during internship. *J. Am. Med. Assoc.* 287, 3143–3146. doi: 10.1001/jama.287.23.3143

Bernhardt, B. C., and Singer, T. (2012). The neural basis of empathy. *Annu. Rev. Neurosci.* 35, 1–23. doi: 10.1146/annurev-neuro-062111-150536

Bird, G., Silani, G., Brindley, R., White, S., Frith, U., and Singer, T. (2010). Empathic brain responses in insula are modulated by levels of alexithymia but not autism. *Brain A J. Neurol.* 133(Pt. 5), 1515–1525. doi: 10.1093/brain/awq060

Cameron, O. G. (2001). Interoception: the inside story - a model for psychosomatic processes. *Psychosomat. Med.* 63, 697–710. doi: 10.1097/00006842-200109000-00001

Cuenen, E., Van Diest, I., and Vlaeyen, J. (2012). Accuracy and awareness of perception: related, yet distinct. *Biol. Psychol.* 92, 426–427. doi: 10.1093/scan/nsr051

Davis, M. H. (1983). Measuring individual differences in empathy: evidence for a multidimensional approach. *J. Pers. Soc. Psychol.* 44, 113–126. doi: 10.1037/0022-3514.44.1.113

Decety, J. (2010). To what extent is the experience of empathy mediated by shared neural circuits? *Emot. Rev.* 2, 204–207. doi: 10.1177/1754073910361981

Decety, J., and Cowell, J. M. (2014). Friends or foes: is empathy necessary for moral behaviour? *Perspect. Psychol. Sci.* 9, 525–537. doi: 10.1177/1745691614545130

Dunn, B., Dalgleish, T., Ogilvie, A., and Lawrence, A. (2007). Heartbeat perception in depression. *Behav. Res. Ther.* 45, 1921–1930. doi: 10.1016/j.brat.2006.09.008

Dunn, B. D., Galton, H. C., Morgan, R., Evans, D., Oliver, C., Meyer, M., et al. (2010). Listening to your heart: how interoception shapes emotion experience and intuitive decision making. *Psychol. Sci.* 21, 1835–1844. doi: 10.1177/0956797610389191

Ehlers, A., Breuer, P., Dohn, D., and Fiegenbaum, W. (1995). Heartbeat perception and panic disorder: possible explanations for discrepant findings. *Behav. Res. Ther.* 33, 69–76. doi: 10.1016/0005-7967(94)E0002-Z

Ernst, J., Northoff, G., Böker, H., Seifritz, E., and Grimm, S. (2013). Interoceptive awareness enhances neural activity during empathy. *Hum. Brain Mapp.* 34, 1615–1624. doi: 10.1002/hbm.22014

Eshkevari, E., Rieger, E., Musiat, P., and Treasure, J. (2014). An investigation of interoceptive sensitivity in eating disorders using a heartbeat detection task and a self-report measure. *Eur. Eat. Dis. Rev.* 22, 383–388. doi: 10.1002/erv.2305

Esysenck, S. B. G., and Eysenck, H. J. (1978). Impulsiveness and venturesomeness: their position in a dimensional system of personalilty description. *Psychol. Rep.* 43, 1247–1255. doi: 10.2466/pr0.1978.43.3f.1247

Fairclough, S. H., and Goodwin, L. (2007). The effect of psychological stress and relaxation on interoceptive accuracy: implications for symptom perception. *J. Psychosom. Res.* 62, 289–295. doi: 10.1016/j.jpsychores.2006.10.017

Fukushima, H., Terasawa, Y., and Umeda, S. (2011). Association between interoception and empathy: evidence from heartbeat-evoked brain potential. *Int. J. Psychophysiol.* 79, 259–265. doi: 10.1016/j.ijpsycho.2010.10.015

Gallese, V. (2007). Before and below "theory of mind": embodied simulation and the neural correlates of social cognition. *Philos. Trans. R. Soc. Lond. B Biol. Sci.* 362, 659–669. doi: 10.1098/rstb.2006.2002

Gallese, V., and Sinigaglia, C. (2011). What is so special about embodied simulation? *Trends Cogn. Sci.* 15, 512–519. doi: 10.1016/j.tics.2011.09.003

Garfinkel, S. N., and Critchley, H. D. (2013). Interoception, emotion and brain: new insights link internal physiology to social behaviour. Commentary on: "Anterior insular cortex mediates bodily sensibility and social anxiety" by Terasawa et al.(2012). *Soc. Cogn. Affect. Neurosci.* 8, 231–234. doi: 10.1093/scan/nss140

Gonzalez-Liencres, C., Shamay-Tsoory, S. G., and Brune, M. (2013). Towards a neuroscience of empathy: ontogeny, phylogeny, brain mechanisms, context and psychpoathology. *Neurosci. Biobehav. Rev.* 37, 1537–1548. doi: 10.1093/scan/nst058

Grynberg, D., and Pollatos, O. (2015). Perceiving one' s body shapes empathy. *Physiol. Behav.* 140, 54–60. doi: 10.1016/j.physbeh.2014.12.026

Handford, C., Lemon, J., Grimm, M. C., and Vollmer-Conna, U. (2013). Empathy as a function of clinical exposure–reading emotion in the eyes. *PLoS ONE* 8:e65159. doi: 10.1371/journal.pone.0065159

Herbert, B. M., Muth, E. R., Pollatos, O., and Herbert, C. (2012). Interoception across modalities: on the relationship between cardiac awareness and the sensitivity for gastric functions. *PLoS ONE* 7:e36646. doi: 10.1371/journal.pone.0036646

Herbert, B. M., Pollatos, O., Flor, H., Enck, P., and Schandry, R. (2010). Cardiac awareness and autonomic cardiac reactivity during emotional picture viewing and mental stress. *Psychophysiology* 47, 342–354. doi: 10.1111/j.1469-8986.2009.00931.x

Hogan, R. (1969). Development of an empathy scale. *J. Consult. Clin. Psychol.* 33, 307–316. doi: 10.1037/h0027580

Jackson, P. L., Meltzoff, A. N., and Decety, J. (2006). Neural circuits involved in imitation and perspective-taking. *NeuroImage* 31, 429–439. doi: 10.1016/j.neuroimage.2005.11.026

Kaplan, J. T., and Iacoboni, M. (2006). Getting a grip on other minds: mirror neurons, intention understanding, and cognitive empathy. *Soc. Neurosci.* 1, 175–183. doi: 10.1080/17470910600985605

Knapp-Kline, K., and Kline, J. P. (2005). Heart rate, heart rate variability, and heartbeat detection with the method of constant stimuli: slow and steady wins the race. *Biol. Psychol.* 69, 387–396. doi: 10.1016/j.biopsycho.2004.09.002

Lamm, C., Batson, C. D., and Decety, J. (2007). The neural substrate of human empathy: effects of perspective-taking and cognitive appraisal. *J. Cogn. Neurosci.* 19, 42–58. doi: 10.1162/jocn.2007.19.1.42

Lamm, C., Porges, E. C., Cacioppo, J. T., and Decety, J. (2008). Perspective taking is associated with specific facial responses during empathy for pain. *Brain Res.* 1227, 153–161. doi: 10.1016/j.brainres.2008.06.066

Lawrence, E. J., Shaw, P., Baker, D., Baron-Cohen, S., and David, S. (2004). Measuring empathy: reliability and validity of the Empathy Quotient. *Psychol. Med.* 34, 911–919. doi: 10.1017/S0033291703001624

Lenggenhager, B., Azevedo, R., Mancini, A., and Aglioti, S. M. (2013). Listening to your heart and feeling yourself: effects of exposure to interoceptive signals during the ultimatum game. *Exp. Brain Res.* 230, 233–241. doi: 10.1007/s00221-013-3647-5

Mussgay, L., Klinkenberg, N., and Rüddel, H. (1999). Heartbeat perception in patients with depressive, somatoform, and personality disorders. *J. Psychophysiol.* 13, 27–36. doi: 10.1027//0269-8803.13.1.27

Panksepp, J. (1998). *Affective Neuroscience: The Foundations of Human and Animal Emotions.* Oxford: Oxford University Press.

Pollatos, O., Füstös, J., and Critchley, H. D. (2012). On the generalised embodiment of pain: how interoceptive sensitivity modulates cutaneous pain perception. *Pain* 153, 1680–1686. doi: 10.1016/j.pain.2012.04.030

Pollatos, O., Kirsch, W., and Schandry, R. (2005). On the relationship between interoceptive awareness, emotional experience, and brain processes. *Cogn. Brain Res.* 25, 948–962. doi: 10.1016/j.cogbrainres.2005.09.019

Pollatos, O., and Schandry, R. (2008). Emotional processing and emotional memory are modulated by interoceptive awareness. *Cogn. Emot.* 22, 272–287. doi: 10.1080/02699930701357535

Preston, S. D., and de Waal, F. B. M. (2002). Empathy: its ultimate and proximate bases. *Behav. Brain Sci.* 25, 1–20. doi: 10.1017/S0140525X02360011

Preston, S. D., and Hofelich, A. J. (2012). The many faces of empathy: parsing empathic phenomena through a proximate, dynamic-systems view of representing the other in the self. *Emot. Rev.* 4, 24–33. doi: 10.1177/1754073911421378

Reniers, R. L., Corcoran, R., Drake, R., Shryane, N. M., and Völlm, B. A. (2011). The QCAE: a questionnaire of cognitive and affective empathy. *J. Pers. Assess.* 93, 84–95. doi: 10.1080/00223891.2010.528484

Santiesteban, I., Shah, P., White, S., Bird, G., and Heyes, C. (2014). Mentalizing or submentalizing in a communication task? Evidence from autism and a camera control. *Psychon. Bull. Rev.* doi: 10.3758/s13423-014-0716-0 [Epub ahead of print].

Santiesteban, I., White, S., Cook, J., Gilbert, S. J., Heyes, C., and Bird, G. (2012). Training social cognition: from imitation to Theory of Mind. *Cognition* 122, 228–235. doi: 10.1016/j.cognition.2011.11.004

Schandry, R. (1981). Heartbeat perception and emotional experience. *Psychophysiology* 18, 483–488. doi: 10.1111/j.1469-8986.1981.tb02486.x

Singer, T. (2006). The neuronal basis and ontogeny of empathy and mind reading: review of literature and implications for future research. *Neurosci. Biobehav. Rev.* 30, 855–863. doi: 10.1016/j.neubiorev.2006.06.011

Singer, T., and Klimecki, O. M. (2013). Empathy and Compassion. *Curr. Biol.* 24, R875. doi: 10.1002/9781118583098.ch7

Singer, T., and Lamm, C. (2009). The social neuroscience of empathy. *Ann. N. Y. Acad. Sci.* 1156, 81–96. doi: 10.1111/j.1749-6632.2009.04418.x

Singer, T., Seymour, B., O'Doherty, J., Kaube, H., Dolan, R. J., and Frith, C. D. (2004). Empathy for pain involves the affective but not sensory components of pain. *Science* 303, 1157–1162. doi: 10.1126/science.1093535

Stevens, S., Gerlach, A. L., Cludius, B., Silkens, A., Craske, M. G., and Hermann, C. (2011). Heartbeat perception in social anxiety before and during speech anticipation. *Behav. Res. Ther.* 49, 138–143. doi: 10.1016/j.brat.2010.11.009

Suzuki, K., Garfinkel, S. N., Critchley, H. D., and Seth, A. K. (2013). Multisensory integration across exteroceptive and interoceptive domains modulates self-experience in the rubber-hand illusion. *Neuropsychologia* 51, 2909–2917. doi: 10.1016/j.neuropsychologia.2013.08.014

Tajadura-Jiménez, A., and Tsakiris, M. (2013). Balancing the "inner" and the "outer" self: interoceptive sensitivity modulates self-other boundaries. *J. Exp. Psychol. Gen.* 143, 736–744. doi: 10.1037/a0033171

Terasawa, Y., Moriguchi, Y., Tochizawa, S., and Umeda, S. (2014). Interoceptive senstivity predcits sensitivity to the emotions of others. *Cogn. Emot.* 28, 37–41. doi: 10.1080/02699931.2014.888988

Tsakiris, M., Tajadura-Jiménez, A., and Costantini, M. (2011). Just a heartbeat away from one's body: interoceptive sensitivity predicts malleability of body-representations. *Proc. R. Soc. B Biol. Sci.* 278, 2470–2476. doi: 10.1098/rspb.2010.2547

Werner, N. S., Kerschreiter, R., Kindermann, N. K., and Duschek, S. (2013). Interoceptive awareness as a moderator of affective responses to social exclusion. *J. Psychophysiol.* 27, 39–50. doi: 10.1027/0269-8803/a000086

Werner, N. S., Peres, I., Duschek, S., and Schandry, R. (2010). Implicit memory for emotional words is modulated by cardiac perception. *Biol. Psychol.* 85, 370–376. doi: 10.1016/j.biopsycho.2010.08.008

Whitehead, W. E., and Drescher, V. M. (1980). Perception of gastric contractions and self-control of gastric motility. *Psychophysiology* 17, 552–558. doi: 10.1111/j.1469-8986.1980.tb02296.x

Wiens, S. (2005). Interoception in emotional experience. *Curr. Opin. Neurol.* 18, 442–447. doi: 10.1097/01.wco.0000168079.92106.99

Wiens, S., Mezzacappa, E., and Katkin, E. S. (2000). Heartbeat detection and the experience of emotions. *Cogn. Emot.* 14, 417–427. doi: 10.1080/026999300378905

Zahavi, D. (2012). Comment: basic empathy and complex empathy. *Emot. Rev.* 4, 81–82. doi: 10.1177/1754073911421387

2

Interoception, Contemplative Practice and Health

Norman Farb[1]*, Jennifer Daubenmier[2], Cynthia J. Price[3], Tim Gard[4], Catherine Kerr[5], Barnaby D. Dunn[6], Anne Carolyn Klein[7], Martin P. Paulus[8] and Wolf E. Mehling[2]*

[1] University of Toronto Mississauga, Mississauga, ON, Canada, [2] University of California San Francisco, San Francisco, CA, USA, [3] University of Washington, Seattle, WA, USA, [4] Maastricht University, Maastricht, Netherlands, [5] Brown University, Providence, RI, USA, [6] University of Exeter, Exeter, UK, [7] Rice University, Houston, TX, USA, [8] University of California San Diego, La Jolla, CA, USA

*Correspondence:
_Norman Farb,
Department of Psychology, University of Toronto Mississauga,
3359 Mississauga Road,
Mississauga,
ON L5L 1C6, Canada
norman.farb@utoronto.ca;
Wolf E. Mehling,
Department of Family and Community Medicine, Osher Center for Integrative Medicine, University of California San Francisco, 1545 Divisadero Street,
San Francisco, CA 94115, USA
wolf.mehling@ucsf.edu_

Interoception can be broadly defined as the sense of signals originating within the body. As such, interoception is critical for our sense of embodiment, motivation, and well-being. And yet, despite its importance, interoception remains poorly understood within modern science. This paper reviews interdisciplinary perspectives on interoception, with the goal of presenting a unified perspective from diverse fields such as neuroscience, clinical practice, and contemplative studies. It is hoped that this integrative effort will advance our understanding of how interoception determines well-being, and identify the central challenges to such understanding. To this end, we introduce an expanded taxonomy of interoceptive processes, arguing that many of these processes can be understood through an emerging predictive coding model for mind–body integration. The model, which describes the tension between expected and felt body sensation, parallels contemplative theories, and implicates interoception in a variety of affective and psychosomatic disorders. We conclude that maladaptive construal of bodily sensations may lie at the heart of many contemporary maladies, and that contemplative practices may attenuate these interpretative biases, restoring a person's sense of presence and agency in the world.

Keywords: interoception, contemplative practice, meditation, body awareness, mindfulness, yoga, mind–body therapies

Introduction

Introducing Interoception

Interoception is the process of receiving, accessing and appraising internal bodily signals. Maintaining desired physiological states is critical for an organism's survival, and so interoception is a powerful motivator of behavior in the pursuit of these states (Craig, 2002, 2009). While interoceptive awareness has been more narrowly defined as the representation of afferent body sensations (Craig, 2003; Critchley et al., 2004), broader definitions cast interoception as a multi-dimensional construct that also takes into account how people attend to, appraise and respond to these sensations (Vaitl, 1996; Cameron, 2001; Verdejo-Garcia et al., 2012). Interoception is supported by an increasingly well-specified neuro-anatomical pathway with dedicated representation cortices akin to the five external senses (Craig, 2002; Critchley et al., 2004; Critchley and Harrison, 2013), although the nature of such representation is decidedly more mysterious and in need of empirical investigation.

Interoception is an iterative process, requiring the interplay between perception of body states and cognitive appraisal of these states to inform response selection. Afferent sensory signals continuously interact with higher order cognitive representations of goals, history, and environment, informing emotional experience and motivating regulatory behavior (Craig, 2009). Together, these iterations create a sense of self laden with motivational context (Damasio, 2003b). Such context may then inform a person's approach or avoidance tendencies, and thus these interactions hold important implications for well-being (Beck, 2008). Conversely, life experience impinges upon interoception: the perceived availability and relevance of body representations are moderated by factors such as attentional cues (Ainley et al., 2012), contemplative training such as meditation, yoga, tai chi, etc. (Bornemann et al., 2014), and age (Khalsa et al., 2009). Understanding how interoceptive processing is shaped by experience is therefore important for efforts to cultivate well-being and stress resilience.

Interoception and Well-Being

One reason that interoception, a perceptual capacity, is introduced in relation to well-being is that it is thought to be intimately connected to self-regulation, having likely evolved to help organisms maintain homeostasis (Paulus, 2007; Herbert and Pollatos, 2012; Craig, 2013; Gu and Fitzgerald, 2014). In modern life, emotionally valenced body signals are also thought to contribute to broader mood states that support emotional balance (Paulus and Stein, 2006; Craig, 2007; Seth, 2013). To the extent that a person is sensitive to interoceptive signals, such signals guide decision making (Dunn et al., 2010). And yet, high sensitivity is not without its price: when body sensation is irregular, as for individuals with joint hypermobility, greater interoceptive sensation may also contribute to feelings of anxiety (Mallorquí-Bagué et al., 2014). Thus, interoceptive sensitivity may either contribute or detract from well-being, suggesting a need for guidance in regulating salient visceral signals.

Paralleling the findings of modern secular clinical science, contemplative science suggests that reflection on interoceptive processes is important for adaptive behavior (Wallace, 2007), disrupting overlearned perceptual, and interpretive habits formed throughout cognitive development and aligning behavior with higher-order intentions (Vago and Silbersweig, 2012). As the embodied self is more fully realized through awareness of ongoing interoceptive interactions, two complimentary senses emerge: *presence,* one's connection to the moment, and *agency,* one's ability to effect change, which are both foundational in determining a person's sense of well-being (Seth et al., 2011). To chart interoception's salutary relevance, research must therefore consider the integration of both early sensory and later affective discriminations into a broader motivational context.

Motivation for the Present Paper

Over the past decade, the mechanisms of interoceptive awareness, the importance of interoceptive cues for physical and psychological health, and the cultivation of healthy interoceptive habits through contemplative practices have become topics of active investigation. However, insights from these three domains are rarely integrated, impeding the progression of understanding their collective significance. In April 2013, the authors met to discuss interoception across a variety of perspectives – including neuroscience, social and clinical psychology, complementary and alternative care, medicine, Buddhist and contemplative studies, and a wealth of personal experiences with contemplative practices. Our goal is to better explain how interoceptive information can be revealed and masked from conscious awareness, how our appraisals of a given level of interoceptive awareness help to determine well-being, and what role contemplative practices may play in these processes.

Here, we attempt to integrate contemporary interoceptive theory from these diverse perspectives, in the hope of better specifying interoceptive constructs and identifying next steps for advancing the field. First, we define interoception from scientific and contemplative perspectives, focusing on a predictive coding model of interoceptive representation that is growing in popularity within the scientific community. Second, we discuss the role of interoception for physical and psychological well-being, with particular regard for emerging clinical evidence. Third, we explore how contemplative practices may interact with interoceptive processes to enhance well-being in oneself and connection to others. Finally, we summarize outstanding issues in the study of interoception and offer recommendations for future research.

Interoception in Psychology, Neuroscience, and Contemplative Traditions

Toward a Science of Interoception

The scientific study of brain, mind, and body provides a framework for objectively modeling interoception. Neuroanatomical studies have identified pathways that support and modulate interoceptive representations, and as such, may be measurably impacted by psychopathology or contemplative practice. This framework is comprised of peripheral receptors, C-fiber afferents, spinothalamic projections, specific thalamic nuclei, posterior and anterior insula as the limbic sensory cortex, and anterior cingulate cortex (ACC) as the limbic motor cortex (Vaitl, 1996; Bush et al., 2000; Vuilleumier, 2005; Critchley and Harrison, 2013). In some cases, definitions of interoception extend beyond visceral afferents (Sherrington, 1906) to include afferents from other deep body receptors, such as C-tactile somatic receptors of the skin that are associated with internal feelings of warmth or pleasure from gentle touch (Björnsdotter et al., 2009), circulating chemicals, proprioceptive inputs, and unexpected or yet unknown sites, such as recently discovered photoreceptor cells within the eyes (Lucas, 2013).

The anatomical pathways for interoception are well-specified, detailing the connections between sensory receptors, spinal cord, brainstem, and brain (Craig, 2002). Neurotransmitter concentrations in the insula and ACC in particular have been associated with subjective interoceptive awareness and subjective well-being (Ernst et al., 2014; Wiebking et al., 2014). However, how such neural representations are constructed and the mechanisms by which they influence cognition remain unclear. Most investigations have focused narrowly on interoceptive awareness, which may only be one aspect of interoceptive processing. Interrelated but distinct constructs such as interoceptive attention tendency, sensitivity, coherence between physiology and subjective experience, accuracy, and regulatory habits may all emerge as important properties of information propagation and integration throughout interoceptive networks (**Box 1**).

Adding to the complexity, multiple interoceptive sub-systems operate concurrently, and may each afford different capacities and tendencies. There is some evidence for general, multimodal interoceptive capacity: heartbeat detection accuracy is correlated with gastric sensitivity (Herbert et al., 2012b), and sometimes with pain sensitivity in some studies (Pollatos et al., 2012), but not with others (Werner et al., 2009). Furthermore, the relationship between interoceptive constructs across domains such as the heartbeat, blood glucose, respiration, temperature and other modalities is largely unknown. This gap in understanding may be due to the fact that most interoception studies gage accuracy within a single modality, and do not measure related constructs such as attention habits, regulation ability, or sensitivity to signal change, all of which converge to support interoceptive awareness (Ceunen et al., 2013).

Acknowledgment of interoception's multimodal and multifaceted nature is important, because consideration of these factors in isolation can lead to spurious inferences. For example, knowledge that meditators have strong interoceptive attention tendencies might lead one to expect them also to possess superior accuracy in heartbeat detection, an idea contradicted by recent research on this topic (Khalsa et al., 2008; Parkin et al., 2013). The benefits of interoceptive attention in a particular domain such as breath or body monitoring therefore seem to be independent from a domain-general enhancement of interoceptive accuracy. Instead, meditation practices appears to promote changes in a

BOX 1 | A taxonomy of interoception terms.

Interoception has been generally construed as the representation of the individual's body at a particular point in time (Craig, 2002), but accounts differ on how interoceptive capacity should be measured. Similar to a recently proposed model (Garfinkel and Critchley, 2013), we suggest several criteria to supplement the measurement of interoception:

• Interoceptive *awareness*, such as the sense of one's heart beating, is the most common measure of interoception, and is usually operationalized in terms of the reportability of interoceptive signals. However, awareness may be a limited criterion, as interoceptive processes may operate implicitly, such as thermoregulation promoting shivering even when one is asleep (Craig, 2003).

• *Coherence* between physiological and subjective states measures the degree to which objectively observable interoceptive signals manifest in reportable experience. For example, low blood sugar (hypoglycemia) is characterized both by interoceptive signals (Herbert et al., 2012a) and irritability (Matthew et al., 1997). However, awareness of low blood sugar is not necessary for it to promote irritability (Taylor and Rachman, 1988; Matthew et al., 1997). Coherence appears to vary widely among individuals, even though both are impacted by environmental stressors (Sze et al., 2010).

• *Attention tendency* refers to whether a person habitually attends to particular interoceptive signals, potentially inhibiting or ignoring others (Pollatos et al., 2005). It can also be expressed as whether a person attends more to interoceptive than exteroceptive sources of information.

• *Sensitivity* refers to the minimum threshold for detecting interoceptive signal change (Holzl et al., 1996); sensitivity may operate at multiple levels of the representational system, culminating in access to conscious awareness. Although not often used in interoception research, specificity is the common counterpart to sensitivity in signal detection theory (Abdi, 2007), the ability to reject competing signals from being classified as interoceptive afferent signals.

• *Accuracy* is perhaps the most commonly investigated measure of interoception, and refers to the ability to reliably discriminate interoceptive signals from noise or competing signals (Vaitl, 1996), and between different levels of signal intensity (Daubenmier et al., 2013). It appears that accuracy varies widely between individuals (Critchley et al., 2004; Ceunen et al., 2013), but may improve with focused training (Brener, 1977; Daubenmier et al., 2013; Mirams et al., 2013). Accuracy is usually thought to be a function of sensitivity and specificity.

• *Sensibility* refers to an individual's personal account of how they experience internal sensations, including a confidence estimate of one's own interoceptive ability and feelings of engagement during interoception (Garfinkel et al., 2015). Sensibility is often gaged using interviews and/or questionnaires, frequently the Porges Body Perception Questionnaire (Porges, 1993) is applied. The usefulness of this instrument for interoception, however, has been questioned as it appears to primarily serve as a proxy measure for anxiety-related symptoms (Mehling et al., 2009). The Scale of Body Connection (Price and Thompson, 2007) and the Multidimensional Assessment of Interoceptive Awareness (MAIA; Mehling et al., 2012) were created in part to help extend interoceptive assessment.

• *Regulation* refers to how well a person can match an interoceptive signal to his or her desired state. Regulation can involve shaping either the signal or the desire. For example, regulation could shape interoceptive signals to meet goals through reappraisal, suppression, or distraction, techniques often cited in modern scientific models of emotion regulation (Gross, 2002). However, regulation could also follow more contemplative traditions, intentionally accepting and examining such signals with curiosity, a strategy that encourages shifts in interoceptive experience without attempting to control unexpected interoceptive signals or create desirable ones.

Concurrent measurement of these constructs may inform a more complete model of interoception, its relationship to well-being, and the effects of contemplative training

It is important to note that these constructs have varied definitions across the literature. Sensitivity and accuracy are combined as interoceptive accuracy in a recent interoception model (Garfinkel et al., 2015), measured as the performance on objective behavioral tests such as the heartbeat detection task, and clearly distinguished from interoceptive awareness and interoceptive sensibility. In the Garfinkel et al. (2015, p. 65) model, interoceptive awareness has been operationalized as "metacognitive awareness of interoceptive accuracy, e.g. confidence-accuracy correspondence." However, the term metacognitive awareness is also a term commonly used in contemplative practice pointing to the possibility of taking awareness itself and the process of thinking as an object of attention and has been defined elsewhere as the ability to reflect or be aware of ongoing thought or mental states (Smallwood and Schooler, 2006; Epel et al., 2009). There, it has been defined as the ability to experience negative thoughts and feelings as mental events that pass through the mind, rather than as a part of the self (Herbert and Forman, 2011). We would therefore avoid using the term "metacognitive awareness" to mean confidence–accuracy correspondence, preferring the term coherence instead. Thus, the terms listed above represent an attempt to bring a common taxonomy to our discussion, but should not be read as universally definitive in this developing field of inquiry.

more specific subset of interoceptive capacities and tendencies, accompanied by non-interoceptive factors such as changes in intentions or goal-orientations unique to each system of practice (Bornemann et al., 2014).

Just as capacity may vary across interoceptive modalities, interoceptive constructs may also vary independently within a given modality. For example, sensitivity may be high without commensurate accuracy or regulation ability, such as with uncontrollable interoceptive 'false alarms' often observed in anxiety disorders (Paulus and Stein, 2006; Domschke et al., 2010). For this reason, rather than arguing for the primacy of any of one interoceptive construct, the present discussion uses the term interoception to describe general processes of perceiving body states, while also acknowledging that a variety of contextual factors influence specific interoceptive capacities. A huge amount of research is needed to begin to catalog the relationships between specific capacities and modalities, let alone their plasticity and effects on well-being.

Integrating Contemplative Insights

If we could establish a working model of interoception that accounts for differences in how interoceptive signals are represented and managed, it is likely that some sub-systems would be more relevant for well-being than others. It is at this point that contemplative science and paradigms of cognitive processing may be extremely helpful, implicating particular interoceptive processes, addressing how interoception ideally functions, and how such function is shaped by experience. In particular, contemplative accounts have much to say on the distinction between maladaptive interoceptive conditioning that may lead to disorders, and adaptive learning that is purportedly engendered by contemplative practices.

Here, we refer to contemplative practice in the broadest sense, i.e., traditions of first-person reflection upon or cultivation of specific modes of experience, and focus on those practices that explicitly involve interoceptive awareness, including types of meditation and mindfulness-based approaches that allocate attention to body sensations (e.g., the breath), or to specific areas of the body (e.g., the center of the abdomen; Kabat-Zinn, 1982), and yoga, tai chi, and other mind-body practices performed in or outside of an explicit spiritual context (Baer, 2003). A wealth of other body-oriented healing and psychotherapy methods may also fall into this classification, such as Dialectical–Behavioral Therapy (Linehan et al., 1999), Acceptance and Commitment Therapy (Hayes et al., 1999), and also body-mind approaches such as Feldenkrais (Buchanan and Ulrich, 2001), Alexander method (Woodman and Moore, 2012), Focusing (Gendlin, 2012), Rosen work (Fogel, 2009), Hakomi (Kurtz, 1997), Sensory Awareness (Selver et al., 2007), Somatic Experiencing (Payne et al., 2015), Breath Therapy (Mehling, 2001), Holotropic Breathwork (Grof and Grof, 2010), and Mindful Awareness in Body-oriented Therapy (MABT; Price, 2005). We discuss primarily practices derived from Asian contemplative traditions, as these have undergone the most research and been the focus for the largest effort to translate contemplative concepts into modern scientific terms. This is not meant to exclude other contemplative traditions, and we hope future research will be extended into other such areas.

Classical contemplative practices such as mindfulness and equanimity seem to speak to the issues of how interoceptive signals are integrated into a complex representation of self and the world beyond, and have models of how they affect health and well-being. There are, however, several outstanding challenges to realizing the potential benefits of these classical traditions in secular modern society. The first issue is the lack of equivalence between traditional contemplative and modern secular practices: although mindfulness has found a strong representation in secular clinical therapies and sciences, it is disputed whether this translation remains faithful to its classical sources (Christopher et al., 2009; Grossman, 2011), and little attention has been given to the diversity of interpretations of mindfulness within the Buddhist tradition itself (Williams and Kabat-Zinn, 2011), which may present competing models for interoceptive representation and function. Second, while both clinical science and contemplative traditions share the goal of reducing human suffering, they differ in scope. In modern secular contexts, practice goals are largely pragmatic, aimed at reducing affective symptoms and improving daily function. By contrast, classical traditions tend to orient toward more extensive change, seeking to generate insight into the fundamental nature of reality, with the intention of liberating individuals from their conditioned states (Grossman and Van Dam, 2011). It is uncertain whether current appropriations of classical contemplative practices are sufficient to promote their historically lauded benefits. Given emerging evidence of contemplative practices' efficacies in secular contexts, there is reason for optimism that at least some beneficial elements have been successfully translated (Farb, 2014). However, our understanding of these practices' true mechanisms of action has lagged behind the scientific validation of their efficacies. There may be central constructs in classical contemplative theory that have yet to be operationalized in scientific discourse, let alone empirically studied.

Despite the great heterogeneity among contemplative traditions, we may begin by selecting a common concept that may contribute to scientific discourse: that of the 'subtle body'. Contemplative practices such as mindfulness are traditionally grounded in traditions, epistemologies, and medical treatises that articulate holistic rather than dualistic models of body and mind (Mehling et al., 2011). Such traditional sources each have their own distinct theories of the psycho-physical complex and invoke concepts of subtle body structures and 'currents' flowing through those structures (Samuel, 2008; Klein, 2013). These structures and 'flows' are amenable to influencing and being influenced by the mind, emotions, posture, and the condition of the grosser (physical) body. Their presence is indicated through awareness of a rich array of internal body sensations and a long phenomenological history of sensate processes that relate events occurring in the outside world to the experience within an individual. Exploring these somatic sensations, their sources, and their modulation has been an important focus of Tibetan, Chinese, and Indian medicine, represented in anatomical maps of channels, meridians, and 'energy centers' (chakras: Sanskrit) through which 'subtle energies' pass known as lung/ch'i/prāṇa, respectively (Loizzo et al., 2009; Klein, 2014). Every mental event – that is, all states of consciousness- are said to ride the 'steed of wind' or 'energy' currents. It is currently unclear how these conceptualizations map onto scientific approaches of interoception. However, these concepts do suggest that attention to embodied experience is significant for self-representation and well-being, and therefore supports the more general hypothesis that over-dependence on top–down, or merely conceptual (in contrast to sensory) awareness significantly limits a human being's potential for relating to self, others, and the world.

Integration Challenges

Asian classical traditions have a rich history of describing the integration of varied interoceptive signals into a common representation, a phenomenon often referred to as the 'subtle body' (Samuel, 2008; Klein, 2013). The subtlety of this body has to do with its typically functioning outside the horizon of ordinary consciousness. However, as already indicated, as the 'steed' or support of consciousness, it impacts perception significantly. We suggest that scientific discourse is just beginning to address this question of integrated representation, creating new possibilities for understanding what differentiates wholesome from maladaptive integration states. And yet, there are few empirical accounts of such integration, and so the potential to operationalize fundamental constructs from these traditions and practices is largely unknown. For example, many contemplative traditions, and many practitioners, speak of changes in the 'flow of energy,' resulting in feelings of lightness or heat, but it is largely unknown whether modern psychophysiological research methods can detect such changes (Loizzo et al., 2009; Kozhevnikov et al., 2013). Physiological arousal itself is inherently ambiguous and highly constrained by cognitive appraisal (Blascovich and Tomaka, 1996), and the science of how interoceptive appraisals translate into distinct emotional experiences is inexact at best (Wilson-Mendenhall et al., 2013). Given this uncertainty, we might begin by searching for the training-related changes described in classical contemplative theories of interoceptive integration. Such descriptions offer a rich set of testable hypotheses for filling this explanatory gap between physiology and experience. Inclusion of first-person experiences, particularly from people with extensive training in directing and reporting on interoceptive attention, may be an important step in developing more comprehensive models of interoception (Gallagher, 1997; Lutz and Thompson, 2003).

As an example of such hypothesis testing, we might consider a classical description of the first steps in mindfulness training. Classical Buddhist texts, such as the fifth century *Path of Purification,* describe the primary goal of early mindfulness training is to develop stable awareness of momentary sensation, distinguishing it from awareness of conceptual thought (Buddhaghosa, 2010). Recent scientific investigation suggests that the conceptualization of an experience, which may include detailed description and analysis of events that are associated with the experience and shaped by our collective cultural socialization, and the actual felt, immediate experience itself, which may arise in the body spontaneously, activate dissociable neural networks, and that the strength of this dissociation is indeed sensitive to mindfulness training (Farb et al., 2007). In this way, contemplative traditions provide a ground for distinction that is ripe for translation into objective, neuroscientific terms. Representation of the momentary, sensory self is relevant to practices of compassion and insight, as well to the classic mindfulness practices that undergird them.

Learning to attend to visceral, momentary sensation is only one step of many in the path to well-being, and likewise a neural distinction between immediate interoceptive and conceptual processing is only an initial step in the development of a scientific model. As will be discussed, access to visceral sensations may be helpful or harmful to a person depending on how such awareness is understood. Under the right conditions, contemplative practices may have therapeutic impact on health and well-being (Farb et al., 2012a), as has been shown in experiential psychotherapy research for some time (for a review, see Hendricks, 2001). On the other hand, interoceptive signals may also be catastrophized in panic and related disorders. Understanding how to skillfully relate to interoceptive sensations, and under what circumstances they should be attended to, is therefore a central question for the study of interoceptive training for well-being. To progress from a general description of interoceptive processing toward characterizing the wholesomeness of such processing requires a greater level of theoretical complexity, the beginnings to which we will attempt to address.

Regulation through Simulation

A scientific model of interoception requires some description of how diffuse sensory signals become integrated into a holistic representation of the body, one that is amenable to integration with higher-order cognition. The concept of a *simulation map* provides one way to characterize such integration. An analog of the contemplative subtle body, the simulation map is also known as an "as–if" representation, a neural representation of the body, a relatively stable abstraction of rapidly fluctuating sensory events (Damasio, 2003a; Seth and Critchley, 2013). From a computational perspective, the simulation map is the ongoing selection of encoded body states into a working memory buffer that serves as the best approximation of the body's current state as predicted by these prior states. The simulation map is layered, with lowest layers being closest to raw sensory afferents from the body, and higher layers representing the aggregation of information at these lower layers into representations that may be accessible to consciousness. As such, the simulation map affords executive brain areas with relatively stable sensory representations from which to interpret experience and coordinate responses. The concept of a simulation map provides a rich canvas on which to observe the nuances and variability in the cognitive representation of interoceptive signals, and furthermore suggests how such representations may be altered, both through the voluntary deployment of attention, and through the relatively involuntary processes of attentional conditioning across the lifespan, such as sensory degradation in aging (Baltes and Lindenberger, 1997).

It is important to stress that the simulation map is not identical to current sensation, but is rather an abstraction from recent sensory experience. Current sensation refers to the current set of sensory inputs arriving at primary interoceptive representation areas of the brain, most likely including the nucleus of the solitary tract, thalamus, posterior insula, and somatosensory cortices (Craig, 2002; Critchley and Harrison, 2013). The simulation map on the other hand is a filtered and integrated representation of these sensory afferents- it is, in effect, an interpreted signal. When considering regulatory motivation, a central question is the degree to which unexpected sensations are viewed as acceptable, as opposed to problematic, deviations from one's expected body state.

Like many neural representations, the contents of the simulation map may not be fully accessible to consciousness. Of the many simulation layers, only a subset will be accessible to conscious experience. Thus, the conscious, *phenomenal map* may not be the same as the broader simulation map, although we view the simulation map as necessary for phenomenal bodily experience. Furthermore, there is large inter-individual variability among phenomenal maps as a function of goals, culture, personal experience, and possibly genetics and other contributing factors (Ferron, 1997; Altabe, 1998; Ma-Kellams et al., 2012; Maister and Tsakiris, 2014). These factors may combine to limit how the simulation map is consciously accessed, and a history of biased simulation map access may override momentary introspective efforts. Given that the simulation map may only be partially accessible to conscious introspection, it may be impractical to rely entirely on subjective reports to gage individual or cultural differences in simulation map composition. However, the simulation map may be tractably examined at the level of neural representation, serving as a substrate for a person's sense of embodiment in the world. Through investigation of the simulation map, the visceral source of our sense of presence and our motivations toward action may be understood.

Neuroscience techniques have helped to move beyond subjective reports in modeling interoceptive simulation maps. By tracking changes in interoceptive signals and subjective body experiences during recordings of brain activity, researchers have begun to create rich models that distinguish between bottom–up sensory signals and top–down sensory predictions. For example, neuroimaging research suggests that attention to the body increases activity at the corresponding level of the spinal cord (Nejad, 2014), suggesting that interoceptive attention may operate even upon relatively distal aspects of the central nervous system. Within the brain, it appears that the middle insula integrates afferent interoceptive information with exteroceptive context into broader motivational space, but individuals differ in the degree to which such information propagates to the prefrontal cortex, and presumably conscious awareness, as a function of interoceptive practice (Singer et al., 2009; Farb et al., 2012b). In addition to neuroimaging, it should be noted that many other psychophysiological indicators of interoceptive processing show promise for revealing interoceptive processing in the absence of participant report, such as heartbeat-evoked potentials (Leopold and Schandry, 2001), respiratory-related potentials (Von Leupoldt et al., 2010), cardiac modulation of startle (Schulz et al., 2009), or EEG-ECG single trial covariation (Mueller et al., 2013). For example, Von Leupoldt et al. (2010) demonstrated that respiratory occlusion produced a reliable respiratory-related evoked potential measurable by EEG, which serves as an index of cortical tracking of interoceptive signals. Mueller et al. (2013) indexed interoceptive processing by the degree to which EEG responses to task feedback predicted subsequent heartbeat acceleration. While it is not our intention to describe all of such techniques in depth, such methods may provide objective indices of the impact of interoceptive signals at different levels of neural representation, such as the brainstem or cortex.

Perhaps the most important application of the simulation map is in explaining how visceral feeling promotes action. This explanatory construct is needed in both scientific and contemplative accounts of bodily health: analogous to the simulation map, contemplative traditions posit the existence of a subtle body, including various quasi-physical and delicate elements of the body that integrate and organize visceral feelings to provide a sense of well-being (Samuel, 2008). Across both scientific and contemplative traditions, motivational salience is afforded to simulation maps or subtle bodies in the form of emotional valence, with aversive or negative valence signaling a need to return to certain adaptive homeostatic ranges. The motivated process of achieving homeostasis, through physiological or behavioral change, has been dubbed *allostasis* (Sterling and Eyer, 1988).

In mammals, allostasis comprises many integrated functions and includes autonomic, neuroendocrine, and behavioral mechanisms. Much of allostasis occurs through autonomic self-regulatory physiology- it occurs internally and automatically. For example, we do not choose nor directly sense the dilation of blood vessels or pupils in response to changing luminance or emotional relevance, and yet such physiological accommodations occur continuously to promote homeostatic ends. A complete understanding of interoception doubtlessly requires the modeling of these allostatic reflexes, especially when their dysregulation acts as a source of suffering. Theories of early contemplative practice are agnostic to the ultimate accessibility of such responses; although direct experience of one's visceral stress response and its triggering conditions are essential for the types of insights that will promote the liberation from motivation driven by conditioned expectations (Goleman, 2008; Hart, 2011). At some point, simulation map theories must account for how even subtle representations can lead to conscious insight that promotes behavior change.

The relationship of the simulation map to Asian (especially Indian and Tibetan) mappings of the 'energy circuits' within the body bears further investigation. At this stage, we note that, like the contents of the simulation map, many 'energy flows' within the body are not available to consciousness. However, these 'energy flows' are *always*, in traditional theorizing, intimately associated with consciousness. These 'flows' are capable of being brought to consciousness through training. As with the simulation map, these 'energies' seem formative of a person's sense of embodiment, emotional orientation, agency, purpose, and sense of self-worth. Unlike the simulation model, perhaps, the mind-energy interaction is less dualistic than the visceral-cognitive dyad that so far seems central to characterization of the simulation map.

Leaving aside the question of how very subtle interoceptive afferents can be shaped to yield conscious insight, interoceptive signals motivate a wealth of unsubtle, overt behavior to satisfy allostatic concerns. Interoceptive dynamics are critical to understanding why identical stimuli can promote divergent behaviors. For example, a person will approach a heat source in a cold environment but avoid it when the ambient temperature is high. Conversely, an ice cube in one's hand can feel pleasant during a hot summer day but aversive in winter. The meaning of the ice cube cannot be modeled based on the sense-perception of the ice cube alone, but rather the coolness of the ice is situated within a broader interoceptive milieu, one that reward or punishes cold sensation in response to thermoregulatory imperatives. Furthermore, allostatic cues can inform cognition and behavior at higher levels of representation- our trust in others is exaggerated by warm sensations and attenuated by cold sensations (Kang et al., 2011).

If an ice cube's value is deeply influenced by allostatic concerns, a complete model of interoception must describe how such influence comes about. While acknowledging that many physiological perturbations are addressed unconsciously, through autonomic control of the internal milieu, conscious regulatory acts seem structured to resolve interactions between the body and its external environment (Gu and Fitzgerald, 2014). In this way, interoceptive signals motivate overt behavior that feeds back upon our physiology. It is also important to note that not all motivated behavior is allostatic: hedonic and pragmatic goals also have a large role to play, such as sensation-seeking to distract or regulate low mood (Taylor and Hamilton, 1997), or self-caffeinating in the face of fatigue (Lorist and Tops, 2003). Thus while homeostatic demands provide a convenient heuristic for predicting regulatory motivation, we must consider that many of our dominant drives sacrifice balance in the body to achieve other goals: riding roller-coasters, watching horror movies, and consuming spicy or sugary foods all fall into this category, in which extremes of arousal are intentionally provoked and enjoyed. Interoceptive regulation thus includes more than allostatic goals, but indeed any action intended to reshape the sensory signals that constitute the interoceptive simulation map. Rather than basing motivation on homeostasis, it may be represented more flexibly by how closely sensations match predicted or desired states.

The Predictive Coding Model of Interoception

In recognition of the complexity of interoceptive processing, the "predictive coding" model of interoception has recently been introduced to the discourse on interoception (Seth et al., 2011; Limanowski and Blankenburg, 2013; Seth, 2013; Apps and Tsakiris, 2014). This model makes use of a final critical attribute of interoception, in that interoceptive processing regularly involves a comparison between immediate sensation and simulated past and future states. Comparison between observed and expected or desired states may then motivate behavior to resolve the discrepancy (Paulus and Stein, 2010; Seth, 2013). From an evolutionary perspective, awareness of interoceptive processes may promote adaptive behavior (Damasio and Carvalho, 2013), and body-focused contemplative practices may support such awareness (Mehling et al., 2011; Price et al., 2012b). We shall argue that the predictive coding model helps to operationalize many of the claims around the role of interoception in contemplative practice, particularly surrounding the deconditioning of maladaptive regulatory habits.

The idea that the simulation map can represent states distinct from one's current interoceptive milieu suggests that comparisons are continuously being made within a given layer of the map, between the current sensation of the body, and the body as it is expected to be based upon past experience (the prior). The simulation is constructed through the consideration of both ongoing sensation and priors which contextualize these sensations. Each layer of the simulation map has its own set of priors, based on that layer's experience with past sensory inputs and adjustments to these priors influenced by top–down expectancy signals. When a prior diverges from incoming sensation, it may update the simulation at that layer, creating a posterior probability that serves as a descending input for lower layers of the simulation map. For example, a person may go outside and be surprised by unexpectedly hot weather. Such surprise may begin at lower layers of the simulation map, promoting a response that is largely unconscious and automatic, targeting local physiology, such as the dilation of blood vessels in response to a rise in temperature. If rapid, automatic regulatory responses fail to reduce prediction error (PE), a prior is updated, forming a posterior probability that updates a lower sensory layer in the map hierarchy to 'expect' hotter temperatures. It is likely that updating of priors at lower layers of the map is therefore itself involuntary and automatic, and not subject to conscious deliberation or intentionality. On the other hand, at higher levels of the simulation map, conscious deliberation may select among several regulatory alternatives for minimizing PE. In the hot weather example, one might choose to seek shelter from the heat, or appraise the heat as welcome and attempt to enjoy the warm air. Whether one selects to regulate sensation to fit the prior, or allows the prior to be updated, is an important regulatory distinction that will be discussed at length below. For now, the critical point is that the ability for sensations to motivate regulatory responses is driven by the magnitude of the PE, the deviation from the priors at a given layer of simulation. The involvement of higher order cognition in this resolution process is determined by how high an error signal processes up through layers of the simulation map before it is resolved, with larger discrepancies being more likely to make their way up through the layers to reach conscious awareness.

The comparison between sensed and expected states, reminiscent of Higgins' self-discrepancy theory (Higgins, 1987), serves two major functions: (i) to orient us to surprising physiological changes that require immediate attention, and

(ii) to allow comparison between current and potential future states as a cue to action. In the face of a mismatch between sensed and expected states, ***active inference*** promotes responses intended to shape the interoceptive milieu to match expected states. For example, given an unexpected headache, one may feel an urge to take a pain-killer, in order to restore one's desired, habitual pain-free state. Not all active inference requires overt behavioral intervention, or even awareness of a PE: some forms of active inference are physiological, wherein the autonomic nervous system regulates physiologically largely automatically and unconsciously to return the body to its expected state (Gu and Fitzgerald, 2014). On the other hand, when exposed to stressors that exceed one's autonomic regulatory capacity, cascading PEs may reach levels of the simulation map that are accessible to conscious awareness. Such awareness prompts overt active inference, by which one interacts with the environment to approach an expected state (Seth, 2013). Critical to our definition, both physiological and overt forms of active inference seek to reduce PE by changing sensation to approach prior expectations rather than updating the priors themselves.

However, active inference is not the only way to reduce the disparity between current and desired states. One may also adjust one's expected state to match the internal milieu. The process of updating the expected simulation map to more accurately reflect immediate sensation is known as ***perceptual inference*** (Seth, 2013). A person's goals and attitudes toward interoceptive sensation may powerfully influence the form of inference made (**Box 2**).

While active and perceptual inference both seek to minimize the disparity between sensed and expected states, they differ in their means of reducing this disparity. Overt active inference is a process by which an organisms acts to confirm/disconfirm attributed causes of unexpected interoceptive sensation, whereas perceptual inference acts to reduce the surprising nature of the sensation by broadening sensory expectations, reducing their inferential weight on the simulation layer. In many ways, this distinction is analogous to the difference between modern psychological and contemplative accounts of emotion regulation. While modern psychological models often discuss suppression, distraction, or reappraisal to alter the characteristics of the interoceptive signal (Gross, 2002), contemplative traditions make use of terms such as acceptance and equanimity, or simply continuous non-interfering observation, as ways of changing ones attitude toward sensation rather than attempting to change sensation itself (Mikulas, 2011). More recent secular adaptations of contemplative traditions almost universally feature integration of these perceptual inference strategies.

We note that this conceptualization of acceptance or equanimity as a 'bottom–up' perceptual rather than 'top–down' cognitive strategy may be controversial, but believe that this distinction is critical for understanding contemplative

BOX 2 | How motivation shapes interoceptive inference.

Given the power of interoceptive signals to alternately capture attention and provoke regulatory behavior, achieving an optimal balance between active and perceptual inference seems central to successful self-regulation. The tension between interrogating perception and formulating a response has been addressed in recent models of interoception processing (Hankin, 2012; Paulus et al., 2012; Seth, 2013), and we elaborate on it here to discuss the role of contemplative practice in addressing the relatively underspecified process of perceptual inference. At the level of perception, interoception requires a mental representation of body state as well as tactile or other related sensory information within the simulation map, which is constrained both by bottom-up sensory inputs, and by top–down expectations based on the stored knowledge of prior interoceptive states and knowledge of one's current context. Neither sensory afferents nor prior representations are necessarily accessible to conscious awareness, although particular layers of the simulation map may become the objects of awareness. Motivation within the model is generated as a response to discrepancies between the simulated and incoming interoceptive signal, which is also referred to as the prediction error or 'surprise' signal (**Figure 1A**). Incoming sensory signals are contrasted against prior information at a given level of the simulation layer. Discrepancies between the sensation and prior feed up through simulation layers, and generally trigger regulatory activities to maintain a homeostatic state. Ultimately, prediction errors are resolved by attributing causes to the unexpected signal, resulting in the creation of a posterior probability that cascades down simulation layers to update priors and reduce prediction error. The posterior probability determines the meaning of the signal and consequently the nature of the regulatory response.

Interoceptive *sensitivity* to unexpected, afferent interoceptive signals describes how readily a prediction error propagates up through simulation layers, serving as a basis for motivation in dynamic interoceptive representation. However, sensitivity alone does not speak to differences in how a person responds to the motivating discrepancy signal. Top–down cognitive factors in the forms of goals, causal appraisals, and contextual cues help to shape the motivated response. The tension between active and perceptual inference is often resolved by how one responds to the surprising mismatch between simulation and sensation. If current goals lead one to value *regulation* over *accuracy* of the interoceptive signal, priors at lower layers will be updated with an inferred explanatory cause (the posterior probability) from the discrepant layer that suggests overt active inference to address this cause. In this situation, surprise is minimized by down-weighting discrepant sensory information in favor of acting to restore a previously expected state (**Figure 1B**). In active inference, the simulation is closely aligned with the prior, and the individual attempts to shape incoming sensation to match the prior, thereby reducing prediction error. Conversely, if a person values accuracy over regulation, he or she will place high weight on sensory relative to prior information, updating the posterior to create a simulation that matches the unexpected afferent sensory information without attempting to restore a prior state (**Figure 1C**). In perceptual inference, the updated simulation departs from the prior and aligns with incoming sensory information; the posterior probability updates the prior rather than prompting efforts to change the sensory input, thereby reducing prediction error. In either situation, the discrepancy is minimized; the concept of sensory precision weighting allows the model to predict how such minimization occurs.

It is unknown how the different forms of inference will affect the distribution of priors in calculating prediction error. If perceptual inference allows for greater updating of priors, then it will likely promote greater variability of priors than active inference, which attempts to constrain the simulation to fit an existing set of priors. For perceptual inference, the distribution of sensory information is rendered more precise instead of the priors. These precise sensory distributions will have greater impact on higher level posterior probabilities, which feed back upon the simulation, making it more accurate while being agnostic to its effects on the prior distribution. If the body shows greater variation in sensation than in prior experience, then the priors will become less precise; if, however, it turns out that the body is displaying a sensory input that is more consistent than in prior experience, then one should expect to see more precise priors following updating. It is our yet untested hypothesis that for most people there is greater variation in the body than we tend to expect, and so the effect of contemplative practice will be to expand the range of anticipated experiences rather than hone in on any one experience, leading to a broadening of the prior distribution. The consequence of such broadening is that the simulation map will more accurately reflect sensory experience as it is less influenced by a rigid prior and more so by incoming sensation.

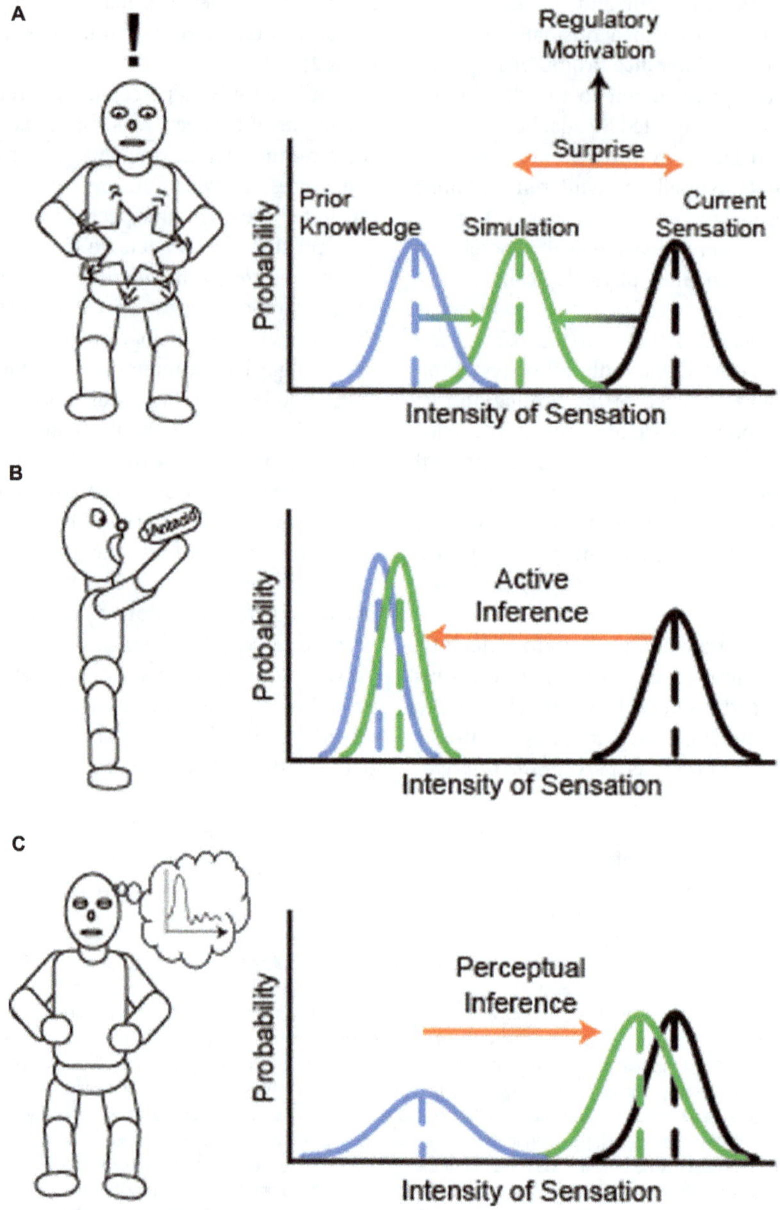

FIGURE 1 | A prediction error (PE) model of interoceptive inference, adapted from (Seth, 2013), in which interoceptive knowledge is represented by three terms: priors (blue lines), i.e., the probable body state as informed by prior events; sensation (black lines), the current sensory input from the body; and simulation (green lines), the current predicted body state based upon integration of current bodily feedback and prior learned contextual influences. Critically, our model suggests that simulation rather than raw sensation is the closest construct to interoceptive awareness. **(A)** Unexpected interoceptive events, such as a stomach ache, are represented as a PE signal that motivates a regulatory response to minimize the error signal. The simulation distribution is displayed as equidistant from the sensation and prior distributions to indicate the potential for updating from both of these information sources. **(B)** Active inference reduces PE by weighting priors over current sensation. A high-specificity range of expected body states leads to large PEs from unexpected sensations, motivating attempts to modify internal states indirectly through cognition or behavior. Successful inference reduces PE by aligning incoming sensation to match the prior. **(C)** Perceptual inference reduces simulation error by weighting current sensation over priors, updating the simulation to fit sensation. A low-specificity range of expected body states lessens immediate PEs from unexpected sensations, lending interoceptive simulations (i.e., future priors) greater sensory accuracy. Successful perceptual inference reduces PE by updating the prior to match incoming sensation. Active and perceptual inference may co-occur dynamically over time, shifting attention between sensory updating and regulatory response.

therapies' intended mechanisms of action. Indeed, contemplative therapies may be valuable precisely because they challenge existing models of 'top–down' emotion regulation by introducing the idea that sometimes attempting to control or regulate emotional experience is itself the problem. In a system where perception and appraisal are seemingly obligatory, iterating steps in human experience, perceptual inference may facilitate emotion regulation by reducing overlearned and seemingly obligatory perception–appraisal associations. Appraisals will still follow perceptions, but they need not be so stereotyped and rigidly constrained.

The distinction between active and perceptual inference matters in everyday life: for example, consider a person experiencing a sense of restlessness, who infers that her unexpected restlessness stems from hunger. In this case, active inference rapidly pairs an unexpected interoceptive signal with an externally directed behavioral response intended to restore the interoceptive milieu to its expected, homeostatic state. In doing so, active inference shifts attention away from interoception itself, at least temporarily. If it turns out that the restlessness stems from some other source, such as workplace stress, active inference's promotion of a shift from interoception to regulation reduces the opportunity to explore unfolding interoceptive signals. By contrast, perceptual inference involves reducing reliance on prior expectations in the face of an unexpected sensation. Through perceptual inference, the state of restlessness becomes the new expected state. From this perspective, the dynamic time course of arousal may be explored, including attention to conditions associated with changing interoceptive states. With time and reflection, perceptual inference might allow the individual to realize that her arousal was greatest when thinking about the workplace, and that the arousal was not actually a hunger signal. Over repeated applications of perceptual inference, prior expectations for arousal may change so that arousal following reflection on the workplace is no longer unexpected, but is instead a familiar consequence of such reflection. Knowledge of such conditioning may then afford new regulatory opportunities, a more adaptive set of priors for anticipating and explaining physiological arousal.

If perceptual inference provides grounds for personal insight, moving directly from an unexpected sensation to regulation (overt active inference) may reduce the opportunity for such insight. It is thought that such 'knee-jerk' regulatory responses to emotional arousal may be a factor that drives emotional eating behavior (Ouwens et al., 2009). In such cases, active inference does not lead to improved interoceptive accuracy, but instead maintains an error state. Indeed, overt active inference may require a masking or abstraction of nuanced and rich interoceptive signals for the purpose of promoting a rapid behavioral response.

It is not the authors' goal to denigrate the importance of active inference in allowing human beings to flexibly and dynamically adapt to the world in which they are inextricably embodied. Many unexpected sensations do require 'doing,' an active regulatory response, rather than simply 'being' with unexpected sensation (Williams, 2008). Indeed the most adaptive behavior may come from iterative cycling between perceptual and active inferences. However, the use of active inference should come with an oft-unappreciated caveat: it seems inevitable that sensory granularity will be lost when one redirects attention away from the body toward response formulation. Sensory granularity here is understood as the ability to notice specific details of internal sensory experience such as subtle changes in sensation. We posit that granularity requires the capacity for sustained attention to sensation, or least the ability to flexibly shift back and forth between sensory monitoring and conceptual inference.

Degradation of sensory granularity is an important tradeoff in shifting to a problem-solving, active inference mode of processing. In many cases, the ability to notice changes in hunger and satiety levels as one eats is adaptive, empowering feelings of satiety to terminate the eating response and maintain homeostatic energy balance. Conversely, in our example above, realization that a sense of restlessness was not alleviated by food consumption is likewise important for ruling out eating as an appropriate regulatory strategy. A failure to return to interoceptive sensation following active inference can be seen as a fixation with a mode of 'doing' rather than 'being.' Such fixation may promote continuation of food intake and potentially disrupt homeostatic energy balances if hedonic eating drives are chronic (Lowe and Butryn, 2007). If one is able to respond to unexpected sensation through perceptual rather than overt active inference, the motivational importance of even powerful interoceptive signals may be mitigated, allowing for deeper inquiry and reflection into the conditioning underlying the interoceptive signal. In modern secular culture, it has been argued that the balance between 'being' and 'doing' has been skewed toward the 'doing' mode; it may therefore take work to rebalance regulatory dynamics to optimize one's regulatory potential (Williams, 2008). Affording a transition from 'doing' to 'being is consequently a primary aim of interventions such a Mindfulness-Based Cognitive Therapy (Segal et al., 2012).

In summary, the integration of active and perceptual inference in the predictive coding model allows for the shared influence of secular and contemplative traditions in accounting for how body sensation are regulated. The idea of overt active inference fits well with the existing scientific literature- the idea that we are the curators of our bodies, needing to regulate interoceptive perturbation. Active inference can serve a variety of ends, be they allostatic, aiming to restore the body to its homeostatic baseline (Craig, 2009; Gu and Fitzgerald, 2014), hedonic, aiming to achieve some desired pleasurable, energetic or tranquil state (Kringelbach, 2005; Paulus et al., 2009; Naqvi and Bechara, 2010), or even nihilistic, aiming for freedom from sensation altogether, as is found in placebo-based analgesia (Büchel et al., 2014). Perceptual inference, by contrast is often less specified in discussions of adaptive self-regulation.

A greater understanding of the mechanisms underlying perceptual inference is important, because high sensory precision weighting may be one avenue by which to effect change in habitual regulatory behavior. Indeed, high levels of sensory granularity may be critical in challenging interpretive biases during the translation of concrete, granular levels of the simulation map into more gross or abstract levels of representation. Through attention to unelaborated emotional or physical sensation, a person may discover significant levels of tension and/or psychological distress associated with the body that were not previously in awareness, revealing a need for greater self-care (Price, 2005; Price et al., 2012a, 2013). Conversely, interoceptive attention may reveal that an allostatic habit such as self-medication is not always necessary in the face of noxious interoceptive sensation (Kabat-Zinn et al., 1985). Recent research suggests that the brains of experienced meditators demonstrate reduced PEs to passively viewed reward (Kirk and Montague, 2015), indicative of less clinging or attachment to positive outcomes that may serve as a basis for addictive behavior. Constructively, perceptual inference may also make people aware of inner resources, affording access to previously hidden capacities for calm, contentment, and appraisals that things are going well just as they are. In this way, every moment requires a choice between perceptual and active inference. And yet, how does a person know which information to use or whether he or she has enough information to act adaptively in response to their goals? When should one switch from perceptual to active inference? The ability to respond appropriately to interoceptive perturbation is a challenging problem that can serve to determine a person's sense of well-being across the lifespan.

With respect to the relevance of contemplative practice, a final point warrants mention surrounding the accessibility of active and perceptual inference. While ideally we are presented with a choice between the two error-minimizing responses, often this is not the case. While active inference appears to be a ubiquitous, evolutionarily conserved regulatory strategy (McBride, 2012), reframing unexpected sensations as a perceptual problem may require the cultivation of attentional capacities, intentions, and attitudes that are less intuitive. Our bodies perform physiological active inference even before we are born (Moor, 1968), and learning to apply overt active inference in the form of clothing, feeding, and otherwise caring for ourselves is a complex but near-universal feature of childhood development. Perceptual inference at these early stages may not be possible given limited cognitive resources or maturity of perspective, and indeed it may be counter-productive to an individual's mastery of overt active inference responses. As we mature, however, we realize that not all unexpected sensations can be corrected, or that our habitual, reflexive corrective habits are actually maladaptive. While behavioral therapies address this problem by working to bring such overt active inference habits to awareness so that they may be fruitfully restructured, perceptual inference may require development of a qualitatively different skill set. Thus it is important to consider that active and perceptual inferences are not equally available, and that perceptual inference tends to be the less available of the two without specific types of learning experiences.

Interoception in Health, Disease, and Well-Being

It has been proposed that mental representations of selfhood are based fundamentally on embodied sensory experience, supporting a sense of the self in the world that is crucial for interacting with the environment (Seth, 2013). From this perspective, greater accuracy of interoceptive self-representation promotes greater moment-by-moment adaptation, whereas dissociation from accurate representation can lead to dysregulation. Accordingly, many contemporary health problems involve dysregulated interoceptive processes, including affective disorders (Paulus and Stein, 2010), addiction (Naqvi and Bechara, 2010), eating disorders (Garner et al., 1983; Pollatos et al., 2008; Herbert and Pollatos, 2014), chronic pain (Schmidt et al., 1989), dissociative disorders (Hankin, 2012; Michal et al., 2014; Sedeño et al., 2014), post-traumatic stress disorder (PTSD; Wald and Taylor, 2008), and somatoform disorders (Mirams et al., 2012; Schaefer et al., 2012). Understanding how interoceptive processes influence representations of the self in the world and self-regulation may lead to improved disease and treatment models (Pollatos et al., 2005).

Presence and Agency

One approach to understanding dysfunctions in interoceptive awareness is to apply the predictive coding framework as an explanatory model distinguishing adaptive and maladaptive processing. Toward this end, two extensions of the model have been proposed, describing how inference processes outlined above operate to minimize PE (Seth et al., 2011). The first of these concepts is *presence*: presence is thought to arise when interoceptive and/or exteroceptive PE signals are successfully minimized. In Seth et al.'s (2011) account, such error minimization has been attributed to either overt or physiological active inference; as one successfully resolves unexpected sensation through autonomic or behavioral responses, a feeling of engagement and connection with one's body and environment ensues. However, we propose that, from a contemplative perspective, presence is accessible through perceptual inference as well.

There are several ways that PEs may be reduced to give rise to presence: in a virtual reality environment, interoceptive signals may be successfully masked by salient external signals, such as seeing oneself walking in the absence of proprioceptive motion signals, leading to a sense of presence through acceptance of the external signal as one's own embodied state. Given predictable and salient external cues in a virtual reality environment, a sense of presence may come more easily than trying to detect and match faint and chaotic interoceptive cues within the noisy internal milieu. By contrast, in contemplative practice the sense of presence through a variety of visceral experiences may broaden the distribution of interoceptive expectations (priors), reducing the precision of those priors, and thereby minimizing the potential for PE. In other words, in a simulation map that allows for great variation in sensory inputs, relatively few visceral sensations are extreme enough to create a PE demanding a regulatory response.

We suggest that with successful iterations of perceptual inference, the influence of prior expectations is thereby weakened, leading to feelings of automatic simulation map updating: effortless presence (Sjölie, 2014). While healthy individuals likely experience such effortless presence as a matter of course, it is likely that such experiences operate across a continuum of effort, with extreme difficulties in achieving presence manifesting as depersonalization and derealization disorders (Seth et al., 2011). However, even within the healthy population, the degree to which presence is experienced and maintained may be dependent upon regulatory strategies and habits. A person whose regulatory habit is overt active inference may spend a great deal of time pursuing idealized interoceptive states, rather than learning to allow autonomic regulation to more subtly achieve and maintain such states. Such a person would require greater effort to achieve feelings of presence. Furthermore, a reliance on active inference in general may reduce the possibility for presence when a sensation cannot be easily mapped to an active regulatory pathway, be it through overt behavior or autonomic regulation. In such situations, effortless presence may seem like an important but unachievable state, similar to the ideal of accepting one's negative thoughts and feelings for individuals with a history of depression or anxiety (Pauley and McPherson, 2010). For such individuals, learning to reduce PE through perceptual rather than active inference may constitute a radical shift in self-processing, the introduction of a new regulation strategy based on acceptance rather than control. It is unknown how variable feelings of presence are in the general population; but we would hypothesize that effortless experiences of presence are fleeting at best in the frenetic, "what's next"-paced endemic of modern Western culture.

As an example of presence functioning adaptively, one might imagine the perception and then acceptance of an unexpected, transient increase in heart rate. Following the initial upset, a person may engage in perceptual inference, exploring the interoceptive signal, which we operationalize as an attempt to update the simulation map to match this signal. Acceptance would then occur when perceptual inference successfully simulates the elevated heart rate as the normal, expected state. By accepting the internal change, the motivational force of the physiological change is minimized without requiring the suppression of either expectations or interoceptive signals. By contrast, the use of overt active inference to resolve unexpected events has been associated with the sense of *agency* (Seth et al., 2011), a feeling of control over one's actions in the world, that one can act to produce particular results. Agency can be inspired both by direct regulation of bodily sensation, as well as through changes in behavioral patterns that indirectly contribute to one's interoceptive state. In the context of our discussion of interoception and well-being, the sense of agency is important for supporting a sense of responsibility for self-regulation. This applies both to conventional forms of overt active inference, such as quitting a stressful job and going for a less paid but more livable employment, but also to contemplative practices such as learning that one can reduce feelings of anxious arousal through breath monitoring or other regulatory activities.

Presence and agency are important in that they establish norms for experience and control that guide behavior and ultimately determine one's sense of well-being. In anxiety disorders, for instance, agency is achieved through withdrawal from stressful situations, a form of active inference that is effective in the short term but ultimately maladaptive, as that person never learns that the feared outcome is unlikely to be true. In anxiety, PEs are augmented- a person may experience a strongly aversive response to a previously neutral stimulus (Paulus and Stein, 2006). This aversive response is then augmented rather than extinguished when the anxious individual makes catastrophic appraisals of the aversive sensation that are paired with active inferences to manage these perceived catastrophes. For example, if a racing heartbeat during exercise is appraised as a heart attack; active inference requires a trip to the hospital. If rapid breathing prior to public speaking is appraised as lack of ability that will lead to embarrassment, active inference requires social withdrawal. Faced with threats to agency, maladaptive behaviors are reinforced, which ultimately interferes with exploration and modification of one's relationship to stressful events. From a predictive coding model, chronic stress may lead to a reduction in the precision of PE coding in favor of strong priors around the recognition of aversive sensation and confirmatory active inference behaviors to alleviate that sensation. While it is unpleasant to experience unexpected aversive arousal, it is this secondary appraisal around agency, the threat to control over the situation that is the cause of deep distress. The idea that higher order constructs such as agency are at the heart of subjective well-being is a central theme of modern appraisal theory (Scherer et al., 2001), just as presence lies at the heart of contemplative accounts of well-being (Brown and Ryan, 2003). These theories serve to reinforce the point that subjective mental health stems from our inferences about coping capacity, for which each attempt at perceptual or active inference serves as a momentary test. If imprecise PEs prompt inefficient or maladaptive active inference behaviors, inferences of coping failure are likely, exacerbating feelings of powerless and inefficacy and contributing to the deleterious impact of the stressor.

In its ability to link higher order appraisals of self-worth to momentary regulatory acts, the predictive coding framework is helpful in understanding many other disorders, because it suggests that well-being requires active sustenance, i.e., that presence and agency must be continuously constructed through successful reduction of PE (Paulus, 2007), regardless of the balance between active or perceptual inference. This perspective stands in contrast to notions that isolation from our bodies and the world is preferential except to alert us about negative events. Instead, it is the lack of integration of accurate interoceptive simulations into higher order representations that is hypothesized to underlie dysregulated cognition and behavior. In this observation there is considerable convergence with contemplative perspectives. For example, following trauma, some individuals may experience dissociation from bodily experience, as is found in some forms of severe depression and PTSD (Feeny et al., 2000). In such cases, the priors associated with body error signals may have been linked to traumatic experiences, which are too powerful to be actively controlled (active inference), and too aversive to be accepted or, as in many contemplative orientations, altered simply by being observed (both instances of perceptual inference). Since the error signals cannot be minimized, the simulation map itself may be suppressed in the form of *experiential avoidance*, a form of protective habituation to powerful and uncontrollable interoceptive signals, or *behavioral avoidance* of provocative situations altogether. Such inhibition may amount to a reduction of the precision of priors in the map, so that only very powerful sensory signals are able to trigger an upward flow of PEs toward higher simulation layers. While avoidance provides momentary relief from the attentional pull of PE signals, inhibition of interoceptive awareness may also create its own host of psychosocial issues, both through the obfuscation of important interoceptive cues, the absence of desirable interoceptive sensations, and the maintenance of dysfunctional beliefs about the world and their relation to it (Paulus and Stein, 2010). Indeed, the severity of somatic dissociation has been linked to greater susceptibility to experimentally induced somatic illusions, suggesting an over-reliance on prior knowledge that interferes with current sensory input (McKenzie and Newport, 2015). Many contemplative traditions begin from the assumption that a suffering individual's interoceptive integration may be dysfunctional or deficient; to rectify this situation, attention to the body may delay cognitive appraisal, creating space to restructure the integration process (Shapiro et al., 2008).

Interoceptive Training in Clinical Practice

Specific examples of dysfunctional interoceptive integration are myriad in psychological disorders. For instance, when an individual with severe attachment disturbances considers close social interaction, there is often an immediate discrepancy between expected and actual feelings of connectedness, providing a new source of interoceptive surprise that may be anxiously appraised (Rector et al., 2006; Gilbert et al., 2012). Similarly, it is not uncommon for a person with a severe history of trauma to be so anxious that it is not possible to experience the positive sensations of close interactions (Fogel, 2009; Frewen et al., 2012). In the absence of new interoceptive information, such fears and concerns may take on the quality of rumination, perseverative thinking that is associated with affective disorder vulnerability (Nolen-Hoeksema, 2000). The predictive coding framework provides an explanation for how rumination can dominate attention: when simulation maps operate with low perceptual weighting, attention is more often drawn to active regulation, leaving little opportunity for experiences of interoceptive presence, a feeling that a sensation is acceptable and tolerable, that may challenge sustained bleak and dysphoric expectations. Contemplative training is largely oriented to increasing one's capacity to de-habituate such coding. Commonly, such cognitive distortions of reality may continue unfettered, leading to a downward spiral that characterizes the chronic and recurrent nature of affective disorders (Raes et al., 2008).

And yet, despite the importance of perceptual inference for well-being, we should stress that such inference alone is not a panacea. Perceptual experience is not easily divorced from subsequent appraisal, and so even accurate and precise interoception may activate powerfully conditioned negative associations and appraisals (Treleaven, 2009). For example, in anorexia nervosa, patients may possess powerful associations between awareness of bodily states associated with starvation and a 'doing' mode that focuses on the control of eating, shape and weight, ultimately culminating in interoceptive suppression once more (Watts et al., 2007; Park et al., 2011). In anorexia, bringing awareness to the body will initially be met with an increase rather than decrease in maladaptive behavior. Thus there is a reason that interoception becomes disrupted, and skillful guidance is particularly important in restoring interoceptive access.

If a regulatory framework is not in place, such as the capacity for acceptance of ambiguous or challenging sensations, restoring interoceptive access- or advocating contemplative intervention- could introduce new trauma. For example, enhancing interoceptive awareness without compensatory regulation may be maladaptive to a severely depressed and suicidal individual or someone with acute pain from a severe burn. In many cases, however, it seems that specifically trained, skillful clinical or contemplative guidance can successfully address the challenges inherent to interoceptive re-engagement. For example, research with women in treatment for substance use disorder suggests that when they are encouraged to discover the connection between physiological and emotional distress, they often learn that they have the capacity and skills to attend and negotiate emotional stress. Enhanced interoceptive awareness provides nuanced cues for self-care that facilitate emotion regulation, reducing conditioned substance-use responses to stress, and allowing such patients to maintain sobriety (Bowen et al., 2007; Price et al., 2012b).

The need for a skillful interpretive framework is apparent in psychometric research on mindfulness of body sensation, such as the 'observe' facet of the five-factor mindfulness questionnaire (Baer et al., 2008). This research on the questionnaire suggests that high levels of this 'observe' facet are associated with mixed health outcomes in the general population, but with more uniformly positive outcomes in those who are relatively trauma-free and have received a background in contemplative training. A recently published report pointed out that the commonly assessed 'observe' facet, becoming aware of bodily changes, may change with these approaches to a much smaller degree than the regulatory aspects of interoceptive awareness, that is, how the body is used for self-regulation in daily life (Bornemann et al., 2014). In this case, a multidimensional self-report measure was sufficient for distinguishing between aspects of interoceptive change, and serves as an important precedent for further clinical research. It should be noted, however, that the bulk of the clinical research discussed in the following section has employed only more global qualitative reports of interoceptive change, a finding which may in fact be driven by training effects on only a subset of interoception's many facets.

Clinical Examples in Patients with PTSD or Chronic Pain

To clarify the relevance of interoception in the promotion of well-being, it may be useful to consider examples where the cultivation of adaptive interoception has yielded clinical success. Emerging research suggests that contemplative practices may powerfully support the process of interoceptive re-discovery. In the treatment of women with a history of interpersonal sexual violence, participants learn to engage in perceptual inference, recognizing when sensory cues of dissociation are triggered emotionally, allowing them to maintain awareness of their bodies instead of dissociating from those sensations into habitual, active regulatory responses.

Over time, participants discover that their bodies can be a helpful informative resource rather than a source of threat signals that should be avoided. Thus, body sensations not previously incorporated into awareness can be allowed into phenomenological experience, and can be more appropriately integrated into self-schemas. These experiences can lead to a greater sense of safety in the world, greater ability to engage in intimate interactions with spouse/partner without dissociating, greater ability to negotiate stressful environments and interactions, and lastly a greater sense of wholeness and empowerment (Price, 2005, 2007). It remains to be determined whether these contemplative practices increase interoceptive accuracy directly, or instead support other elements of our interoceptive taxonomy, such as increasing sensitivity to interoceptive events and moderating subsequent regulatory habits.

A further example of contemplative practices' potential lies in the psychological management of chronic pain, in which the most commonly applied approach is currently a combination of cognitive reframing by cognitive behavioral therapy and attentional distraction (Hoffman et al., 2007). Recently, however, it has been questioned whether distraction works for chronic pain as well as it does for acute experimental pain (Goubert et al., 2004). One reason for reconsidering distraction techniques is that part of chronic pain pathology may actually revolve around maladaptive fear associations that impoverish interoceptive processing and exacerbate pain syndromes, conditioned withdrawal associations which distraction would only serve to reinforce rather than challenge (Zaman et al., 2015). Instead, "interoceptive exposure" may be an alternative (Craske et al., 1997; Boswell et al., 2013). Yoga includes interoceptive training, is associated with decreased prefrontal brain activity and has shown benefits for pain management (Villemure et al., 2013). Keeping in mind concerns that interoceptive awareness is not always beneficial, helpful, or tolerable, it would seem that these and other examples suggest that in many cases, careful guidance of attention, informed by contemplative theory, and with case-sensitive guidance, may allow people to reconnect with their bodies with great therapeutic potential. With appropriate support, patients may learn to tolerate negatively conditioned experiences of the body, eventually accepting their prepotent aversive response, which frees cognitive resources for reconstruction of the appraisal process (Park et al., 2012).

In summary, interoception is important for well-being, at a pragmatic level of maintaining desired physiological states within the body, but also at an epistemic level for its contributions to perceptions of presence and agency. Indeed, many psychological disorders are characterized by disruptions to presence and agency in the form of dissociation or hopelessness, or by maladaptive solutions to agency violations, as is found in anxious withdrawal. Contemplative training may be effective to restore adaptive interoceptive dynamics that address these violations, keeping in mind the caveat that the reframing of interoceptive signals must be carried out with care and support to increase tolerance for aversively conditioned sensation. How such practices may accomplish this restorative feat is the topic of the next section.

Contemplative Practices for Revitalizing Interoception: The Example of Mindfulness

Given compelling evidence that interoceptive processes are integral to many forms of affective and related disorders, we might ask how interoception can be restored to an optimal state. Clinical experience suggests that trusting in body signals as potentially decision-guiding information and valuing the body as an important resource in directing one's behavior may be key conditional precursors for this change (Mehling et al., 2011). In the predictive coding framework, trust or acceptance of interoceptive signals fits with our discussion of increased sensory weighting, perceptual inference, and the cultivation of feelings of presence. While modern scientific traditions have much to say about the mechanics of the body, how such mechanics can be used to cultivate a more embodied phenomenological state has not historically been a focus of these traditions. It is in this effort to understand the cultivation and sustenance of presence and agency that contemplative science may be of value.

Many body-based contemplative practices involve explicit direction of attention to interoceptive sensations. The exploration of interoceptive awareness under many other names is central to Asian contemplative, medical, and philosophical traditions. Importantly, these are presented in a philosophical context of exploring errors about the self and individual subjectivity that result in part from ignoring, misinterpreting, or missing more subtle levels of an individual's interoceptive experience. Such contemplative practices, although they may have numerous other goals in addition to training interoceptive awareness, offer a method for training in such awareness and, in many cases, for reorienting experience from, for example, distraction to attentional control, effort to ease, separateness to connection. All these shifts are deemed beneficial. In this section, we will focus on mindfulness meditation traditions as they contribute to a sense of presence through largely stationary interoceptive attention practices. It is however important to note that other, movement-based traditions such as yoga or tai chi may be especially well-suited toward the cultivation of agency. A great deal of research is needed to explore the effects of particular practices on the predictive coding of bodily awareness.

Among contemplative traditions, mindfulness has recently received a particularly strong representation in modern sciences. There is an extensive literature on the health benefits of mindfulness approaches- for overviews see (Grossman et al., 2004; Chiesa and Serretti, 2009; Hofmann et al., 2010), particularly in the area of preventing relapse in those vulnerable to depression (Goyal et al., 2014). It should be noted that interoceptive training is only one aspect of mindfulness interventions, which also emphasize changing one's relationship to thought content. Nonetheless, such interventions may represent the most prominent introduction of interoceptive training in the West. Mindfulness approaches may be particularly efficacious in reducing somatic dissociation from chronic pain and sexual trauma (Price, 2007) but also in other mental health disorders such as substance use (Price et al., 2012b) depression (Williams et al., 2013), anxiety (Hoge et al., 2013) and eating dysregulation (Daubenmier et al., 2011). There are suggestions that interoceptive mindfulness practices yield changes across a variety of cognitive domains, including self-reference (Brown and Ryan, 2003; Farb et al., 2007; Ingram et al., 2011), attention (Jha et al., 2007; De Raedt and Koster, 2010), emotion regulation (Ortner et al., 2007; Jha et al., 2010), pain perception (Zeidan et al., 2010a; Gard et al., 2012), agency (Allen et al., 2009), and feelings of social connectedness (Hutcherson et al., 2008; Neff and Germer, 2013); see (Hölzel et al., 2011) for an integrative review.

Modern scientific definitions conceptualize mindfulness as an open, engaged, and non-judgmental awareness of the ongoing flux of present moment experience, including internal experiences of sensations, thoughts, and feelings, as well as exteroceptive sensations. It has been argued that one of the primary and early means by which mindfulness benefits its practitioners is by anchoring attention to interoceptive signals such as the breath or body sensation (Mehling et al., 2011; Kerr et al., 2013; Farb et al., 2014), an idea supported by training-related reports of both increased subjective body awareness (Mehling, 2001) and increased strength of brain networks dedicated to interoceptive processing (Farb et al., 2013). *Focal attention* to bodily sensations is at least one method by which mindfulness training shapes cognition (Hölzel et al., 2011); with stability of interoceptive anchoring, practitioners also engage in *open monitoring* of experience, which, from the modern secular perspective of using contemplative practice for therapeutic effects, reveals and weakens maladaptive reactive patterns so that they may be adaptively modified (Lutz et al., 2008; van den Hurk et al., 2010). From a predictive coding framework, body-focused contemplative practices may alter interoceptive processing by shifting regulatory habits from active to perceptual inference, increasing bottom–up integration of what is happening in the body rather than attempting to alter body sensation to fit top–down expectations of what should happen in the body (Pagnoni and Porro, 2014). Since even low level perceptions of interoceptive sensations commonly rely on the integration of both sensory and expectation processes, engaging in a contemplative interoceptive practice may reduce the precision of prior expectations, reducing the ability of priors to trigger PEs in response to incoming sensations. The consequence of this widening of acceptable sensory inputs may therefore allow for more accurate and dynamic representations of sensation, yielding more nuanced, and adaptive behavioral responses.

The wealth of empirical studies on mindfulness training mechanisms makes it a useful starting point in discussing how contemplative practices may impact interoceptive processing. Below, we describe several domains in which contemplative traditions such as mindfulness may positively impact interoceptive processing, acknowledging that these domains likely interact and support each other in an iterative and dynamic (rather than linear) manner.

Enhanced Sensitivity

Contrary to the perceptions of even experienced meditators, mindfulness does not appear to generally increase interoceptive sensitivity when assessed in laboratory settings, at least where the most popular metrics of interoceptive accuracy, heartbeat perception, is concerned (Khalsa et al., 2008; Parkin et al., 2013). Mindfulness may, however, increase interoceptive sensitivity in domains that are the foci of meditative practice, such as sensation of the breath (Daubenmier et al., 2013), or interoceptive cues indicating the presence of subtle reactive patterns. Even brief body-scan meditations reduce errors in a subtle somatic signal detection task (Mirams et al., 2013), similar to the enhanced tactile acuity related to movement-based tai chi practices (Kerr et al., 2008). Furthermore, mindfulness training appears to alter interoceptive attention tendencies, focusing attention on interoceptive sensations rather than cognitive appraisals of such sensations (Garland et al., 2012), in keeping with a model in which contemplative practices shift regulatory strategy from active to perceptual inference. As such, sensitivity enhancements following mindfulness training may constitute a kind of 'embodied ethic' that promotes interoceptive attention deployment, a daily commitment to perceptual inference rather than an enhancement of sensitivity itself (Grossman, 2015).

One consequence of increased perceptual inference is increased *granularity* of interoceptive experience, reducing the self-appraised emotional impact of experience in favor of enhanced clarity and sensitivity to subtle emotion provocation (Nielsen and Kaszniak, 2006). Such sensitivity may manifest at multiple levels of the simulation map, allowing practitioners to recognize subtle, temporally extended dynamics of physiological arousal, dynamics that would ordinarily be obfuscated by definitive cognitive appraisals in response to early perturbations. The benefit of enhanced granularity is an opportunity to learn more about one's body and its conditioning in the world. Indeed, mindfulness practitioners show increased accuracy between subjective and objective measures of body sensitivity, an index of the relative sensitivity to different body regions (Fox et al., 2012), and increased coherence between physiological and subjective states (Sze et al., 2010). Such coherence allows for greater appreciation of how stressful situations impact both mind and body, increasing the chances of adaptive regulatory action. Furthermore, because interoception engages the same neural pathways supporting pro-social emotions such as empathy (Singer et al., 2009), increased interoceptive sensitivity may also lead to improved social function, although direct evidence for this relationship is still needed.

Enhanced Non-Reactivity

The deeper benefits of contemplative practices lie in leveraging non-reactivity to generate adaptive regulatory insights. The predictive-coding model can account for the availability of such insights through the idea of *precision weighting*, i.e., how much a person divides finite metabolic and attentional resources between (i) representing, exploring, and accepting unexpected sensory signals, i.e., PEs, and (ii) maintaining prior expectations. As discussed, attention can lead to greater precision weighting of sensation over priors, which in turn promotes perceptual as opposed to overt active inference (**Figure 1**). Both perceptual and overt active inferences require the use of metabolic resources, as evidenced by the feelings of effort when one first engages in either process. However, only overt active inference engages cognitive elaboration, requiring cognitive appraisal of the many action affordances available in response to interoceptive perturbations. Given that overt active inference is likely to consume considerably greater metabolic resources, individuals must attempt to optimize the precision of incoming sensory data and priors to minimize wasted regulatory effort (Fotopoulou, 2013). Unfortunately, for individuals with strongly conditioned regulatory tendencies, i.e., highly precise tuning of priors, this minimization may prove difficult in the face of seemingly automatic and obligatory regulatory responses. Resolving the disconnection between the ideal of regulatory efficiency and the reality of conditioned active inference is a non-trivial problem.

We hypothesize that one major benefit of contemplative interoceptive practice is that it provides a pathway toward perceptual inference in the face of such conditioning, effectively relaxing appraisal tendencies. While still demanding attention, perceptual inference frees higher order cognition to arise and pass without particular attachment to a given appraisal of the simulated state. In other words, perceptual inference frees higher order resources to allow metacognition, representation of the mind's current state. The simple act of observation itself, as mentioned, can create a shift, possibly even without setting in motion higher-order resources. The idea that one's thoughts can be viewed as transient mental events rather than as cues to immediate action is often referred to as *decentering* or *reperceiving* in the mindfulness literature (Shapiro et al., 2006; Fresco et al., 2007), and may be an adaptive consequence of responding to arousing experiences using perceptual rather than active inference. Decentering may also be related to the intentional arc, which sets mindfulness practice in motion and sustains it. Mindfulness training seems particularly well-suited for promoting decentering relative to progressive muscle relaxation or loving kindness interventions (Feldman et al., 2010), perhaps due to the explicit enhancement of sensory precision weighting at the expense of active inference. Such decentering may not be an end unto itself, but also afford flexible 're-centering' on habitually ignored sensations that constitute meaningful, constructive, and positive experiences. Accordingly, mindfulness experience has been associated with increased cognitive flexibility (Moore and Malinowski, 2009). In this way, dedicated perceptual processing of interoceptive signals may serve to enhance conceptual non-reactivity to experience that in turn promotes well-being.

Enhanced Regulation

In the metacognitive space that is afforded by enhanced non-reactivity, multiple interoceptive appraisals may be observed as arising and passing. From a decentered perspective, a person may become aware both of her most prepotent, i.e., frequently and powerfully occurring, appraisal tendencies, but also of alternative appraisal options. From awareness of multiple options, flexibility of choice can be directly experienced, thereby allowing for novel, creative, and potentially more adaptive appraisals and actions over time. As an example, a person in a stressful public speaking situation may use mindful attention to notice her elevated physiological response. From this perspective, appraisals of threat and incompetence may rapidly arise, but with continued attention toward sensory perception, such appraisals may also pass, interspersed with weaker, and less frequent appraisals of potential failure, support from loved ones, and feelings of determination. With this greater tableau of appraisals before her, the speaker may elect to appraise her arousal as a surmountable challenge rather than an imminent threat that requires withdrawal from the situation. It is through this process of forming insight that interoceptive attention may iteratively shift engagement away from habitual appraisals to promote experiences of insight and choice.

With practice, iterations of weighting attention toward perceptual inference may allow an individual to more permanently decondition and eventually replace appraisals of environmental stressors as threat with exploration of such stressors, appraising them instead as challenges to be explored, a distinction with important consequences for reducing physiological stress (Tomaka et al., 1993). Over time, this process of overcoming habitual dysphoric or catastrophic appraisals in favor of a less rigidly deterministic simulation map may result in positive trajectories for transformation, upward spirals in the promotion of well-being (Garland et al., 2010, 2011).

One example of adaptive sensory precision weighting in mindfulness can be found in a study of pain perception. Pain perception is theorized to have both a sensory component, representing the intensity and location of the interoceptive signal, and an affective component, representing the signal's motivational relevance in term of pleasantness or unpleasantness (Melzack, 1975), a distinction supported by neuroimaging research (Tölle et al., 1999). In one recent study (Farb et al., 2013), mindful attention in experienced meditators was associated with increased attention to bottom-up signals as reflected in increased posterior insula activation, a region of primary interoceptive representation (Craig, 2002; Farb et al., 2012b), but decreased top–down processing as reflected in decreased lateral prefrontal cortical activity, a region involved in cognitive appraisal (Ochsner et al., 2004). These neural changes were linked to a change in the pain experience, such that pain was perceived as similarly intense but less unpleasant during a state of mindful attention. In the absence of cognitive judgment, cognition may be freed to consider alternative interpretations of sensory states. From a regulatory perspective, such freedom allows an individual to explore different forms of active inference, and may perhaps even reveal that active regulation of interoceptive signals is no longer necessary.

Enhanced Insight

While perceptual inference, by definition, precludes immediate cognitive elaboration on sensation, in the long term, a greater corpus of interoceptive information provides a richer set of data from which to investigate habitual sources of interoceptive perturbation, to identify the relationship between inner somatic experience and cognitive experience, and one's internal responses to outside events and stimuli. This meta-cognitive awareness may lead to *insight*, recognition of how events, emotions, thoughts, and bodily sensations relate to each other (Lavie et al., 2003; Sze et al., 2010). For example, recent studies in body ownership show that a sense of one's body can be manipulated by experimentally induced visual and tactile feedback (Ainley et al., 2013; Suzuki et al., 2013) but those with greater interoceptive accuracy (as assessed by heartbeat detection) are less susceptible to illusions of body-ownership (Tsakiris et al., 2011). This 'rubber hand illusion' is a good example of prioritizing attention toward priors for the interpretation of visual input over attention toward PEs stemming from afferent interoceptive signals from the hand. For insight to occur one need not have perfect accuracy in such interoceptive signal representation; however, insight could emerge as a process of weighting sensory PEs over priors, leading to a higher-fidelity simulation of body state. In theory, contemplative practices could similarly reduce false inferences about the relationship between one's body and the world.

As interoceptive signals inform emotional experience, contemplative practice may promote a cycle of awareness of the contingencies between environmental triggers, bodily responses, cognitive appraisals, and emotional experiences, knowledge which can then be leveraged to regulate cognition and behavior in the service of emotional well-being. We argue that this process optimally occurs when interoception is viewed as foundational to emotional experience, and thus interoceptive attention becomes a basis for engaging in emotional processing, enhancing awareness, and regulation of rapidly escalating emotional responses to stress. For example, in a study for women in substance use disorder treatment, those taught interoceptive awareness and related skills for self-care perceived such awareness as facilitating their ability to identify, accept, and process their emotions, key regulating factors that they attributed to successfully preventing relapse to substance use (Price et al., 2012a). Through these introspective cycles, insight is fostered that radically modifies inference of interoceptive signals, toward perceptual over active inference, exploring causal factors rather than requiring an immediate inference and response to interoceptive causality. Despite such enticing claims, more research is needed to substantiate the idea that mindfulness practice helps to alter the extent to which particular interoceptive signals are attended to and received with finer granularity in daily life.

Enhanced Presence and Agency

If the link between perceptual inference and insight can be empirically validated, there are many downstream benefits that may ensue. For example, mindfulness' ability to curtail habitual reactive tendencies may enhance one's sense of *presence* and *agency*, and indeed mindfulness training has been associated with increased motor control during perceptual-motor conflicts (Teper and Inzlicht, 2013). Enhanced perceptual-motor integration reflecting increased agency may impact self-representations related to one's ability to control the environment, which may have important consequences for one's sense of well-being. Minimally, the increased weighting of interoceptive signal may increase interoceptive accuracy, decreasing the impact of dysfunctional simulations and self-representations on cognitive and behavior, as previously proposed (Farb et al., 2012a). For example, depression is characterized by external locus of control, learned helplessness, and low self-efficacy, and some evidence suggests depressed people have poorer interoceptive accuracy (Ehlers and Breuer, 1992; Pessoa and Ungerleider, 2004; Dunn et al., 2007; Pollatos et al., 2009), whereas mindfulness training appears to bolster self-efficacy (Chang et al., 2004). Furthermore, mindfulness training has been associated with normalization of the gait pattern of depressed individuals, suggesting concurrent changes in proprioceptive and emotional responses (Michalak et al., 2011). Presence and agency appears to matter for physiological self-regulation: the rubber hand illusion, in which visual-tactile illusions decrease a sense of ownership over one's arm, results in decreased skin perfusion and temperature (Moseley et al., 2008), and increased stress hormone release in that arm (Barnsley et al., 2011), without providing commensurate pain relief in that arm (Mohan et al., 2012). These findings suggest that a strategy for minimizing interoceptive PEs through substitution of interoceptive inputs with exteroceptive inputs leads to dysfunctional, and, in the long run, unhealthy physiological changes in the respective bodily region. Conversely, we would hypothesize that minimization of interoceptive PE by increased precision weighting of sensation over priors may create a sense of embodied presence and be able to reverse such dysfunctional physiological changes.

Over time, an increased sense of presence and agency may also begin to enhance higher order self-representations such as self-esteem. Through this process, a person may hold greater confidence in his engagement with interoceptive inference, which may in turn reinforce the exploratory cycling between perceptual and active inference, again promoting upward spirals in the promotion of well-being (Garland et al., 2010, 2011). It remains to be tested whether enhanced interoceptive attention may produce differential cognitive-emotional-physiological effects when placed in different parts of the body and contemplative theories may guide the formulation of such hypotheses. Future research could examine to what extent contemplative practices enhance a sense of presence or agency that can account for improvements in self-efficacy and related aspects of mental health. In particular, movement practices such as yoga and tai chi may more readily promote an increased sense of agency in which the opportunity to explore sensorimotor signals in slow, controlled movements is readily apparent.

Increased Positive Experiences

Another downstream consequence of enhanced interoceptive capacity may be the ability to engage with and appreciate *pleasant sensations*. In the case of mindful eating interventions, for example, practices involve bringing greater attention to the pleasure of seeing, smelling, tasting, and eating of palatable foods and noticing when the satisfaction subsides, as taste-specific satiety mechanisms become apparent (Kristeller and Wolever, 2010; Daubenmier et al., 2011). Indeed, many mindfulness training programs involve a raisin-eating exercise (Kabat-Zinn, 1990), in which increased sensory precision weighting is encouraged to decenter habitual interpretations of the consumptive act, replaced with re-centering on unexpected positive aspects of the experience. The ability to enter into such sensory states and appreciate positive sensations is important, because reduced response in the brain's reward regions to consumption of palatable foods such as milkshakes is associated with increased, unwanted weight gain (Stice et al., 2010). Such reduced activation may account for the subjective experience of "chasing the flavor" and continued eating. It remains to be determined whether increased awareness of the pleasurable taste of palatable foods increases reward activation and ultimately reduces food consumption, in particular among those who are overweight. Similarly, mindfulness-based interventions during recovery from drug addiction appear to increase interoceptive responsiveness to natural reward while decreasing responsiveness to drug cues, an effect correlated with reductions in drug craving (Garland et al., 2014). As interoceptive attention broadens to allow reactions to less conditioned cues, it appears that freedom from maladaptive cycles of craving may follow (Khanna and Greeson, 2013). Furthermore, classical accounts of mindfulness training suggest that the continued exploration of experience, including interoceptive conditioning, can itself lead to feelings of joy and rapture (Brewer et al., 2013); the importance of such feelings is made concrete in consideration of mindfulness' impact in these and other clinical disorders.

Embodied Effects

Finally, it is important to note that because mindfulness practice itself takes place in an embodied context, engagement with interoceptive processes may promote further physiological effects. One example common to all Buddhist traditions is the importance of sitting posture in cultivating attentional stability and emotional equipoise. As traditionally stated in the Tibetan Buddhist tradition: "When the body is straight, the channels are straight, when the channels are straight, the energies are straight, when the energies are straight the mind is straight" (Rinpoche, 1998). An upright sitting posture is thought to affect the movement of energy through the channels thereby enhancing the effectiveness of the meditation practice. One could hypothesize that just sitting in this posture without engaging in a meditative practice may have beneficial effects. Upright postures are associated with enhanced physiological outcomes in hospital patients (Convertino, 2003). And indeed, sitting in an upright posture without actually engaging in meditation practice, referred to as "sham meditation," is associated with slowed respiration rate which predicts decreased pain unpleasantness ratings (Zeidan et al., 2010b). Another example is mindfulness meditation's association with increased heart rate variability (HRV; Ditto et al., 2006), particularly in the high frequency band, an indicator of parasympathetic activation (Wu and Lo, 2008; Krygier et al., 2013). Greater resting HRV is associated with a host of cognitive benefits; including greater sustained attention, working memory, and motor-response control (Thayer et al., 2009). Thus, as the mindfulness practice itself unfolds, mindful attention impacts physiological systems, which in turn may feed back to influence cognitive processes in a self-reinforcing cycle. In other words, the process of attending to interoceptive sensations provides time for autonomic processes to restore homeostasis, rather than perpetuating inefficient or maladaptive regulatory habits that rely on overt behavioral intervention. Awareness of homeostasis may generate feelings of calm, peace, and satisfaction, greater connection to others, and a decreased desire to seek out externally rewarding stimuli to maintain a hedonic set point. Unlike external reward, the reward of increasingly accurate perceptual predictions may be continuous. However, this is not to say that mindfulness meditation predisposes one to passive monitoring states, but such training may increase the ability to respond more adaptively to environmental challenges and return to homeostasis more quickly.

Taken together, an interoceptive consequence of contemplative practice is the decoupling of hedonic stimulus response arcs, i.e., an increased capacity to refrain from automatically responding to aversion with avoidance and to pleasure with approach. This relative freedom from interoceptive appraisal habits that drive behavior may optimize homeostasis that becomes a self-reinforcing process. Thus, this model may serve as a working hypothesis for the mechanisms of action underlying how changes in interoceptive awareness resulting from contemplative practice may enhance health and well-being including affective and dissociative disorders, pain, and addiction. Research is needed to better understand what difference variations in the particular foci of interoceptive attention make toward promoting salutary effects.

A Caveat on the Primacy of Interoceptive Processing

Admittedly, it seems unlikely that contemplative training can or should aim to engender continual and total awareness of the panoply of interoceptive sensations present in each moment. Instead, skillful attention to interoceptive sensation may improve self-regulation (Bornemann et al., 2014), allowing an individual to operate more closely to his or her optimal homeostatic state. We have argued that increased interoceptive attention represents one way to return to 'being' in a world that prioritizes 'doing,' restoring balance between the two. This balance may then serve to promote regulatory flexibility, with the ideal of bringing 'flow' to experience, the alignment of being with doing in living a meaningful and satisfying life. It may be that one reason focusing on the body has such positive potential is that rumination, already shown to be largely a negative factor in experience, operates, by and large, by shifting attention to past or future (Killingsworth and Gilbert, 2010). By contrast, the body is always in the present. Attending to the body therefore anchors the mind in the present and away from rumination. Admittedly, there are other forms of present-centered foci that would distract from rumination such as watching television, and the research literature supports many short-term regulatory benefits of distraction strategies (Sheppes and Meiran, 2007; Denson et al., 2012). However, distraction may lead to later 'rebound effects' when a person is faced with an unresolved stressor compared to conditions in which that stressor was directly attended to (Thiruchselvam et al., 2011).

While more work is needed to properly distinguish the effects of attention-based regulatory strategies such as mindful attention and distraction, skillfully attending to the stressor itself may be the more adaptive strategy (Kross and Ayduk, 2008), and in any case an improvement over rumination.

The Road Ahead

Other Elements of Traditional Contemplative Approaches to Interoception

Continued dialog with representatives of traditional contemplative practices may further explore how they understand and modulate the flow of sensations through the body to enhance health and healing. From many Asian medical/contemplative perspectives, disturbances or blockages in the 'flow' of *lung/ch'i/prāṇa* sensations are related to disease, and free movement of *chi* is related to health, as well as to insight, kindheartedness, and other positive qualities, a central principle of Tibetan medical literature (Loizzo et al., 2009). Their understanding of the body is not only a matter of interoceptive awareness; it is most explicitly a pathway for exploring and deeply revising the sense of self, agency, and substantiality that bears on virtually every activity in life. The import of these issues, and their potential significance for mental and physical health outcomes could be a fruitful endeavor for future research.

In these Asian traditions, the interior of the body, deep in the belly, or the heart, or deep inside the head, are all significant areas for focusing the mind. The mapping of channels and the "steeds of wind" moving through them further illustrate the body's interiority as characterized in traditional Asian or Buddhist perspectives. We also see that varying the locus of attention, for example shifting one's focus from breath to sound, or to a particular area of the body, may produce different results.

Contemplative traditions may have theoretical insights on how placement of attention in certain parts of the body affects perception, cognitive, and emotional responses. Practices vary according to whether the body is alone or in a group, with or without direct eye contact, whether the scope of attention includes others, whether one sits in silence or performs movement, whether one attends to sound or focuses on other sensory stimuli. There is also variation in the types of intentional and cognitive framework surrounding the practice. How these variations in interoceptive training may influence cognitive, emotional, and behavioral processes could be tested in empirical research.

While the focus of this discussion has largely been on the use of contemplative practices in general, and mindfulness in particular, to increase awareness of interoceptive processes as a means to enhance self-regulation, other practices, such as those found in the Tibetan Buddhist tradition, involve the use of interoceptive awareness, sometimes in combination with mental imagery or somatic manipulations, to intentionally *modulate* the 'flow' of sensations through the channels. Importantly, these practices take place within specific philosophical and ethical understandings, such as believing the nature of self and all of existence to be impermanent and interdependent. From an integrated mind–body perspective, in which 'currents flowing through the body' are conceived as the energetic support for consciousness, such "body-oriented" practices can be a direct means to impact psychological functioning and achieve meditative states of realization. One example is the practice of *tummo*, which has been studied by scientists showing dramatic increases in core body temperature (Kozhevnikov et al., 2013). While the short-term physiological outcomes of this practice have been studied for years (Benson et al., 1982), further attention could be devoted to understanding traditional theories of how these practices work and how conscious modulation of interoceptive sensations, especially when supported by accompanying intentional body-imagery, may impact physiological and psychological functioning more generally by promoting autonomic control of physiological processes. Equally interesting in this regard would be a consideration of shifts in 'energetic flows' and related sensations that accompany the cultivation of love or compassion, or the use of simple sounds and their impact on mind and body. First-person descriptions can be useful here as well, since they typically will describe both physical and mental responses to such training.

Future Research Directions

Although we have rich contemplative traditions to draw from, as well as new interoceptive awareness-enhancing approaches and emerging theory from a psychology perspective, we are facing a shortage of tools, paradigms, and appropriate measures to rigorously interrogate and integrate them. With this challenge in mind, we derive the following topics for further research from our considerations:

(1) Qualitative exploration of body-based practices from traditional contemplative traditions, and greater understanding of Asian (and eventually other) contemplative and medical models of the body that involve the 'flow' of 'energetic currents' through the body.
(2) Further refinement of scientific conceptualizations of mindfulness, including exploration of first-person accounts, to see how these match the conceptual framework for interoceptive awareness provided here. Concurrently, we must continue to explore traditional contemplative texts and practice traditions for new perspectives, and to deepen our understanding of constructs already appropriated into scientific models of interoception.
(3) Developing and/or refining appropriate quantitative measures, objective and self-report, able to capture cognitive, behavioral, and physiological changes occurring with interoceptive awareness training, moving beyond measures of interoceptive accuracy to include attention habits, sensitivity to interoceptive cues, coherence of physiological and subjective changes, and regulatory strategies.
(4) Developing research designs that integrate first, second and third-person perspectives, such as neurofeedback, in which third-person objective measures of physiological or neural change provide second-person reports to the participant in real time, potentially modulating the quality of first

person experience during interoceptive attention (Lutz and Thompson, 2003).

(5) Finding appropriate ways to operationalize behavioral interventions representing mindful interoceptive awareness, e.g., to the breath.

(6) Defining clear criteria for when or under which conditions interoceptive awareness is beneficial in treating patients with psychiatric, psychosomatic, and pain conditions.

(7) Using neuroscience tools and technique to follow the process/flow of interoceptive information through the brain, e.g., by EEG oscillations or with fMRI with its top-down modulations. For example, interoceptive cultivation should be apparent in terms of increased correspondence between feedback-evoked EEG potentials and subsequent physiological change. Conversely manipulation of interoceptive signals such as respiratory occlusion following enhanced interoception should provoke stronger and more reliable evoked potentials.

(8) Develop longitudinal studies applying interoceptive awareness training to patients with clinical conditions.

(9) Consider testing effects of mindfulness training and other contemplative practices on existing measures of presence and agency (rubber hand illusion, virtual environment experiments) to see if they mediate improvements in health outcomes.

By improving the measurement of interoception at subjective and physiological levels, and seeking to better understand the coherence (and lack thereof) between these levels of representation, it may be possible to broaden our understanding of interoception and its connection to motivation and well-being. Furthermore, such paradigms will help to test basic properties of information processing in the brain, investigating how we both predict and adjust to changing sensory stimuli, and how such stimuli are incorporated into a broader motivational milieu.

Concluding Remarks

Interoception has in many ways been a hidden sense, perhaps due to the challenges involved in measuring and manipulating interoceptive signals. And yet, interoception is arguably at least as important as the external senses, providing a sense of embodiment in the world that is foundational to a person's subjective sense of well-being. Contemporary scientific frameworks largely lack the ability to articulate contemplative models of how energy within the body helps to determine well-being, but, concepts such as the simulation map offer an avenue for bridging this cultural divide. Indeed, the idea of energy or prana may one day be expressed as the allocation of metabolic and cognitive resources to afford neural representation of body states. While genetic, social, and environmental factors clearly have great impact on a person's quality of life, our goal in writing this paper is to bring to light the process by which health and suffering are revealed to us in the moment, in the hope of providing a more integrated psychological account.

Fortunately, interest in interoception appears to be growing, and its researchers are privileged to have access to a rich repository of conceptual models from neuroscience, clinical, and contemplative research traditions. The challenge for the field will be to pay careful attention to the claims made in each of these traditions, to better characterize their predictions, and in doing so to more even-handedly advance modern secular interoceptive science while respecting the validity of all of these perspectives. While perhaps no single researcher will hold sufficient expertise in all of these domains, through collaboration such integration is possible- we hope that the current paper and other articles contained in the special issue is testament to that lofty but achievable goal. It is our hope that we may all benefit from the continued study of interoception, in paralleling the advances modern science has already made in understanding the external world.

Acknowledgments

We are very grateful to the reviewers and their intense discussion of many details that clearly strengthened this manuscript. The project was funded by supportive grants from the Mind and Life Institute, Amherst MN, and by the Peter Baumann Foundation, San Francisco CA. We would also like to thank Claire Mehling for her administrative assistance at our initial meeting and the Mind and Life Institute with Wendy Hasenkamp for providing the meeting space.

References

Abdi, H. (2007). "Signal detection theory (SDT)," in *Encyclopedia of Measurement and Statistics,* ed. N. J. Salkind (Thousand Oaks: Sage), 886–889.

Ainley, V., Maister, L., Brokfeld, J., Farmer, H., and Tsakiris, M. (2013). More of myself: manipulating interoceptive awareness by heightened attention to bodily and narrative aspects of the self. *Conscious. Cogn.* 22, 1231–1238. doi: 10.1016/j.concog.2013.08.004

Ainley, V., Tajadura-Jiménez, A., Fotopoulou, A., and Tsakiris, M. (2012). Looking into myself: changes in interoceptive sensitivity during mirror self-observation. *Psychophysiology* 49, 1672–1676. doi: 10.1111/j.1469-8986.2012.01468.x

Allen, M., Bromley, A., Kuyken, W., and Sonnenberg, S. J. (2009). Participants' experiences of mindfulness-based cognitive therapy: "it changed me in just about every way possible". *Behav. Cogn. Psychother.* 37, 413–430. doi: 10.1017/S135246580999004X

Altabe, M. (1998). Ethnicity and body image: quantitative and qualitative analysis. *Int. J. Eat. Disord.* 23, 153–159. doi: 10.1002/(SICI)1098-108X(199803)23:2<153::AID-EAT5>3.0.CO;2-J

Apps, M. A., and Tsakiris, M. (2014). The free-energy self: a predictive coding account of self-recognition. *Neurosci. Biobehav. Rev.* 41, 85–97. doi: 10.1016/j.neubiorev.2013.01.029

Baer, R. A. (2003). Mindfulness training as a clinical intervention: a conceptual and empirical review. *Clin. Psychol. Sci. Pract.* 10, 125–143. doi: 10.1093/clipsy.bpg015

Baer, R. A., Smith, G. T., Lykins, E., Button, D., Krietemeyer, J., Sauer, S., et al. (2008). Construct validity of the five facet mindfulness questionnaire in meditating and nonmeditating samples. *Assessment* 15, 329–342. doi: 10.1177/1073191107313003

Baltes, P. B., and Lindenberger, U. (1997). Emergence of a powerful connection between sensory and cognitive functions across the adult life span: a new window to the study of cognitive aging? *Psychol. Aging* 12, 12–21. doi: 10.1037/0882-7974.12.1.12

Barnsley, N., Mcauley, J., Mohan, R., Dey, A., Thomas, P., and Moseley, G. (2011). The rubber hand illusion increases histamine reactivity in the real arm. *Curr. Biol.* 21, R945–R946. doi: 10.1016/j.cub.2011.10.039

Beck, A. T. (2008). The evolution of the cognitive model of depression and its neurobiological correlates. *Am. J. Psychiatry* 165, 969–977. doi: 10.1176/appi.ajp.2008.08050721

Benson, H., Lehmann, J. W., Malhotra, M., Goldman, R. F., Hopkins, J., and Epstein, M. D. (1982). Body temperature changes during the practice of g Tum-mo yoga. *Nature* 295, 234–236. doi: 10.1038/295234a0

Björnsdotter, M., Löken, L., Olausson, H., Vallbo, Å., and Wessberg, J. (2009). Somatotopic organization of gentle touch processing in the posterior insular cortex. *J. Neurosci.* 29, 9314–9320. doi: 10.1523/JNEUROSCI.0400-09.2009

Blascovich, J., and Tomaka, J. (1996). The biopsychosocial model of arousal regulation. *Adv. Exp. Soc. Psychol.* 28, 1–51. doi: 10.1016/s0065-2601(08) 60235-x

Bornemann, B., Herbert, B. M., Mehling, W. E., and Singer, T. (2014). Differential changes in self-reported aspects of interoceptive awareness through 3 months of contemplative training. *Front. Psychol.* 5:1504. doi: 10.3389/fpsyg.2014.01504

Boswell, J. F., Farchione, T. J., Sauer-Zavala, S., Murray, H. W., Fortune, M. R., and Barlow, D. H. (2013). Anxiety sensitivity and interoceptive exposure: a transdiagnostic construct and change strategy. *Behav. Ther.* 44, 417–431. doi: 10.1016/j.beth.2013.03.006

Bowen, S., Witkiewitz, K., Dillworth, T. M., and Marlatt, G. A. (2007). The role of thought suppression in the relationship between mindfulness meditation and alcohol use. *Addict. Behav.* 32, 2324–2328. doi: 10.1016/j.addbeh.2007.01.025

Brener, J. (1977). "Visceral perception," in *Biofeedback and Behavior*, eds J. Beatty and H. Legewie (New York, NY: Springer), 235–259. doi: 10.1007/978-1-4684-2526-0_15

Brewer, J. A., Davis, J. H., and Goldstein, J. (2013). Why is it so hard to pay attention, or is it? Mindfulness, the factors of awakening and reward-based learning. *Mindfulness* 4, 1–6. doi: 10.1007/s12671-012-0164-8

Brown, K. W., and Ryan, R. M. (2003). The benefits of being present: mindfulness and its role in psychological well-being. *J. Pers. Soc. Psychol.* 84, 822–848. doi: 10.1037/0022-3514.84.4.822

Buchanan, P. A., and Ulrich, B. D. (2001). The Feldenkrais Method: a dynamic approach to changing motor behavior. *Res. Q. Exerc. Sport.* 72, 315–323. doi: 10.1080/02701367.2001.10608968

Büchel, C., Geuter, S., Sprenger, C., and Eippert, F. (2014). Placebo analgesia: a predictive coding perspective. *Neuron* 81, 1223–1239. doi: 10.1016/j.neuron.2014.02.042

Buddhaghosa, B. (2010). *Visuddhimagga: The Path of Purification*. Onalaska, WA: Pariyatti

Bush, G., Luu, P., and Posner, M. I. (2000). Cognitive and emotional influences in anterior cingulate cortex. *Trends Cogn. Sci.* 4, 215–222. doi: 10.1016/S1364-6613(00)01483-2

Cameron, O. (2001). Interoception: the inside story - a model for psychosomatic processes. *Psychosom. Med.* 63, 697–710. doi: 10.1097/00006842-200109000-00001

Ceunen, E., Van Diest, I., and Vlaeyen, J. W. (2013). Accuracy and awareness of perception: related, yet distinct (commentary on Herbert et al., 2012). *Biol. Psychol.* 92, 426–427. doi: 10.1016/j.biopsycho.2012.09.012

Chang, V. Y., Palesh, O., Caldwell, R., Glasgow, N., Abramson, M., Luskin, F., et al. (2004). The effects of a mindfulness-based stress reduction program on stress, mindfulness self-efficacy, and positive states of mind. *Stress Health* 20, 141–147. doi: 10.1002/smi.1011

Chiesa, A., and Serretti, A. (2009). Mindfulness-based stress reduction for stress management in healthy people: a review and meta-analysis. *J. Altern. Complement. Med.* 15, 593–600. doi: 10.1089/acm.2008.0495

Christopher, M. S., Charoensuk, S., Gilbert, B. D., Neary, T. J., and Pearce, K. L. (2009). Mindfulness in Thailand and the United States: a case of apples versus oranges? *J. Clin. Psychol.* 65, 590–612. doi: 10.1002/jclp.20580

Convertino, V. A. (2003). Value of orthostatic stress in maintaining functional status soon after myocardial infarction or cardiac artery bypass grafting. *J. Cardiovasc. Nurs.* 18, 124–130. doi: 10.1097/00005082-200304000-00008

Craig, A. D. (2002). How do you feel? Interoception: the sense of the physiological condition of the body. *Nat. Rev. Neurosci.* 3, 655–666. doi: 10.1038/nrn894

Craig, A. D. (2003). Interoception: the sense of the physiological condition of the body. *Curr. Opin. Neurobiol.* 13, 500–505. doi: 10.1016/S0959-4388(03)00090-4

Craig, A. D. (2007). "Interoception and emotion: a Neuaroanatomical perspective," in *Handbook of Emotions*, eds M. Lewis, J. M. Haviland-Jones, and L. Feldman Barrett (New York, NY: Guilford Press), 272–290.

Craig, A. D. (2009). How do you feel–now? The anterior insula and human awareness. *Nat. Rev. Neurosci.* 10, 59–70. doi: 10.1038/nrn2555

Craig, A. (2013). An interoceptive neuroanatomical perspective on feelings, energy, and effort. *Behav. Brain Sci.* 36, 685–686. doi: 10.1017/S0140525X13001489

Craske, M. G., Rowe, M., Lewin, M., and Noriega-Dimitri, R. (1997). Interoceptive exposure versus breathing retraining within cognitive-behavioural therapy for panic disorder with agoraphobia1. *Br. J. Clin. Psychol.* 36, 85–99. doi: 10.1111/j.2044-8260.1997.tb01233.x

Critchley, H. D., and Harrison, N. A. (2013). Visceral influences on brain and behavior. *Neuron* 77, 624–638. doi: 10.1016/j.neuron.2013.02.008

Critchley, H. D., Wiens, S., Rotshtein, P., Ohman, A., and Dolan, R. J. (2004). Neural systems supporting interoceptive awareness. *Nat. Neurosci.* 7, 189–195. doi: 10.1038/nn1176

Damasio, A. (2003a). Feelings of emotion and the self. *Ann. N. Y. Acad. Sci.* 1001, 253–261. doi: 10.1196/annals.1279.014

Damasio, A. (2003b). Mental self: the person within. *Nature* 423, 227. doi: 10.1038/423227a

Damasio, A., and Carvalho, G. B. (2013). The nature of feelings: evolutionary and neurobiological origins. *Nat. Rev. Neurosci.* 14, 143–152. doi: 10.1038/nrn3403

Daubenmier, J., Kristeller, J., Hecht, F. M., Maninger, N., Kuwata, M., Jhaveri, K., et al. (2011). Mindfulness intervention for stress eating to reduce cortisol and abdominal fat among overweight and obese women: an exploratory randomized controlled study. *J. Obes.* 2011:651936. doi: 10.1155/2011/651936

Daubenmier, J., Sze, J., Kerr, C. E., Kemeny, M. E., and Mehling, W. (2013). Follow your breath: respiratory interoceptive accuracy in experienced meditators. *Psychophysiology* 50, 777–789. doi: 10.1111/psyp.12057

De Raedt, R., and Koster, E. H. (2010). Understanding vulnerability for depression from a cognitive neuroscience perspective: a reappraisal of attentional factors and a new conceptual framework. *Cogn. Affect. Behav. Neurosci.* 10, 50–70. doi: 10.3758/CABN.10.1.50

Denson, T. F., Moulds, M. L., and Grisham, J. R. (2012). The effects of analytical rumination, reappraisal, and distraction on anger experience. *Behav. Ther.* 43, 355–364. doi: 10.1016/j.beth.2011.08.001

Ditto, B., Eclache, M., and Goldman, N. (2006). Short-term autonomic and cardiovascular effects of mindfulness body scan meditation. *Ann. Behav. Med.* 32, 227–234. doi: 10.1207/s15324796abm3203_9

Domschke, K., Stevens, S., Pfleiderer, B., and Gerlach, A. L. (2010). Interoceptive sensitivity in anxiety and anxiety disorders: an overview and integration of neurobiological findings. *Clin. Psychol. Rev.* 30, 1–11. doi: 10.1016/j.cpr.2009.08.008

Dunn, B. D., Dalgleish, T., Ogilvie, A. D., and Lawrence, A. D. (2007). Heartbeat perception in depression. *Behav. Res. Ther.* 45, 1921–1930. doi: 10.1016/j.brat.2006.09.008

Dunn, B. D., Galton, H. C., Morgan, R., Evans, D., Oliver, C., Meyer, M., et al. (2010). Listening to your heart. How interoception shapes emotion experience and intuitive decision making. *Psychol. Sci.* 21, 1835–1844. doi: 10.1177/0956797610389191

Ehlers, A., and Breuer, P. (1992). Increased cardiac awareness in panic disorder. *J. Abnorm. Psychol.* 101, 371–382. doi: 10.1037/0021-843X.101.3.371

Epel, E., Daubenmier, J., Moskowitz, J. T., Folkman, S., and Blackburn, E. (2009). Can meditation slow rate of cellular aging? Cognitive stress, mindfulness, and telomeres. *Ann. N. Y. Acad. Sci.* 1172, 34–53. doi: 10.1111/j.1749-6632.2009.04414.x

Ernst, J., Böker, H., Hättenschwiler, J., Schüpbach, D., Northoff, G., Seifritz, E., et al. (2014). The association of interoceptive awareness and alexithymia with neurotransmitter concentrations in insula and anterior cingulate. *Soc. Cogn. Affect. Neurosci.* 9, 5857–863. doi: 10.1093/scan/nst058

Farb, N. A. (2014). From retreat center to clinic to boardroom? Perils and promises of the modern mindfulness movement. *Religions* 5, 1062–1086. doi: 10.3390/rel5041062

Farb, N. A., Anderson, A. K., and Segal, Z. V. (2012a). The mindful brain and emotion regulation in mood disorders. *Can. J. Psychiatry* 57, 70–77.

Farb, N. A., Segal, Z. V., and Anderson, A. K. (2012b). Attentional modulation of primary interoceptive and exteroceptive cortices. *Cereb. Cortex* 23, 114–126. doi: 10.1093/cercor/bhr385

Farb, N., Anderson, A., Irving, J., and Segal, Z. (2014). "Mindfulness interventions and emotion regulation," in *Handbook of Emotion Regulation*, 2nd Edn, ed. J. J. Gross (New York, NY: Guilford Press), 548–567.

Farb, N. A., Segal, Z. V., and Anderson, A. K. (2013). Mindfulness meditation training alters cortical representations of interoceptive attention. *Soc. Cogn. Affect. Neurosci.* 8, 15–26. doi: 10.1093/scan/nss066

Farb, N. A., Segal, Z. V., Mayberg, H., Bean, J., Mckeon, D., Fatima, Z., et al. (2007). Attending to the present: mindfulness meditation reveals distinct neural modes of self-reference. *Soc. Cogn. Affect. Neurosci.* 2, 313–322. doi: 10.1093/scan/nsm030

Feeny, N. C., Zoellner, L. A., Fitzgibbons, L. A., and Foa, E. B. (2000). Exploring the roles of emotional numbing, depression, and dissociation in PTSD. *J. Trauma. Stress* 13, 489–498. doi: 10.1023/A:1007789409330

Feldman, G., Greeson, J., and Senville, J. (2010). Differential effects of mindful breathing, progressive muscle relaxation, and loving-kindness meditation on decentering and negative reactions to repetitive thoughts. *Behav. Res. Ther.* 48, 1002–1011. doi: 10.1016/j.brat.2010.06.006

Ferron, C. (1997). Body image in adolescence: cross-cultural research—results of the preliminary phase of a quantitative survey. *Adolescence* 32, 735–745.

Fogel, A. (2009). *The Psychophysiology of Self-Awareness: Rediscovering the Lost Art of Body Sense.* New York, NY: W. W. Norton.

Fotopoulou, A. (2013). Beyond the reward principle: consciousness as precision seeking. *Neuropsychoanalysis* 15, 33–38. doi: 10.1080/15294145.2013.10773715

Fox, K. C., Zakarauskas, P., Dixon, M., Ellamil, M., Thompson, E., and Christoff, K. (2012). Meditation experience predicts introspective accuracy. *PLoS ONE* 7:e45370. doi: 10.1371/journal.pone.0045370

Fresco, D. M., Moore, M. T., Van Dulmen, M. H., Segal, Z. V., Ma, S. H., Teasdale, J. D., et al. (2007). Initial psychometric properties of the experiences questionnaire: validation of a self-report measure of decentering. *Behav. Ther.* 38, 234–246. doi: 10.1016/j.beth.2006.08.003

Frewen, P. A., Dozois, D. J., and Lanius, R. A. (2012). Assessment of anhedonia in psychological trauma: psychometric and neuroimaging perspectives. *Eur. J. Psychotraumatol.* 3:8587. doi: 10.3402/ejpt.v3i0.8587

Gallagher, S. (1997). Mutual enlightenment: recent phenomenology in cognitive science. *J. Conscious. Stud.* 4, 195–214.

Gard, T., Holzel, B. K., Sack, A. T., Hempel, H., Lazar, S. W., Vaitl, D., et al. (2012). Pain attenuation through mindfulness is associated with decreased cognitive control and increased sensory processing in the brain. *Cereb. Cortex* 22, 2692–2702. doi: 10.1093/cercor/bhr352

Garfinkel, S. N., and Critchley, H. D. (2013). Interoception, emotion and brain: new insights link internal physiology to social behaviour. Commentary on: "Anterior insular cortex mediates bodily sensibility and social anxiety" by Terasawa et al. (2012). *Soc. Cogn. Affect. Neurosci.* 8, 231–234. doi: 10.1093/scan/nss140

Garfinkel, S. N., Seth, A. K., Barrett, A. B., Suzuki, K., and Critchley, H. D. (2015). Knowing your own heart: distinguishing interoceptive accuracy from interoceptive awareness. *Biol. Psychol.* 104, 65–74. doi: 10.1016/j.biopsycho.2014.11.004

Garland, E. L., Fredrickson, B., Kring, A. M., Johnson, D. P., Meyer, P. S., and Penn, D. L. (2010). Upward spirals of positive emotions counter downward spirals of negativity: insights from the broaden-and-build theory and affective neuroscience on the treatment of emotion dysfunctions and deficits in psychopathology. *Clin. Psychol. Rev.* 30, 849–864. doi: 10.1016/j.cpr.2010.03.002

Garland, E. L., Froeliger, B., and Howard, M. O. (2014). Effects of mindfulness-oriented recovery enhancement on reward responsiveness and opioid cue-reactivity. *Psychopharmacology (Berl.)* 231, 3229–3238. doi: 10.1007/s00213-014-3504-7

Garland, E. L., Gaylord, S. A., and Fredrickson, B. L. (2011). Positive reappraisal mediates the stress-reductive effects of mindfulness: an upward spiral process. *Mindfulness* 2, 59–67. doi: 10.1007/s12671-011-0043-8

Garland, E. L., Gaylord, S. A., Palsson, O., Faurot, K., Mann, J. D., and Whitehead, W. E. (2012). Therapeutic mechanisms of a mindfulness-based treatment for IBS: effects on visceral sensitivity, catastrophizing, and affective processing of pain sensations. *J. Behav. Med.* 35, 591–602. doi: 10.1007/s10865-011-9391-z

Garner, D. M., Olmstead, M. P., and Polivy, J. (1983). Development and validation of a multidimensional eating disorder inventory for anorexia nervosa and bulimia. *Int. J. Eat. Disord.* 2, 15–34. doi: 10.1002/1098-108X(198321)2:2<15::AID-EAT2260020203>3.0.CO;2-6

Gendlin, E. T. (2012). *Focusing.* New York, NY: Bantam Book.

Gilbert, P., Mcewan, K., Gibbons, L., Chotai, S., Duarte, J., and Matos, M. (2012). Fears of compassion and happiness in relation to alexithymia, mindfulness, and self-criticism. *Psychol. Psychother. Theory Res. Pract.* 85, 374–390. doi: 10.1111/j.2044-8341.2011.02046.x

Goleman, D. (2008). *Destructive Emotions: A Scientific Dialogue with the Dalai Lama.* New York City, NY: Bantam Books

Goubert, L., Crombez, G., Eccleston, C., and Devulder, J. (2004). Distraction from chronic pain during a pain-inducing activity is associated with greater post-activity pain. *Pain* 110, 220–227. doi: 10.1016/j.pain.2004.03.034

Goyal, M., Singh, S., Sibinga, E. M., Gould, N. F., Rowland-Seymour, A., Sharma, R., et al. (2014). Meditation programs for psychological stress and well-being: a systematic review and meta-analysis. *JAMA Int. Med.* 174, 357–368. doi: 10.1001/jamainternmed.2013.13018

Grof, S., and Grof, C. (2010). *Holotropic Breathwork: A New Approach to Self-Exploration and Therapy.* Albany, NY: SUNY Press.

Gross, J. J. (2002). Emotion regulation: affective, cognitive, and social consequences. *Psychophysiology* 39, 281–291. doi: 10.1017/S0048577201393198

Grossman, P. (2011). Defining mindfulness by how poorly I think I pay attention during everyday awareness and other intractable problems for psychology's (re)invention of mindfulness: comment on Brown et al. (2011). *Psychol. Assess.* 23, 1034–1040. doi: 10.1037/a0022713

Grossman, P. (2015). Mindfulness: awareness informed by an embodied ethic. *Mindfulness* 6, 17–22. doi: 10.1007/s12671-014-0372-5

Grossman, P., Niemann, L., Schmidt, S., and Walach, H. (2004). Mindfulness-based stress reduction and health benefits: a meta-analysis. *J. Psychosom. Res.* 57, 35–43. doi: 10.1016/S0022-3999(03)00573-7

Grossman, P., and Van Dam, N. T. (2011). Mindfulness, by any other name: trials and tribulations of sati in western psychology and science. *Contemp. Buddhism* 12, 219–239. doi: 10.1080/14639947.2011.564841

Gu, X., and Fitzgerald, T. (2014). Interoceptive inference: homeostasis and decision-making. *Trends Cogn. Sci.* 18, 269–270. doi: 10.1016/j.tics.2014.02.001

Hankin, B. L. (2012). Future directions in vulnerability to depression among youth: integrating risk factors and processes across multiple levels of analysis. *Clin. Child Adolesc. Psychol.* 41, 695–718. doi: 10.1080/15374416.2012.711708

Hart, W. (2011). *The Art of Living: Vipassana meditation: As Taught by SN Goenka.* Onalaska, WA: Pariyatti.

Hayes, S. C., Strosahl, K. D., and Wilson, K. G. (1999). *Acceptance and Commitment Therapy: An Experiential Approach to Behavior Change.* New York City, NY: Guilford Press.

Hendricks, M. (2001). "Focusing-oriented/experiential psychotherapy," in *Humanistic Psychotherapy*, eds D. Cain and J. Seeman (Washington, DC: Handbook of Research and Practice American Psychological Association). doi: 10.1016/j.biopsycho.2011.09.004

Herbert, B. M., Herbert, C., Pollatos, O., Weimer, K., Enck, P., Sauer, H., et al. (2012a). Effects of short-term food deprivation on interoceptive awareness, feelings and autonomic cardiac activity. *Biol. Psychol.* 89, 71–79. doi: 10.1016/j.biopsycho.2011.09.004

Herbert, B. M., Muth, E. R., Pollatos, O., and Herbert, C. (2012b). Interoception across modalities: on the relationship between cardiac awareness and the sensitivity for gastric functions. *PLoS ONE* 7:e36646. doi: 10.1371/journal.pone.0036646

Herbert, B. M., and Pollatos, O. (2012). The body in the mind: on the relationship between interoception and embodiment. *Top. Cogn. Sci.* 4, 692–704. doi: 10.1111/j.1756-8765.2012.01189.x

Herbert, B. M., and Pollatos, O. (2014). Attenuated interoceptive sensitivity in overweight and obese individuals. *Eat. Behav.* 15, 445–448. doi: 10.1016/j.eatbeh.2014.06.002

Herbert, J. D., and Forman, E. M. (2011). *Acceptance and Mindfulness in Cognitive Behavior Therapy: Understanding and Applying the New Therapies.* Hoboken, NJ: John Wiley & Sons. doi: 10.1002/9781118001851

Higgins, E. T. (1987). Self-discrepancy: a theory relating self and affect. *Psychol. Rev.* 94, 319–340. doi: 10.1037/0033-295X.94.3.319

Hoffman, B. M., Papas, R. K., Chatkoff, D. K., and Kerns, R. D. (2007). Meta-analysis of psychological interventions for chronic low back pain. *Health Psychol.* 26, 1–9. doi: 10.1037/a0018555

Hofmann, S. G., Sawyer, A. T., Witt, A. A., and Oh, D. (2010). The effect of mindfulness-based therapy on anxiety and depression: a meta-analytic review. *J. Consult. Clin. Psychol.* 78, 169–183. doi: 10.1037/a0018555

Hoge, E. A., Bui, E., Marques, L., Metcalf, C. A., Morris, L. K., Robinaugh, D. J., et al. (2013). Randomized controlled trial of mindfulness meditation for generalized anxiety disorder: effects on anxiety and stress reactivity. *J. Clin. Psychiatry* 74, 786–792. doi: 10.4088/JCP.12m08083

Hölzel, B. K., Lazar, S. W., Gard, T., Schuman-Olivier, Z., Vago, D. R., and Ott, U. (2011). How does mindfulness meditation work? Proposing mechanisms of action from a conceptual and neutral perspective. *Perspect. Psychol. Sci.* 6, 537–559. doi: 10.1177/1745691611419671

Holzl, R., Erasmus, L. P., and Moltner, A. (1996). Detection, discrimination and sensation of visceral stimuli. *Biol. Psychol.* 42, 199–214. doi: 10.1016/0301-0511(95)05155-4

Hutcherson, C. A., Seppala, E. M., and Gross, J. J. (2008). Loving-kindness meditation increases social connectedness. *Emotion* 8, 720. doi: 10.1037/a0013237

Ingram, R. E., Atchley, R. A., and Segal, Z. V. (2011). *Vulnerability to Depression: From Cognitive Neuroscience to Prevention and Treatment.* New York, NY: Guilford Press.

Jha, A. P., Krompinger, J., and Baime, M. J. (2007). Mindfulness training modifies subsystems of attention. *Cogn. Affect. Behav. Neurosci.* 7, 109–119. doi: 10.3758/CABN.7.2.109

Jha, A. P., Stanley, E. A., Kiyonaga, A., Wong, L., and Gelfand, L. (2010). Examining the protective effects of mindfulness training on working memory capacity and affective experience. *Emotion* 10, 54–64. doi: 10.1037/a0018438

Kabat-Zinn, J. (1982). An outpatient program in behavioral medicine for chronic pain patients based on the practice of mindfulness meditation: theoretical considerations and preliminary results. *Gen. Hosp. Psychiatry* 4, 33–47. doi: 10.1016/0163-8343(82)90026-3

Kabat-Zinn, J. (1990). *Full Catastrophe Living: Using the Wisdom of Your Body and Mind to Face Stress, Pain and Illness.* New York, NY: Delacorte.

Kabat-Zinn, J., Lipworth, L., and Burney, R. (1985). The clinical use of mindfulness meditation for the self-regulation of chronic pain. *J. Behav. Med.* 8, 163–190. doi: 10.1007/BF00845519

Kang, Y., Williams, L. E., Clark, M. S., Gray, J. R., and Bargh, J. A. (2011). Physical temperature effects on trust behavior: the role of insula. *Soc. Cogn. Affect. Neurosci.* 6, 507–515. doi: 10.1093/scan/nsq077

Kerr, C. E., Sacchet, M. D., Lazar, S. W., Moore, C. I., and Jones, S. R. (2013). Mindfulness starts with the body: somatosensory attention and top-down modulation of cortical alpha rhythms in mindfulness meditation. *Front. Hum. Neurosci.* 7:12. doi: 10.3389/fnhum.2013.00012

Kerr, C. E., Shaw, J. R., Wasserman, R. H., Chen, V. W., Kanojia, A., Bayer, T., et al. (2008). Tactile acuity in experienced Tai Chi practitioners: evidence for use dependent plasticity as an effect of sensory-attentional training. *Exp. Brain Res.* 188, 317–322. doi: 10.1007/s00221-008-1409-6

Khalsa, S. S., Rudrauf, D., Damasio, A. R., Davidson, R. J., Lutz, A., and Tranel, D. (2008). Interoceptive awareness in experienced meditators. *Psychophysiology* 45, 671–677. doi: 10.1111/j.1469-8986.2008.00666.x

Khalsa, S. S., Rudrauf, D., and Tranel, D. (2009). Interoceptive awareness declines with age. *Psychophysiology* 46, 1130–1136. doi: 10.1111/j.1469-8986.2008.00666.x

Khanna, S., and Greeson, J. M. (2013). A narrative review of yoga and mindfulness as complementary therapies for addiction. *Complement. Ther. Med.* 21, 244–252. doi: 10.1016/j.ctim.2013.01.008

Killingsworth, M. A., and Gilbert, D. T. (2010). A wandering mind is an unhappy mind. *Science* 330, 932–932. doi: 10.1126/science.1192439

Kirk, U., and Montague, R. (2015). Mindfulness meditation modulates reward prediction errors in the striatum in a passive conditioning task. *Front. Psychol.* 6:90. doi: 10.3389/fpsyg.2015.00090

Klein, A. (2014). *The Knowing Body: Currents of Connection and Women in Religious Dialogue. Catherine Cornille (edit): Women and Inter-Religious Dialogue.* Columbia, MD: Crossroads Press.

Klein, A. C. (2013). Seeing mind, being body. *Contemplat. Pract. Buddh. Epistemol.* 572–584. doi: 10.1002/9781118324004.ch37

Kozhevnikov, M., Elliott, J., Shephard, J., and Gramann, K. (2013). Neurocognitive and somatic components of temperature increases during g-tummo meditation: legend and reality. *PLoS ONE* 8:e58244. 10.1371/journal.pone.0058244 doi: 10.1371/journal.pone.0058244

Kringelbach, M. L. (2005). The human orbitofrontal cortex: linking reward to hedonic experience. *Nat. Rev. Neurosci.* 6, 691–702. doi: 10.1038/nrn1747

Kristeller, J. L., and Wolever, R. Q. (2010). Mindfulness-based eating awareness training for treating binge eating disorder: the conceptual foundation. *Eat. Disord.* 19, 49–61. doi: 10.1080/10640266.2011.533605

Kross, E., and Ayduk, O. (2008). Facilitating adaptive emotional analysis: distinguishing distanced-analysis of depressive experiences from immersed-analysis and distraction. *Personal. Soc. Psychol. Bull.* 34, 924–938. doi: 10.1177/0146167208315938

Krygier, J. R., Heathers, J. A., Shahrestani, S., Abbott, M., Gross, J. J., and Kemp, A. H. (2013). Mindfulness meditation, well-being, and heart rate variability: a preliminary investigation into the impact of intensive Vipassana meditation. *Int. J. Psychophysiol.* 89, 305–313. doi: 10.1016/j.ijpsycho.2013.06.017

Kurtz, R. (1997). *Body-Centered Psychotherapy: The Hakomi Method: The Integrated Use of Mindfulness, Nonviolence and the Body.* Mendocino, CA: LifeRhythm.

Lavie, N., Ro, T., and Russell, C. (2003). The role of perceptual load in processing distractor faces. *Psychol. Sci.* 14, 510–515. doi: 10.1111/1467-9280.03453

Leopold, C., and Schandry, R. (2001). The heartbeat-evoked brain potential in patients suffering from diabetic neuropathy and in healthy control persons. *Clin. Neurophysiol.* 112, 674–682. doi: 10.1016/S1388-2457(01)00480-1

Limanowski, J., and Blankenburg, F. (2013). Minimal self-models and the free energy principle. *Front. Hum. Neurosci.*7:547. doi: 10.3389/fnhum.2013.00547

Linehan, M. M., Schmidt, H., Dimeff, L. A., Craft, J. C., Kanter, J., and Comtois, K. A. (1999). Dialectical behavior therapy for patients with borderline personality disorder and drug-dependence. *Am. J. Addict.* 8, 279–292. doi: 10.1080/105504999305686

Loizzo, J. J., Blackhall, L. J., and Rapgay, L. (2009). Tibetan medicine: a complementary science of optimal health. *Ann. N. Y. Acad. Sci.* 1172, 218–230. doi: 10.1196/annals.1393.008

Lorist, M. M., and Tops, M. (2003). Caffeine, fatigue, and cognition. *Brain Cogn.* 53, 82–94. doi: 10.1016/S0278-2626(03)00206-9

Lowe, M. R., and Butryn, M. L. (2007). Hedonic hunger: a new dimension of appetite? *Physiol. Behav.* 91, 432–439. doi: 10.1016/j.physbeh.2007.04.006

Lucas, R. J. (2013). Mammalian inner retinal photoreception. *Curr. Biol.* 23, R125–R133. doi: 10.1016/j.cub.2012.12.029

Lutz, A., Slagter, H. A., Dunne, J. D., and Davidson, R. J. (2008). Attention regulation and monitoring in meditation. *Trends Cogn. Sci.* 12, 163–169. doi: 10.1016/j.tics.2008.01.005

Lutz, A., and Thompson, E. (2003). Neurophenomenology integrating subjective experience and brain dynamics in the neuroscience of consciousness. *J. Conscious. Stud.* 10, 31–52.

Ma-Kellams, C., Blascovich, J., and Mccall, C. (2012). Culture and the body: East–West differences in visceral perception. *J. Pers. Soc. Psychol.* 102:718. doi: 10.1037/a0027010

Maister, L., and Tsakiris, M. (2014). My face, my heart: cultural differences in integrated bodily self-awareness. *Cogn. Neurosci.* 5, 10–16. doi: 10.1080/17588928.2013.808613

Mallorquí-Bagué, N., Garfinkel, S. N., Engels, M., Eccles, J. A., Pailhez, G., Bulbena, A., et al. (2014). Neuroimaging and psychophysiological investigation of the link between anxiety, enhanced affective reactivity and interoception in people with joint hypermobility. *Front. Psychol.* 5:1162. doi: 10.3389/fpsyg.2014.01162

Matthew, T., Sherwin, R. S., Murphy, J., and Kerr, D. (1997). Importance of cerebral blood flow to the recognition of and physiological responses to hypoglycemia. *Diabetes Metab. Res. Rev.* 46, 829–833. doi: 10.2337/diab.46.5.829

McBride, R. (2012). A framework for error correction under prediction. *Front. Psychol.* 3:411. doi: 10.3389/fpsyg.2012.00411

McKenzie, K. J., and Newport, R. (2015). Increased somatic sensations are associated with reduced limb ownership. *J. Psychosom. Res.* 78, 88–90. doi: 10.1016/j.jpsychores.2014.11.005

Mehling, W. E. (2001). The experience of breath as a therapeutic intervention – psychosomatic forms of breath therapy. A descriptive study about the actual situation of breath therapy in Germany, its relation to medicine, and its application in patients with back pain. *Forsch. Komplementarmed. Klass. Naturheilkd.* 8, 359–367. doi: 10.1159/000057253

Mehling, W. E., Gopisetty, V., Daubenmier, J., Price, C. J., Hecht, F. M., and Stewart, A. (2009). Body awareness: construct and self-report measures. *PLoS ONE* 4:e5614. doi: 10.1371/journal.pone.0005614

Mehling, W. E., Price, C., Daubenmier, J. J., Acree, M., Bartmess, E., and Stewart, A. (2012). The multidimensional assessment of interoceptive awareness (MAIA). *PLoS ONE* 7:e48230. doi: 10.1371/journal.pone.0048230

Mehling, W. E., Wrubel, J., Daubenmier, J. J., Price, C. J., Kerr, C. E., Silow, T., et al. (2011). Body Awareness: a phenomenological inquiry into the common ground of mind-body therapies. *Philos. Ethics Humanit. Med.* 6:6. doi: 10.1186/1747-5341-6-6

Melzack, R. (1975). The McGill Pain Questionnaire: major properties and scoring methods. *Pain* 1, 277–299. doi: 10.1016/0304-3959(75)90044-5

Michal, M., Reuchlein, B., Adler, J., Reiner, I., Beutel, M. E., Vögele, C., et al. (2014). Striking discrepancy of anomalous body experiences with normal interoceptive accuracy in depersonalization-derealization disorder. *PLoS ONE* 9:e89823. doi: 10.1371/journal.pone.0089823

Michalak, J., Troje, N., and Heidenreich, T. (2011). The effects of mindfulness-based cognitive therapy on depressive gait patterns. *J. Cogn. Behav. Psychother.* 11, 13–27.

Mikulas, W. L. (2011). Mindfulness: significant common confusions. *Mindfulness* 2, 1–7. doi: 10.1007/s12671-010-0036-z

Mirams, L., Poliakoff, E., Brown, R. J., and Lloyd, D. M. (2012). Interoceptive and exteroceptive attention have opposite effects on subsequent somatosensory perceptual decision making. *Q. J. Exp. Psychol.* 65, 926–938. doi: 10.1080/17470218.2011.636823

Mirams, L., Poliakoff, E., Brown, R. J., and Lloyd, D. M. (2013). Brief body-scan meditation practice improves somatosensory perceptual decision making. *Conscious. Cogn.* 22, 348–359. doi: 10.1016/j.concog.2012.07.009

Mohan, R., Jensen, K. B., Petkova, V. I., Dey, A., Barnsley, N., Ingvar, M., et al. (2012). No pain relief with the rubber hand illusion. *PLoS ONE* 7:e52400. doi: 10.1371/journal.pone.0052400

Moor, R. (1968). Foetal homeostasis: conceptus-ovary endocrine balance. *Proc. R. Soc. Med.* 61, 1217–1226.

Moore, A., and Malinowski, P. (2009). Meditation, mindfulness and cognitive flexibility. *Conscious. Cogn.* 18, 176–186. doi: 10.1016/j.concog.2008.12.008

Moseley, G. L., Olthof, N., Venema, A., Don, S., Wijers, M., Gallace, A., et al. (2008). Psychologically induced cooling of a specific body part caused by the illusory ownership of an artificial counterpart. *Proc. Natl. Acad. Sci. U.S.A.* 105, 13169–13173. doi: 10.1073/pnas.0803768105

Mueller, E. M., Stemmler, G., Hennig, J., and Wacker, J. (2013). 5-HTTLPR and anxiety modulate brain-heart covariation. *Psychophysiology* 50, 441–453. doi: 10.1111/psyp.12016

Naqvi, N. H., and Bechara, A. (2010). The insula and drug addiction: an interoceptive view of pleasure, urges, and decision-making. *Brain Struct. Funct.* 214, 435–450. doi: 10.1007/s00429-010-0268-7

Neff, K. D., and Germer, C. K. (2013). A pilot study and randomized controlled trial of the mindful self-compassion program. *J. Clin. Psychol.* 69, 28–44. doi: 10.1002/jclp.21923

Nejad, K. K. (2014). Spinal fMRI of interoceptive attention/awareness in experts and novices. *Neural Plast.* 679509, 1–7. doi: 10.1155/2014/679509

Nielsen, L., and Kaszniak, A. W. (2006). Awareness of subtle emotional feelings: a comparison of long-term meditators and nonmeditators. *Emotion* 6, 392–405. doi: 10.1037/1528-3542.6.3.392

Nolen-Hoeksema, S. (2000). The role of rumination in depressive disorders and mixed anxiety/depressive symptoms. *J. Abnorm. Psychol.* 109, 504–511. doi: 10.1037/0021-843X.109.3.504

Ochsner, K. N., Ray, R. D., Cooper, J. C., Robertson, E. R., Chopra, S., Gabrieli, J. D., et al. (2004). For better or for worse: neural systems supporting the cognitive down-and up-regulation of negative emotion. *Neuroimage* 23, 483–499. doi: 10.1016/j.neuroimage.2004.06.030

Ortner, C. N., Kilner, S. J., and Zelazo, P. D. (2007). Mindfulness meditation and reduced emotional interference on a cognitive task. *Motiv. Emot.* 31, 271–283. doi: 10.1007/s11031-007-9076-7

Ouwens, M. A., Van Strien, T., Van Leeuwe, J. F., and Van Der Staak, C. P. (2009). The dual pathway model of overeating. Replication and extension with actual food consumption. *Appetite* 52, 234–237. doi: 10.1016/j.appet.2008.07.010

Pagnoni, G., and Porro, C. A. (2014). Cognitive modulation of pain and predictive coding: comment on "Facing the experience of pain: a neuropsychological perspective" by Fabbro & Crescentini. *Phys. Life Rev.* 11, 555–557. doi: 10.1016/j.plrev.2014.05.006

Park, R. J., Dunn, B. D., and Barnard, P. J. (2011). Schematic models and modes of mind in anorexia nervosa I: a novel process account. *Int. J. Cogn. Ther.* 4, 415–437. doi: 10.1521/ijct.2011.4.4.415

Park, R. J., Dunn, B. D., and Barnard, P. J. (2012). Schematic models and modes of mind in anorexia nervosa II: implications for treatment and course. *Int. J. Cogn. Ther.* 5, 86–98. doi: 10.1521/ijct.2012.5.1.86

Parkin, L., Morgan, R., Rosselli, A., Howard, M., Sheppard, A., Evans, D., et al. (2013). Exploring the relationship between mindfulness and cardiac perception. *Mindfulness* 5, 1–16. doi: 10.1007/s12671-012-0181-7

Pauley, G., and McPherson, S. (2010). The experience and meaning of compassion and self-compassion for individuals with depression or anxiety. *Psychol. Psychother.* 83, 129–143. doi: 10.1348/147608309X471000

Paulus, M. P. (2007). Decision-making dysfunctions in psychiatry—altered homeostatic processing? *Science* 318, 602–606. doi: 10.1126/science.1142997

Paulus, M. P., Flagan, T., Simmons, A. N., Gillis, K., Kotturi, S., Thom, N., et al. (2012). Subjecting elite athletes to inspiratory breathing load reveals behavioral and neural signatures of optimal performers in extreme environments. *PLoS ONE* 7:e29394. doi: 10.1371/journal.pone.0029394

Paulus, M. P., and Stein, M. B. (2006). An insular view of anxiety. *Biol. Psychiatry* 60, 383–387. doi: 10.1016/j.biopsych.2006.03.042

Paulus, M. P., and Stein, M. B. (2010). Interoception in anxiety and depression. *Brain Struct. Funct.* 214, 451–463. doi: 10.1007/s00429-010-0258-9

Paulus, M. P., Tapert, S. F., and Schulteis, G. (2009). The role of interoception and alliesthesia in addiction. *Pharmacol. Biochem. Behav.* 94, 1–7. doi: 10.1016/j.pbb.2009.08.005

Payne, P., Levine, P. A., and Crane-Godreau, M. A. (2015). Somatic experiencing: using interoception and proprioception as core elements of trauma therapy. *Front. Psychol.* 6:93. doi: 10.3389/fpsyg.2015.00093

Pessoa, L., and Ungerleider, L. G. (2004). Neuroimaging studies of attention and the processing of emotion-laden stimuli. *Prog. Brain Res.* 144, 171–182. doi: 10.1016/s0079-6123(03)14412-3

Pollatos, O., Füstös, J., and Critchley, H. D. (2012). On the generalised embodiment of pain: how interoceptive sensitivity modulates cutaneous pain perception. *Pain* 153, 1680–1686. doi: 10.1016/j.pain.2012.04.030

Pollatos, O., Kirsch, W., and Schandry, R. (2005). On the relationship between interoceptive awareness, emotional experience, and brain processes. *Brain Res. Cogn. Brain Res.* 25, 948–962. doi: 10.1016/j.cogbrainres.2005.09.019

Pollatos, O., Kurz, A.-L., Albrecht, J., Schreder, T., Kleemann, A. M., Schöpf, V., et al. (2008). Reduced perception of bodily signals in anorexia nervosa. *Eat. Behav.* 9, 381–388. doi: 10.1016/j.eatbeh.2008.02.001

Pollatos, O., Traut-Mattausch, E., and Schandry, R. (2009). Differential effects of anxiety and depression on interoceptive accuracy. *Depress. Anxiety* 26, 167–173. doi: 10.1002/da.20504

Porges, S. (1993). *Body Perception Questionnaire*. College Park, MD: Laboratory of Developmental Assessment, University of Maryland.

Price, C. (2005). Body-oriented therapy in recovery from child sexual abuse: an efficacy study. *Altern. Ther. Health Med.* 11, 46–57.

Price, C. (2007). Dissociation reduction in body therapy during sexual abuse recovery. *Complement Ther. Clin. Pract.* 13, 116–128. doi: 10.1016/j.ctcp.2006.08.004

Price, C. J., Diana, T. M., Smith-Dijulio, K. L., and Voss, J. G. (2013). Developing compassionate self-care skills in persons living with HIV: a pilot study to examine mindful awareness in body-oriented therapy feasibility and acceptability. *Int. J. Ther. Massage Bodywork* 6, 9–19.

Price, C. J., and Thompson, E. A. (2007). Measuring dimensions of body connection: body awareness and bodily dissociation. *J. Altern. Complement. Med.* 13, 945–953. doi: 10.1089/acm.2007.0537

Price, C. J., Wells, E. A., Donovan, D. M., and Brooks, M. (2012a). Implementation and acceptability of mindful awareness in body-oriented therapy in women's substance use disorder treatment. *J. Altern. Complement. Med.* 18, 454–462. doi: 10.1089/acm.2011.0126

Price, C. J., Wells, E. A., Donovan, D. M., and Rue, T. (2012b). Mindful awareness in body-oriented therapy as an adjunct to women's substance use disorder treatment: a pilot feasibility study. *J. Subst. Abuse Treat.* 43, 94–107. doi: 10.1016/j.jsat.2011.09.016

Raes, F., Hermans, D., Williams, J. M. G., Bijttebier, P., and Eelen, P. (2008). A "Triple W"-Model of rumination on sadness: why am i feeling sad, what's the meaning of my sadness, and wish i could stop thinking about my sadness (but i can't!). *Cogn. Ther. Res.* 32, 526–541. doi: 10.1007/s10608-007-9137-y

Rector, N. A., Kocovski, N. L., and Ryder, A. G. (2006). Social anxiety and the fear of causing discomfort to others: conceptualization and treatment. *J. Soc. Clin. Psychol.* 25, 906–918. doi: 10.1521/jscp.2006.25.8.906

Rinpoche, P. (1998). *The Words of My Perfect Teacher: A Complete Translation of a Classic Introduction to Tibetan Buddhism.* Lanham, MA: Rowman Altamira.

Samuel, G. (2008). *The Origins of Yoga and Tantra: Indic Religions to the Thirteenth Century.* New York, NY: Cambridge University Press. doi: 10.1017/CBO9780511818820

Schaefer, M., Egloff, B., and Witthöft, M. (2012). Is interoceptive awareness really altered in somatoform disorders? Testing competing theories with two paradigms of heartbeat perception. *J. Abnorm. Psychol.* 121:719. doi: 10.1037/a0028509

Scherer, K. R., Schorr, A., and Johnstone, T. (2001). *Appraisal Processes in Emotion: Theory, Methods, and Research.* Oxford: Oxford University Press.

Schmidt, A. J., Gierlings, R. E., and Peters, M. L. (1989). Environmental and interoceptive influences on chronic low back pain behavior. *Pain* 38, 137–143. doi: 10.1016/0304-3959(89)90231-5

Schulz, A., Lass-Hennemann, J., Nees, F., Blumenthal, T. D., Berger, W., and Schachinger, H. (2009). Cardiac modulation of startle eye blink. *Psychophysiology* 46, 234–240. doi: 10.1111/j.1469-8986.2008.00768.x

Sedeño, L., Couto, B., Melloni, M., Canales-Johnson, A., Yoris, A., Baez, S., et al. (2014). How do you feel when you can't feel your body? Interoception, functional connectivity and emotional processing in depersonalization-derealization disorder. *PLoS ONE* 9:e98769. doi: 10.1371/journal.pone.0098769

Segal, Z. V., Williams, J. M. G., and Teasdale, J. D. (2012). *Mindfulness-Based Cognitive Therapy for Depression.* New York, NY: Guilford Press.

Selver, C., Brooks, C. V. W., Lowe, R., and Laeng-Gilliatt, S. (2007). *Reclaiming Vitality and Presence: Sensory Awareness as a Practice for Life.* Berkeley, CA: North Atlantic Books.

Seth, A. K. (2013). Interoceptive inference, emotion, and the embodied self. *Trends Cogn. Sci.* 17, 565–573. doi: 10.1016/j.tics.2013.09.007

Seth, A. K., and Critchley, H. D. (2013). Extending predictive processing to the body: emotion as interoceptive inference. *Behav. Brain Sci.* 36, 227–228. doi: 10.1017/S0140525X12002270

Seth, A. K., Suzuki, K., and Critchley, H. D. (2011). An interoceptive predictive coding model of conscious presence. *Front. Psychol.* 2:395. doi: 10.3389/fpsyg.2011.00395

Shapiro, S. L., Carlson, L. E., Astin, J. A., and Freedman, B. (2006). Mechanisms of mindfulness. *J. Clin. Psychol.* 62, 373–386. doi: 10.1002/jclp.20237

Shapiro, S. L., Oman, D., Thoresen, C. E., Plante, T. G., and Flinders, T. (2008). Cultivating mindfulness: effects on well-being. *J. Clin. Psychol.* 64, 840–862. doi: 10.1002/jclp.20491

Sheppes, G., and Meiran, N. (2007). Better late than never? On the dynamics of online regulation of sadness using distraction and cognitive reappraisal. *Pers. Soc. Psychol. Bull.* 33, 1518–1532. doi: 10.1177/0146167207305537

Sherrington, C. S. (1906). *The Integrative Action of the Nervous System.* New Haven, CT: Yale University Press. 412.

Singer, T., Critchley, H. D., and Preuschoff, K. (2009). A common role of insula in feelings, empathy and uncertainty. *Trends Cogn. Sci.* 13, 334–340. doi: 10.1016/j.tics.2009.05.001

Sjölie, D. (2014). "Measuring presence in the simulating brain," in *Interacting with Presence*, eds G. Riva, J. Waterworth, and D. Murray (Berlin: De Gruyter), 46–56.

Smallwood, J., and Schooler, J. W. (2006). The restless mind. *Psychol. Bull.* 132, 946. doi: 10.1037/0033-2909.132.6.946

Sterling, P., and Eyer, J. (1988). "Allostasis: a new paradigm to explain arousal pathology," in *Handbook of Life Stress, Cognition and Health*, eds S. Fisher and J. Reason (New York, NY: John Wiley & Sons), 629–649.

Stice, E., Yokum, S., Blum, K., and Bohon, C. (2010). Weight gain is associated with reduced striatal response to palatable food. *J. Neurosci.* 30, 13105–13109. doi: 10.1523/JNEUROSCI.2105-10.2010

Suzuki, K., Garfinkel, S. N., Critchley, H. D., and Seth, A. K. (2013). Multisensory integration across exteroceptive and interoceptive domains modulates self-experience in the rubber-hand illusion. *Neuropsychologia* 51, 2909–2917. doi: 10.1016/j.neuropsychologia.2013.08.014

Sze, J. A., Gyurak, A., Yuan, J. W., and Levenson, R. W. (2010). Coherence between emotional experience and physiology: does body awareness training have an impact? *Emotion* 10, 803–814. doi: 10.1037/a0020146

Taylor, L. A., and Rachman, S. (1988). The effects of blood sugar level changes on cognitive function, affective state, and somatic symptoms. *J. Behav. Med.* 11, 279–291. doi: 10.1007/BF00844433

Taylor, R. L. III, and Hamilton, J. C. (1997). Preliminary evidence for the role of self-regulatory processes in sensation seeking. *Anxiety Stress Coping* 10, 351–375. doi: 10.1080/10615809708249309

Teper, R., and Inzlicht, M. (2013). Meditation, mindfulness and executive control: the importance of emotional acceptance and brain-based performance monitoring. *Soc. Cogn. Affect. Neurosci.* 8, 85–92. doi: 10.1093/scan/nss045

Thayer, J. F., Hansen, A. L., Saus-Rose, E., and Johnsen, B. H. (2009). Heart rate variability, prefrontal neural function, and cognitive performance: the neurovisceral integration perspective on self-regulation, adaptation, and health. *Ann. Behav. Med.* 37, 141–153. doi: 10.1007/s12160-009-9101-z

Thiruchselvam, R., Blechert, J., Sheppes, G., Rydstrom, A., and Gross, J. J. (2011). The temporal dynamics of emotion regulation: an EEG study of distraction and reappraisal. *Biol. Psychol.* 87, 84–92. doi: 10.1016/j.biopsycho.2011.02.009

Tölle, T. R., Kaufmann, T., Siessmeier, T., Lautenbacher, S., Berthele, A., Munz, F., et al. (1999). Region-specific encoding of sensory and affective components of pain in the human brain: a positron emission tomography correlation analysis. *Ann. Neurol.* 45, 40–47. doi: 10.1002/1531-8249(199901)45:1<40::AID-ART8>3.0.CO;2-L

Tomaka, J., Blascovich, J., Kelsey, R. M., and Leitten, C. L. (1993). Subjective, physiological, and behavioral effects of threat and challenge appraisal. *J. Pers. Soc. Psychol.* 65, 248–260. doi: 10.1037/0022-3514.65.2.248

Treleaven, D. (2009). Meditation, trauma, and contemplative dissociation. *Somatics* 16:20.

Tsakiris, M., Tajadura-Jiménez, A., and Costantini, M. (2011). Just a heartbeat away from one's body: interoceptive sensitivity predicts malleability of body-representations. *Proc. Biol. Sci.* 278, 2470–2476. doi: 10.1098/rspb.2010.2547

Vago, D. R., and Silbersweig, D. A. (2012). Self-awareness, self-regulation, and self-transcendence (S-ART): a framework for understanding the neurobiological mechanisms of mindfulness. *Front. Hum. Neurosci.* 6:296. doi: 10.3389/fnhum.2012.00296

Vaitl, D. (1996). Interoception. *Biol. Psychol.* 42, 1–27. doi: 10.1016/0301-0511(95)05144-9

van den Hurk, P. A., Janssen, B. H., Giommi, F., Barendregt, H. P., and Gielen, S. C. (2010). Mindfulness meditation associated with alterations in bottom-up processing: psychophysiological evidence for reduced reactivity. *Int. J. Psychophysiol.* 78, 151–157. doi: 10.1016/j.ijpsycho.2010.07.002

Verdejo-Garcia, A., Clark, L., and Dunn, B. D. (2012). The role of interoception in addiction: a critical review. *Neurosci. Biobehav. Rev.* 36, 1857–1869. doi: 10.1016/j.neubiorev.2012.05.007

Villemure, C., Čeko, M., Cotton, V. A., and Bushnell, M. C. (2013). Insular cortex mediates increased pain tolerance in yoga practitioners. *Cereb. Cortex* 24, 2732–2740. doi: 10.1093/cercor/bht124

Von Leupoldt, A., Vovk, A., Bradley, M. M., Keil, A., Lang, P. J., and Davenport, P. W. (2010). The impact of emotion on respiratory-related evoked potentials. *Psychophysiology* 47, 579–586. doi: 10.1111/j.1469-8986.2009.00956.x

Vuilleumier, P. (2005). How brains beware: neural mechanisms of emotional attention. *Trends Cogn. Sci.* 9, 585–594. doi: 10.1016/j.tics.2005.10.011

Wald, J., and Taylor, S. (2008). Responses to interoceptive exposure in people with posttraumatic stress disorder (PTSD): a preliminary analysis of induced anxiety reactions and trauma memories and their relationship to anxiety sensitivity and PTSD symptom severity. *Cogn. Behav. Ther.* 37, 90–100. doi: 10.1080/16506070801969054

Wallace, B. A. (2007). *Contemplative Science: Where Buddhism and Neuroscience Converge*. New York, NY: Columbia University Press.

Watts, A. G., Salter, D. S., and Neuner, C. M. (2007). Neural network interactions and ingestive behavior control during anorexia. *Physiol. Behav.* 91, 389–396. doi: 10.1016/j.physbeh.2007.04.010

Werner, N. S., Duschek, S., Mattern, M., and Schandry, R. (2009). The relationship between pain perception and interoception. *J. Psychophysiol.* 23, 35–42. doi: 10.1027/0269-8803.23.1.35

Wiebking, C., Duncan, N. W., Tiret, B., Hayes, D. J., Marjańska, M., Doyon, J., et al. (2014). GABA in the insula—a predictor of the neural response to interoceptive awareness. *Neuroimage* 86, 10–18. doi: 10.1016/j.neuroimage.2013.04.042

Williams, J. M. G. (2008). Mindfulness, depression and modes of mind. *Cogn. Ther. Res.* 32, 721–733. doi: 10.1007/s10608-008-9204-z

Williams, J. M., Crane, C., Barnhofer, T., Brennan, K., Duggan, D. S., Fennell, M. J., et al. (2013). Mindfulness-Based cognitive therapy for preventing relapse in recurrent depression: a randomized dismantling trial. *J. Consult. Clin. Psychol.* 82, 275–286. doi: 10.1037/a0035036

Williams, J. M. G., and Kabat-Zinn, J. (2011). Mindfulness: diverse perspectives on its meaning, origins, and multiple applications at the intersection of science and dharma. *Contemp. Buddhism* 12, 1–18. doi: 10.1080/14639947.2011.564811

Wilson-Mendenhall, C. D., Barrett, L. F., and Barsalou, L. W. (2013). Neural evidence that human emotions share core affective properties. *Psychol. Sci.* 24, 947–956. doi: 10.1177/0956797612464242

Woodman, J. P., and Moore, N. R. (2012). Evidence for the effectiveness of Alexander Technique lessons in medical and health-related conditions: a systematic review. *Int. J. Clin. Pract.* 66, 98–112. doi: 10.1111/j.1742-1241.2011.02817.x

Wu, S.-D., and Lo, P.-C. (2008). Inward-attention meditation increases parasympathetic activity: a study based on heart rate variability. *Biomed. Res.* 29, 245–250. doi: 10.2220/biomedres.29.245

Zaman, J., Vlaeyen, J. W., Van Oudenhove, L., Wiech, K., and Van Diest, I. (2015). Associative fear learning and perceptual discrimination: a perceptual pathway in the development of chronic pain. *Neurosci. Biobehav. Rev.* 51, 118–125. doi: 10.1016/j.neubiorev.2015.01.009

Zeidan, F., Gordon, N. S., Merchant, J., and Goolkasian, P. (2010a). The effects of brief mindfulness meditation training on experimentally induced pain. *J. Pain* 11, 199–209. doi: 10.1016/j.jpain.2009.07.015

Zeidan, F., Johnson, S. K., Gordon, N. S., and Goolkasian, P. (2010b). Effects of brief and sham mindfulness meditation on mood and cardiovascular variables. *J. Altern. Complement. Med.* 16, 867–873. doi: 10.1089/acm.2009.0321

3

Investigating the Relationship between Interoceptive Accuracy, Interoceptive Awareness and Emotional Susceptibility

Giuseppe Calì[1†], *Ettore Ambrosini*[1,2†], *Laura Picconi*[3], *Wolf E. Mehling*[4] *and Giorgia Committeri*[1,2*]

[1] *Laboratory of Neuropsychology and Cognitive Neuroscience, Department of Neuroscience, Imaging and Clinical Sciences, University G. d'Annunzio, Chieti, Italy,* [2] *Institute for Advanced Biomedical Technologies, University G. d'Annunzio, Chieti, Italy,* [3] *Department of Psychological, Health and Territory Sciences, University G. d'Annunzio, Chieti, Italy,* [4] *Department of Family and Community Medicine, Osher Center for Integrative Medicine, University of California, San Francisco, San Francisco, CA, USA*

Correspondence:
Giorgia Committeri, Laboratory of Neuropsychology and Cognitive Neuroscience, Department of Neuroscience, Imaging and Clinical Sciences, University G. d'Annunzio, via dei Vestini 33, 66013 Chieti Scalo, Italy
gcommitteri@unich.it

[†] *These authors have contributed equally to this work.*

Interoception, the sense of the physiological condition of the body, provides a basis for subjective feelings and emotions. Anterior insular cortex activity represents the state of the body and varies according to personality traits, such as emotional susceptibility (ES)—the tendency to experience feelings of discomfort and vulnerability when facing emotionally-laden stimuli. The accuracy of perceiving one's own bodily signals, or interoceptive accuracy (IAc), can be assessed with the heartbeat perception task (HPT), which is the experimental measure used by most of the existing research on interoception. However, IAc is only one facet of interoception. Interoceptive awareness (IAw) is the conscious perception of sensations from inside the body, such as heart beat, respiration, satiety, and the autonomic nervous system sensations related to emotions, which create the sense of the physiological condition of the body. We developed an Italian version of the recent self-report Multidimensional Assessment of Interoceptive Awareness (MAIA), tested its psychometric properties (reliability, dimensionality, and construct validity), and examined its relationship to ES, as assessed using the Emotional Susceptibility Scale, in a sample (n = 321) of healthy Italian psychology students (293 females, mean age: 20.5 years). In a subgroup of females (n = 135), we measured IAc with the HPT. We used a series of correlation/regression analyses to examine the complex interplay between the three constructs. We provide further evidence for a substantial independence of the IAc and IAw measures, confirming previous reports and current theoretical models that differentiate between IAc and IAw. Our analyses elucidate the complex relationship between distinct dimensions of IAw and ES, highlighting the need for continued efforts to shed more light on this topic.

Keywords: interoception, personality, heartbeat, MAIA, interoceptive awareness, interoceptive accuracy, emotional susceptibility

Introduction

Interoception has been classically conceived of as the sense of the physiological condition specifically of the viscera, as opposed to the five senses (see also Sherrington, 1948). However, subsequent research has led to a redefinition of the classical conception of interoception as the sense of the physiological condition of the entire body (Craig et al., 1996, 2000; Craig, 2002; Saper, 2002). The neurological correlates of interoceptive processes have been well-defined. They convey information essential to the maintenance of an optimal physiological balance in the body—the physiological system's homeostasis—through autonomic, neuroendocrine, and behavioral responses. Interoception has thus been proposed as a core facet of motivational regulation of behavior and cognition (Craig, 2002; Singer et al., 2009). The progressive meta-representation of the physiological condition of the body from posterior to anterior insula (Farb et al., 2013) enables an integration of homeostatic afferent inputs with information from other limbic regions involved in motivation and emotions and from the prefrontal cortex involved in planning (Chikama et al., 1997; Adolphs, 2002; Brooks et al., 2002; Craig, 2002; Olausson et al., 2002). Perception and feedback of interoceptive signals is considered an essential aspect in many theories of emotion (Darwin, 1873; James, 1884; Schachter and Singer, 1962; Damasio, 1994) and has become the subject of research exploring the relationship between interoception and emotional experience.

Interestingly, activity of the anterior insular cortex also appears to be modulated by specific personality traits, such as neuroticism or emotional susceptibility (ES). This latter personality trait, measured by self-report through the Emotional Susceptibility Scale (Caprara et al., 1985), relates to the tendency to experience feelings of discomfort, helplessness, inadequacy, and vulnerability after exposure to emotionally salient stimuli. ES is viewed as reflecting an individual's inclination to experience a state of negative affect and as a tendency to place oneself in a defensive position due to the inability to control excitement, arousal, and reactions in situations—real or imagined—of danger, offense, threat, or attack. Therefore, this personality trait may be of particular relevance for individual differences in reactions to emotional stimuli, particularly to the experienced intensity of affective states, or their arousal dimensions (Bradley and Lang, 1994), which is reflected in the activity of the autonomic nervous system. In an fMRI study, the ES personality trait has been shown to be associated with insular activity (Iaria et al., 2008): participants with high ES scores, as compared to those with low ES scores, showed greater activity within the anterior insula during processing of visual stimuli with emotional content, independent of the stimuli's valence (i.e., positive vs. negative affect). In another study (Ebisch et al., submitted), high ES participants showed greater anterior insular activity in response to neutral gustatory stimuli when interspersed between distasteful and pleasant-tasting affective stimuli. These findings suggest the implication of interoceptive processes in the modulation of basic emotional processing by the ES trait. Although a potential association between ES and interoception has yet to be explored, there is a growing body of evidence supporting the idea that the insula is a key region involved in interoception and emotion (e.g., Zaki et al., 2012; Ernst et al., 2013). Moreover, how interoception is involved in emotional processing has been studied over the last several years (e.g., Wiens, 2005; Craig, 2008; Pollatos and Schandry, 2008; Garfinkel et al., 2014). These studies have suggested that individuals with stronger interoceptive ability report more intense emotions (e.g., Wiens et al., 2000), place a stronger emphasis on the arousal dimension in reporting their emotional experience (Barrett et al., 2004), and show a stronger link between their body reactions to emotional stimuli and their subjective arousal ratings (Dunn et al., 2010). Moreover, interoceptive accuracy has been shown to be positively correlated to measures of central and peripheral processing of emotional stimuli (Pollatos et al., 2005, 2007; Herbert et al., 2007).

However, investigations of the relationship between interoceptive ability and emotional processing have used a variety of measures of interoception, which has triggered recent attempts at more clearly defining the different constructs and measures related to interoception, such as interoceptive accuracy (IAc) and awareness (IAw) (Farb et al., 2015; Garfinkel et al., 2015). Ceunen et al. (2013), for example, defined IAc as the ability to accurately perceive changes in the homeostatic function and IAw as the conscious perception of these body signals, expressing disagreement with the idea that cardiac interoceptive awareness is reflected in and appropriately assessed as the sensorial accuracy in perceiving one's own cardiac signals (e.g., Herbert et al., 2012). Ceunen et al. (2013) proposed that the equivalence between the two concepts probably derives from the "interoceptive sensitivity hypothesis" (Tyrer, 1973; Ehlers and Breuer, 1992), according to which highly anxious individuals and patients with panic disorder—conditions associated with a high symptom awareness level (IAw)—have highly accurate perceptions of anxiety-related bodily sensations (IAc). Ceunen et al. suggested: "IAw should be taken to mean the cognizant, mindful perception of bodily signals [...]. Although IAw can be accompanied by an accurate perception of bodily sensations, such accuracy is not necessarily implied" (Ceunen et al., 2013). This suggestion is clearly supported by recent research, which has led to a more differentiated theoretical model of interoception processes (Garfinkel et al., 2015). However, the definition of IAw we are using here, which was developed in focus groups of clinicians and patients and applied in the recently developed Multidimensional Assessment of Interoceptive Awareness (MAIA; Mehling et al., 2012), does not exactly correspond to that provided by Garfinkel and Critchley (2013) and Garfinkel et al. (2015). The latter was operationalized as "a metacognitive measure that quantifies individuals' explicit knowledge of (and confidence in) their interoceptive accuracy" (Garfinkel and Critchley, 2013, p. 233), or "the correspondence between objective IAc and subjective report, i.e., metacognition" (Garfinkel et al., 2015). This pragmatic operationalization of IAw is based on signal-detection theory and reduces the IAw construct to a measure for the extent to which self-reported confidence of perception predicts the accuracy of perception. Our view of the IAw construct, as applied to this study, is also broader and more complex than the conceptualization of interoceptive sensibility, defined as the "self-perceived dispositional tendency to be internally self-focused and interoceptively cognizant" (Garfinkel et al., 2015) and assessed by self-report questionnaires regarding specific bodily sensations or self-rated confidence in one's perceptual ability. We incorporated the earlier, broader definition by Cameron of IAw as awareness of "the afferent information that arises from anywhere and everywhere within the body ... [involving] higher mental processes such as emotions, conscious awareness, and behavior" (Cameron, 2001), differentiated further through qualitative analyses from clinically-oriented focus groups (Mehling et al., 2011). By including these attitudinal, appraisal, and self-regulatory aspects of IAw, our conceptualization of IAw extends beyond interoceptive sensibility, which nevertheless remains as one of the dimensions ("Noticing") in the multidimensional self-report MAIA (Mehling et al., 2012). The pragmatic operationalization of IAw as metacognition is clearly innovative in that it extends beyond self-report, but it appears to strip the construct of some of its psychological, particularly its regulatory, aspects (Bornemann et al., 2014).

The most commonly applied objective measures of interoceptive ability are two different types of Heartbeat Perception Tasks (HPT): the tracking task (Schandry, 1981), in which the perception of cardiac activity is measured by asking the participant to count his/her heartbeats within a defined time interval, and the discrimination task (Whitehead et al., 1977), in which the participant is asked to discriminate between exteroceptive (auditory or visual) and interoceptive (heartbeat) signals. Both methods provide a measure of the individual's interoceptive accuracy (IAc), which in turn has been taken in various earlier studies as an index of one's awareness of one's own bodily sensations, that is, a measure of interoceptive awareness (IAw). Yet early findings showed that IAw, as assessed by self-report measures, does not necessarily predict actual IAc, as assessed experimentally by HPT (e.g., McFarland, 1975). Consequently, in recent years, numerous authors (Wiens, 2005; Mehling et al., 2009; Ceunen et al., 2013; Garfinkel and Critchley, 2013; Bornemann et al., 2014; Farb et al., 2015; Garfinkel et al., 2015) have highlighted the need to discriminate between IAc and IAw and to better define interoception-related constructs. Still, IAc is currently viewed as the central construct underpinning other interoceptive measures (Garfinkel et al., 2015).

In reviewing existing self-report questionnaires on body awareness, which includes both proprioceptive and IAw, and their psychometric properties, Mehling et al. showed that most of the questionnaires are based on a conceptualization of body awareness that considers the attentional focus on bodily symptoms as a maladaptive expression of anxiety, depression, or somatization (Mehling et al., 2009), which may be associated with ES. Instead, a more recent (and alternative) conceptualization of body awareness considers the ability to recognize subtle bodily signals as a controlled state of sustained attention to events that happen within the body here-and-now, which has been integrated into numerous therapeutic approaches geared to improve this ability. An example is mindfulness: the training of a specific style of attention aimed at a state of mind characterized by non-judgmental acceptance and a sense of self that is rooted in experiencing physical sensations in the present moment with equanimity (Thompson and Varela, 2001; Carruthers, 2008; Fogel, 2009; Grossman, 2015). These more recent considerations formed the basis for the development of the MAIA (Mehling et al., 2012; see also Mehling et al., 2009, 2011), a multidimensional self-report instrument designed to measure interoceptive body awareness. It consists of 32 items evaluating eight aspects grouped into five dimensions. The MAIA attempts to overcome the limitations of previous accounts of IAw (described above) by integrating what appear to be the hallmarks of interoceptive body awareness: (1) its multidimensional nature; (2) the distinction between the anxiety-related hypervigilance and the attentional style to bodily sensations in the present moment (which would represent a step away from the assumptions of the interoceptive sensitivity hypothesis); (3) the differential assessment of essential psychological aspects of the perception and evaluation of bodily sensations. To date, the MAIA has been translated into 13 languages (http://www.osher.ucsf.edu/maia/) but, to our knowledge, only two adaptations with adequate validation have been published: the German (Bornemann et al., 2014) and Chilean (Valenzuela Moguillansky and Reyes Reyes, 2015) versions, which both replicated the global structure of the original MAIA.

The first aim of the present work was to provide a preliminary validation of the Italian MAIA, by analyzing the psychometric properties (reliability, dimensionality, and construct validity) of the Italian translation of the MAIA. To this aim, we applied it to a population mainly composed of female psychology undergraduate students.

Our second aim was to investigate the relationship between the multiple facets composing the IAw construct—as assessed by the MAIA—and both the IAc as assessed by the HPT and the ES personality trait. To this aim, we used a series of correlational and regression analyses to examine the complex interplay between the three measures of IAw, IAc, and ES, which are involved in shaping the way individuals experience their "bodily" feelings and emotional states. At the time of the survey (early 2013), it was still an open question whether IAw, as measured by the 8 MAIA scales, would potentially be correlated with IAc or not. Therefore, we aimed at investigating such a relationship with the hypothesis that these constructs are not correlated, supporting the idea that they are distinct and dissociable dimensions. Second, we hypothesized that IAc may be positively correlated with ES, possibly mediated by the arousal component of emotions. Our third exploratory hypothesis was that different scales of the MAIA for the IAw construct would correlate in a complex way with ES, which may add new details to the relationship between interoception, emotion processing, and personality traits.

Materials and Methods

Participants

A total of 321 participants, aged between 19 and 27 years (M = 20.53, SD = 0.88), were recruited for the study. Almost all of them were students of the University G. d'Annunzio of Chieti. The majority of the participants (293, representing approximately 91% of the total sample) were female (mean age = 20.49 years,

$SD = 0.85$); 28 were male (mean age = 20.96 years, $SD = 1.10$). This disproportion in the participants' gender reflects the gender distribution in the psychology student population from which the experimental sample was mainly recruited. Of this sample, 135 female participants (mean age = 20.40 years, $SD = 0.72$) volunteered to perform the Heartbeat Perception Task (HPT). Participants gave informed consent prior to their inclusion in the study, which was approved by the ethical committee of the University of Chieti and conducted according to the guidelines of the Declaration of Helsinki.

Instruments and Procedures

Italian Translation of the MAIA

We assessed the multiple aspects of our participants' interoceptive awareness (IAw) by using a new Italian version of the original English-language MAIA (Mehling et al., 2012). Since no Italian version of the MAIA was available, we first systematically translated it. In a first step, three independent forward translations were produced, two by native Italian speakers proficient in English, and one by a native English speaker proficient in Italian. These translators were not familiar with the IAw construct, but have a background in Psychology. The three resulting provisional Italian versions were then compared, item by item, by two of the authors (EA and GCo) and two other researchers from the Laboratory of Neuropsychology and Cognitive Neuroscience of the University of Chieti (Marcello Costantini and Gianluca Finotti), who were familiar with the IAw construct. Following consensus, a provisional Italian version was drafted. Next, an English native bilingual translator, who had a background in Psychology but who was not familiar with the IAw construct, performed the back-translation into English. Finally, differences between the original English version and the back-translation were identified and discussed with the first author of the original MAIA (WM), and further small corrections were applied to improve and finalize the items. The final Italian version was approved by consensus among the same four researchers from the Laboratory of Neuropsychology and Cognitive Neuroscience named above, as well as by the first author of the original MAIA.

The MAIA is composed of 32 items on a 6-points Likert scale, in which the participant has to rate "how often each statement applies to you generally in daily life," with ordinal responses coded from 0 ("never") to 5 ("always"). This multidimensional instrument measures IAw on eight scales: (1) Noticing, the awareness of one's body sensations (4 items); (2) Not-distracting, the tendency not to ignore or distract oneself from sensations of pain or discomfort (3 items); (3) Not-worrying, the tendency not to experience emotional distress or worry with sensations of pain or discomfort (3 items); (4) Attention regulation, the ability to sustain and control attention to body sensation (7 items); (5) Emotional awareness, the awareness of the connection between body sensations and emotional states (5 items); (6) Self-regulation, the ability to regulate psychological distress by attention to body sensations (4 items); (7) Body listening, the tendency to actively listen to the body for insight (3 items); and (8) Trusting: the experience of one's body as safe and trustworthy (3 items). The score for each scale is calculated by averaging the scores of its individual items, and thus can vary in the 0–5 range.

Emotional Susceptibility Scale

We assessed the participants' ES by using the Italian version of the Emotional Susceptibility Scale, a self-administered, unidimensional instrument developed by Caprara et al. (1985; Caprara et al., 1983) to measure the participants' "tendency to experience feelings of discomfort, inadequacy, and vulnerability" after the exposure to emotionally salient stimuli. This instrument is composed of 40 items on a 6-points Likert scale, in which the participant has to rate "how true" each statement is for him/her, with ordinal responses coded from 1 ("completely false for me") to 6 ("completely true for me"). The ES scale includes 10 control items that do not contribute to the final score, which is obtained by summing the scores of the 30 non-control items.

Heartbeat Perception Task

From our 321 participants, we recruited volunteers to perform the HPT ($n = 135$) in the laboratory. After the participants had arrived, they were informed about the general procedure and the measures. The participants were seated in a comfortable chair in a sound-attenuated room. The HPT procedure was performed according to the mental tracking method proposed by Schandry (1981), which was modified by adding a fourth time interval (Pollatos et al., 2008). The four time intervals were 25, 35, 45, and 100 s and were presented in random order. During each time interval, participants were asked to count and keep track of their heartbeats by focusing on their heart activity and bodily feelings. The participants were explicitly instructed not to take their own pulse in any way or to try any other physical manipulations; they were also discouraged from simply tracking the number of seconds of each time interval and using that information to estimate the number of heartbeats. After each interval, the participants were asked to verbally report the count or estimated number of heartbeats, which was manually recorded by the experimenter.

Participants were distributed in three groups. In the first two groups, the actual number of the participant's heartbeats was recorded by one of the authors (GCa) and a trained assistant by means of either tactile radial arterial palpation at the wrist or direct chest auscultation through a stethoscope. The experimenters kept track of time using a digital stopwatch. The beginning and end of each time interval was indicated to the participants by the acoustic signal emitted by the stopwatch. In the tactile arterial palpation procedure, particular care was taken by the experimenters to avoid applying excessive pressure to the participants' wrist skin, which could have made it easier for the participants to perceive their heartbeats. In the third group, the participant's heartbeats were recorded by another author (EA) and a trained assistant using a three-lead electrocardiogram: three Ag/AgCl pre-gelled electrodes (ADInstruments, UK) were placed on the participant's chest in an Einthoven's triangle configuration to record the electrocardiogram (sampling rate: 1 KHz; Powerlab, ADInstruments, UK); the heartbeats were identified by the R peaks using the BioSig package (Vidaurre et al., 2011) for Matlab (MathWorks, Natick, MA). The beginning and the end of each time interval were indicated to the participant through acoustic signals presented via headphones. The entire procedure was controlled by a computer running the Matlab Psychophysics Toolbox (Brainard, 1997). The participants of this third group were part of a larger study investigating the physiological responses to affective stimuli.

General Procedure

Participants were asked to complete the MAIA and the ES scale. Both scales were administered in paper-and-pencil versions in classroom settings during psychology courses or, for a small number of participants recruited from campus facilities, in the laboratory setting. Data collection was conducted by the first author (GCa) and trained assistants. The purpose of the study was briefly explained to the participants, who were informed that they were free to respond (but asked to respond truthfully), and that their collaboration would potentially be requested again at a later stage of the study. The HPT was performed in a separate session as detailed in Section Heartbeat Perception Task.

Data Analysis

Missing values for items of the MAIA and ES scale were imputed using the series mean method, that is, by replacing them with the mean of all participants' values for the same item. For the MAIA, there were on average 0.97 missing values ($SD = 2.63$) for each item, whereas the missing values were, on average, 1.18 ($SD = 1.20$) for each item of the ES scale (out of the total of 321 responses per item). The scores of the eight scales of the MAIA were computed by averaging the values of the items of each scale, according to the final factorial structure we obtained (see Section Psychometric Properties and Factorial Structure of the Italian MAIA). The ES score was computed for each participant by averaging the score of the 30 non-control items of the ES scale, thus obtaining a value ranging from 1 to 6.

The participants' performance in perceiving their heartbeats during each time interval of the HPT was calculated as a relative error score, that is, the absolute difference between reported and actual number of heartbeats divided by the actual number of heartbeats. Next, in line with standard practice (e.g., Koch and Pollatos, 2014), the participants' interoceptive accuracy (IAc) score was computed according to the formula: $\text{IAc} = 1/4\ \Sigma\ [1 - (|\text{recorded heartbeats} - \text{counted heartbeats}|/\text{recorded heartbeats})]$. We then tested for differences in IAc scores between the three groups of participants (see Section Heartbeat Perception Task). We carried out pairwise comparisons with two-tailed independent-samples t-test and the corresponding Bayesian t-test (Rouder et al., 2009) for accepting the null hypothesis of no differences between groups.

We first investigated the preliminary psychometric properties and the dimensionality of the Italian version of the MAIA, in order to evaluate whether the factor structure of the original version would replicate in the Italian version. To this aim, we carried out an exploratory factor analysis (EFA) and an analysis of covariances within the framework of confirmatory factor analysis (CFA), and assessed the reliability of the MAIA scales. The hypothesized model was estimated via maximum likelihood (ML). For the evaluation of covariance structure models we used the chi-square goodness of fit supplemented by the comparative fit index (CFI), the root-mean-square error of approximation (RMSEA) and the standardized root-mean-square residual (SRMR). The CFI (Bentler, 1990) assesses the reduction in misfit of a population target model relative to a population baseline model in which no structure is specified (i.e., all correlations among variables are equal to zero). Values of at least 0.90 are considered adequate for good models (Bentler, 1990). The RMSEA is a measure of the discrepancy of the variance covariance matrix of fitted model from the starting variance covariance matrix per degree of freedom. Values lower than 0.05 reflect a small error of approximation and values between 0.05 and 0.08 reflect an acceptable error of approximation. Values greater than 0.10 constitute poor model fit (Browne and Cudeck, 1993). The SRMR is an absolute index of the discrepancy between reproduced and observed correlations. Hu and Bentler (1998) suggested a cutoff criterion of 0.08, with higher values indicating poorer fit to the empirical data and values lower than 0.05 indicating an excellent fit. Finally, the Cronbach's alpha coefficient and corrected item-scale correlations were used to assess the reliability of the scales.

Next, we explored the complex interplay between the three constructs of ES, IAw, and IAc. We first calculated the Pearson correlations between each pair of variables, including ES score, scores of the eight scales of MAIA (as derived from the factorial analyses, see Section Psychometric Properties and Factorial Structure of the Italian MAIA)[1], and IAc score, with pairwise deletion of cases with missing data. For a more precise investigation of the actual relationship between these measures, partial correlations were calculated between each pair of variables, controlling for all the other variables. We carried out multiple regression analyses to investigate in more detail the relationship between the participants' ES scores (the dependent variable) and the eight MAIA scores and the IAc score (continuous predictors), as well as all two-way interactions. The model that best explained the ES scores, with the appropriate number of predictors, was selected by the method of the best subset as implemented in Statistica (StatSoft). In brief, all possible regression models with up to a defined number (subset) of predictors were evaluated; among those with the same number of predictors, the model with the highest percentage of explained variance (R^2) was provisionally selected as best explaining the ES scores. This procedure identifies the best model for any number of predictors (Neter et al., 1989; as cited in StatSoft Inc, 2013). Finally, models with different numbers of predictors were compared by testing for the significance of the R^2 difference taking into account the tolerance index, a measure of multicollinearity among the predictors included in the model. Statistical analyses were conducted through Statistical Software for Social Science (SPSS), Statistica (StatSoft), and Structural Equation Modeling (EQS).

[1]Note that when we applied the original structure of the MAIA the analyses yielded similar results, and thus are not reported here for the sake of brevity.

Results

We first describe the preliminary psychometric properties and the dimensionality of the Italian MAIA and compare them with those of both the original version and the two other published German (Bornemann et al., 2014) and Chilean (Valenzuela Moguillansky and Reyes Reyes, 2015) translations.

Psychometric Properties and Factorial Structure of the Italian MAIA

The appropriateness of the factor analysis as a model for analyzing the data was supported by the Bartlett's test of sphericity ($\chi^2 = 3129.50p < 0.001$) and the Kaiser-Meyer-Olkin measure of sampling adequacy (0.822). Assessment of distributions showed that all but one item had skewness and kurtosis values in the (-1, 1) range (item 4 had a skewness value of -1.054, see **Table 1**).

For the EFA, we extracted factors using the principal axis factoring method and factor loadings higher than $\pm$ 0.30 were considered as building criteria for the EFA model. In order to determine the number of factors to retain, we examined the eigenvalues (Cattell and Vogelmann, 1977) (extraction criterion: eigenvalue > 1; varimax rotation). After the successive deletion of three items (7, 10, and 12) that did not load significantly on the theoretical factor, and after the interpretation of the factor scores, the eight factors solution, accounting for the 47.74% of the total variance, was also corroborated by an inspection of the scree-plot of eigenvalues (the first 10 eigenvalues were 6.566, 2.542, 1.942, 1.679, 1.520, 1.283, 1.170, 1.089, 0.931, and 0.883). This eight factors solution clusters the remaining items as in the original version, with the exception of items 19 ("When something is wrong in my life, I can feel it in my body."), which loaded on both its original factor Emotional Awareness (0.469) and on Body Listening (0.377) (for a similar result, see Bornemann et al., 2014), and 4 ("I notice changes in my breathing, such as whether it slows down or speeds up."), which loaded on a different factor (Emotional Awareness: 0.387) than its original factor (Noticing: 0.147). We then carried out a CFA, which showed that the eight factors solution obtained with the EFA provided an adequate fit to the data [$\chi^{2(349)} = 408.99$; $p = 0.015$; RMSEA = 0.023 (90% CI = 0.011, 0.032); SRMR = 0.057, CFI = 0.974].

TABLE 1 | Univariate descriptive item statistics (n = 321).

Item	Mean	SD	Skewness	Kurtosis
Item1	2.47	1.23	−0.01	−0.43
Item2	2.94	1.30	−0.34	−0.46
Item3	3.09	1.20	−0.45	−0.28
Item4	3.69	1.24	−0.05	0.72
Item5	2.74	1.36	−0.12	−0.76
Item6	2.79	1.21	0.05	−0.71
Item7	2.43	1.37	0.13	−0.67
Item8	2.32	1.26	0.13	−0.78
Item9	2.70	1.15	−0.02	−0.56
Item10	2.19	1.20	0.10	−0.68
Item11	2.57	1.27	−0.26	−0.58
Item12	2.83	1.13	−0.26	−0.27
Item13	2.77	1.32	−0.32	−0.56
Item14	2.48	1.12	−0.10	−0.31
Item15	2.68	1.02	−0.27	0.07
Item16	2.63	1.16	−0.04	−0.45
Item17	3.01	1.01	−0.03	0.11
Item18	3.00	1.27	−0.40	−0.54
Item19	2.79	1.28	−0.13	−0.57
Item20	3.41	1.14	−0.67	0.14
Item21	3.74	1.16	−0.92	0.44
Item22	3.57	1.16	−0.93	0.58
Item23	2.60	1.19	−0.09	−0.51
Item24	2.44	1.15	−0.03	−0.47
Item25	3.02	1.21	−0.47	−0.22
Item26	2.45	1.21	0.02	−0.56
Item27	2.95	1.08	−0.38	−0.19
Item28	1.98	1.17	0.32	−0.34
Item29	2.02	1.21	0.29	−0.43
Item30	2.80	1.39	−0.19	−0.70
Item31	2.88	1.32	−0.14	−0.56
Item32	3.35	1.16	−0.52	−0.05

Table 2 shows mean values, standard deviation, average item-scale correlations, and Cronbach's alphas of the Italian MAIA. The Cronbach's alpha values varied between 0.53 and 0.80. Compared to the alpha values of the original MAIA (Mehling et al., 2012) by means of the Feldt's test for independent samples (Feldt, 1986), the alpha values of the Italian MAIA were significantly lower than in the original MAIA for five scales (Attention regulation, Self-regulation, Body listening, Not-worrying, and Not-distracting), and not statistically different for the remaining three scales. Similar results were obtained when comparing the alpha values of the Italian MAIA with those of both the German and Chilean versions of the MAIA. Notwithstanding these differences, it should be noted that our Cronbach's alpha values were highly correlated to those of the English, German, and Chilean versions (r = 0.855, 0.973, and 0.928; all $ps \leq 0.007$). Moreover, the pattern of values we observed was very similar to that of the other MAIA versions, with the Noticing, Not-worrying, and Not-distracting scales consistently showing the lower Cronbach's alpha values among the scales (all $ps \leq 0.0001$, Feldt's test for dependent samples; Feldt et al., 1987).

Table 2 also shows the Pearson's interscale correlations. The results revealed a complex pattern of intercorrelations between the scores of the eight scales of MAIA, with a high percentage of significant correlations (16 out of 28 possible pairs, representing approximately 57%, with α = 0.05). The only two scales that seem to be relatively independent from the other scales were Not-worrying and Not-distracting, which composed the dimension "Emotional Reactions and Attentional Response to a Sensation" in the original MAIA (Mehling et al., 2012). Again, this pattern of results was consistent with that found in the other available versions of the MAIA. In fact, the Pearson's correlations between the interscale correlation matrices of the Italian MAIA, on the one side, and those of the English, German, and Chilean MAIA, on the other side, were very high (respectively, 0.896, 0.925, and 0.925; all ps < 0.001), highlighting a high second-order

TABLE 2 | Descriptive statistics, reliability indices, and interscale/inter-construct correlations for IAw, ES, and IAc.

	Measures	Mean	SD	Cronbach's alpha	Item-scale correlation	No. of items	Interscale/Inter-construct Pearson's correlations[a]									
							1	2	3	4	5	6	7	8	9	10
	IAw (MAIA, *n* = 321)															
1	Noticing	2.84	0.97	0.68	0.495	3		0.704	0.278	**<0.001**	**<0.001**	**<0.001**	**<0.001**	**<0.001**	0.316	0.543
2	Not-distracting	2.77	1.06	0.53	0.361	2	0.021		0.107	0.087	0.599	0.887	0.910	0.967	0.695	0.495
3	Not-worrying	2.51	1.01	0.59	0.418	2	−0.061	−0.090		0.344	**<0.001**	0.171	0.214	0.959	**<0.001**	0.352
4	Attention regulation	2.69	0.77	0.75	0.495	6	**0.329**	−0.096	0.053		**<0.001**	**<0.001**	**<0.001**	**<0.001**	**<0.001**	**0.020**
5	Emotional awareness	3.37	0.85	0.79	0.550	6	**0.335**	0.029	**−0.196**	**0.315**		**<0.001**	**<0.001**	**0.002**	**0.001**	0.534
6	Self-regulation	2.63	0.90	0.75	0.540	4	**0.310**	−0.008	−0.076	**0.361**	**0.428**		**<0.001**	**<0.001**	0.103	0.713
7	Body listening	2.32	0.94	0.74	0.569	3	**0.364**	0.006	−0.069	**0.327**	**0.429**	**0.571**		**<0.001**	0.299	0.974
8	Trusting	3.01	1.09	0.80	0.653	3	**0.123**	0.002	0.003	**0.375**	**0.170**	**0.414**	**0.281**		**<0.001**	0.338
9	*ES* (*n* = 321)	4.02	0.92	0.90	0.508	30	0.056	−0.022	**−0.359**	**−0.244**	**0.187**	−0.091	0.058	**−0.331**		0.412
10	*IAc* (HPT, *n* = 135)	0.47	0.19	0.89	0.852	4	−0.053	−0.059	0.081	**0.200**	−0.054	−0.032	−0.003	0.083	−0.071	

[a]*Values of the Pearson's correlation coefficients (r) are shown in the lower triangle, and the corresponding p-values are shown in the upper triangle. Significant correlations are indicated in bold.*

isomorphism (Shepard and Chipman, 1970) of the correlational structure of the MAIA across these three languages and cultures.

Heartbeat Perception Task

There were no significant differences in IAc scores between the three groups of participants (see Section Heartbeat Perception Task), suggesting that the method of measurement of the heartbeats (ECG, stethoscope, wrist palpation) did not influence the participants' IAc. Mean IAc scores in the three groups were 0.503 ($SD = 0.191$), 0.542 ($SD = 0.182$), and 0.540 ($SD = 0.185$), respectively. These were not different in pairwise comparisons by t-tests [ECG vs. stethoscope: $t_{(90)} = 0.987$, $p = 0.327$; ECG vs. wrist palpation: $t_{(90)} = 0.721$, $p = 0.473$; stethoscope vs. wrist palpation: $t_{(84)} = 0.939$, $p = 0.350$]. More importantly, Bayesian t-tests (Rouder et al., 2009) provided support for the null hypothesis across the three methods with JZS Bayes factors of 3.96, 4.90, and 4.02, respectively, for the same pairwise comparisons. Therefore, we felt justified in pooling the data across heartbeat recording methods for our subsequent analyses.

Inter-constructs Relationships

Table 2 shows the Pearson's correlations between the eight scales of the Italian MAIA and both the ES and IAc scores of the participants. The analysis revealed significant correlations between the ES scale and four of the eight scales of the MAIA: Not-worrying, Attention regulation, Emotional awareness, and Trusting (respectively, $r = -0.359, -0.244, 0.187$, and -0.331; all $ps \leq 0.001$; $n = 321$). The IAc score was weakly but statistically significantly correlated with only one MAIA scale: Attention regulation ($r = 0.200$, $p = 0.020$, $n = 135$).

Given the complex pattern of intercorrelations among the MAIA scales, we also calculated partial correlations between each pair of variables, controlling for all other variables. This analysis essentially confirms the results of the previous one and revealed weak but statistically significant partial correlations ($r_{partial}$) between the ES scale and the same four MAIA scales that emerged from the regular correlation analysis (respectively, $r_{partial} = -0.339, -0.222, 0.212, -0.289$; all $ps \leq 0.001$; $n = 321$), with an additional significant partial correlation for the Body Listening scale ($r_{partial} = 0.128$, $p = 0.023$; **Table 3**). Moreover, the weak but statistically significant correlation between the IAc score and the MAIA Attention Regulation scale was confirmed by the partial correlation analysis ($r_{partial} = 0.226$, $p = 0.011$, $n = 135$).

Since the correlational analyses showed that the IAc score was not reliably related to the ES score, and since a first multiple regression analyses showed that IAc did not reliably predict ES score (neither individually nor in interaction with other predictors), we chose to exclude it from the list of continuous predictors for the multiple regression analysis, in order to have a larger sample of valid cases ($n = 321$, the participants who completed both the MAIA and ES scales, instead of $n = 135$ for the subsample of participants who also performed the HPT) and, thus, increase the statistical power. The subsequent multiple regression analyses, performed with the best subset method as detailed in the Data Analysis Section, revealed a

TABLE 3 | Partial correlations between IAw scales, ES, and IAc.

	Measures	Interscale/Inter-construct partial correlations[a]									
	IAw (MAIA, *n* = 321)	1	2	3	4	5	6	7	8	9	10
1	Noticing		0.424	0.970	**<0.001**	**0.022**	0.189	**0.003**	0.350	0.271	0.236
2	Not-distracting	0.045		0.076	**0.021**	0.364	0.658	0.713	0.846	0.094	0.939
3	Not-worrying	0.002	−0.100		0.624	0.075	0.351	0.403	0.095	**<0.001**	0.628
4	Attention regulation	**0.217**	**−0.130**	0.028		**0.001**	0.329	0.112	**0.001**	**<0.001**	**0.011**
5	Emotional awareness	**0.129**	0.051	−0.100	**0.194**		**<0.001**	**0.005**	0.868	**<0.001**	0.644
6	Self regulation	0.074	−0.025	−0.053	0.055	**0.214**		**<0.001**	**<0.001**	0.093	0.487
7	Body listening	**0.165**	0.021	0.047	0.090	**0.157**	**0.403**		0.159	**0.023**	0.685
8	Trusting	−0.053	0.011	−0.094	**0.194**	0.009	**0.242**	0.079		**<0.001**	0.583
9	*ES* (*n* = 321)	0.062	−0.095	**−0.339**	**−0.222**	**0.212**	−0.095	**0.128**	**−0.289**		0.710
10	*IAc* (HPT, *n* = 135)	−0.106	−0.007	0.043	**0.226**	−0.041	−0.062	0.036	0.049	0.033	

[a]*Partial correlations are shown in the lower triangle with corresponding p-values in the upper triangle. Significant correlations are indicated in bold.*

model that included three predictors as the best model to explain the participants' ES scores. This model included, ordered by the percentage of explained variance, the interaction between Attention regulation and Trusting [$\beta = -0.413$, $SE = 0.049$, $t_{(317)} = 8.142$, $p < 0.001$], Not-worrying [$\beta = -0.295$, $SE = 0.048$, $t_{(317)} = 6.123$, $p < 0.001$], and Emotional awareness ($\beta = 0.241$, $SE = 0.050$, $t_{(317)} = 4.829$, $p < 0.001$) as predictors, explaining 29.87% of the variance [$F_{(3, 317)} = 45.01$, $p < 0.001$], with good tolerance values (≥ 0.883). Therefore, participants with higher ES scores reported more emotional distress or worry with sensations of pain or discomfort (Not-worrying), higher awareness of the connection between body sensations and emotional states (Emotional awareness), and were either less able to sustain and control attention to body sensation (Attention regulation) or less prone to experience their own body as a safe and trustworthy place (Trusting). In particular, the interaction between Attention regulation and Trusting revealed that the participants who had a score near zero in at least one of these scales also had higher ES scores, that is, were more prone to experience feelings of discomfort, helplessness, inadequacy, and vulnerability due to the inability to control their reactions in negative situations (either real or imagined), irrespective of their score in the other scale; conversely, only participants who showed high scores in both these scales reported to have low ES. It is important here to note that the significant effects revealed in this multiple correlation analysis concern the same variables for which we found significant regular and partial correlations with the ES score, thus highlighting the stability of these findings but, at the same time, revealing a more complex pattern of relationships between the ES and the multiple facets of the IAw, as evaluated by the MAIA.

Discussion

In the present study, we aimed to test the psychometric properties of the new Italian translation of the MAIA and explore the complex interplay between three constructs that contribute in shaping how we experience our "bodily" feelings and emotional states, that is: interoceptive accuracy (IAc), interoceptive awareness (IAw), and emotional susceptibility (ES).

First, analyses of the preliminary psychometric properties and the factorial structure of the Italian version of the MAIA revealed that it has acceptable reliability and a dimensionality that is comparable to that of the other available MAIA versions (Mehling et al., 2012; Bornemann et al., 2014; Valenzuela Moguillansky and Reyes Reyes, 2015). However, this validation is limited, and the present data should be considered as a first, preliminary validation of the MAIA in the Italian population, as we tested it in a sample of 321 mostly female undergraduate Psychology students that is hardly representative of the Italian population. Therefore, further examinations on a more representative sample are needed.

Second, our analyses elucidate the complex relationship between distinct dimensions of IAw and ES and add new data to the discriminatory and convergent validity of the MAIA scales. Our results from correlational analyses revealed that the participants' scores on four of the eight MAIA scales, namely Not-worrying, Emotional awareness, Attention regulation, and Trusting were reliably correlated with the participants' ES. Multiple regression analyses further elucidated the specific contribution of each MAIA scale in explaining the variability in the participants' ES. These analyses revealed that, while both the Not-worrying and Emotional awareness scales were independent significant predictors of the participants' ES scores, the Attention regulation and Trusting scales interacted in explaining the (larger portion of the) variability in the participants' ES scores. Our results thus showed that two specific aspects of IAw independently contributed in explaining the variability in participants' ES. First, Not-worrying was significantly and negatively related to ES: participants who reported to be more prone to experience emotional distress or worry in response to negative body sensations (i.e., with a score near zero in the Not-worrying scale) also reported to be more prone to experience discomfort, vulnerability, and inadequacy in controlling their excitement, arousal, and reactions in situations of danger, offense, or attack (Caprara et al., 1983), as evidenced by high ES scores. This result suggests a tight link between these two constructs, which makes sense, as both are related to measures of trait anxiety (see Caprara et al., 1983; Mehling et al., 2012), and situations of danger, threat, offense, or attack all lead to the activation of the sympathetic nervous system, with subsequent strong bodily signals of alarm. As shown by Mehling et al. (2012), the MAIA Not-worrying scale was related to the Physical Concern subscale of the Anxiety Sensitivity Index (Zinbarg et al., 1997), which assesses the proneness or enduring tendency to worry when experiencing bodily sensations of discomfort, such as quickened respiration or heartbeat and chest constriction. These sensations can be triggered by "emotionally salient stimuli" and the "situations of danger, offense, or attack" that contribute to the definition of the ES construct and, therefore, are all phenomena that individuals with high ES experience as a state of negative affect. Thus, to not worry would mean to accept and tolerate these negative body signals, to have a mindful, nonjudgmental acceptance in experiencing the physical sensations in the present moment (Mehling et al., 2012), and not to "burden oneself with them," while experiencing feelings of helplessness, inadequacy, and vulnerability (Caprara et al., 1985).

Next, although the Emotional awareness scale was significantly related to the participants' ES, those who reported to have high awareness of the connection between their emotional states and body sensations also had high ES. At first glance, this finding may seem counterintuitive: one could expect to find the opposite relation between the Emotional awareness and the ES. However, a recent study provides findings in line with our results. Indeed, Lichev et al. (2015) found that individuals with high emotional awareness showed stronger (implicit) affective reactivity. Moreover, the results of an analysis conducted by Mehling et al. (2012) may provide some insights to help us better understand how the multiple aspects of the IAw relate to anxiety. The authors showed that Emotional awareness was the only scale that showed a specific—even if marginally significant—positive relation with the trait anxiety measure, despite its negative association with anxiety measures in a simple correlational analysis. Follow-up regression analyses showed that a possible cause of this inversion of the association between anxiety and Emotional awareness was the portion of shared variance between the latter measure and another MAIA measure, Self-regulation. In other words, even if Emotional awareness was negatively related to anxiety when taken alone, this relation became positive after having removed the portion of variance shared with Self-regulation. This suggests that the Emotional awareness scale would assess distinct aspects of this construct that may be negatively and positively related to anxiety (respectively, those shared with the Self-regulation scale and those that are specific to the Emotional awareness). This interpretation led the authors to conclude that "mere awareness of how body sensations correspond to emotional states [i.e., the Emotional awareness], without the ability to use awareness of those sensations to reduce distress [i.e., the Self-regulation], could increase anxiety" (Mehling et al., 2012). Therefore, it is possible that the positive relationship we found between Emotional awareness and ES was mostly driven by the same specific aspects that Mehling and colleagues found to be related to an increased trait anxiety, which in turn is strongly related to ES (Caprara et al., 1983). Albeit speculative, this possible explanation highlights the need to conduct further studies and intensify the efforts to clarify how specific aspects of body awareness affect anxiety.

Finally, and more importantly, the multiple regression analysis showed an even stronger relationship between the participants' ES and the interaction between Attention regulation and Trusting. This result is of particular interest, as it reveals a pattern of relationships between the ES and these facets of the IAw that is more complex than it would at first appear by relying on the correlational analyses. The significant interaction between Attention regulation and Trusting reveals their non-additive effect in explaining the ES variability: participants who had a score near zero in either the Attention regulation or the Trusting scale also had higher ES scores, that is, reported to be highly prone to experiencing feelings of discomfort, helplessness, inadequacy, and vulnerability due to the inability to control their reactions in negative situations, irrespective of their score in the other scale; on the contrary, only participants who showed high scores in both these scales reported to have a low ES. This result indicates that, when taken alone, neither of these two abilities is enough to protect oneself against the negative feelings implied by a high ES score. In other words, the inclination to experience the body as a safe and trustworthy place (i.e., a high Trusting score) also needs the concomitant ability to actively direct and maintain the attention on the body signals (i.e., a high Attention regulation score) to allow one to control feelings and reactivity in negative situations. In line with this result, different authors considered the skills in attention regulation as a precondition for the capacity to be nonreactive and tolerant of body sensations, which are key elements of a more general mindfulness approach to experience (e.g., Shapiro et al., 2006; Mehling et al., 2009, 2012; Hölzel et al., 2011). Our data emphasize that focusing attention on body signals can be maladaptive if one does not "feel at home" in his/her own body. Therefore, good attention regulation skills (as assessed by the MAIA) should be considered as just one necessary—but not sufficient—precondition of a beneficial mind-body relation and emotional stability, even when it is viewed as related to a "positive" mode of mind, or mindful presence (Mehling et al., 2009), (i.e., "focusing attention directly on immediately experienced feelings," instead of "[having an] abstract ruminative self-focus," Mehling et al., 2012; see also Watkins and Teasdale, 2004; Shapiro et al., 2006).

Furthermore, our correlational analyses revealed that Attention regulation was the only facet of the IAw, as assessed by the MAIA, to be significantly, albeit weakly, related to the participants' IAc, i.e., their ability to accurately perceive their bodily signals, as assessed by the performance in the HPT. This result is consistent with the fact that the Attention regulation scale assesses the participants' perception of their ability to actively direct and maintain attention on body signals, which is the very ability that was required to have a good performance in the HPT. This result is consistent with the report that Attention regulation was found to be the MAIA scale with the strongest correlation (Mehling et al., 2012) with the Private Body Consciousness subscale of the Body Consciousness Questionnaire (Miller et al., 1981), which specifically relates to the ability to notice bodily sensations such as the heart beating, thus involving IAc skills. Apart from this single, albeit interesting, significant result, the results of our analyses suggest the substantial independence of the IAc and IAw measures, supporting the notion that the assessment of IAc alone cannot provide a full comprehension of IAw. This independence is in line with more recent findings showing that subjective self-report measures of interoception ability, such as those provided by the MAIA, may diverge significantly from objective measures of IAc (Garfinkel et al., 2015). As a limitation of our IAc measures we need to note that three different measures of heart rate measures were used including manual palpation and chest auscultation that would potentially interfere with the participants' self-assessment. However, our analyses supported equivalence of the three methods. In addition, using exclusively self-report to assess IAw is a limitation that in future studies can—at least partially—be addressed following the method suggested by Garfinkel et al. (2015).

The complex pattern of correlations between the IAw, IAc, and ES revealed by our analyses would induce another important consideration supporting the above conclusion. While the correlational analyses showed that ES was not significantly related to the ability to accurately perceive bodily signals (IAc), multiple regression analyses showed that it was significantly and specifically related to different facets of IAw, particularly to the interaction between Trusting and Attention Regulation, which is the only MAIA scale to be related to the IAc. This pattern of results appears to provide additional support against the interoceptive sensitivity hypothesis (Ehlers and Breuer, 1992), which equates the IAw and IAc constructs, and, again, in favor of a conceptual distinction between these two constructs, as proposed by Ceunen et al. (2013) and Garfinkel et al. (2015). Moreover, the lack of a significant relation between IAc and ES would suggest that this latter construct does not simply correspond to the subjective experience of the intensity of emotional states. It has been shown that individuals with high IAc experience emotions as more intense (e.g., Wiens et al., 2000) and report them with a greater emphasis on the arousal dimension (Barrett et al., 2004). Based on these findings, we hypothesized that individuals with high IAc may be more emotionally susceptible, as they would tend to have greater arousal/excitement, especially "in situations, real, or imagined, of danger, offense, threat, or attack." Contrary to our hypothesis, our finding indicates that this presumably greater activation in response to emotional stimuli does not automatically lead to the "feelings of discomfort, helplessness, inadequacy, and vulnerability," and the "state of negative affect and a tendency to place oneself in a defensive position" that characterize individuals with high ES, as these feelings/emotions triggered by highly arousing situations are critically "due to the inability [of the individual] to control" this arousal/excitement (Caprara et al., 1983). This explanation is supported by reports (e.g., Füstös et al., 2013) that, although individuals with high IAc experienced negative emotions as more arousing, they were at the same time better able to actively down-regulate affect-related states, suggesting that the accurate perception of bodily states may be essential for adequate emotion regulation.

Alternatively, it has also been shown (Van't Wout et al., 2013) that IAc is not related to self-reported levels of the habitual use of emotion regulatory strategies, as assessed by the Emotion Regulation Questionnaire (ERQ: Gross and John, 2003). Therefore, it is possible that a positive association between IAc and the subjective experience of the intensity of emotions, combined with no association between IAc and emotion regulation ability canceled out any association between IAc and ES. Regrettably, we did not check our participants' habitual use of emotion regulation strategies, and thus we cannot verify this possible interpretation. Taken together, these findings suggest that the relationship between IAc and ES/emotion regulation appears to be more complex than we initially thought, and further investigation is needed to shed light on this issue.

Support for the conceptual distinction of the IAw and IAc constructs also comes from the analysis of the pattern of relations that these two constructs have with anxiety-related traits, which resembles a sort of "double dissociation." First, as showed by Mehling et al. (2012), different MAIA scales are negatively correlated with trait anxiety, a general stable tendency to respond with anxiety to perceived threats in the environment, and to a lesser degree with anxiety sensitivity, a dispositional trait characterized by the fear of feelings related to anxiety, based on the belief that these symptoms have physical, psychological, or social consequences (Reiss and McNally, 1985; Reiss, 1991). Conversely, the HPT shows the opposite pattern, with stronger positive correlations with the anxiety sensitivity and weaker correlations with the trait anxiety (Domschke et al., 2010), suggesting that individuals with high anxiety sensitivity are generally more accurate in detecting their own heartbeat. Second, an increased IAc has been proposed as a risk factor for the development of state anxiety, trait anxiety or clinical disorders related to anxiety due to an attentional bias involving catastrophic misinterpretations of somatic signals (McNally, 2002; Schmidt et al., 2008; Perez Benitez et al., 2009; Domschke et al., 2010). Conversely, mindful meditation, and thus potentially improved IAw, is associated with the transition from (a) a form of conceptual and narrative thinking that is susceptible to evaluative interferences to (b) a form of experiential, non-evaluative, immediate sensing, in which the attention is focused directly on present-moment feelings and bodily sensations (Mehling et al., 2009; Daubenmier et al., 2013). These distinct modes of mind can thus determine whether body awareness is beneficial or maladaptive: "focusing attention directly on immediately experienced feelings appears to be adaptive, whereas an abstract ruminative self-focus appears to be maladaptive" (Mehling et al., 2012).

These distinct modes of minds correspond to distinct forms of self-awareness that are habitually integrated but, as shown with fMRI by Farb et al. (2007), "can be dissociated through attentional training: the self across time [i.e., the narrative focus] and the self in the present moment [i.e., the experiential focus]." Following 8 weeks of mindfulness meditation, an experiential focus was associated with three fMRI findings: (i) reduced activity of the medial prefrontal cortex, which supports narrative self-awareness by linking subjective experiences across time (Northoff and Bermpohl, 2004), providing narrative self-reference that preserves the identity stability across time (Gallagher, 2004); (ii) increased engagement of a right lateralized network (including the insula) for higher-order representations of the self (Critchley et al., 2004) through moment-to-moment self-reference and "experiential" focus, and (iii) decoupling of the functional connectivity between these two sets of areas. These findings are also in line with other studies investigating how meditation affects IAc and IAw measures. It has been shown that meditation, contemplative practice, or other mind–body interventions can longitudinally improve different facets of the IAw construct (e.g., Bornemann et al., 2014), but they seem to have no effect on the ability to accurately perceive bodily signals, as assessed by the HPT (Nielsen and Kaszniak, 2006; Khalsa et al., 2008; Melloni et al., 2013; Parkin et al., 2013).

To sum up, our analyses confirmed the factorial structure of the original MAIA in the new Italian version of the instrument. Moreover, they (a) confirmed the substantial independence of the IAc from self-reported IAw, (b) showed independence of IAc from ES measure, and (c) revealed a complex pattern of relationships between ES and distinct dimensions of IAw, such as Not-worrying, Emotional awareness, and the interplay between Attention regulation and Trusting. Taken together, these findings further highlight the need for a multidimensional assessment of the IAw construct and for a more in-depth analysis of its relationship with personality traits, such as ES or neuroticism.

Acknowledgments

We thank Marcello Costantini and Gianluca Finotti for the MAIA translation and preliminary data discussion. Research was funded by G. d'Annunzio University grants to GC.

References

Adolphs, R. (2002). Trust in the brain. *Nat. Neurosci.* 5, 192–193. doi: 10.1038/nn0302-192

Barrett, L. F., Quigley, K. S., Bliss-Moreau, E., and Aronson, K. R. (2004). Interoceptive sensitivity and self-reports of emotional experience. *J. Pers. Soc. Psychol.* 87:684. doi: 10.1037/0022-3514.87.5.684

Bentler, P. M. (1990). Comparative fit indexes in structural models. *Psychol. Bull.* 107:238. doi: 10.1037/0033-2909.107.2.238

Bornemann, B., Herbert, B. M., Mehling, W. E., and Singer, T. (2014). Differential changes in self-reported aspects of interoceptive awareness through 3 months of contemplative training. *Front. Psychol.* 5:1504. doi: 10.3389/fpsyg.2014.01504

Bradley, M. M., and Lang, P. J. (1994). Measuring emotion: the self-assessment manikin and the semantic differential. *J. Behav. Ther. Exp. Psychiatry* 25, 49–59. doi: 10.1016/0005-7916(94)90063-9

Brainard, D. H. (1997). The psychophysics toolbox. *Spat. Vis.* 10, 433–436. doi: 10.1163/156856897X00357

Brooks, J. C., Nurmikko, T. J., Bimson, W. E., Singh, K. D., and Roberts, N. (2002). fMRI of thermal pain: effects of stimulus laterality and attention. *Neuroimage* 15, 293–301. doi: 10.1006/nimg.2001.0974

Browne, M. W., and Cudeck, R. (1993). "Alternative ways of assessing model fit," in *Testing Structural Equation Models*, eds K. A. Bollen and J. S. Long (Beverly Hills, CA: Sage), 136–162.

Cameron, O. G. (2001). Interoception: the inside story—a model for psychosomatic processes. *Psychosom. Med.* 63, 697–710. doi: 10.1097/00006842-200109000-00001

Caprara, G. V., Cinanni, V., D'Imperio, G., Passerini, S., Renzi, P., and Travaglia, G. (1985). Indicators of impulsive aggression: present status of research on irritability and emotional susceptibility scales. *Pers. Individ. Dif.* 6, 665–674. doi: 10.1016/0191-8869(85)90077-7

Caprara, G. V., Cinanni, V., Passerini, S., Renzi, P., and Zelli, A. (1983). Caratteristiche e sviluppi di due scale per la misura dell'irritabilità e della suscettibilità emotiva: indicazioni di ricerca in ambito clinico. *Psicologia clinica* 3, 323–345.

Carruthers, G. (2008). Types of body representation and the sense of embodiment. *Conscious. Cogn.* 17, 1302–1316. doi: 10.1016/j.concog.2008.02.001

Cattell, R. B., and Vogelmann, S. (1977). A comprehensive trial of the scree and KG criteria for determining the number of factors. *Multivariate Behav. Res.* 12, 289–325. doi: 10.1207/s15327906mbr1203_2

Ceunen, E., van Diest, I., and Vlaeyen, J. W. S. (2013). Accuracy and awareness of perception: related, yet distinct (commentary on Herbert et al., 2012). *Biol. Psychol.* 92, 426–427. doi: 10.1016/j.biopsycho.2012.09.012

Chikama, M., McFarland, N. R., Amaral, D. G., and Haber, S. N. (1997). Insular cortical projections to functional regions of the striatum correlate with cortical cytoarchitectonic organization in the primate. *J. Neurosci.* 17, 9686–9705.

Craig, A. D. (2002). How do you feel? Interoception: the sense of the physiological condition of the body. *Nat. Rev. Neurosci.* 3, 655–666. doi: 10.1038/nrn894

Craig, A. D. (2008). "Interoception and emotion: a neuroanatomical perspective," in *Handbook of Emotions*, 3rd Edn., eds M. Lewis, J. M. Haviland-Jones, and L. F. Barrett (New York, NY: Guilford Press), 272–292.

Craig, A. D., Chen, K., Bandy, D., and Reiman, E. M. (2000). Thermosensory activation of insular cortex. *Nat. Neurosci.* 3, 184–190. doi: 10.1038/72131

Craig, A. D., Reiman, E. M., Evans, A., and Bushnell, M. C. (1996). Functional imaging of an illusion of pain. *Nature* 384, 258–260. doi: 10.1038/384258a0

Critchley, H. D., Wiens, S., Rotshtein, P., Ohman, A., and Dolan, R. J. (2004). Neural systems supporting interoceptive awareness. *Nat. Neurosci.* 7, 189–195. doi: 10.1038/nn1176

Damasio, A. R. (1994). *Descartes Error: Emotion, Reason and the Human Brain*. New York, NY: Avon.

Darwin, C. (1873). *The Expression of the Emotions in Man and Animals*. New York, NY: Appleton.

Daubenmier, J., Sze, J., Kerr, C. E., Kemeny, M. E., and Mehling, W. (2013). Follow your breath: respiratory interoceptive accuracy in experienced meditators. *Psychophysiology* 50, 777–789. doi: 10.1111/psyp.12057

Domschke, K., Stevens, S., Pfleiderer, B., and Gerlach, A. L. (2010). Interoceptive sensitivity in anxiety and anxiety disorders: an overview and integration of neurobiological findings. *Clin. Psychol. Rev.* 30, 1–11. doi: 10.1016/j.cpr.2009.08.008

Dunn, B. D., Galton, H. C., Morgan, R., Evans, D., Oliver, C., Meyer, M., et al. (2010). Listening to your heart how interoception shapes emotion experience and intuitive decision making. *Psychol. Sci.* 1, 1835–144. doi: 10.1177/0956797610389191

Ehlers, A., and Breuer, P. (1992). Increased cardiac awareness in panic disorder. *J. Abnorm. Psychol.* 101, 371–382. doi: 10.1037/0021-843X.101.3.371

Ernst, J., Böker, H., Hättenschwiler, J., Schüpbach, D., Northoff, G., Seifritz, E., et al. (2013). The association of interoceptive awareness and alexithymia with neurotransmitter concentrations in insula and anterior cingulate. *Soc. Cogn. Affect. Neurosci.* 9, 857–863. doi: 10.1093/scan/nst058

Farb, N. A., Segal, Z. V., and Anderson, A. K. (2013). Attentional modulation of primary interoceptive and exteroceptive cortices. *Cereb. Cortex* 23, 114–126. doi: 10.1093/cercor/bhr385

Farb, N. A., Segal, Z. V., Mayberg, H., Bean, J., McKeon, D., Fatima, Z., et al. (2007). Attending to the present: mindfulness meditation reveals distinct neural modes of self-reference. *Soc. Cogn. Affect. Neur.* 2, 313–322. doi: 10.1093/scan/nsm030

Farb, N., Daubenmier, J., Price, C. J., Gard, T., Kerr, C., Dunn, B. D., et al. (2015). Interoception, contemplative practice, and health. *Front. Psychol.* 6:763. doi: 10.3389/fpsyg.2015.00763

Feldt, L. S. (1986). "Statistical tests and confidence intervals for Cronbach's coefficient alpha," in *Iowa Testing Programs Occasional Papers Number 33* (Iowa City, IA: Iowa Testing Programs).

Feldt, L. S., Woodruff, D. J., and Salih, F. A. (1987). Statistical inference for coefficient alpha. *Appl. Psychol. Meas.* 11, 93–103. doi: 10.1177/014662168701100107

Fogel, A. (2009). *The Psychophysiology of Self-awareness: Rediscovering the Lost Art of Body Sense*. New York, NY: W. W. Norton.

Füstös, J., Gramann, K., Herbert, B. M., and Pollatos, O. (2013). On the embodiment of emotion regulation: interoceptive awareness facilitates reappraisal. *Soc. Cogn. Affect. Neurosci.* 8, 911–917. doi: 10.1093/scan/nss089

Gallagher, S. (2004). Philosophical conceptions of the self: implications for cognitive science. *Trends Cogn. Sci.* 4, 14–21. doi: 10.1016/S1364-6613(99)01417-5

Garfinkel, S. N., and Critchley, H. D. (2013). Interoception, emotion and brain: new insights link internal physiology to social behaviour. Commentary on: "Anterior insular cortex mediates bodily sensibility and social anxiety" by Terasawa et al. (2012). *Soc. Cogn. Affect. Neurosci.* 8, 231–234. doi: 10.1093/scan/nss140

Garfinkel, S. N., Minati, L., Gray, M. A., Seth, A. K., Dolan, R. J., and Critchley, H. D. (2014). Fear from the heart: sensitivity to fear stimuli depends on individual heartbeats. *J. Neurosci.* 34, 6573–6582. doi: 10.1523/jneurosci.3507-13.2014

Garfinkel, S. N., Seth, A. K., Barrett, A. B., Suzuki, K., and Critchley, H. D. (2015). Knowing your own heart: distinguishing interoceptive accuracy from interoceptive awareness. *Biol. Psychol.* 104, 65–74. doi: 10.1016/j.biopsycho.2014.11.004

Gross, J. J., and John, O. P. (2003). Individual differences in two emotion regulation processes: implications for affect, relationships, and well-being. *J. Pers. Soc. Psychol.* 85:348. doi: 10.1037/0022-3514.85.2.348

Grossman, P. (2015). Mindfulness: awareness informed by an embodied ethic. *Mindfulness* 6, 17–22. doi: 10.1007/s12671-014-0372-5

Herbert, B. M., Herbert, C., Pollatos, O., Weimer, K., Enck, P., Sauer, H., et al. (2012). Effects of short-term food deprivation on interoceptive awareness, feelings and autonomic cardiac activity. *Biol. Psychol.* 89, 71–79. doi: 10.1016/j.biopsycho.2011.09.004

Herbert, B. M., Pollatos, O., and Schandry, R. (2007). Interoceptive sensitivity and emotion processing: an EEG study. *Int. J. Psychophysiol.* 65, 214–227. doi: 10.1016/j.ijpsycho.2007.04.007

Hölzel, B. K., Lazar, S. W., Gard, T., Schuman-Olivier, Z., Vago, D. R., and Ott, U. (2011). How does mindfulness meditation work? Proposing mechanisms of action from a conceptual and neural perspective. *Perspect. Psychol. Sci.* 6, 537–559. doi: 10.1177/1745691611419671

Hu, L. T., and Bentler, P. M. (1998). Fit indices in covariance structure modeling: sensitivity to underparameterized model misspecification. *Psychol. Methods* 3:424. doi: 10.1037/1082-989x.3.4.424

Iaria, G., Committeri, G., Pastorelli, C., Pizzamiglio, L., Watkins, K. E., and Carota, A. (2008). Neural activity of the anterior insula in emotional processing depends on the individuals' emotional susceptibility. *Hum. Brain Mapp.* 29, 363–373. doi: 10.1002/hbm.20393

James, W. (1884). What is emotion? *Mind* 19, 188–205. doi: 10.1093/mind/os-IX.34.188

Khalsa, S. S., Rudrauf, D., Damasio, A. R., Davidson, R. J., and Lutz, A. (2008). Interoceptive awareness in experienced meditators. *Psychophysiology* 45, 671–677. doi: 10.1111/j.1469-8986.2008.00666.x

Koch, A., and Pollatos, O. (2014). Interoceptive sensitivity, body weight and eating behavior in children: a prospective study. *Front. Psychol.* 5:1003. doi: 10.3389/fpsyg.2014.01003

Lichev, V., Sacher, J., Ihme, K., Rosenberg, N., Quirin, M., Lepsien, J., et al. (2015). Automatic emotion processing as a function of trait emotional awareness: an fMRI study. *Soc. Cogn. Affect. Neurosci.* 10, 680–689. doi: 10.1093/scan/nsu104

McFarland, R. A. (1975). Heart rate perception and heart rate control. *Psychophysiology* 12, 402–405. doi: 10.1111/j.1469-8986.1975.tb00011.x

McNally, R. J. (2002). Anxiety sensitivity and panic disorder. *Biol. Psychiatry.* 52, 938–946. doi: 10.1016/S0006-3223(02)01475-0

Mehling, W. E., Gopisetty, V., Daubenmier, J., Price, C. J., Hecht, F. M., and Stewart, A. (2009). Body awareness: construct and self-report measures. *PLoS ONE* 4:e5614. doi: 10.1371/journal.pone.0005614

Mehling, W. E., Price, C., Daubenmier, J. J., Acree, M., and Bartmess, E. (2012). The Multidimensional Assessment of Interoceptive Awareness (MAIA). *PLoS ONE* 7:e48230. doi: 10.1371/journal.pone.0048230

Mehling, W. E., Wrubel, J., Daubenmier, J. J., Price, C. J., Kerr, C. E., Silow, T., et al. (2011). Body Awareness: a phenomenological inquiry into the common ground of mind-body therapies. *Philos. Ethics Humanit. Med.* 6:6. doi: 10.1186/1747-5341-6-6

Melloni, M., Sedeño, L., Couto, B., Reynoso, M., Gelormini, C., Favaloro, R., et al. (2013). Preliminary evidence about the effects of meditation on interoceptive sensitivity and social cognition. *Behav. Brain Funct.* 9, 10–1186. doi: 10.1186/1744-9081-9-47

Miller, L. C., Murphy, R., and Buss, A. H. (1981). Consciousness of body: private and public. *J. Pers. Soc. Psychol.* 41, 397–406. doi: 10.1037/0022-3514.41.2.397

Neter, J., Wasserman, W., and Kutner, M. H. (1989). *Applied Linear Regression Models, 2nd Edn.* Homewood, IL: Irwin.

Nielsen, L., and Kaszniak, A. W. (2006). Awareness of subtle emotional feelings: a comparison of long-term meditators and nonmeditators. *Emotion* 6, 392–405. doi: 10.1037/1528-3542.6.3.392

Northoff, G., and Bermpohl, F. (2004). Cortical midline structures and the self. *Trends Cogn. Sci.* 8, 102–107. doi: 10.1016/j.tics.2004.01.004

Olausson, H., Lamarre, Y., Backlund, H., Morin, C., Wallin, B. G., Starck, G., et al. (2002). Unmyelinated tactile afferents signal touch and project to insular cortex. *Nat. Neurosci.* 5, 900–904. doi: 10.1038/nn896

Parkin, L., Morgan, R., Rosselli, A., Howard, M., Sheppard, A., Evans, D., et al. (2013). Exploring the relationship between mindfulness and cardiac perception. *Mindfulness* 5, 298–313. doi: 10.1007/s12671-012-0181-7

Perez Benitez, C. I., Shea, M. T., Raffa, S., Rende, R., Dyck, I. R., Ramsawh, H. J., et al. (2009). Anxiety sensitivity as a predictor of the clinical course of panic disorder: a 1-year follow-up study. *Depress. Anxiety* 26, 335–342. doi: 10.1002/da.20423

Pollatos, O., Herbert, B. M., Matthias, E., and Schandry, R. (2007). Heart rate response after emotional picture presentation is modulated by interoceptive awareness. *Int. J. Psychophysiol.* 63, 117–124. doi: 10.1016/j.ijpsycho.2006.09.003

Pollatos, O., Kirsch, W., and Schandry, R. (2005). On the relationship between interoceptive awareness, emotional experience, and brain processes. *Brain Res. Cogn. Brain Res.* 25, 948-962. doi: 10.1016/j.cogbrainres.2005.09.019

Pollatos, O., Kurz, A. L., Albrecht, J., Schreder, T., Kleemann, A. M., Schopf, V., et al. (2008). Reduced perception of bodily signals in anorexia nervosa. *Eat. Behav.* 9, 381–388. doi: 10.1016/j.eatbeh.2008.02.001

Pollatos, O., and Schandry, R. (2008). Emotional processing and emotional memory are modulated by interoceptive awareness. *Cogn. Emot.* 22, 272–287. doi: 10.1080/02699930701357535

Reiss, S. (1991). Expectancy model of fear, anxiety, and panic. *Clin. Psychol. Rev.* 11, 141–153. doi: 10.1016/0272-7358(91)90092-9

Reiss, S., and McNally, R. J. (1985). "Expectancy model of fear," in *Theoretical Issues in Behavior Therapy*, eds S. Reiss and R. R. Bootzin (San Diego, CA: Academic Press), 107–121.

Rouder, J. N., Speckman, P. L., Sun, D., Morey, R. D., and Iverson, G. (2009). Bayesian t tests for accepting and rejecting the null hypothesis. *Psychon. Bull. Rev.* 16, 225–237. doi: 10.3758/PBR.16.2.225

Saper, C. B. (2002). The central autonomic nervous system: conscious visceral perception and autonomic pattern generation. *Annu. Rev. Neurosci.* 25, 433–469. doi: 10.1146/annurev.neuro.25.032502.111311

Schachter, S., and Singer, J. E. (1962). Cognitive, social, and physiological determinants of emotional state. *Psychol. Rev.* 69, 379–399. doi: 10.1037/h0046234

Schandry, R. (1981). Heart beat perception and emotional experience. *Psychophysiology* 18, 483–488. doi: 10.1111/j.1469-8986.1981.tb02486.x

Schmidt, N. B., Mitchell, M. A., and Richey, J. A. (2008). Anxiety sensitivity as an incremental predictor of later anxiety symptoms and syndromes. *Compr. Psychiatry* 49, 407–412. doi: 10.1016/j.comppsych.2007.12.004

Shapiro, S. L., Carlson, L. E., Astin, J. A., and Freedman, B. (2006). Mechanisms of mindfulness. *J. Clin. Psychol.* 62, 373–386. doi: 10.1002/jclp.20237

Shepard, R. N., and Chipman, S. (1970). Second-order isomorphism of internal representations: shapes of states. *Cogn. Psychol.* 1, 1–17. doi: 10.1016/0010-0285(70)90002-2

Sherrington C. S. (1948). *The Integrative Action of the Nervous System*. Cambridge, UK: Cambridge University Press.

Singer, T., Critchley, H. D., and Preuschoff, K. (2009). A common role of insula in feelings, empathy and uncertainty. *Trends Cogn. Sci.* 13, 334–340. doi: 10.1016/j.tics.2009.05.001

StatSoft Inc., (2013). *Electronic Statistics Textbook*. Tulsa, OK. Available online at: http://www.statsoft.com/textbook/

Thompson, E., and Varela, F. J. (2001). Radical embodiment: neural dynamics and consciousness. *Trends Cogn. Sci.* 5, 418–425. doi: 10.1016/S1364-6613(00)01750-2

Tyrer, P. J. (1973). Relevance of bodily feelings in emotion. *Lancet* 301, 915–916.

Valenzuela Moguillansky, C., and Reyes Reyes, A. (2015). Psychometric properties of the Multidimensional Assessment of Interoceptive Awareness (MAIA) in a Chilean population. *Front. Psychol.* 6:120. doi: 10.3389/fpsyg.2015.00120

Van't Wout, M., Faught, S., and Menino, D. (2013). Does interoceptive awareness affect the ability to regulate unfair treatment by others? *Front. Psychol.* 4:880. doi: 10.3389/fpsyg.2013.00880

Vidaurre, C., Sander, T. H., and Schlögl, A. (2011). BioSig: the free and open source software library for biomedical signal processing. *Comput. Intell. Neurosci.* 2011:935364. doi: 10.1155/2011/935364

Watkins, E., and Teasdale, J. D. (2004). Adaptive and maladaptive self-focus in depression. *J. Affect. Disord.* 82, 1–8. doi: 10.1016/j.jad.2003.10.006

Whitehead, W. E., Drescher, V. M., Heiman, P., and Blackwell, B. (1977). Relation of heart rate control to heartbeat perception. *Biofeedback Self Regul.* 2, 371–392. doi: 10.1007/BF00998623

Wiens, S. (2005). Interoception in emotional experience. *Curr. Opin. Neurol.* 18, 442–447. doi: 10.1097/01.wco.0000168079.92106.99

Wiens, S., Mezzacappa, E. S., and Katkin, E. S. (2000). Heartbeat detection and the experience of emotions. *Cogn. Emot.* 14, 417–427. doi: 10.1080/026999300378905

Zaki, J., Davis, J. I., and Ochsner, K. N. (2012). Overlapping activity in anterior insula during interoception and emotional experience. *Neuroimage* 62, 493–499. doi: 10.1016/j.neuroimage.2012.05.012

Zinbarg, R. E., Barlow, D. H., and Brown, T. A. (1997). Hierarchical structure and general factor saturation of the anxiety sensitivity index: evidence and implications. *Psychol. Assess.* 9, 277–284. doi: 10.1037/1040-3590.9.3.277

4

Effects of Mindfulness-Based Cognitive Therapy on Body Awareness in Patients with Chronic Pain and Comorbid Depression

Marasha de Jong[1,2,3†], *Sara W. Lazar*[4†], *Kiran Hug*[5,6], *Wolf E. Mehling*[7], *Britta K. Hölzel*[4,8], *Alexander T. Sack*[9], *Frenk Peeters*[2], *Heidi Ashih*[1], *David Mischoulon*[1] *and Tim Gard*[4,10,11*]

[1] *Depression Clinical and Research Program, Massachusetts General Hospital, Harvard Medical School, Boston, MA, USA,* [2] *Department of Psychiatry and Psychology, School of Mental Health and Neuroscience, Maastricht University Medical Center, Maastricht, Netherlands,* [3] *Mondriaan, Maastricht, Netherlands,* [4] *Department of Psychiatry, Massachusetts General Hospital, Harvard Medical School, Boston, MA, USA,* [5] *Department of Psychosomatic Medicine, Medical Faculty, Medical Center-University of Freiburg, Freiburg, Germany,* [6] *Division of Clinical Psychology and Psychotherapy, Faculty of Psychology, University of Basel, Basel, Switzerland,* [7] *Department of Family and Community Medicine, Osher Center for Integrative Medicine, University of California, San Francisco, San Francisco, CA, USA,* [8] *Department of Neuroradiology, Klinikum rechts der Isar, Technical University of Munich, Munich, Germany,* [9] *Department of Cognitive Neuroscience, Faculty of Psychology and Neuroscience, Maastricht University, Maastricht, Netherlands,* [10] *Institute for Complementary and Integrative Medicine, University Hospital Zurich, University of Zurich, Zurich, Switzerland,* [11] *Translational Neuromodeling Unit, Institute for Biomedical Engineering, University of Zurich and Swiss Federal Institute of Technology (ETH), Zurich, Switzerland*

Correspondence:
Tim Gard
tgard@mgh.harvard.edu

† *These authors have contributed equally to this work.*

Body awareness has been proposed as one of the major mechanisms of mindfulness interventions, and it has been shown that chronic pain and depression are associated with decreased levels of body awareness. We investigated the effect of Mindfulness-Based Cognitive Therapy (MBCT) on body awareness in patients with chronic pain and comorbid active depression compared to treatment as usual (TAU; $N = 31$). Body awareness was measured by a subset of the Multidimensional Assessment of Interoceptive Awareness (MAIA) scales deemed most relevant for the population. These included: Noticing, Not-Distracting, Attention Regulation, Emotional Awareness, and Self-Regulation. In addition, pain catastrophizing was measured by the Pain Catastrophizing Scale (PCS). These scales had adequate to high internal consistency in the current sample. Depression severity was measured by the Quick Inventory of Depressive Symptomatology—Clinician rated (QIDS-C_{16}). Increases in the MBCT group were significantly greater than in the TAU group on the "Self-Regulation" and "Not Distracting" scales. Furthermore, the positive effect of MBCT on depression severity was mediated by "Not Distracting." These findings provide preliminary evidence that a mindfulness-based intervention may increase facets of body awareness as assessed with the MAIA in a population of pain patients with depression. Furthermore, they are consistent with a long hypothesized mechanism for mindfulness and emphasize the clinical relevance of body awareness.

Keywords: mindfulness-based cognitive therapy, mindfulness meditation, interoceptive awareness, body awareness, pain catastrophizing, chronic pain, depression, mediation

INTRODUCTION

Chronic pain is a highly prevalent and disabling condition with major impact on individuals, their significant others, and society (Turk et al., 2011). Prevalence rates for chronic pain range from 10 to 30% (Reid et al., 2011), and Major Depressive Disorder (MDD) is the most frequent psychiatric disorder in patients with chronic pain, with a 12-month prevalence ranging from 18% in population based settings up to 85% in specialized pain clinics (Bair et al., 2003). Since patients who suffer from both chronic pain and depression are particularly difficult to treat (Tunks, 2008), more effective interventions for this population are needed.

Mindfulness-based interventions have recently been shown to be effective for the treatment of chronic pain with small to moderate effect sizes on pain and depression (Veehof et al., 2011). Mindfulness-based therapies, and particularly mindfulness based cognitive therapy (MBCT), also have been shown to be effective for relapse prevention in recurrent depression and the treatment of active depression (Hofmann et al., 2010; Piet and Hougaard, 2011; Marchand, 2012; Sipe and Eisendrath, 2012). Results from a recent pilot randomized controlled trial (RCT) suggested that MBCT may be an effective intervention for the treatment of active depression in a population with chronic pain (de Jong et al., in press).

Because mindfulness-based interventions seem beneficial for chronic pain and depression, the question arises how mindfulness exerts its effects. Mindfulness entails paying attention to present moment experience, including thoughts, emotions, and bodily sensations. Training body awareness is a significant component of most mindfulness-based interventions, including the body scan, in which individuals specifically pay attention to all parts of the body; and yoga, which entails paying attention to movements of the body (Kabat-Zinn, 1990; Segal et al., 2013). Body awareness has been proposed as a potential mechanism for the therapeutic effects of mindfulness and is considered an integral part of the mindfulness construct (Mehling et al., 2009; Hölzel et al., 2011; Farb et al., 2015). The definition of body awareness emphasizes the fact that this is a complex multi-dimensional construct: "the sensory awareness that originates from the body's physiological states, processes (including pain and emotion), and actions (including movement), and functions as an interactive process that includes a person's appraisal and is shaped by attitudes, beliefs, and experience in their social and cultural context" (Mehling et al., 2012). In this article, the terms body awareness and interoceptive awareness are used interchangeably.

Several studies lend support to the notion of enhanced body awareness through mindfulness training. For example, meditators have been reported to show greater coherence between objective physiological data and their subjective experience—in regard to both emotional experience (Sze et al., 2010) and sensitivity of body regions (Fox et al., 2012). With regard to the heart beat perception task, which assesses the ability of subjects to accurately determine their heartbeat rate by comparing the subjectively counted heartbeats to heartbeats measured by an electrocardiogram, a number of studies with small sample sizes did not find increased interoceptive accuracy in meditators (Nielsen and Kaszniak, 2006; Khalsa et al., 2008; Melloni et al., 2013; Parkin et al., 2014). However, a large ($N = 160$), recent longitudinal study revealed that heart beat accuracy was increased after 39 weeks of a mindfulness-based contemplative intervention (Bornemann and Singer, 2015). Neuroimaging studies indicate mindfulness training-related changes in brain function and structure in regions that are thought to be involved in body awareness (Lazar et al., 2005; Hölzel et al., 2008; Farb et al., 2010, 2013; Gard et al., 2012).

It has been long postulated that interoceptive awareness plays an important role in the experience of emotions (James, 1984) and there is empirical evidence that the extent to which one can accurately perceive bodily functions has a positive relationship with the intensity of emotions (Herbert et al., 2007). Former studies on this topic in clinical populations have mainly focused on anxiety disorders, which demonstrated a close association with increased interoceptive awareness (Ehlers and Breuer, 1996). Depression often entails anhedonia and blunted emotions. In fact, body awareness has been found to be reduced in individuals with depression (Ehlers and Breuer, 1992; Dunn et al., 2007), and higher levels of depressive symptoms are associated with decreased body awareness in healthy subjects (Pollatos et al., 2009). Reduction in interoceptive awareness in depression is also supported by a recent neuro-imaging study, which shows reduced effective connectivity in networks involved in interoception in patients with melancholia (Hyett et al., 2015). A recent study revealed that body awareness therapy resulted in decreased self-rated depressive symptoms, but no changes in body awareness were found (Danielsson et al., 2014). Whether improvements in body awareness lead to reduced depression has yet to be established.

Variations in body awareness appear to be particularly important in patients with chronic pain. Mehling et al. (2013), for example, reported differences in some dimensions of interoceptive awareness between patients with current or past low back pain and mind-body trained individuals. Neuroscientific evidence indicates that some of the brain regions activated during pain are also activated when engaging in interoceptive awareness (Craig, 2003). Attention styles toward chronic pain sensations are of key importance for psychological pain management (Johnston et al., 2012), and fMRI studies suggest that mindfulness meditation facilitates a reduction of pain through increased sensory processing (Gard et al., 2012). Thus, body awareness and pain perception are closely linked on a neuro-biological level, such that the enhancement of specific styles or dimensions of body awareness may facilitate the self-regulation of pain.

Although body awareness is considered an integral part of the mindfulness construct, there has been a paucity of instruments that measure body awareness (Mehling et al., 2009). Previous body awareness questionnaires either measured non-adaptive forms of body awareness (as indicated in disorders such as panic disorder), were uni-dimensional, lacked systematic development, or did not measure body awareness specifically, but rather a more general observation ability (Mehling et al., 2012). The Multidimensional Assessment of Interoceptive Awareness

(MAIA; Mehling et al., 2012) scale is a relatively new, multifaceted body awareness questionnaire that intends to fill the apparent gap.

Bornemann et al. (2015), who recently translated the MAIA into German and demonstrated good scale properties, found that a 3-month contemplative training that included the bodyscan and breath meditation techniques lead to changes on several scales in a sample of individuals with good psychological and physical health. Values on all scales increased following the training, and changes were significantly greater compared to a retest control group for most of the scales.

While there is evidence for an effect of a mind-body intervention on body awareness as measured with the MAIA in a healthy sample, no intervention studies have been reported on effects in chronic pain patients or in depressed individuals. In the present pilot RCT we investigated the longitudinal effects of MBCT on body awareness, as measured by the MAIA in a population of patients with chronic pain and comorbid depression. It was hypothesized that MBCT enhances aspects of body awareness in this population. Furthermore, as this is the first study using the MAIA in a sample of patients with chronic pain and comorbid active depression, we investigated the reliability of the MAIA scales in this population.

METHODS

Participants

Participants in this add-on study were part of a larger clinical trial reported elsewhere (de Jong et al., in press; clinicaltrials.gov id: NCT01473615). Patients were recruited from different outpatient clinics through introduction of the study by their physicians, as well as via web-based advertisements and through several online mailing lists. After phone screening and a subsequent in-person screening visit, eligible participants were offered enrollment and were randomly assigned to treatment as usual (TAU) or TAU plus mindfulness-based cognitive therapy (MBCT). English-language literate individuals aged 18 or older were eligible if they (a) had persistent chronic pain for a minimum of 3 months; (b) met the DSM-IV criteria for MDD, Dysthymic Disorder, or Depressive disorder Not Otherwise Specified (NOS); and (c) a score $\geq$10 on the QIDS-C_{16} scale. After initiating the study the cutoff was reduced to a QIDS-C_{16} score $\geq$6 (indicative of at least mild depressive symptoms) to allow more ample recruitment.

Exclusion criteria were: (a) serious suicide or homicide risk; (b) current or past bipolar disorder, current psychotic symptoms, or a current or past primary psychotic disorder; (c) diagnosis of substance abuse or dependence disorder during the last 3 months; (d) general condition that impedes attendance in group interventions, such as severe personality disorders, cognitive impairment, or tendencies toward physical aggression; (e) severe and unstable medical illness including cardiovascular, hepatic, renal, respiratory, endocrine, neurological, or hematological disease; and (f) significant present meditation practice with more than 3 h of mindfulness, insight, or vipassana meditation per week. Patients were requested to keep their psychological and pharmacological treatment as stable as possible from 8 weeks before the beginning of the study until its conclusion.

Seventy-one participants were screened, of which 40 were randomized to TAU or MBCT + TAU in a 1:2 ratio (**Figure 1**). Participants received $40 for completed study participation and provided written informed consent. The study was approved by the Partners Human Research Committee, Massachusetts General Hospital (protocol 2011-P-001699/1).

Measures

Multidimensional Assessment of Interoceptive Awareness (MAIA)

The MAIA is a 32-item instrument that assesses body awareness on 6-point Likert-type scales that range from 0 (Never) to 5 (Always). It comprises eight scales, namely Noticing, Not-Distracting, Not-Worrying, Attention Regulation, Emotional Awareness, Self-Regulation, Body Listening, and Trusting (Mehling et al., 2012). The eight scales have been shown to have adequate to excellent internal-consistency reliabilities, with Cronbach's alphas from 0.66 to 0.87, and above 0.70 for five of the eight scales (Mehling et al., 2012). Because the current study was an add-on to a larger study, subject burden had to be kept to a minimum. For this reason only the scales deemed most relevant for the specific population were administered, namely Noticing, Not-Distracting, Attention Regulation, Emotional Awareness, and Self-Regulation.

The Noticing scale assesses the awareness of comfortable, neutral, and uncomfortable body sensations. Not-Distracting refers to not ignoring or distracting oneself from uncomfortable body sensations such as pain. Attention Regulation is the ability to maintain and regulate attention to body sensations, and Emotional Awareness is defined as consciousness of the interrelation of emotions and body sensations. Self-Regulation refers to the ability to control psychological distress by consciously attending to body sensations (Mehling et al., 2013). The MAIA Not Worrying scale assesses worrying or feeling emotionally distressed in response to uncomfortable body sensations including pain and was not administered, since its items are similar to the Pain Catastrophizing Scale, which was included in the larger study.

Pain Catastrophizing Scale (PCS)

The PCS is a 13-item scale, comprised of three subscales, that measures pain catastrophizing. Pain catastrophizing is defined as "an exaggerated negative mental set brought to bear during (actual or anticipated) painful experience" (Sullivan et al., 2001). The PCS total score can range from 0 to 52 and has been shown to have excellent reliability (Cronbach's alpha = 0.87; Sullivan et al., 2001).

Quick Inventory of Depressive Symptomatology—Clinician Rated (QIDS-C_{16})

The QIDS-C is a widely used clinician rated instrument to assess depression symptom severity. The instrument is comprised of 16 items, has good psychometric properties, score range from 0 to 27 and is sensitive to changes in depressive severity (Rush et al., 2003; Trivedi et al., 2004).

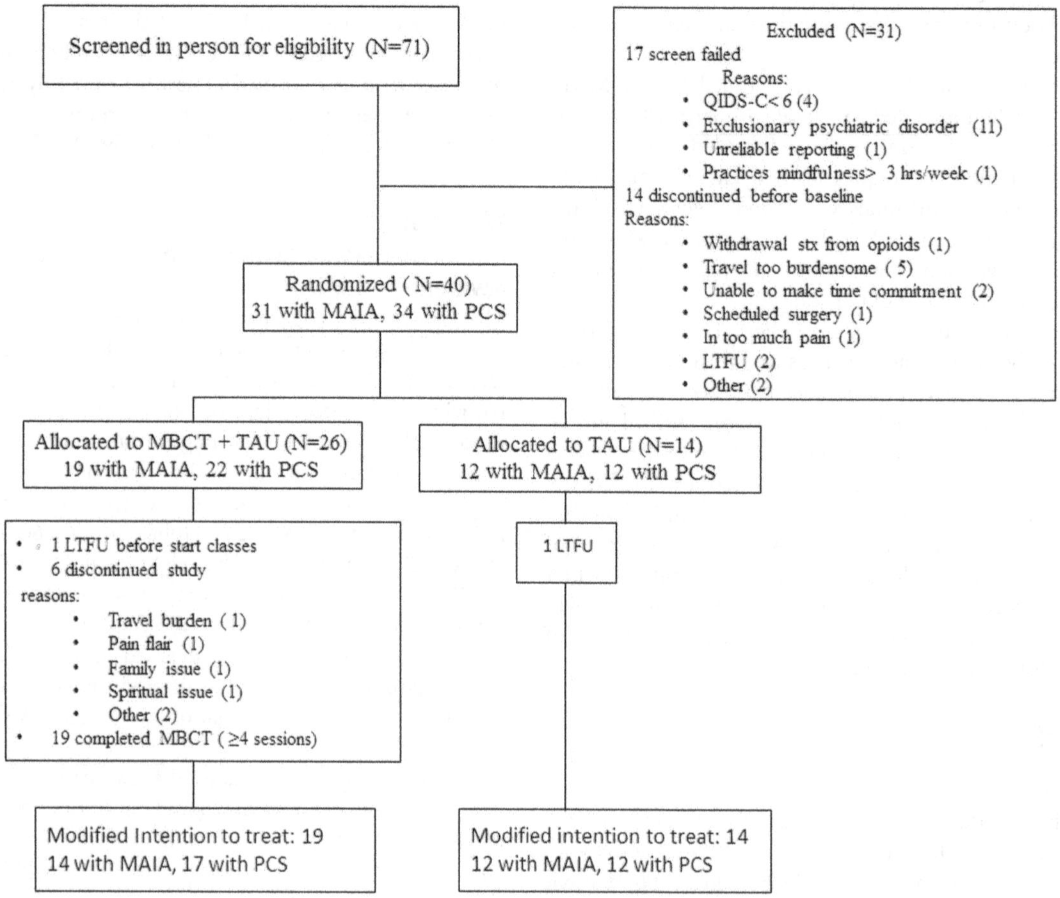

FIGURE 1 | Patient flow. LTFU, lost to follow up; MAIA, Multidimensional Assessment of Interoceptive Awareness; MBCT, Mindfulness Based Cognitive Therapy; PCS, Pain Catastrophizing Scale; QIDS-C, Quick Inventory of Depressive Symptoms-Clinician Rated; TAU, Treatment As Usual.

Procedure

Eligible participants were randomly assigned to TAU (control group) or MBCT + TAU (intervention group) in a 1:2 ratio, which allowed the main project to fill the MBCT groups with participants more quickly. An independent researcher not involved in the project generated the randomization sequence in blocks of five (using the sequence generator on www.random.org). In order to assure equal gender distribution in both groups we stratified for gender. The intervention group received an 8-week MBCT group skills program. Interoceptive awareness, pain catastrophizing, and depression were assessed at baseline (week 0), midpoint (week 4), and endpoint (week 8).

Intervention

The intervention consisted of an 8-week group skills program with one 2-h mindfulness training session each week and individual exercises for homework practice. It was modeled on the MBCT program developed by Segal et al. (2013), which was developed as a program to address recurrent depressive episodes and combines elements of cognitive behavioral therapy (CBT), such as psycho-education, with experiential mindfulness practices. The program is intended to teach and foster a non-judgmental, accepting attitude toward one's internal, and external experience. For the current study the original program was adapted to our specific population by modifying the psychoeducation and CBT elements to a depressed CP population. This included psycho-education linking CP, negative thoughts, negative emotions, and depressive behaviors such as withdrawal; identifying automatic thoughts related to CP; and paying attention to behavioral elements such as pacing of activities. We also included meditations that specifically focused on cultivating mindfulness in relationship to CP. The MBCT program was led by two instructors, an experienced licensed independent clinical social worker (LICSW) and a fellow in psychology and was provided free of charge. Subjects that were assigned to TAU were waitlisted and offered the MBCT treatment after completion of the study. TAU included all regular visits with the pain physician, psychiatrist, psychotherapist and prescribed pain and/or antidepressant medications.

Statistical Analyses

Differences in patient characteristics at baseline were assessed by performing independent-samples t-tests for continuous variables and Chi-square tests for categorical variables. To evaluate internal consistency/reliability, Cronbach's alphas were assessed. Inter-scale correlations were obtained for the MAIA scales and the PCS total score based on the data of all subjects at baseline. The effects of intervention/group (MBCT + TAU vs. TAU) and time (baseline vs. endpoint) on the dependent variables body awareness and pain catastrophizing were assessed by conducting repeated measures analysis of variance (rmANOVA), with time as repeated measure, treatment group as between-subjects factor, and MAIA scales and PCS as dependent variables. Assumptions of normality and homogeneity were met. Paired samples t-tests were conducted to compare baseline and endpoint scores on the MAIA scales and the PCS within each group. As a measure of effect size, Cohen's d was calculated for each pre-post change. Analyses were conducted according to a modified intention-to-treat (ITT) principle with the last observation carried forward (LOCF). When endpoint measures were missing, midpoint measures were imputed, and if midpoint data were missing, baseline data were used. Only participants who attended at least four of the eight classes were included in the analyses.

The main study, to which this study was added on, revealed a significant effect of MBCT (group by time interaction) on depression as measured with the Quick Inventory of Depressive Symptomatology-Clinician rated (QIDS-C_{16}) (de Jong et al., in press). To explore if and how this effects is mediated by the MAIA, a multiple mediator model was tested. The model comprised group (MBCT + TAU vs. TAU) as independent variable, depression measured at week 8 as dependent variable and MAIA scales (measured at week 8) that revealed significant group by time interactions, as mediators. Depression and respective MAIA scales measured at baseline (week 0) were included as covariates. Mediation analyses were conducted with a macro by Preacher and Hayes (2008) that implements a bootstrapping procedure to create confidence intervals for partial and total indirect effects. For mediation analyses only participants who participated in at least 4 classes and who had week 0 and week 8 data available were included in the analyses and 10,000 bootstrap iterations were used. All analyses were conducted with SPSS 21 (SPSS Inc., Chicago, IL, USA).

RESULTS

Participant Characteristics

For the main study, 71 patients were screened, of which 40 were randomized to TAU ($n = 14$) or MBCT + TAU ($n = 26$) in a 1:2 ratio. Of those 40, 34 completed the pain catastrophizing scale (PCS) and 31 the MAIA at baseline. Of the 14 patients in the TAU, 12 had PCS and MAIA data and were included in modified ITT analyses. Of the 26 patients in MBCT + TAU, 22 had PCS and 19 MAIA data. Five of the patients with MAIA and PCS data participated in <4 sessions of MBCT and were excluded from further analyses, resulting in 17 patients with PCS data and 14 with MAIA data in the modified ITT analyses (**Figure 1**). For the mediation analyses 11 subjects with MAIA data were in MBCT + TAU and 7 in TAU.

Characteristics of participants ($N = 29$) who attended at least four classes (MBCT+TAU) or had four clinic visits (TAU), are shown in **Table 1**. Types of chronic pain included: chronic back pain, neuropathic pain, osteoarthritis, fibromyalgia, and migraines. There were no significant differences in demographics and patient characteristics (see **Table 1**) or in average baseline scores on the five MAIA scales or total PCS scores between the two groups at baseline (see **Table 2**). For the collapsed sample that completed the MAIA, regardless of class attendance, participants ($N = 31$) were on average 50 years old ($M = 49.45$, $SD = 10.58$), and college graduates ($M = 16.30$, $SD = 2.55$ years of education). Most of the participants were female (74.2%), Caucasian (90.3%), and non-Hispanic (83.9%). Three (9.7%) participants were African American and one (3.2%) was Hispanic. The largest proportion of participants was married (46.7%) or never married (36.7%), and five were separated or divorced (16.7%). Employed participants comprised 32.3% of the sample, and disabled ones comprised 32.3% of the sample. Most participants had (MDD; 83.9%), with the remaining 16.1% suffering from Depressive Disorder not otherwise specified (NOS). 46.4% of the participants were taking Anti-Depressant medication.

Scale Properties

Table 3 summarizes scale means with standard deviations, range of observed values, Cronbach's alphas, and range of item-scale

TABLE 1 | Participant characteristics ($N = 29$).

	TAU		MBCT + TAU		t/χ^2-test		
	M/%	*SD*	*M*	*SD*	t/χ^2	*df*	*p*
Age (years)	51.67	10.08	50.06	11.68	0.39	27	0.703
Education (years)[a]	16.58	2.61	15.94	2.56	0.65	26	0.519
Gender (% female)	66.7		76.5		0.34	1	0.561
Race (%)					0.88	1	0.348
African-American	16.7		5.9				
Caucasian	83.3		94.1				
Ethnicity (%)					3.28	2	0.194
Hispanic	0.0		5.9				
Non-hispanic	100.0		76.5				
Unknown/Not reported			17.6				
Marital status (%)[a]					0.933	2	0.627
Never married	25.0		37.5				
Married/Live together	50.0		50.0				
Separated/Divorced	25.0		12.5				
Employment status (%)[a]					1.67	2	0.435
Employed	16.7		37.5				
Disabled	41.7		37.5				
Other/Not reported	41.7		25.0				
Type depression (%)					0.20	1	0.653
NOS	16.7		23.5				
MDD	83.3		76.5				
ADM (% taking)[b]	50.0		35.3		0.564	1	0.453

ADM, Anti-Depressant Medication; MDD, Major Depressive Disorder; MBCT, Mindfulness Based Cognitive Therapy; NOS, Depressive Disorder Not Otherwise Specified; SD, Standard Deviation; TAU, Treatment As Usual.

[a]Based on N = 28 due to missing value.

[b]Based on N = 27 due to missing values.

TABLE 2 | Baseline scores on MAIA and PCS.

	TAU		MBCT + TAU		*t*-test		
	M	*SD*	*M*	*SD*	*t*	*df*	*p*
MAIA							
Noticing	3.13	1.26	2.86	1.01	0.60	24	0.553
Attention regulation	2.08	1.34	2.37	1.05	0.60	24	0.550
Emotional awareness	2.65	1.63	2.71	1.29	0.11	24	0.912
Self-regulation	2.10	1.42	1.98	1.24	0.23	24	0.817
Not distracting	2.14	0.80	1.95	1.23	0.45	24	0.657
PCS	27.17	10.67	31.82	12.28	1.06	27	0.299

MAIA, Multidimensional Assessment of Interoceptive Awareness; PCS, Pain Catastrophizing Scale.

TABLE 3 | Scale properties (N).

Scale	Number of items	Cronbach's Alpha	Range of item-scale correlations	Mean (SD)	Observed *N*	
MAIA						
Noticing	4	0.67	0.65–0.74	3.10 (1.15)	1–5	31
Attention regulation	7	0.92	0.64–0.91	2.42 (1.22)	0.1–4.57	31
Emotional awareness	5	0.94	0.85–0.93	2.73 (1.50)	0–5	31
Self-regulation	4	0.93	0.89–0.96	1.92 (1.37)	0–4.5	31
Not distracting	3	0.72	0.49–0.84	2.04 (1.06)	0.3–5	31
PCS	13	0.93	0.51–0.85	30.18 (11.63)	4–52	34

MAIA, Multidimensional Assessment of Interoceptive Awareness; PCS, Pain Catastrophizing Scale; SD, Standard Deviation.

correlations of the MAIA and the PCS for the entire sample ($N = 31$ and 34 respectively) at baseline. Cronbach's alphas for three of the five administered MAIA scales were excellent and ranged from 0.92 (Attention Regulation) to 0.94 (Emotional Awareness). The Not-Distracting and Noticing scales had alphas of 0.72 and 0.67 respectively. The PCS had a Cronbach's alpha of 0.93.

Table 4 shows Pearson correlations between the MAIA scales. The correlations ranged from 0.76 for Self-regulation and Emotional awareness and 0.71 for Emotional awareness and Noticing, to ($r \leq |0.29|$) for Not Distracting, which did not correlate significantly with any other MAIA scale.

Effects of MBCT on Body Awareness

For Noticing, no significant group-by-time interaction [$F_{(1, 24)} = 0.18$, $p = 0.676$, $\eta_p^2 = 0.007$] and no significant main effect of time [$F_{(1, 24)} = 2.59$, $p = 0.121$, $\eta_p^2 = 0.097$] was found. No significant within group changes between the pre- and the post-treatment measurements were found in either group with effect

TABLE 4 | Scale-scale correlations.

Scale	Noticing	Attention regulation	Emotional awareness	Self-regulation	Not distracting
Noticing	–				
Attention regulation	0.53**	–			
Emotional awareness	0.71**	0.62**	–		
Self-regulation	0.52**	0.50**	0.76**	–	
Not distracting	0.02	0.17	−0.29	−0.09	–

Pearson product moment correlations among MAIA scales in the total sample (N = 31).
***Correlations are significant at p < 0.01.*
MAIA, Multidimensional Assessment of Interoceptive Awareness.

sizes of the pre-post change being medium for the treatment group but small for the control group (**Figure 2A, Table 5**).

For the Attention Regulation scale, no significant group-by-time interaction effect [$F_{(1, 24)} = 0.03$, $p = 0.863$, $\eta_p^2 \leq 0.001$] was found, but a significant main effect of time was revealed [$F_{(1, 24)} = 5.34$, $p = 0.030$, $\eta_p^2 = 0.182$]. Paired sample *t*-tests showed significant, medium size increases of Attention Regulation scores in the control group. The increase in Attention Regulation in the MBCT group did not reach statistical significance, despite its large effect size (**Figure 2B, Table 5**).

For Emotional Awareness, no significant group-by-time interaction effect [$F_{(1, 24)} = 2.17$, $p = 0.153$, $\eta_p^2 = 0.083$] was found, but a significant main effect of time was revealed [$F_{(1, 24)} = 4.63$, $p = 0.042$, $\eta_p^2 = 0.162$]. Paired samples *t*-tests revealed a large and significant increase in Emotional Awareness over time within the treatment group, while the change within the control group was small and not significant (**Figure 2C, Table 5**).

A rmANOVA revealed a significant group-by-time interaction for Self-Regulation [$F_{(1, 24)} = 5.93$, $p = 0.023$, $\eta_p^2 = 0.198$]. This interaction was driven by large and significant increases in Self-Regulation over time in the treatment group and small but significant increases in the control group. Furthermore, the analyses revealed a significant main effect of time [$F_{(1, 24)} = 22.86$, $p < 0.001$, $\eta_p^2 = 0.488$] (**Figure 2D, Table 5**).

For the Not Distracting scale, a significant group-by-time interaction was revealed [$F_{(1, 24)} = 4.87$, $p = 0.037$, $\eta_p^2 = 0.169$]. This interaction was driven by an increase in Not Distracting scores in the treatment group and a decrease within the control group, none of which reached statistical significance. No main effect of time [$F_{(1, 24)} = 0.45$, $p = 0.511$, $\eta_p^2 = 0.018$] was found (**Figure 2E, Table 5**).

For pain catastrophizing, no significant group-by-time interaction effect [$F_{(1, 27)} = 1.15$, $p = 0.294$, $\eta_p^2 = 0.041$] was revealed. The main effect of time approached significance [$F_{(1, 27)} = 3.60$, $p = 0.069$, $\eta_p^2 = 0.118$]. Analysis showed a large and significant decrease of pain catastrophizing within the MBCT group, but only a small and not significant decrease in the control group (**Figure 2F, Table 5**).

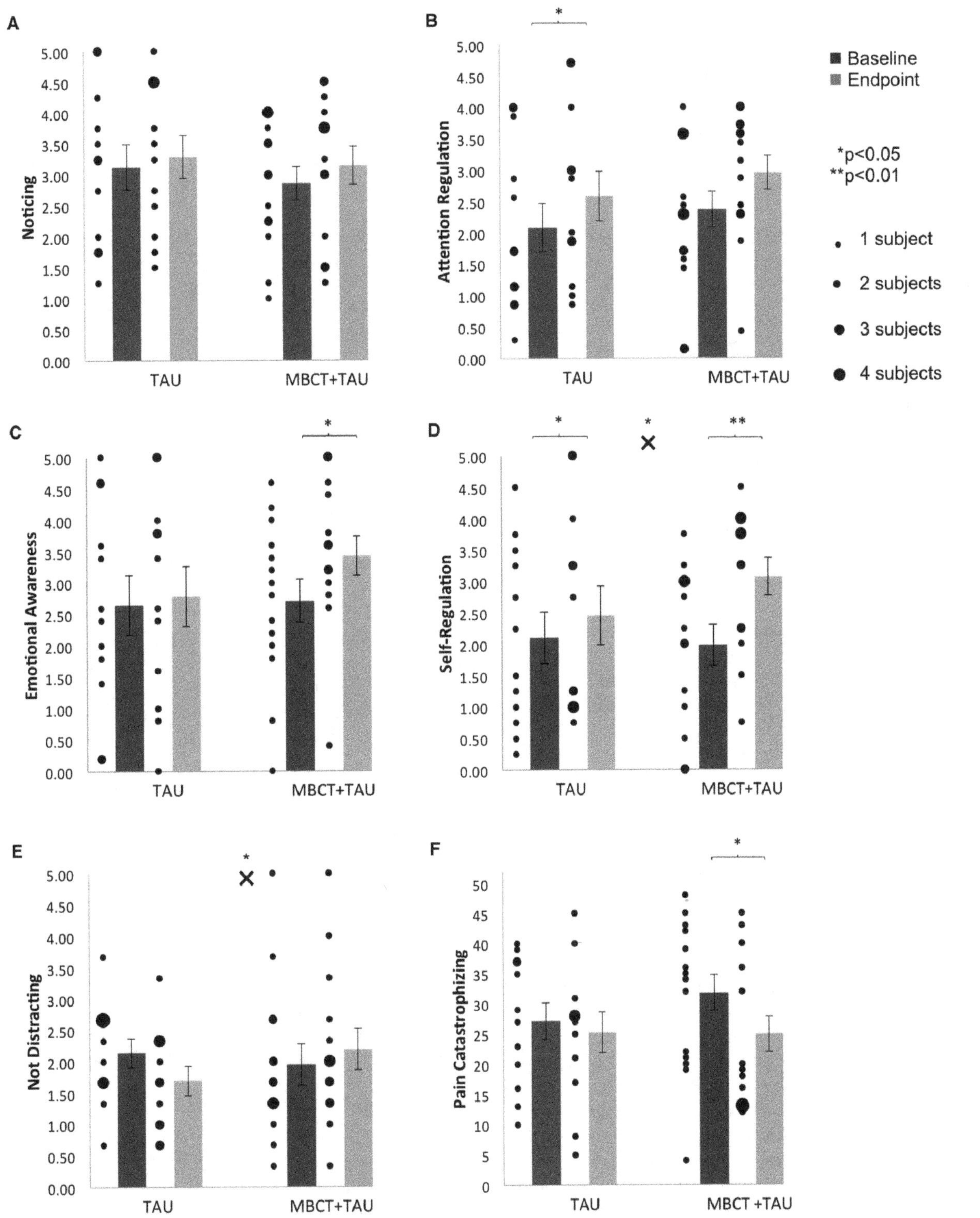

FIGURE 2 | Individual data and averages for baseline and endpoint. (A) Noticing, **(B)** Attention Regulation, **(C)** Emotional Awareness, **(D)** Self-regulation, **(E)** Not Distracting, **(F)** Pain Catastrophizing. Error bars are ± 1 SEM; Asterisks above the × symbols indicate significant *p*-values based on group by time interaction effects as revealed by repeated measures ANOVAs; Asterisks above the bars indicate significant *p*-values based on pre-post treatment changes as revealed by paired-samples *t*-tests.

TABLE 5 | Changes in scores for body awareness and pain catastrophizing.

	TAU						MBCT + TAU					
	Pre *M (SD)*	Post *M (SD)*	*t*	*df*	*p*	*d*	Pre *M (SD)*	Post *M (SD)*	*t*	*df*	*p*	*d*
MAIA												
Noticing	3.13 (1.26)	3.29 (1.20)	1.30	11	0.220	+0.134	2.86 (1.01)	3.14 (1.15)	1.21	13	0.247	+0.262
Attention regulation	2.08 (1.34)	2.58 (1.36)	2.52	11	0.029	+0.371	2.37 (1.05)	2.95 (1.01)	1.46	13	0.167	+0.564
Emotional awareness	2.65 (1.63)	2.78 (1.66)	0.73	11	0.482	+0.081	2.71 (1.29)	3.43 (1.19)	2.17	13	0.049	+0.573
Self-regulation	2.10 (1.42)	2.46 (1.62)	2.49	11	0.030	+0.216	1.98 (1.24)	3.07 (1.12)	4.34	13	0.001	+0.913
Not distracting	2.14 (0.80)	1.69 (0.81)	−1.85	11	0.092	−0.553	1.95 (1.23)	2.19 (1.23)	1.20	13	0.253	+0.194
PCS	27.17 (10.67)	25.25 (11.52)	−0.57	11	0.580	−0.172	31.82 (12.28)	24.94 (12.16)	−2.23	16	0.041	−0.564

Results of paired-samples t-tests and Cohen's d effect sizes for pre-post changes for treatment and control group.
MAIA, Multidimensional Assessment of Interoceptive Awareness; PCS, Pain Catastrophizing Scale; d, Cohen's d; SD, Standard Deviation.

Body Awareness As Mediator

A multiple mediator model as described by Preacher and Hayes (2008) was tested. The model (**Figure 3**) was comprised of group (MBCT + TAU vs. TAU) as independent variable, depression measured with QIDS-C_{16} at week 8 as dependent variable, and the MAIA Not Distracting and Self-Regulation scales, measured at week 8, as mediators. QIDS-C_{16}, MAIA Not Distracting and Self-Regulation scales measured at week 0 were included as covariates.

Analyses resulting in bias-corrected confidence intervals (CI) based on 10,000 bootstrap iterations, revealed a significant indirect effect of group on depression through the MAIA scale Not Distracting ($a_1 \times b_1 = -3.584$, 95% CI −8.880 to −0.357), but not through the Self-Regulation scale ($a_2 \times b_2 = -2.317$, 95% CI −8.733 to 0.284). There also was a significant direct effect of group on depression ($c' = 4.817$, $p = 0.0485$) independent of Self-Regulation or Not Distracting. These findings indicate that the effect of MBCT on depression was partially mediated by Not Distracting but not by Self-Regulation.

DISCUSSION

In the present study, we investigated the effects of MBCT on body awareness as measured with the MAIA in patients with chronic pain and comorbid active depression. To our knowledge, this is the first RCT that investigates the effects of MBCT on body awareness as measured with the MAIA in patients with chronic pain and depression. The MAIA appears a reliable instrument with scales of adequate consistency in this newly studied population. In accordance with our hypothesis, MBCT resulted in increases for several dimensions of body awareness in the studied patient population. More specifically, a significantly greater increase in Self-Regulation and Not Distracting, but no changes in Noticing in the MBCT group as compared to the TAU group were observed. In addition, participants in the MBCT group, but not in the control group, had increases in Emotional Awareness. For Pain Catastrophizing, we found significant decreases within the treatment group, but not within the control group. Furthermore, mediation analyses revealed that the effect of MBCT on depression was mediated by Not Distracting, but not by Self-Regulation. We discuss these results for each dimension of body awareness separately in more detail below.

Self-Regulation

The finding of a significantly greater increase in Self-Regulation ratings in the treatment than in the control group is in line with findings by Mehling et al. (2013, 2014). In two cross-sectional studies they showed that patients with chronic low back pain who were practicing mind-body therapies (Mehling et al., 2013), or yoga and meditation (Mehling et al., 2014) had greater self-regulatory body awareness than patients without such practice. In the cohort study (Mehling et al., 2014), this difference was more pronounced for the Self-Regulation dimension than for any of the other dimensions of body awareness. Similarly, Bornemann et al. (2015) also found that increases following 3 months of contemplative training including the bodyscan and breath awareness meditation were largest on the Self-Regulation scale (effect size $d = 0.72$). Our study extends the previous findings by demonstrating effects through a well-established intervention (MBCT) in a patient population (chronic pain with active depression).

The present findings of a change in Self-Regulation through a mindfulness intervention are consistent with a hypothesized link between mindfulness and enhanced self-management (Baer, 2003), a concept closely related to general self-regulation (Vohs and Baumeister, 2011). Our findings also converge with empirical evidence for better general self-regulation of chronic pain through a mindfulness intervention (Kabat-Zinn et al., 1985), as well as with evidence from paradigms with acute pain induction, in which mindfulness practice enhanced pain tolerance (Kingston et al., 2007; Gard et al., 2012), thereby having potential clinical implications.

Emotional Awareness

Paired samples *t*-tests revealed a large and significant increase in Emotional Awareness over time within the treatment group, while the change within the control group was small and not significant However, no significant group by time interaction was found. The significant increase in Emotional Awareness over

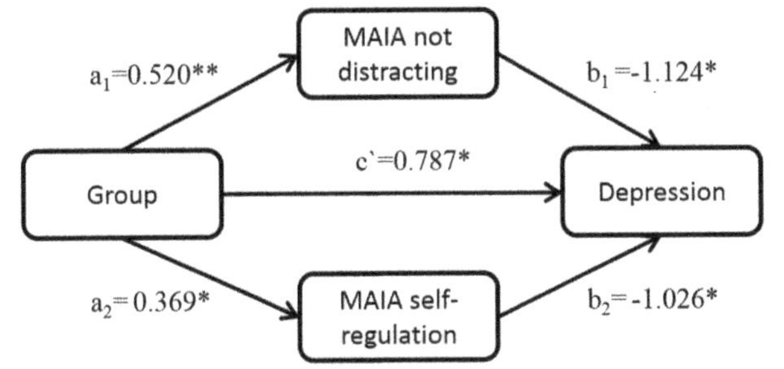

FIGURE 3 | Mediation model. Mediation model for the effect of group [Mindfulness Based Cognitive Therapy + treatment as usual (TAU), vs. TAU alone] on depression (measured with the Quick Inventory of Depressive Symptomatology—Clinician rated) through the Not Distracting and Self-Regulation scales of body awareness as measured with the Multidimensional Assessment of Interoceptive Awareness (MAIA). Analyses are based on $N = 18$ and numbers are standardized regression coefficients. $^*p < 0.05$, $^{**}p < 0.01$.

time in the treatment group is in line with the expectation that the MBCT intervention increases the awareness of the connection between body sensations and emotional states and corresponds to results by Mehling et al. (2013, 2014). Their group found that levels of Emotional Awareness were higher in chronic pain patients with mind-body practice than in patients without such practice (Mehling et al., 2013). Bornemann et al. (2015) also found significant increases of scores of the Emotional Awareness scale following 3 months of bodyscan and breath awareness training, but the effect size was rather small (<0.20). Beyond the body awareness specific construct of Emotional Awareness, our findings confirm theory (Bishop et al., 2004; Phillipot and Segal, 2009) and previous evidence that suggests an association between mindfulness and general emotional awareness (Sze et al., 2010; Boden et al., 2015).

In light of evidence that demonstrates that emotional awareness is linked to better clinical outcomes and quality of life (Williams et al., 2010; Boden et al., 2015), as well as to reductions in depression symptomatology and in depression-related affective and cognitive outcomes (Goldman et al., 2005; Arch and Craske, 2006; Farb et al., 2010), the present findings of increased Emotional Awareness in the MBCT group may be relevant for clinical practice.

In the present study, the Emotional Awareness and Self-Regulation scores were strongly correlated ($r = 0.76$). Similarly, Mehling et al. (2013) found a relatively high correlation between these two scales ($r = 0.60$) in patients with chronic pain, as well as in body-mind practitioners, and Bornemann et al. (2015) also found a considerable interscale correlation between these two scales ($r = 0.46$). These findings may correspond to the notion of the general self-regulation concept (Baumeister et al., 1998; Muraven and Baumeister, 2000) as being closely related to emotion regulation (Vohs and Baumeister, 2011; Vago and Silbersweig, 2012).

Not Distracting

The significant group-by-time interaction on Not Distracting was driven by a decrease in the TAU group and an increase in the MBCT + TAU group. Distracting oneself from uncomfortable body sensations or ignoring them, is a common coping strategy in chronic pain (Reid et al., 2002; Peres and Lucchetti, 2010), as well as in depression (Matheson and Anisman, 2003). In the intervention group, however, subjects learned to pay mindful attention to all body sensations including pain and depression-related symptoms, independent of their valence. In the control group, subjects did not learn to use this alternative approach and may have practiced the more common coping strategy of distraction so that they became "better" at ignoring and distracting themselves from uncomfortable body sensations.

Correspondingly, when a sample of patients with chronic lower-back pain was compared to healthy mind-body therapy practitioners, the mind-body sample had significantly higher Not Distracting scores (Mehling et al., 2013). However, contrary to our findings, Mehling et al. (2013) found no significant difference between the Not Distracting capacities of those patients with past or current pain that did have mind-body therapy experience and those that did not have such practice. Bornemann et al. (2015) did not find that changes on the Not Distracting scale in healthy participants, following bodyscan and breath awareness training were significantly higher than those in the retest control group. These discrepancies between the studies may be due to methodological differences, differences in the interventions, and differences in the studied population. Our findings are further aligned with prior studies showing a negative association between mindfulness and experiential avoidance (Riley, 2014), and general distraction (Jain et al., 2007).

In the present study, we did not find significant correlations between the Not Distracting scale and any other MAIA scales. This result corresponds to the finding by Mehling et al. (2013) that Not Distracting scores were not correlated with the other scales in their large ($N = 301–434$) chronic pain sample. Bornemann et al. (2015) on the other hand found that the Not Distracting scale was significantly correlated with all other scales in their sample of healthy patients. This difference might be due to their very large sample size ($N = 1076$) or due to the fact that they investigated a sample of healthy participant. These findings may indicate that in patients with chronic pain, a coping style of not distracting from pain is independent of the other aspects of body awareness.

A striking finding of the present study is that Not Distracting mediates the effect of MBCT on depressive symptom reduction. This finding supports the clinical importance of the Not Distracting aspect of body awareness. Furthermore, while it has long been hypothesized that the effects of MBCT on depression are mediated by body awareness, this is to our knowledge the first study supporting such a relationship (Michalak et al., 2010, 2011, 2012; van Der Velden et al., 2015).

Although previous research has shown that for depression, distraction may be an effective coping style, mindfulness is even more effective in reducing negative mood (Broderick, 2005; Huffziger and Kuehner, 2009). Furthermore, mindfulness has been shown to be better in reducing acute pain than distraction in individuals with high levels in pain catastrophizing (Prins et al., 2014). Similarly, in patients with chronic pain that had high health anxiety, paying attention to sensations resulted in greater pain relief than distraction (Hadjistavropoulos et al., 2000), which is in line with the notion that "one problem in chronic pain is not only the pain itself, but the 'turning away' from, the averting of attention from the regions that give rise to painful sensations, either through deliberate distraction, or by thinking *about* the pain (conceptually) rather than experiencing the sensations directly" (Williams, 2010). More in general, experiential avoidance is associated with a wide variety of psychopathology (Hayes et al., 1996) and coping strategies that work contrary to avoidance, such as mindfulness are related to better mental health outcomes (Williams et al., 2010).

Together these findings suggest that the Not Distracting aspect of body awareness may be an important predictor of depressive symptoms and potentially of mental health in general, and that it can be cultivated through mindfulness based interventions.

Noticing and Attention Regulation

The present study did not detect significant changes in the awareness of uncomfortable, comfortable, and neutral body sensations as a result of the MBCT intervention. However, the effect sizes of the pre-post changes in the treatment group were of medium size, whereas the increases in the control group only had a small effect size. These findings contrast earlier findings where large differences in MAIA Noticing scores were found between individuals with chronic pain who had mind-body practice and those patients who did not have such practice (Mehling et al., 2012). Our findings are in line with those of Bornemann et al. (2015), who despite a very large sample size did not find a significantly greater increase in the contemplative training group than in the rest control group.

No significant group by time interaction was revealed for Attention Regulation, indicating that MBCT did not result in increases of the ability to sustain and control attention to body sensations. The absence of a significant increase of Attention Regulation in the MBCT group is surprising, given that previous studies found that mind-body therapy practicing patients with back pain had higher scores of Attention Regulation than non-practicing patients (Mehling et al., 2012, 2013), and that bodyscan and breath awareness training lead to large effects on Attention Regulation in healthy participants (effect size $d = 0.54$; Bornemann et al., 2015).

It remains an open question whether the absence of significantly greater increases in Noticing and Attention Regulation in MBCT + TAU than in TAU is specific for the studied patient population and intervention, or just a lack of power. Larger studies are warranted.

Pain Catastrophizing

The absence of a group by time interaction for pain catastrophizing indicates that there were no greater changes in pain catastrophizing in the MBCT group than in the control group. However, the main effect of time was approaching significance and was driven by a large and significant decrease within the MBCT group and a small, non-significant decrease in the control group.

Consistent with the significant decrease in pain catastrophizing within the MBCT group, Mehling et al. (2013) showed that pain patients who practiced mind-body therapies worried less about their pain or other uncomfortable body sensations than individuals without mind-body practice. Furthermore, our findings are aligned with a growing body of evidence showing that in patients with chronic pain, mindfulness and participation in mindfulness-based interventions are related to decreases in pain catastrophizing (e.g., Schutze et al., 2010; Cassidy et al., 2012; Garland et al., 2012; Day et al., 2014) but see (de Boer et al., 2014).

Limitations

The present study has several limitations. First, this pilot study had a relatively small sample size, resulting in low power to detect true effects. Yet despite this limitation, some significant effects were found. However, it must be mentioned that because of the small sample size we chose a less conservative statistical approach and did not correct for multiple comparisons. Second, because this was a pilot study, we used TAU as a control group, which does not adequately control for "placebo effects" secondary to non-specific factors like attention in patient-clinician interactions (Goyal et al., 2014). This means that, theoretically, the clinical improvement could be due to non-specific factors like attention and expectations about improvement and not due to the specific effects of MBCT. We controlled for attention as much as possible by providing equal numbers of office and phone visits with the clinician for both the MBCT group and TAU group. In future larger studies it would be imperative to compare MBCT to an active non-specific control group, such as group sessions including psycho-education and gentle stretching exercises without any mindfulness components, as in the study of Hoge et al. (2013). Third, the MAIA is a relatively new scale that requires further refinement and validation in different populations. Fourth, because the intervention produced skills and understanding of concepts that were relatively new to participants at baseline, subjects may have reached a different understanding of the concepts enquired by the MAIA items post-treatment. While the first three limitations can be addressed by larger studies with active control groups and future versions of the MAIA, the last limitation is probably inherent to interventions that result in a change of perspective and thus more difficult to address. Using behavioral data might help to validate self-report instruments in these circumstances. Finally, because there were too many missing midpoint data, we conducted mediation analysis only on the endpoint data, which decreases the degree of causal specificity (Kazdin, 2007). However, Hayes argues that mediation analysis is still valuable, even when data collection of the mediator and outcome variable is at the same time point (Hayes, 2013). The findings on the mediation analysis in this study should be interpreted as tentative and purely hypothesis generating. Future larger studies with MBCT in this population with the assessment of the MAIA scales of potential mediators at multiple time points are warranted, to permit more conclusive statements.

CONCLUSION

In summary, our data suggest that MBCT can increase several dimensions of body awareness as measured with the MAIA in patients with chronic pain and comorbid active depression. In particular, the Not Distracting aspect of body awareness mediated the positive effect of MBCT on depressive symptoms. Furthermore, the MAIA appears to be a reliable instrument for measuring self-reported body awareness in this population. This finding is important because the reliability of this new and increasingly used instrument for measuring body awareness has so far not been assessed in a population of patients with chronic pain and comorbid depression. Our finding that MBCT can increase self-reported body awareness represents the first preliminary evidence in support of a causal link between a mindfulness-based intervention and increased body awareness (as measured with the MAIA) in this population. Finally, the fact that body awareness mediated the effect of MBCT on depressive symptoms provides preliminary first evidence for this long-hypothesized relationship and indicates that body awareness may have clinical relevance as an element of mindfulness approaches in the studied population.

AUTHOR CONTRIBUTIONS

All authors listed, have made substantial, direct and intellectual contribution to the work, and approved it for publication.

ACKNOWLEDGMENTS

This study was funded by an anonymous bequest to the DCRP, and the Mind and Life Francisco J. Varela (Grant 2010-01-010) to TG. We would like to thank Narayan Brach for his help with Red Cap, Jim Doorley, Rosemary Walker, and Thomas Calahan for logistics, Laurie Rhoades for teaching the MBCT program, Ronald Kulich, Karsten Kueppenbender, and Egilius Spiering for patient recruitment and Lee Baer, and Alisabet J. Clain, for assistance with data base conversion and general statistical questions, and Elizabeth Hoge, Willoughby Britton, Jonathan Alpert, and Maurizio Fava for advice. We also would like to thank all participants for their efforts.

REFERENCES

Arch, J. J., and Craske, M. G. (2006). Mechanisms of mindfulness: emotion regulation following a focused breathing induction. *Behav. Res. Ther.* 44, 1849–1858. doi: 10.1016/j.brat.2005.12.007

Baer, R. A. (2003). Mindfulness training as a clinical intervention: a conceptual and empirical review. *Clin. Psychol.* 10, 125–143. doi: 10.1093/clipsy.bpg015

Bair, M. J., Robinson, R. L., Katon, W., and Kroenke, K. (2003). Depression and pain comorbidity: a literature review. *Arch. Intern. Med.* 163, 2433–2445. doi: 10.1001/archinte.163.20.2433

Baumeister, R. F., Bratslavsky, E., Muraven, M., and Tice, D. M. (1998). Ego depletion: is the active self a limited resource? *J. Pers. Soc. Psychol.* 74, 1252. doi: 10.1037/0022-3514.74.5.1252

Bishop, S. R., Lau, M., Shapiro, S., Carlson, L., Anderson, N. D., Carmody, J., et al. (2004). Mindfulness: a proposed operational definition. *Clin. Psychol.* 11, 230–241. doi: 10.1093/clipsy.bph077

Boden, M. T., Irons, J. G., Feldner, M. T., Bujarski, S., and Bonn-Miller, M. O. (2015). An investigation of relations among quality of life and individual facets of emotional awareness and mindfulness. *Mindfulness* 6, 700–707. doi: 10.1007/s12671-014-0308-0

Bornemann, B., Herbert, B. M., Mehling, W. E., and Singer, T. (2015). Differential changes in self-reported aspects of interoceptive awareness through three months of contemplative training. *Front. Psychol.* 5:1504. doi: 10.3389/fpsyg.2014.01504

Bornemann, B., and Singer, T. (2015). "Taking time to get in touch with our body: heart beat perception accuracy increases linearly over nine months of contemplative training," in *International Convention of Psychological Science* (Amsterdam).

Broderick, P. C. (2005). Mindfulness and coping with dysphoric mood: contrasts with rumination and distraction. *Cognit. Ther. Res.* 29, 501–510. doi: 10.1007/s10608-005-3888-0

Cassidy, E. L., Atherton, R. J., Robertson, N., Walsh, D. A., and Gillett, R. (2012). Mindfulness, functioning and catastrophizing after multidisciplinary pain management for chronic low back pain. *Pain* 153, 644–650. doi: 10.1016/j.pain.2011.11.027

Craig, A. D. (2003). A new view of pain as a homeostatic emotion. *Trends Neurosci.* 26, 303–307. doi: S0166223603001231

Danielsson, L., Papoulias, I., Petersson, E.-L., Carlsson, J., and Waern, M. (2014). Exercise or basic body awareness therapy as add-on treatment for major depression: a controlled study. *J. Affect. Disord.* 168, 98–106. doi: 10.1016/j.jad.2014.06.049

Day, M. A., Thorn, B. E., Ward, L. C., Rubin, N., Hickman, S. D., Scogin, F., et al. (2014). Mindfulness-based cognitive therapy for the treatment of headache pain: a pilot study. *Clin. J. Pain* 30, 152–161. doi: 10.1097/AJP.0b013e318287a1dc

de Boer, M. J., Steinhagen, H. E., Versteegen, G. J., Struys, M. M. R. F., and Sanderman, R. (2014). Mindfulness, acceptance and catastrophizing in chronic pain. *PLoS ONE* 9:e87445. doi: 10.1371/journal.pone.0087445

de Jong, M., Peeters, F., Gard, T., Ashih, H., Doorley, J., Walker, R., et al. (in press). A randomized controlled pilot study on mindfulness-based cognitive therapy for unipolar depression in patients with chronic pain. *J. Clin. Psychiatry*.

Dunn, B. D., Dalgleish, T., Lawrence, A. D., and Ogilvie, A. D. (2007). The accuracy of self-monitoring and its relationship to self-focused attention in dysphoria and clinical depression. *J. Abnorm. Psychol.* 116, 1–15. doi: 10.1037/0021-843X.116.1.1

Ehlers, A., and Breuer, P. (1992). Increased cardiac awareness in panic disorder. *J. Abnorm. Psychol.* 101, 371–382. doi: 10.1037/0021-843X.101.3.371

Ehlers, A., and Breuer, P. (1996). How good are patients with panic disorder at perceiving their heartbeats? *Biol. Psychol.* 42, 165–182.

Farb, N., Daubenmier, J. J., Price, C. J., Gard, T., Kerr, C., Dunn, B., et al. (2015). Interoception, contemplative practice, and health. *Front. Psychol.* 6:763. doi: 10.3389/fpsyg.2015.00763

Farb, N. A., Anderson, A. K., Mayberg, H., Bean, J., McKeon, D., and Segal, Z. V. (2010). Minding one's emotions: mindfulness training alters the neural expression of sadness. *Emotion* 10, 25–33. doi: 10.1037/a0017151

Farb, N. A., Segal, Z. V., and Anderson, A. K. (2013). Mindfulness meditation training alters cortical representations of interoceptive attention. *Soc. Cogn. Affect. Neurosci.* 8, 15–26. doi: 10.1093/scan/nss066

Fox, K. C., Zakarauskas, P., Dixon, M., Ellamil, M., Thompson, E., and Christoff, K. (2012). Meditation experience predicts introspective accuracy. *PLoS ONE* 7:e45370. doi: 10.1371/journal.pone.0045370

Gard, T., Hölzel, B. K., Sack, A. T., Hempel, H., Lazar, S. W., Vaitl, D., et al. (2012). Pain attenuation through mindfulness is associated with decreased cognitive control and increased sensory processing in the brain. *Cereb. Cortex* 22, 2692–2702. doi: 10.1093/cercor/bhr352

Garland, E. L., Gaylord, S. A., Palsson, O., Faurot, K., Mann, J. D., and Whitehead, W. E. (2012). Therapeutic mechanisms of a mindfulness-based treatment for IBS: effects on visceral sensitivity, catastrophizing, and affective processing of pain sensations. *J. Behav. Med.* 35, 591–602. doi: 10.1007/s10865-011-9391-z

Goldman, R. N., Greenberg, L. S., and Pos, A. E. (2005). Depth of emotional experience and outcome. *Psychother. Res.* 15, 248–260. doi: 10.1080/10503300512331385188

Goyal, M., Singh, S., Sibinga, E. M., Gould, N. F., Rowland-Seymour, A., Sharma, R., et al. (2014). Meditation programs for psychological stress and well-being: a systematic review and meta-analysis. *JAMA Intern. Med.* 174, 357–368. doi: 10.1001/jamainternmed.2013.13018

Hadjistavropoulos, H. D., Hadjistavropoulos, T., and Quine, A. (2000). Health anxiety moderates the effects of distraction versus attention to pain. *Behav. Res. Ther.* 38, 425–438. doi: 10.1016/S0005-7967(99)00044-3

Hayes, A. F. (2013). *Introduction to Mediation, Moderation, and Conditional Process Analysis*. New York, NY: The Guilford Press.

Hayes, S. C., Wilson, K. G., Gifford, E. V., Follette, V. M., and Strosahl, K. (1996). Experiential avoidance and behavioral disorders: a functional dimensional approach to diagnosis and treatment. *J. Consult. Clin. Psychol.* 64, 1152.

Herbert, B. M., Pollatos, O., and Schandry, R. (2007). Interoceptive sensitivity and emotion processing: an EEG study. *Int. J. Psychophysiol.* 65, 214–227. doi: 10.1016/j.ijpsycho.2007.04.007

Hofmann, S. G., Sawyer, A. T., Witt, A. A., and Oh, D. (2010). The effect of mindfulness-based therapy on anxiety and depression: a meta-analytic review. *J. Consult. Clin. Psychol.* 78, 169–183. doi: 10.1037/a0018555

Hoge, E. A., Bui, E., Marques, L., Metcalf, C. A., Morris, L. K., Robinaugh, D. J., et al. (2013). Randomized controlled trial of mindfulness meditation for generalized anxiety disorder: effects on anxiety and stress reactivity. *J. Clin. Psychiatry* 74, 786–792. doi: 10.4088/JCP.12m08083

Hölzel, B. K., Lazar, S. W., Gard, T., Schuman-Olivier, Z., Vago, D. R., and Ott, U. (2011). How does mindfulness meditation work? Proposing mechanisms of action from a conceptual and neural perspective. *Pers. Psychol. Sci.* 6, 537–559. doi: 10.1177/1745691611419671

Hölzel, B. K., Ott, U., Gard, T., Hempel, H., Weygandt, M., Morgen, K., et al. (2008). Investigation of mindfulness meditation practitioners with voxel-based morphometry. *Soc. Cogn. Affect. Neurosci.* 3, 55–61. doi: 10.1093/scan/nsm038

Huffziger, S., and Kuehner, C. (2009). Rumination, distraction, and mindful self-focus in depressed patients. *Behav. Res. Ther.* 47, 224–230. doi: 10.1016/j.brat.2008.12.005

Hyett, M. P., Breakspear, M. J., Friston, K. J., Guo, C. C., and Parker, G. B. (2015). Disrupted effective connectivity of cortical systems supporting attention and interoception in melancholia. *JAMA Psychiatry* 72, 350–358. doi: 10.1001/jamapsychiatry.2014.2490

Jain, S., Shapiro, S., Swanick, S., Roesch, S., Mills, P., Bell, I., et al. (2007). A randomized controlled trial of mindfulness meditation versus relaxation training: effects on distress, positive states of mind, rumination, and distraction. *Ann. Behav. Med.* 33, 11–21. doi: 10.1207/s15324796abm3301_2

James, W. (1984). What is an emotion? *Mind* 9, 188–205.

Johnston, N. E., Atlas, L. Y., and Wager, T. D. (2012). Opposing effects of expectancy and somatic focus on pain. *PLoS ONE* 7:e38854. doi: 10.1371/journal.pone.0038854

Kabat-Zinn, J. (1990). *Full Catastrophe Living: Using the Wisdom of Your Body and Your Mind to Face Stress, Pain and Illness*. New York, NY: Delta.

Kabat-Zinn, J., Lipworth, L., and Burney, R. (1985). The clinical use of mindfulness meditation for the self-regulation of chronic pain. *J. Behav. Med.* 8, 163–190. doi: 10.1007/BF00845519

Kazdin, A. E. (2007). Mediators and mechanisms of change in psychotherapy research. *Annu. Rev. Clin. Psychol.* 3, 1–27. doi: 10.1146/annurev.clinpsy.3.022806.091432

Khalsa, S. S., Rudrauf, D., Damasio, A. R., Davidson, R. J., Lutz, A., and Tranel, D. (2008). Interoceptive awareness in experienced meditators. *Psychophysiology* 45, 671–677. doi: 10.1111/j.1469-8986.2008.00666.x

Kingston, J., Chadwick, P., Meron, D., and Skinner, T. C. (2007). A pilot randomized control trial investigating the effect of mindfulness practice on pain tolerance, psychological well-being, and physiological activity. *J. Psychosom. Res.* 62, 297–300. doi: 10.1016/j.jpsychores.2006.10.007

Lazar, S. W., Kerr, C. E., Wasserman, R. H., Gray, J. R., Greve, D. N., Treadway, M. T., et al. (2005). Meditation experience is associated with increased cortical thickness. *Neuroreport* 16, 1893–1897. doi: 10.1097/01.wnr.0000186598.66243.19

Marchand, W. R. (2012). Mindfulness-based stress reduction, mindfulness-based cognitive therapy, and Zen meditation for depression, anxiety, pain, and psychological distress. *J. Psychiatr. Pract.* 18, 233–252. doi: 10.1097/01.pra.0000416014.53215.86

Matheson, K., and Anisman, H. (2003). Systems of coping associated with dysphoria, anxiety and depressive illness: a multivariate profile perspective. *Stress* 6, 223–234. doi: 10.1080/10253890310001594487

Mehling, W. E., Daubenmier, J., Price, C. J., Acree, M., Bartmess, E., and Stewart, A. L. (2013). Self-reported interoceptive awareness in primary care patients with past or current low back pain. *J. Pain Res.* 6, 403–418. doi: 10.2147/JPR.S42418

Mehling, W. E., Gopisetty, V., Daubenmier, J., Price, C. J., Hecht, F. M., and Stewart, A. (2009). Body awareness: construct and self-report measures. *PLoS ONE* 4:e5614. doi: 10.1371/journal.pone.0005614

Mehling, W. E., Price, C. J., Daubenmier, J., Mike, A., Bartmess, E., and Stewart, A. (2014). Body awareness and the practice of yoga or meditation in 435 primary care patients with past or current low back pain. *J. Altern. Complement. Med.* 20, A63–A64. doi: 10.1089/acm.2014.5165.abstract

Mehling, W. E., Price, C., Daubenmier, J. J., Acree, M., Bartmess, E., and Stewart, A. (2012). The multidimensional assessment of interoceptive awareness (MAIA). *PLoS ONE* 7:e48230. doi: 10.1371/journal.pone.0048230

Melloni, M., Sedeno, L., Couto, B., Reynoso, M., Gelormini, C., Favaloro, R., et al. (2013). Preliminary evidence about the effects of meditation on interoceptive sensitivity and social cognition. *Behav. Brain Funct.* 9:47. doi: 10.1186/1744-9081-9-47

Michalak, J., Burg, J., and Heidenreich, T. (2012). Don't forget your body: mindfulness, embodiment, and the treatment of depression. *Mindfulness* 3, 190–199. doi: 10.1007/s12671-012-0107-4

Michalak, J., Troje, N. F., and Heidenreich, T. (2010). Embodied effects of mindfulness-based cognitive therapy. *J. Psychosom. Res.* 68, 312–313. doi: 10.1016/j.jpsychores.2010.01.004

Michalak, J., Troje, N., and Heidenreich, T. (2011). The effects of mindfulness-based cognitive therapy on depressive gait patterns. *J. Cogni. Behav. Psychother.* 11, 13–27.

Muraven, M., and Baumeister, R. F. (2000). Self-regulation and depletion of limited resources: does self-control resemble a muscle? *Psychol. Bull.* 126:247. doi: 10.1037/0033-2909.126.2.247

Nielsen, L., and Kaszniak, A. W. (2006). Awareness of subtle emotional feelings: a comparison of long-term meditators and nonmeditators. *Emotion* 6, 392. doi: 10.1037/1528-3542.6.3.392

Parkin, L., Morgan, R., Rosselli, A., Howard, M., Sheppard, A., Evans, D., et al. (2014). Exploring the relationship between mindfulness and cardiac perception. *Mindfulness* 5, 298–313. doi: 10.1007/s12671-012-0181-7

Peres, M. F., and Lucchetti, G. (2010). Coping strategies in chronic pain. *Curr. Pain Headache Rep.* 14, 331–338. doi: 10.1007/s11916-010-0137-3

Phillipot, P., and Segal, Z. (2009). Mindfulness based psychological interventions: developing emotional awareness for better being. *J. Conscious. Stud.* 16, 10–12.

Piet, J., and Hougaard, E. (2011). The effect of mindfulness-based cognitive therapy for prevention of relapse in recurrent major depressive disorder: a systematic review and meta-analysis. *Clin. Psychol. Rev.* 31, 1032–1040. doi: 10.1016/j.cpr.2011.05.002

Pollatos, O., Traut-Mattausch, E., and Schandry, R. (2009). Differential effects of anxiety and depression on interoceptive accuracy. *Depress. Anxiety* 26, 167–173. doi: 10.1002/da.20504

Preacher, K. J., and Hayes, A. F. (2008). Asymptotic and resampling strategies for assessing and comparing indirect effects in multiple mediator models. *Behav. Res. Methods* 40, 879. doi: 10.3758/BRM.40.3.879

Prins, B., Decuypere, A., and Van Damme, S. (2014). Effects of mindfulness and distraction on pain depend upon individual differences in pain catastrophizing: an experimental study. *Eur. J. Pain* 18, 1307–1315. doi: 10.1002/j.1532-2149.2014.491.x

Reid, K. J., Harker, J., Bala, M. M., Truyers, C., Kellen, E., Bekkering, G. E., et al. (2011). Epidemiology of chronic non-cancer pain in Europe: narrative review of prevalence, pain treatments and pain impact. *Curr. Med. Res. Opin.* 27, 449–462. doi: 10.1185/03007995.2010.545813

Reid, M. C., Barry, L. C., Kerns, R. D., Duong, B. D., and Concato, J. (2002). Coping strategies and their associations with levels of disability or pain, among older veterans receiving primary care. *J. Clin. Epidemiol.* 55, 629. doi: 10.1016/s0895-4356(02)00420-1

Riley, B. (2014). Experiential avoidance mediates the association between thought suppression and mindfulness with problem gambling. *J. Gambl. Stud.* 30, 163–171. doi: 10.1007/s10899-012-9342-9

Rush, A. J., Trivedi, M. H., Ibrahim, H. M., Carmody, T. J., Arnow, B., Klein, D. N., et al. (2003). The 16-item quick inventory of depressive symptomatology (QIDS), clinician rating (QIDS-C), and self-report (QIDS-SR): a psychometric evaluation in patients with chronic major depression. *Biol. Psychiatry* 54, 573–583. doi: 10.1016/S0006-3223(02)01866-8

Schutze, R., Rees, C., Preece, M., and Schutze, M. (2010). Low mindfulness predicts pain catastrophizing in a fear-avoidance model of chronic pain. *Pain* 148, 120–127. doi: 10.1016/j.pain.2009.10.030

Segal, Z. V., Williams, J. M. G., and Teasdale, J. D. (2013). *Mindfulness-Based Cognitive Therapy for Depression*. New York, NY: The Guilford Press.

Sipe, W. E., and Eisendrath, S. J. (2012). Mindfulness-based cognitive therapy: theory and practice. *Can. J. Psychiatry* 57, 63–69. doi: 10.1177/070674371205700202

Sullivan, M. J. L., Thorn, B., Haythornthwaite, J. A., Keefe, F., Martin, M., Bradley, L. A., et al. (2001). Theoretical perspectives on the relation between catastrophizing and pain. *Clin. J. Pain* 17, 52–64. doi: 10.1097/00002508-200103000-00008

Sze, J. A., Gyurak, A., Yuan, J. W., and Levenson, R. W. (2010). Coherence between emotional experience and physiology: does body awareness training have an impact? *Emotion* 10, 803. doi: 10.1037/a0020146

Trivedi, M. H., Rush, A., Ibrahim, H., Carmody, T., Biggs, M., Suppes, T., et al. (2004). The inventory of depressive symptomatology, clinician rating (IDS-C) and self-report (IDS-SR), and the quick inventory of depressive symptomatology, clinician rating (QIDS-C) and self-report (QIDS-SR) in public sector patients with mood disorders: a psychometric evaluation. *Psychol. Med.* 34, 73–82. doi: 10.1017/S0033291703001107

Tunks, E. R. (2008). Epidemiology of chronic pain with psychological comorbidity: prevalence, risk, course, and prognosis. *Can. J. Psychiatry* 53, 224–234. doi: 10.1177/070674370805300403

Turk, D. C., Wilson, H. D., and Cahana, A. (2011). Treatment of chronic non-cancer pain. *Lancet* 377, 2226–2235. doi: 10.1016/S0140-6736(11)60402-9

Vago, D. R., and Silbersweig, D. A. (2012). Self-awareness, self-regulation, and self-transcendence (S-ART): a framework for understanding the neurobiological mechanisms of mindfulness. *Front. Hum. Neurosci.* 6:296. doi: 10.3389/fnhum.2012.00296

van Der Velden, A. M., Kuyken, W., Wattar, U., Crane, C., Pallesen, K. J., Dahlgaard, J., et al. (2015). A systematic review of mechanisms of change in mindfulness-based cognitive therapy in the treatment of recurrent major depressive disorder. *Clin. Psychol. Rev.* 37, 26–39. doi: 10.1016/j.cpr.2015.02.001

Veehof, M. M., Oskam, M. J., Schreurs, K. M., and Bohlmeijer, E. T. (2011). Acceptance-based interventions for the treatment of chronic pain: a systematic review and meta-analysis. *Pain* 152, 533–542. doi: 10.1016/j.pain.2010.11.002

Vohs, K. D., and Baumeister, R. F. (2011). *Handbook of Self-Regulation: Research, Theory, and Applications*. New York, NY: The Guilford Press.

Williams, J. M. G. (2010). Mindfulness and psychological process. *Emotion* 10, 1–7. doi: 10.1037/a0018360

Williams, V., Ciarrochi, J., and Deane, F. P. (2010). On being mindful, emotionally aware, and more resilient: longitudinal pilot study of police recruits. *Aust. Psychol.* 45, 274–282. doi: 10.1080/00050060903573197

5

The Effect of Meditation on Regulation of Internal Body States

Sahib S. Khalsa[1,2,3]*, *David Rudrauf*[1,4], *Richard J. Davidson*[5] *and Daniel Tranel*[1,6]

[1] *Department of Neurology, University of Iowa, Iowa City, IA, USA,* [2] *Laureate Institute for Brain Research, Tulsa, OK, USA,* [3] *Faculty of Community Medicine, University of Tulsa, Tulsa, OK, USA,* [4] *Laboratory of Functional Imaging, Institut National de la Santé et de la Recherche Médicale U678s/Université Pierre et Marie Curie, Paris, France,* [5] *Department of Psychology, University of Wisconsin-Madison, Madison, WI, USA,* [6] *Department of Psychology, University of Iowa, Iowa City, IA, USA*

Correspondence:
Sahib S. Khalsa,
Laureate Institute for Brain Research,
6655 S. Yale Ave, Tulsa, OK 74136,
USA
skhalsa@laureateinstitute.org

Meditation is commonly thought to induce physiologically quiescent states, as evidenced by decreased autonomic parameters during the meditation practice including reduced heart rate, respiratory rate, blood pressure, skin conductance, and increased alpha activity in the electroencephalogram. Preliminary empirical support for this idea was provided in a case report by Dimsdale and Mills (2002), where it was found that meditation seemed to regulate increased levels of cardiovascular arousal induced by bolus isoproterenol infusions. In that study, while meditating, a self-taught meditator exhibited unexpected decreases in heart rate while receiving moderate intravenous doses of the beta adrenergic agonist isoproterenol. This effect was no longer observed when the individual received isoproterenol infusions while not meditating. The current study was designed to explore this phenomenon empirically in a group of formally trained meditators. A total of 15 meditators and 15 non-meditators individually matched on age, sex, and body mass index were recruited. Participants received four series of infusions in a pseudorandomized order: isoproterenol while meditating (or during a relaxation condition for the non-meditators), isoproterenol while resting, saline while meditating (or during a relaxation condition for the non-meditators), and saline while resting. Heart rate was continuously measured throughout all infusions, and several measures of heart rate were derived from the instantaneous cardiac waveform. There was no evidence at the group or individual level suggesting that meditation reduced the cardiovascular response to isoproterenol, across all measures. These results suggest that meditation is not associated with increased regulation of elevated cardiac adrenergic tone.

Keywords: meditation, adrenergic, regulation, sympathetic, parasympathetic, isoproterenol, body

Introduction

Meditation is a form of mental training that has been practiced for thousands of years, and that can be conceptualized as a family of complex emotional and attentional regulatory training regimens developed for various ends, including the cultivation of well-being and emotional balance (Davidson et al., 1976; Ekman et al., 2005; Brefczynski-Lewis et al., 2007). Meditation has also been defined as involving a process of intentional self-regulation of attention, in which attention is directed from a combination of external and internal stimuli to a primarily internally perceptive state (Astin et al., 2003; Bonadonna, 2003). Traditional philosophies emphasize that anyone can learn to meditate (Taimni, 1961), and that through repeated practice meditation provides long-term effects that outlast the confines of individual meditative states (Nyanaponika, 1969; Ahir, 1999; Burley, 2000).

A central tenet of early investigations into the effects of meditation has been that meditation induces a physiologically quiescent bodily state. This was based on initial reports of decreases in autonomic parameters such as heart rate, respiratory rate, blood pressure, skin conductance, and adrenergic reactivity, as well as increased levels of alpha activity in the electroencephalogram during the practice of meditation [mostly Transcendental Meditation (TM)] (Wallace et al., 1971; Orme-Johnson, 1973; Beary and Benson, 1974; Farrow and Hebert, 1982; Hoffman et al., 1982; Jevning et al., 1992; Travis and Wallace, 1997; Aftanas and Golocheikine, 2002; Aftanas and Golosheykin, 2005). Early researchers described this pattern of autonomic responses as an integrated "relaxation response" (Benson et al., 1974). More recently, investigations have continued finding reductions in physiological parameters such as heart rate and blood pressure occur during the acute meditation practice (Telles et al., 1995; Barnes et al., 1999; Solberg et al., 2004) as well as following practice over longer periods of time (Barnes et al., 2004; Harinath et al., 2004). Similar effects have also been reported for more secular and abbreviated protocols such as an 8 week training in Mindfulness Based Stress Reduction, which incorporates instruction in meditation as well as teaches cognitive and other approaches to enhance stress reduction (Hughes et al., 2013).

Although the notion that meditation results in physiologically quiescent states is well established, the extent to which meditative practices exert such effects is unknown. Early investigations of yogis in India who claimed to be able to exert considerable voluntary control of the heart (some claimed to be able to stop the heart) revealed limited support for this idea. At best, some individuals were able to exert transient bradycardia, or reductions in heart rate by engaging in combinations of posture manipulation, muscular contraction and breath holding (including the Valsalva maneuver, or "bearing down," which elicits a complex pattern of reflexes including tachy/bradycardia and hyper/hypotension) (Wenger et al., 1961). Subsequent studies have reported mixed results, primarily for different yoga adepts (Fenz and Plapp, 1970; Kothari et al., 1973). More recently, shorter term changes in the ability to reduce the resting heart rate have been reported (Telles et al., 2004). The perception that meditation and yoga confers an increased cardiac regulatory capability has persisted, despite this heterogeneous literature.

Preliminary empirical support via a case study approach suggested a novel effect of meditation on the ability to regulate increased levels of cardiovascular arousal. Dimsdale and Mills (2002) reported a case study in which a female meditator was randomly recruited to participate in a standardized isoproterenol challenge as a control subject in their studies of sympathetic nervous system regulation in hypertension (Dimsdale et al., 1988; Mills et al., 1998). Isoproterenol stimulates peripheral beta 1 and beta 2 adrenergic receptors equally, and when administered intravenously results in rapid and transient increases in heart rate, contractility and bronchodilation, as well as decreases in diastolic blood pressure. The protocol involved administration of a standard isoproterenol sensitivity test, involving sequentially increasing doses of intravenous isoproterenol. Halfway into the infusions the woman spontaneously started meditating, and continued to meditate during the infusions. At the time of the isoproterenol challenge the investigators did not know she was a meditator, but noticed afterwards that after the fourth dose (1 mcg) her heart rate had begun to decrease, in stark contrast to the expected increase. They report that at the highest dose (4 mcg), when the expected heart rate response was a 21 beats per minute (bpm) increase above baseline, they observed a 17 bpm *decrease* below baseline. When this unexpected result was mentioned, the participant stated that she had decided to start meditating halfway into the experiment. Suspecting that her meditation practice had interfered with the isoproterenol response, she was asked to return 2 weeks later and repeat the challenge, with explicit instructions to avoid meditating during the infusions. Her heart rate response to the subsequent infusion protocol appeared entirely consistent with the typical laboratory response to isoproterenol at rest (i.e., an approximately 20 bpm increase above baseline at the highest dose). This individual reported a self-taught meditation practice for many years, and that she did not practice any particular tradition of meditation.

The current study was designed to investigate whether there is an effect of meditation on attenuating adrenergically mediated increases in sympathetic arousal, using an empirical group study approach. We recruited formally trained meditators with at least several years of experience, and examined autonomic parameters of sympathetic arousal during meditation and resting conditions, during the application of isoproterenol and saline infusions. To examine whether such a putative effect was specific to a meditation practice or was a generic ability inherent to humans more generally, we also recruited a group of individually matched non-meditators, and tested them under relaxation and resting conditions.

We hypothesized that the practice of meditation would be specifically associated with an enhanced ability to regulate body states occurring in the context of acute physiological arousal. To test this hypothesis, the impact of a meditation practice on the cardiovascular response to isoproterenol was assessed. Several specific predictions derive from our hypothesis: (1) In the face of isoproterenol infusions, during a meditation practice meditators would display lower isoproterenol induced heart rate increases than during a non-meditation condition. (2) Meditators would demonstrate lower heart rate increases during a meditation practice than non-meditators during a relaxation practice. (3) Heart rate reductions in the meditators would be specific to meditation, i.e., meditators and non-meditators would have equivalent heart rate responses to isoproterenol at rest (when not engaging in meditation or relaxation strategies).

Methods

Participants

Fifteen meditators and fifteen non-meditators participated in the study. Each non-meditator was individually matched to a corresponding meditator based on three criteria: age, gender, and body mass index (see **Table 1**). Meditators were considered eligible for participation if they reported a continuous (daily or

TABLE 1 | Meditator and non-meditator demographic data.

	Meditators (M)	Non-meditators (NM)
Sex	10 Men, 5 Women	10 Men, 5 Women
Age (yrs)	44.7 ± 13.2	44.0 ± 13.7
Body Mass Index	24.5 ± 4.6	25.5 ± 4.0
Race	15 Caucasian American	14 Caucasian American, 1 Asian American
Education (years)	17.3 ± 2.2	15.9 ± 2.3
Beck Anxiety Inventory score	5.1 ± 3.4	3.5 ± 2.9
Beck Depression Inventory score	4.3 ± 4.6	3.7 ± 5.3
Meditation practice (years)*	10.8 ± 10.8	0 ± 0
Cumulative meditation practice (hours)*	4947 ± 6251	0 ± 0
Retreat experience (days)**	19 ± 14	0 ± 0
CD25 (micrograms)	4.48 ± 1.5	4.72 ± 2.2
Practice similarity: iso + meditation/relaxation	3.5 ± 1.1	3.4 ± 1.1
Practice similarity: saline + meditation/relaxation	3.7 ± 1.1	3.8 ± 0.8

*Chronotropic Dose 25 (CD25): isoproterenol dose (mcg) required to increase the heart rate by 25 beats per minute. Mean ± SD. *p < 0.01, **p < 0.05. Means ± standard deviation. Iso, isoproterenol.*

near daily) meditation practice during the previous 2 years, and if they had also attended one or more weeklong meditation retreats within the previous year. Using this criteria, 11 of the recruited meditators were Vipassana practitioners, and the other four meditators were Kundalini practitioners. Non-meditators were considered eligible for participation if they had never received formal meditation training in meditation or yoga and did not practice self-taught meditation.

All participants were screened for the presence of any neurological, psychiatric, cardiac or respiratory disease during a detailed phone interview, and were excluded if they reported a history of disease in any of these categories. None of the study participants were smokers, and none of the women took oral contraceptives or were pregnant, as assessed via urine pregnancy test. Each participant demonstrated a normal 12 lead electrocardiogram (EKG), as assessed by a board certified cardiologist or neurologist.

This study was approved by the University of Iowa's Institutional Review Board, and all participants provided informed consent prior to participation.

Tasks

Participants were informed that they would be asked to rest quietly with their eyes open on two occasions, and to meditate (or alternatively, to relax) with their eyes closed on two occasions. The instruction to keep the eyes open or closed was chosen to reflect the manipulation reported in Dimsdale and Mills (2002). It was also chosen to improve the face validity of the meditation and relaxation conditions, as these activities are typically taught and practiced with the eyes closed. Meditators were asked to engage in their usual daily meditation practice for a period of 15–20 min during the meditation condition. Since the Kundalini tradition offers many different types of meditations deriving different types of effects, Kundalini meditators were asked to select a specific meditation practice that they had found helped them decrease their level of bodily arousal in the past. Since the Vipassana tradition does not provide such an approach, Vipassana meditators were simply asked to engage in their usual meditation practice. Non-meditators were instructed in the performance of a cognitive relaxation strategy for a period of 15–20 min during the relaxation condition. Specifically, they were asked to engage in a relaxation practice they had found useful for themselves in the past, one that would help them to "relax your mind and body, help you slow your thoughts, slow your breathing, and slow your heart rate." If a non-meditator reported no such a strategy, they were given the option to perform one or any combination of several relaxation strategies including (1) reliving a pleasant memory, such as a warm day at the beach, (2) replaying their favorite song internally, (3) completely relaxing their musculature and/or (4) slowing their breathing. These strategies were selected with the aim of facilitating elicitation of the relaxation response (Benson et al., 1974). Non-meditators were specifically instructed to avoid falling asleep during the relaxation practice, despite any inclinations that might arise.

During the rest conditions both groups were instructed to avoid engaging in their meditation/relaxation practice, and instead, to explore their everyday thoughts (for example, thoughts about activities from the recent past such as what they had done the day before, or thoughts about potential future activities such as what they would do once the study had been concluded).

All participants were informed that they would be receiving isoproterenol or saline at some points during the entire session, but that neither they nor the experimenter would know when the nurse administered a particular agent. The room in which the study took place was kept quiet and was dimly lit throughout duration of the study to minimize distractions. Participants were seated in a padded chair, with a curtain draped over the arm containing the intravenous line, in order to reduce distraction and to preserve the blinding.

To assess task validity, after each meditation condition was complete, participants were asked to "rate how similar your meditation practice just now compares to your meditation practice in general," on a scale from 1 ("not at all the same") to 5 ("completely the same"). Non-meditators were also asked to rate their relaxation practice in a similar fashion, using the same scale. If they reported never having practiced the specific instructed relaxation condition, they were asked to relate their experience to other events where they had felt a state of relaxation.

Infusion Protocol

Two sets of standard bolus isoproterenol infusion protocols and two sets of matched saline infusion protocols were administered. The isoproterenol protocols consisted of sequentially increasing bolus isoproterenol doses of 0.1, 0.5, 1.0, 2.0, and 4.0 (mcg), delivered 3.5 min apart. The saline protocols consisted of five identically delivered bolus infusions of saline.

Infusion Delivery

Each infusion (isoproterenol and saline) consisted of two 3 milliliter (ml) bolus infusions delivered sequentially through an intravenous catheter. During isoproterenol infusions, a 3 ml bolus containing the specified dose was delivered, immediately followed by a 3 ml bolus of saline to flush the line. During saline infusions, a 3 ml bolus of saline was delivered, immediately followed by an additional 3 ml bolus of saline. Both bolus volumes were administered in entirety within a 15 s period by a nurse from the General Clinical Research Center. This method of delivery minimized the participant's ability to use external cues to distinguish between the different infusion types, and ensured rapid and standardized systemic introduction of isoproterenol.

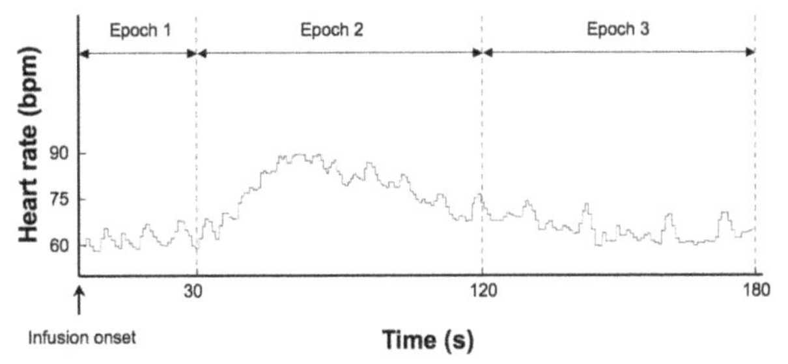

FIGURE 1 | Epochs used to derive different measures of heart rate. This example shows a typical heart rate response to a 2.0 mcg dose of isoproterenol in a single subject.

Procedure

The infusion protocol order was pseudorandomized and single blinded. This was necessitated by practical considerations: in order to ensure the safety of isoproterenol administration, a physician was required to be present during the first round of infusions. Due to constraints in the physician's schedule, a majority of the time the first condition consisted of isoproterenol infusions. However, the remaining task selection process (for example, isoproterenol plus rest vs. isoproterenol plus meditation) remained randomized, and was determined by the nurse administering the infusions. The remaining three infusion conditions were completely randomized.

Psychophysiological Measures

Heart rate was continuously measured throughout all infusions, from a lead II electrocardiogram (MP100 acquisition unit, Biopac Systems, Inc.), at a sampling rate of 200 Hz. All artifacts affecting the cardiac waveform (e.g., movement related and cardiac, such as premature ventricular contractions) were visually identified and manually removed.

Several measures of heart rate were derived from the instantaneous cardiac waveform. These measures were divided across three epochs of specific relevance to the timeline of heart rate changes induced by isoproterenol (**Figure 1**). The first epoch consisted of a 30 s interval starting immediately after the onset of each infusion. This reflects a period when bolus isoproterenol induced heart rate changes are unlikely to occur. The second epoch consisted of a 90 s interval beginning 30 s after infusion administration. This reflects a period when bolus isoproterenol induced heart rate changes are most likely to occur. The third epoch consisted of a 60 s period beginning immediately after the end of the second epoch. This interval reflects a period when the heart rate is nearing baseline or has already returned back to baseline. These three epochs combine to represent the 3 min following each infusion onset, when the probability of isoproterenol induced changes in heart rate are maximal. Because the participant could hear the nurse was preparing the next infusion during the final 30 s period (of the 3.5 min separating each infusion), this period was not included in the analysis to remove any potential influence of this preparatory period on the heart rate.

Within each epoch, four measures were obtained. (1) Mean heart rate change. This was determined for each participant by subtracting the average heart rate during epoch 1 from epoch 2. This controls for any potential residual elevations in baseline heart rate that might occur following increasing doses, and provided a reliable estimate of changes induced by isoproterenol administration. (2) Mean heart rates for each participant were determined across epochs 1, 2, and 3. This provided a measure of average heart rate over the entire window of response to isoproterenol. (3) The lowest heart rates occurring for each participant within a 3 s period throughout epochs 1, 2, and 3. This provided a measure of the floor effect for the heart rate response to isoproterenol. (4) The highest heart rates occurring for each participant within a 3-s period were identified across epochs 1, 2, and 3. This provided a measure of the ceiling effect of the heart rate response to isoproterenol.

Finally, heart rate and blood pressure were measured with an automated non-invasive blood pressure monitoring device (inflatable cuff wrapped around the dominant arm) similar to what was likely used during the Dimsdale and Mills (2002) study[1]. These latter measures were initiated 30 s after the start of each infusion, in an effort to mirror the type of measurement used by Dimsdale and Mills (2002). Due to variability in the amount of time taken for the machine to generate a blood pressure reading, each measure with this device was obtained approximately 60–80 s after each infusion onset (maximum range observed: 58–93 s after initiating measurement).

Data Analysis

Continuous data were analyzed using general linear models (GLM) with repeated measures, with group (meditator, non-meditator) as the between subjects factor and with condition (isoproterenol plus meditation or relaxation, isoproterenol plus rest, saline plus meditation or relaxation, saline plus rest) and dose (isoproterenol, saline) as the within subjects factors. In the GLM analysis, the meditation vs. rest contrast is explicitly evaluated in the condition term. Therefore, the pertinent measures determining whether meditators show differences in their autonomic responses during meditation vs. rest were tested by the group x condition, and by the group x condition x dose interactions. All repeated measures tests were assessed for

[1]These measures were incorporated after several participants had been tested. As a result, data for the full sample are unavailable (measures were collected from 11 meditators and 13 non-meditators).

violations of the sphericity assumption, and when violated, were corrected with the Huynh-Feldt method. In these instances the corrected p-values are reported, along with the Huynh-Feldt epsilon (ε) correction.

Results

Participants

Meditators reported significantly more years of meditation practice $t_{(28)} = 3.90, p = 0.002$, hours of cumulative meditation practice $t_{(28)} = 3.07, p = 0.008$, and days of retreat experience $t_{(28)} = 2.31, p = 0.037$, than non-meditators. The groups did not differ with respect to age $t_{(28)} = 0.15, p = 0.88$, body mass index $t_{(28)} = -0.63, p = 0.53$, or education $t_{(28)} = 1.64, p = 0.11$. The groups also did not differ with respect to baseline levels of anxiety (assessed via the Beck Anxiety Inventory) $t_{(28)} = 1.39, p = 0.18$, or depression (assessed via the Beck Depression Inventory) $t_{(28)} = 0.33, p = 0.74$.

Mean Heart Rate Change

As expected, we observed a significant effect of condition $F_{(1, 3)} = 140.22, p < 0.0001, \eta_p^2 = 0.83, \varepsilon = 0.780$, observed power = 1.00, and dose $F_{(1, 4)} = 68.5, p < 0.0001, \eta_p^2 = 0.71, \varepsilon = 0.554$, observed power = 1.00, on the mean heart rate response to isoproterenol. There was a significant interaction between condition and dose $F_{(1, 12)} = 68.54, p < 0.0001, \eta_p^2 = 0.59, \varepsilon = 0.737$, observed power = 1.00, such that increases in heart rate occurred at increasing doses of isoproterenol (but not saline) administration. However, despite these changes, there were no group differences in the heart rate response to isoproterenol. There was no effect of group $F_{(1, 28)} = 2.97, p = 0.10, \eta_p^2 = 0.10$, observed power = 0.384, and there were no interactions between condition and group $F_{(1, 3)} = 0.21, p = 0.84, \eta_p^2 = 0.01$, observed power = 0.088, between dose and group $F_{(1, 4)} = 0.21, p = 0.84, \eta_p^2 = 0.01$, observed power = 0.093, or between condition and group and dose $F_{(1, 12)} = 0.82, p = 0.60, \eta_p^2 = 0.03$, observed power = 0.482, suggesting that the heart rate increases induced by isoproterenol were not statistically different between the groups (**Figure 2A**). Mean average heart rate changes and associated 95% confidence intervals are displayed in **Table 2**.

Mean Heart Rate

Epoch 1

We did not find an effect of condition on the average heart rate during epoch 1, $F_{(1, 3)} = 0.39, p = 0.76, \eta_p^2 = 0.01$, observed power = 0.12, but did observe a significant effect of dose $F_{(1, 4)} = 4.74, p = 0.008, \eta_p^2 = 0.15, \varepsilon = 0.616$, observed power = 0.80. There was also a significant interaction between condition and dose $F_{(1, 12)} = 4.41, p < 0.0001, \eta_p^2 = 0.14, \varepsilon = 0.772$, observed power = 0.99, suggesting that increases in mean heart rate during epoch 1 occurred at increasing doses of isoproterenol (but not saline) administration. However, despite these changes, there were no group differences in heart rate during this period. There was no effect of group $F_{(1, 28)} = 1.20, p = 0.28, \eta_p^2 = 0.04$, observed power = 0.19, and there were no interactions between condition and group $F_{(1, 3)} = 2.00, p = 0.12, \eta_p^2 = 0.07$, observed power = 0.50, between dose and group $F_{(1, 4)} = 0.33, p = 0.76, \eta_p^2 = 0.01$, observed power = 0.12, or between condition and group and dose $F_{(1, 12)} = 0.74, p = 0.68, \eta_p^2 = 0.03$, observed power = 0.43, suggesting that the mean heart rate increases observed for increasing doses of isoproterenol during epoch 1 were not statistically different between groups. Average heart rate changes and associated 95% confidence intervals are displayed in **Table 3**. The finding of dose related increases during this epoch reflects that fact that some of the heart rate changes induced by isoproterenol (particularly at the higher doses) began occurring early, as was observed in several participants.

Epoch 2

We observed a significant effect of condition on the average heart rate during epoch 2, $F_{(1, 3)} = 84.19, p < 0.0001, \eta_p^2 = 0.75$, observed power = 1.00, as well as for dose $F_{(1, 4)} = 84.19, p < 0.0001, \eta_p^2 = 0.75, \varepsilon = 0.402$, observed power = 1.00. There was also a significant interaction between condition and dose $F_{(1, 12)} = 72.91, p < 0.0001, \eta_p^2 = 0.72, \varepsilon = 0.330$, observed power = 1.00, suggesting that increases in mean heart rate during epoch 2 occurred at increasing doses of isoproterenol (but not saline) administration. However, despite these changes, there were again no group differences in heart rate during this period. There was no effect of group $F_{(1, 28)} = 0.65, p = 0.43, \eta_p^2 = 0.02$, observed power = 0.12, and there were no interactions between condition and group $F_{(1, 3)} = 2.00, p = 0.12, \eta_p^2 = 0.07$, observed power = 0.37, between dose and group $F_{(1, 4)} = 1.44, p = 0.24, \eta_p^2 = 0.05$, observed power = 0.16, or between condition and group and dose $F_{(1, 12)} = 0.70, p = 0.59, \eta_p^2 = 0.02$, observed power = 0.41, suggesting that the mean heart rate increases observed for increasing doses of isoproterenol during epoch 2 were not statistically different between groups (**Figure 2B** and **Table 4**).

Epoch 3

We observed a significant effect of condition on the average heart rate during epoch 3, $F_{(1, 3)} = 8.30, p < 0.0001, \eta_p^2 = 0.23$, observed power = 0.98, as well as for dose $F_{(1, 4)} = 30.66, p < 0.0001, \eta_p^2 = 0.52, \varepsilon = 0.440$, observed power = 1.00. There was also a significant interaction between condition and dose $F_{(1, 12)} = 20.72, p < 0.0001, \eta_p^2 = 0.43, \varepsilon = 0.418$, observed power = 1.00, suggesting that increases in mean heart rate during epoch 3 occurred at increasing doses of isoproterenol (but not saline) administration. However, despite these changes, there were no group differences in heart rate during this period. There was no effect of group $F_{(1,28)} = 0.92, p = 0.35, \eta_p^2 = 0.03$, observed power = 0.15, and there were no interactions between condition and group $F_{(1, 3)} = 1.68, p = 0.18, \eta_p^2 = 0.06$, observed power = 0.43, between dose and group $F_{(1, 4)} = 0.93, p = 0.39, \eta_p^2 = 0.03$, observed power = 0.29, or between condition and group and dose $F_{(1, 12)} = 0.65, p = 0.67, \eta_p^2 = 0.02$, observed power = 0.38, suggesting that the mean heart rate increases observed for increasing doses of isoproterenol during epoch 3 were not statistically different between groups. The finding of dose related increases during this epoch reflects that fact that some of the heart rate changes induced by isoproterenol (particularly at the higher doses) were still present and in the

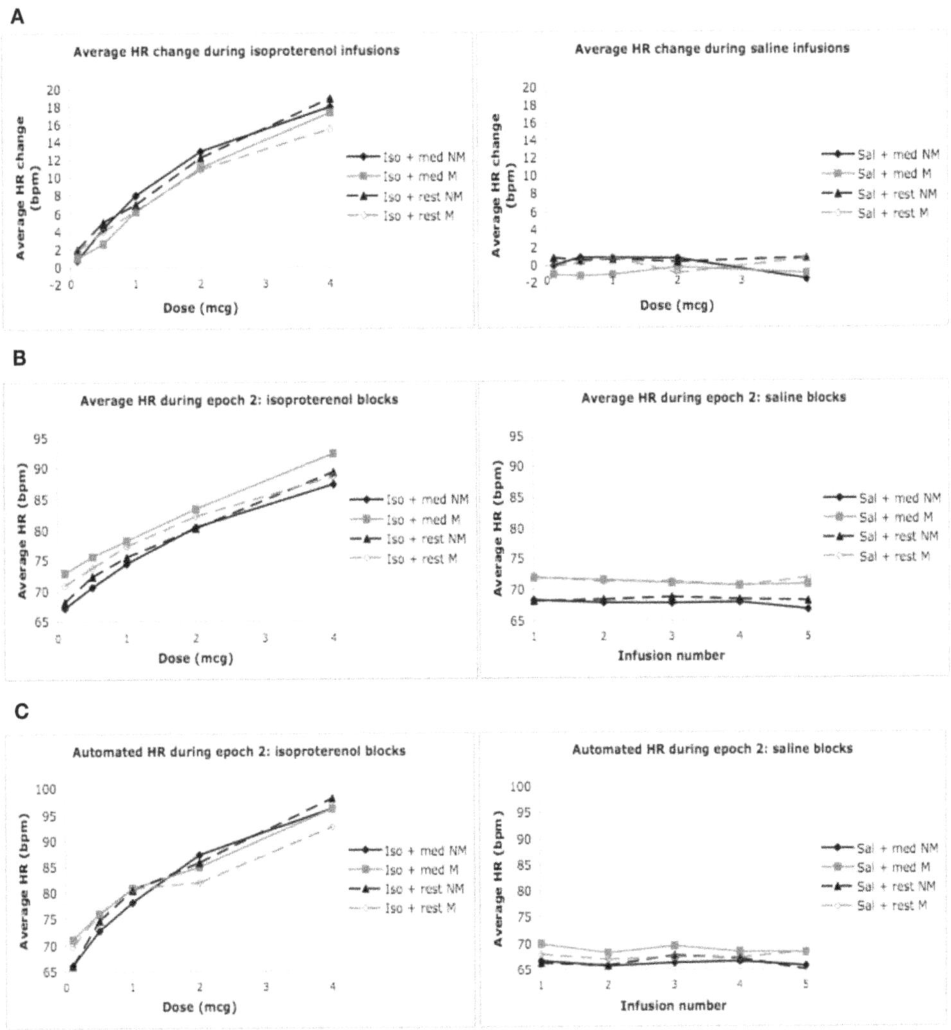

FIGURE 2 | Mean heart rates for both groups during mediation/relaxation and rest. (A) Mean heart rate change (epoch 2 minus epoch 1). **(B)** Mean heart rate during epoch 2. **(C)** Mean heart rates measured via automated blood pressure monitor. For purposes of clarity, mean values are displayed without error bars, and the relaxation condition for the non-meditators is labeled as meditation.

process of returning to baseline, as was observed in several participants (**Table 5**).

Highest and Lowest Heart Rates

The pattern of findings reported above also held true for the highest and lowest heart rates observed during epochs 1, 2, and 3. Specifically, increases in the highest (and lowest) heart rates occurred for conditions containing increasing doses of isoproterenol (but not saline), in the absence of any group differences whatsoever. In the interest of brevity the relevant statistics have not been reported here, but rather, the reader is referred to **Figure 3**.

Automated Heart Rate and Blood Pressure

Automated Heart Rate

We observed a significant effect of condition on the automated measure of heart rate $F_{(1,\ 3)} = 115.88$, $p < 0.0001$, $\eta_p^2 = 0.84$, $\varepsilon = 0.813$, observed power = 1.00, as well as dose $F_{(1,\ 4)} = 74.42$, $p < 0.0001$, $\eta_p^2 = 0.77$, $\varepsilon = 0.909$, observed power = 1.00. There was also a significant interaction between condition and dose $F_{(1,\ 12)} = 40.31$, $p < 0.0001$, $\eta_p^2 = 0.65$, $\varepsilon = 0.438$, observed power = 1.00, suggesting that increases in automated heart rate occurred at increasing doses of isoproterenol (but not saline) administration. There were again no group differences in heart rate during this period. There was no effect of group $F_{(1,\ 28)} = 0.06$, $p = 0.80$, $\eta_p^2 = 0.00$, observed power = 0.06, and there were no interactions between condition and group $F_{(1,\ 3)} = 1.02$, $p = 0.38$, $\eta_p^2 = 0.04$, observed power = 0.27, between dose and group $F_{(1,\ 4)} = 2.12$, $p = 0.09$, $\eta_p^2 = 0.09$, observed power = 0.61, or between condition and group and dose $F_{(1,\ 12)} = 1.19$, $p = 0.32$, $\eta_p^2 = 0.05$, observed power = 0.68, suggesting that the mean heart rate increases observed for increasing doses of isoproterenol with the automated measure were similar between the groups (**Figure 2C**).

TABLE 2 | Average heart rate changes and 95% confidence intervals across all conditions.

Average HR change								
Dose (mcg)	Non-meditators		95% CI		Meditators		95% CI	
	Mean	SE	Lower	Upper	Mean	SE	Lower	Upper
	ISO + RELAXATION				ISO + MEDITATION			
0.1	0.74	0.84	−0.99	2.47	0.97	0.84	−0.76	2.7
0.5	4.33	0.77	2.74	5.91	2.6	0.77	1.03	4.2
1	8.02	1.03	5.91	10.13	6.2	1.03	4.09	8.3
2	12.91	1.63	9.58	16.25	11.11	1.63	7.79	14.45
4	17.93	2.26	13.3	22.56	17.27	2.26	12.64	21.89
	ISO + REST				ISO + REST			
0.1	1.97	0.65	0.65	3.31	1.82	0.65	0.48	3.15
0.5	5.01	0.85	3.28	6.74	3.97	0.85	2.24	5.7
1	6.99	0.76	5.44	8.53	6.28	0.76	4.73	7.82
2	12.22	1.39	9.38	15.06	10.89	1.39	8.05	13.73
4	18.85	1.54	15.71	22.02	15.44	1.54	12.28	18.59
	SALINE + RELAXATION				SALINE + MEDITATION			
0.1	0	0.96	−1.96	1.97	−1.04	0.96	−3	0.92
0.5	0.89	0.66	−0.47	2.2	−1.17	0.66	−2.52	0.18
1	0.86	0.51	−0.19	1.91	−1.03	0.51	−2.08	0.02
2	0.76	0.52	−0.31	1.84	−0.22	0.52	−1.29	0.85
4	−1.54	1.04	−3.67	0.6	−0.88	1.04	−3.02	1.25
	SALINE + REST				SALINE + REST			
0.1	0.85	0.59	−0.37	2.06	−0.13	0.59	−1.35	1.08
0.5	0.58	0.52	−0.49	1.65	0.33	0.52	−0.74	1.4
1	0.69	0.65	−0.62	2	0.88	0.64	−0.42	2.19
2	0.39	0.9	−1.47	2.24	−0.9	0.9	−2.75	0.95
4	0.82	0.61	−0.44	2.07	0.74	0.61	−0.51	2

Systolic Blood Pressure

We did not observe a significant effect of condition $F_{(1,\ 3)} = 0.03$, $p = 0.99$, $\eta_p^2 = 0.00$, observed power $= 0.06$, but did observe a significant effect of dose $F_{(1,\ 4)} = 4.35$, $p = 0.006$, $\eta_p^2 = 0.17$, $\varepsilon = 0.820$, observed power $= 0.82$, on the automated measure of systolic blood pressure. There was no interaction between condition and dose $F_{(1,\ 12)} = 1.74$, $p < 0.08$, $\eta_p^2 = 0.07$, observed power $= 0.87$, suggesting that changes in systolic blood pressure were not influenced by isoproterenol administration. No group differences in systolic blood pressure were observed. There was no effect of group $F_{(1,28)} = 0.71$, $p = 0.41$, $\eta_p^2 = 0.03$, observed power $= 0.13$, and there were no interactions between condition and group $F_{(1,\ 3)} = 0.80$, $p = 0.50$, $\eta_p^2 = 0.04$, observed power $= 0.21$, between dose and group $F_{(1,\ 4)} = 1.52$, $p = 0.21$, $\eta_p^2 = 0.07$, observed power $=$ 0.45, or between condition and group and dose $F_{(1,\ 12)} = 1.03$, $p = 0.42$, $\eta_p^2 = 0.05$, observed power $= 0.59$. These decreases in systolic blood pressure were similar for both groups and appeared to occur during the first few doses, irrespective of condition, suggesting that these effects were likely not related to isoproterenol (**Figure 4A**).

Diastolic Blood Pressure

We observed a significant effect of condition $F_{(1,\ 3)} = 25.37$, $p <$ 0.0001, $\eta_p^2 = 0.54$, observed power $= 1.00$, and dose $F_{(1,\ 4)} =$ 28.58, $p < 0.0001$, $\eta_p^2 = 0.57$, $\varepsilon = 0.653$, observed power $= 1.00$ on the automated measure of diastolic blood pressure. There was also a significant interaction between condition and dose $F_{(1,\ 12)} = 12.43$, $p < 0.0001$, $\eta_p^2 = 0.36$, $\varepsilon = 0.692$, observed power $= 1.00$, suggesting that changes in diastolic blood pressure occurred at increasing doses of isoproterenol (but not saline) administration. There were again no group differences in diastolic blood pressure. There was no effect of group $F_{(1,\ 28)} =$ 0.65, $p = 0.43$, $\eta_p^2 = 0.03$, observed power $= 0.12$, and there were no interactions between condition and group $F_{(1,\ 3)} = 0.29$, $p = 0.82$, $\eta_p^2 = 0.01$, observed power $= 0.10$, between dose and

TABLE 3 | Average heart rates during epoch 1, including 95% confidence intervals.

Average HR—epoch 1								
Dose (mcg)	**Non-meditators**		**95% CI**		**Meditators**		**95% CI**	
	Mean	**SE**	**Lower**	**Upper**	**Mean**	**SE**	**Lower**	**Upper**
	ISO + RELAXATION				**ISO + MEDITATION**			
0.1	66.47	2.54	61.27	71.77	71.86	2.54	66.67	77.06
0.5	66.26	2.83	60.46	72.07	72.92	2.83	67.11	78.72
1	66.37	2.68	60.89	71.86	71.97	2.68	66.49	77.46
2	67.48	2.76	61.83	73.12	72.22	2.76	66.57	77.87
4	69.4	3.12	63	75.8	75.05	3.12	68.65	81.45
	ISO + REST				**ISO + REST**			
0.1	66.18	2.52	61.03	71.35	69.09	2.52	63.92	74.25
0.5	67.32	2.59	62.02	72.2	69.74	2.59	64.44	75.04
1	68.46	2.6	64.12	73.79	70.98	2.6	65.65	76.32
2	67.98	2.65	62.58	73.29	71.28	2.64	65.87	76.69
4	70.45	2.97	64.37	76.53	73.11	2.97	67.03	79.19
	SALINE + RELAXATION				**SALINE + MEDITATION**			
0.1	68.29	3.3	61.54	75.04	72.96	3.3	66.2	79.71
0.5	66.91	2.72	61.33	72.49	72.72	2.72	67.14	78.3
1	66.87	2.71	61.32	72.43	72.01	2.71	66.46	77.57
2	67.07	2.77	61.39	72.74	70.81	2.77	65.13	76.49
4	68.24	3.1	61.88	74.59	71.7	3.1	65.34	78.06
	SALINE + REST				**SALINE + REST**			
0.1	67.26	2.78	61.56	72.96	72.16	2.78	66.46	77.86
0.5	67.73	2.67	62.27	73.19	70.91	2.67	65.45	76.37
1	68.00	2.73	62.42	73.59	70.42	2.73	64.83	76
2	67.94	2.87	62.05	73.83	71.43	2.87	65.54	77.32
4	67.31	2.58	62.02	72.59	71.1	2.58	65.81	76.38

group $F_{(1, 4)} = 0.36, p = 0.76, \eta_p^2 = 0.02$, observed power = 0.13, or between condition and group and dose $F_{(1, 12)} = 0.45$, $p = 0.90$, $\eta_p^2 = 0.02$, observed power = 0.25. These findings suggest decreases in diastolic blood pressure were similar for both groups at increasing doses of isoproterenol (**Figure 4B**). Such a decrease in diastolic blood pressure is consistent with the vasodilatory effects of isoproterenol.

Individual Heart Rates

Given the absence of group effects, an examination of individual heart rate changes during the isoproterenol plus meditation condition was performed. The goal was to identify whether any reductions in heart rate similar those reported in Dimsdale and Mills (2002) had occurred in individual meditators or non-meditators. At the individual level, we examined mean heart rate changes during epoch 2, mean heart rate during epoch 2, and lowest and highest heart rate during epoch 2 derived from the continuous heart rate waveform during the isoproterenol plus meditation/relaxation and isoproterenol plus rest conditions. Based on the hypothesis that meditation would result in a reduced response to isoproterenol, individual meditators and non-meditators displaying the lowest responses were selected for comparison with their respective groups. As **Figure 5A** indicates, there were individuals in each group who displayed a reduced heart rate response and reduced average heart rates relative to their group averages. Similarly, there were individuals in each group who displayed reduced heart rate maxima and minima when compared with their group averages (**Figure 5B**). However, although these reflect large differences in the magnitude of the response to isoproterenol (compared to the respective group mean), only one individual (a meditator) appeared to display a reduced heart rate at increasing doses, and only when using the criterion for lowest heart rate observed during a 3 s period.

Subjective Experience

Similarity ratings of the meditation and relaxation practices did not differ between the groups $F_{(1, 25)} = 0.03$, $p = 0.86$, $\eta_p^2 =$

TABLE 4 | Average heart rates during epoch 2, including 95% confidence intervals.

Dose (mcg)	Non-meditators		95% CI		Meditators		95% CI	
	Mean	SE	Lower	Upper	Mean	SE	Lower	Upper
	ISO + RELAXATION				**ISO + MEDITATION**			
0.1	67.21	2.55	61.98	72.43	72.84	2.55	67.61	78.06
0.5	70.59	2.71	65.04	76.13	75.53	2.71	69.98	81.08
1	74.39	2.95	68.35	80.43	78.17	2.95	72.13	84.21
2	80.39	3.5	73.22	87.55	83.34	3.5	76.17	90.5
4	87.33	4.14	78.87	95.8	92.32	4.14	83.85	100.79
	ISO + REST				**ISO + REST**			
0.1	68.17	2.51	63.03	73.31	70.9	2.51	65.76	76.04
0.5	72.34	2.64	66.93	77.74	73.71	2.64	68.31	79.12
1	75.44	2.84	69.62	81.26	77.26	2.84	71.44	83.08
2	80.21	3.19	73.66	86.75	82.17	3.19	75.63	88.71
4	89.31	3.51	82.13	96.49	88.55	3.51	81.37	95.73
	SALINE + RELAXATION				**SALINE + MEDITATION**			
0.1	68.39	2.9	62.36	74.23	71.92	2.9	65.98	77.85
0.5	67.79	2.69	62.23	73.3	71.55	2.69	66.04	77.06
1	67.73	2.8	62.01	73.45	70.99	2.8	65.26	76.71
2	67.83	2.75	62.2	73.46	70.59	2.75	65.96	76.22
4	66.7	2.73	61.11	72.29	70.82	2.73	65.23	76.4
	SALINE + REST				**SALINE + REST**			
0.1	68.10	2.77	62.44	73.77	72.03	2.77	66.36	77.7
0.5	68.31	2.63	62.93	73.7	71.24	2.63	65.86	76.63
1	68.69	2.69	63.19	74.19	71.3	2.66	65.8	76.8
2	68.33	2.66	62.89	73.77	70.53	2.66	65.09	75.97
4	68.12	2.8	62.38	73.86	71.84	2.8	66.1	77.58

0.001, observed power = 0.054[2]. The meditation and relaxation practices appeared to be rated as more similar to the usual practice during the saline infusion condition, but this difference was not statistically significant $F_{(1, 1)} = 2.18$, $p = 0.15$, $\eta_p^2 = 0.08$, observed power = 0.30.

Post Analysis

Since a heterogeneously recruited sample of meditators could potentially result in differential heart rate responses during meditation, we subsequently evaluated the responses of the Kundalini and Vipassana practitioners separately, and in comparison to each other.

Vipassana vs. Healthy Comparison

During the meditation condition there was no effect of group $F_{(1, 24)} = 1.31$, $p = 0.26$, $\eta_p^2 = 0.05$, observed power = 0.20, and there were no interactions between condition and group $F_{(1, 1)} = 0.002$, $p = 0.98$, $\eta_p^2 = 0.00$, observed power = 0.05, or between condition and group and dose $F_{(1, 4)} = 0.10$, $p = 0.98$, $\eta_p^2 = 0.004$, observed power = 0.07. This suggests that during both meditation conditions (saline and isoproterenol) the Vipassana meditators displayed similar average heart rate increases as healthy comparisons.

Kundalini vs. healthy Comparison

During the meditation condition there was no effect of group $F_{(1, 17)} = 2.1$, $p = 0.17$, $\eta_p^2 = 0.11$, observed power = 0.27, and there were no interactions between condition and group $F_{(1, 1)} = 0.006$, $p = 0.94$, $\eta_p^2 = 0.00$, observed power = 0.05, or between condition and group and dose $F_{(1, 4)} = 1.58$, $p = 0.19$, $\eta_p^2 = 0.09$, observed power = 0.46. This suggests that during both meditation conditions (saline and isoproterenol) the Kundalini meditators displayed similar average heart rate increases as healthy comparisons.

[2]Data were missing from 3 healthy comparison participants.

TABLE 5 | Average heart rates during epoch 3, including 95% confidence intervals.

Average HR—epoch 3								
Dose (mcg)	Non-meditators		95% CI		Meditators		95% CI	
	Mean	SE	Lower	Upper	Mean	SE	Lower	Upper
	ISO + RELAXATION				ISO + MEDITATION			
0.1	66.58	2.58	61.29	71.87	71.57	2.58	66.28	76.85
0.5	66.28	2.71	60.73	71.82	71.7	2.71	66.16	77.25
1	68.1	2.74	62.49	73.71	72.5	2.74	66.89	78.11
2	70.59	3.15	64.13	77.05	75.11	3.15	68.66	81.57
4	73.94	3.67	66.43	81.45	81.28	3.67	73.77	88.79
	ISO + REST				ISO + REST			
0.1	67.88	2.55	62.66	73.1	71.8	2.55	66.57	77.02
0.5	68.38	2.63	62.99	73.77	70.67	2.63	65.28	76.06
1	68.75	2.65	63.32	74.19	70.86	2.65	65.43	76.3
2	71.47	3.04	65.25	77.69	75.57	3.04	69.35	81.79
4	77.07	3.33	70.24	83.9	80.35	3.33	73.52	87.18
	SALINE + RELAXATION				SALINE + MEDITATION			
0.1	67.93	2.74	62.31	73.54	71.97	2.74	66.34	77.57
0.5	68.06	2.67	62.6	73.52	71.85	2.67	66.39	77.31
1	68.58	2.7	63.04	74.12	71.2	2.7	65.66	76.74
2	67.89	2.78	62.2	73.58	70.39	2.78	64.7	76.08
4	66.48	2.69	60.97	72	70.09	2.69	65.57	75.6
	SALINE + REST				SALINE + REST			
0.1	68.12	2.6	62.8	73.44	71.78	2.6	66.46	77.1
0.5	68.67	2.67	63.19	74.14	71.24	2.67	65.77	76.72
1	69.73	2.53	64.54	74.93	71.04	2.53	65.85	76.23
2	68.38	2.66	62.94	73.83	71.83	2.66	66.39	77.27
4	69.81	2.77	64.14	75.47	72.15	2.77	66.49	77.82

Kundalini vs. Vipassana

During the meditation condition there was no effect of group $F_{(1,\ 13)} = 0.11, p = 0.75, \eta_p^2 = 0.008$, observed power = 0.06, and there were no interactions between condition and group $F_{(1,\ 1)} = 0.001, p = 0.98, \eta_p^2 = 0.00$, observed power = 0.05, or between condition and group and dose $F_{(1,\ 4)} = 0.88, p = 0.48, \eta_p^2 = 0.06$, observed power = 0.26. This suggests that during both meditation conditions (saline and isoproterenol) the two groups of meditators displayed similar average heart rate increases.

Discussion

As expected, bolus isoproterenol infusions produced dose-dependent increases in heart rate in both groups during a condition of rest. However, meditators did not display lowered heart rate responses to isoproterenol while practicing meditation. Both meditators and non-meditators displayed similar dose dependent increases in heart rate during conditions of isoproterenol plus rest, and during isoproterenol plus meditation/relaxation. The lack of group differences in heart rate was reliable. It was observed for five different measures of heart rate (mean heart rate change, mean heart rate, lowest heart rate during a 3 s period, highest heart rate during a 3 s period, and via automated heart rate monitor), and throughout three epochs that captured the entire time span over which isoproterenol induced changes occur. Equivalent decreases in diastolic blood pressure were also observed in both groups following increasing doses of isoproterenol. These decreases are consistent with the known vasodilatory effects of isoproterenol on lowering diastolic blood pressure. We do not consider these to be related to the meditation intervention, as they occurred in both isoproterenol conditions. Non-meditators did not display lowered heart rates when practicing a relaxation condition, which was predicted.

Since the Dimsdale and Mills (2002) study reported a finding in a single individual, we examined individual heart rate responses within in each group of participants. Of the measures utilized, there was only one participant who demonstrated an appreciable reduction in heart rate during the meditation plus

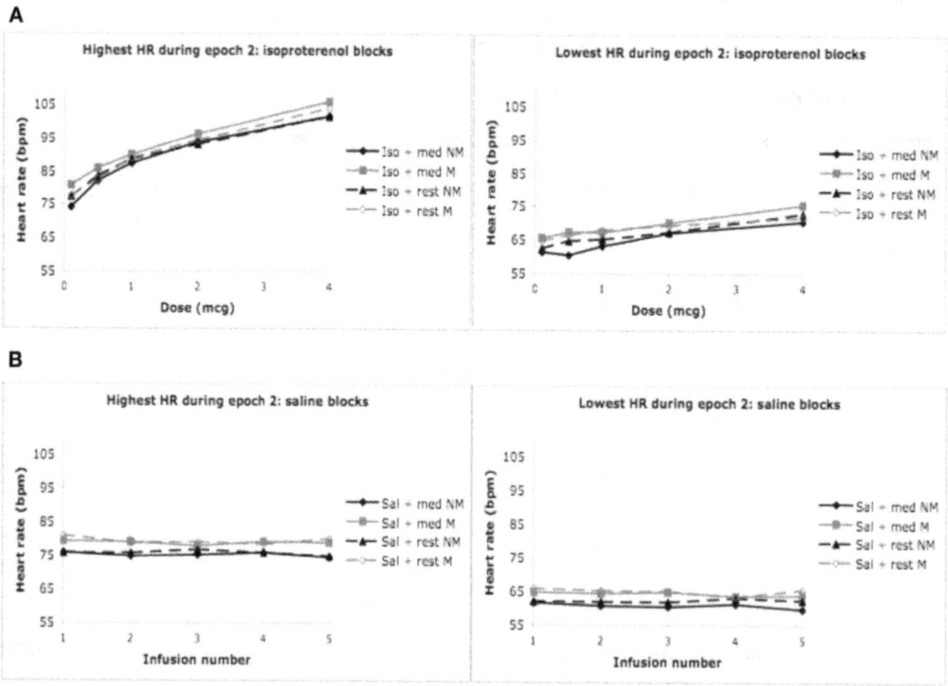

FIGURE 3 | Highest and lowest heart rates for both groups during epoch 2. (A) Highest heart rates observed during a 3 s period, averaged across each group (ceiling effect). **(B)** Lowest heart rates observed during a 3 s period, averaged across each group (floor effect). For purposes of clarity, mean values are displayed without error bars, and the relaxation condition for the non-meditators is labeled as meditation.

isoproterenol condition similar to that reported by Dimsdale and Mills (2002) (i.e., a 17 bpm decrease in heart rate from the lowest to the highest dose). This finding occurred in a meditator, but it was only observed with one measure, the lowest heart rate, which is also the least reliable index of the response to isoproterenol. This finding was no longer present when more robust measures of the heart rate response to isoproterenol, such as the average heart rate and average heart rate change, were examined in the same individual. Thus, it seems unlikely that the individual effect reported by Dimsdale and Mills (2002), although intriguing, generally applies to individuals practicing meditation.

Several potential factors might explain the discrepancy in findings between the two studies. First, individual variability in heart rate responses might explain their observation of lowered heart rate in one subject. We observed considerable variability in heart rate responses to isoproterenol even when using a continuous measure, with select individuals from both groups exhibiting attenuated heart rate responses to isoproterenol. A second and related reason could be measurement error. Dimsdale and Mills (2002) did not utilize a continuous measure of heart rate in their original study, and it is unclear when the automated monitor they utilized actually captured their meditator's heart rate. Although they report that heart rate was measured 60 s after the infusion was delivered, it is possible that there was enough measurement variability to result in missing the heart rate epoch of interest. In the present study, we observed large variability in the output of the automated heart rate measurement when using the blood pressure monitoring device (e.g., 58–93 s after initiating measurement), which is long enough to potentially miss capturing an individual's heart rate response. It is also possible that the window of measurement of the automated heart rate monitor was too brief. In the current study, we observed the largest reductions in heart rate response during a momentary measure, the lowest heart rate during a 3 s period, which provides some plausibility to this theory. Third, their meditator was not formally trained, and indicated that she would often enter such deep states that she would need to set an alarm clock to rouse her from meditation. Thus, it is possible that this individual likened sleep (or some other altered state of consciousness) with meditation, and that the observed effect had nothing to do with meditation but a different process worth understanding further. However, despite these potential methodological and conceptual considerations, the participant in Dimsdale and Mills' study did demonstrate a heart rate increase at the 1.0 mcg dose, followed by two successive decreases below her resting heart rate. Assuming this was not a spurious finding, such a profound decrease in heart rate could also occur during an episode of increased vagal output (as suggested by the authors). Neurocardiogenic syncope is one example of such an increase in vagal output. These episodes are usually preceded by an increase in sympathetic tone (as was observed in the meditator at the 1.0 mcg dose) and can even be triggered by isoproterenol (Kikushima et al., 1999). These episodes are often foreshadowed by symptoms such as lightheadedness, nausea, warmth, pallor and/or sweating, and although the meditator denied experiencing a subset of these (nausea, dizziness or fainting), it is possible that she was not

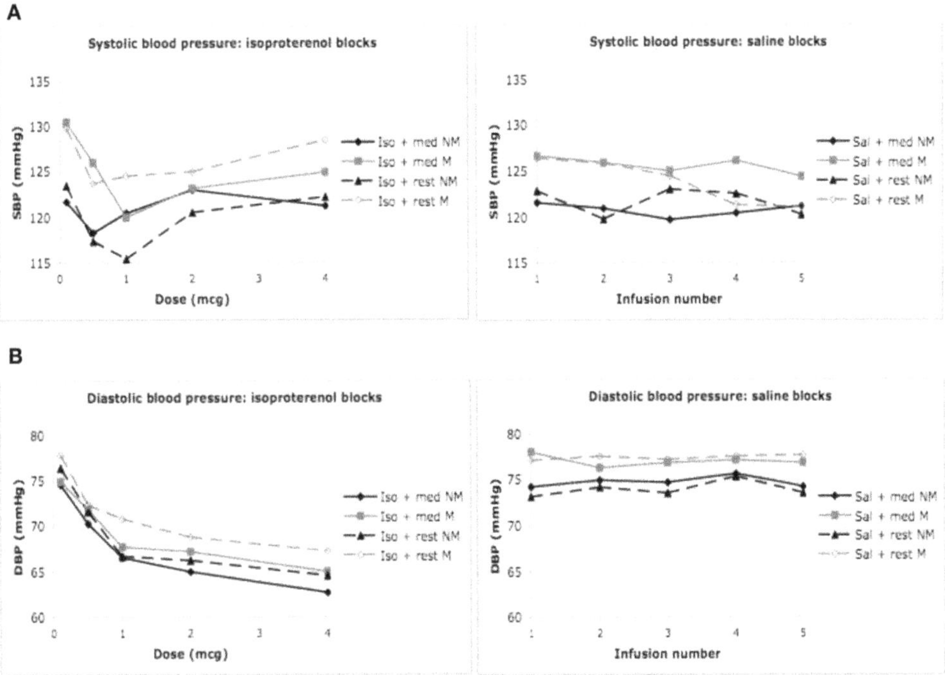

FIGURE 4 | Blood pressure for both groups during meditation/relaxation and rest. (A) Systolic blood pressure measured via automated blood pressure monitor. **(B)** Diastolic blood pressure measured via automated blood pressure monitor. For purposes of clarity, mean values are displayed without error bars, and the relaxation condition for the non-meditators is labeled as meditation.

aware of these symptoms given the fact that she was in a meditative state described as so deep as to require rousing with an alarm clock. Since heart rate was the only autonomic measure reported in that study, it is difficult to determine whether such a reflex occurred. Evaluating this particular meditator's (or others the future) cardiovascular responses to upright tilt table testing (perhaps even incorporating isoproterenol) could be useful to exclude this as a possibility. Another possibility is that respiratory modulation could have played a role in the observed heart rate changes, as respiratory induced changes in heart rate variability have been observed in meditators (Peng et al., 2004). We did not examine this possibility as it was outside the scope of inquiry in our study.

There are several limitations associated with the present study. First, the meditation practice was investigated under highly artificial circumstances. Meditators were meditating in a novel environment, with an intravenous line placed, were attached to various recording devices and were seated next to two individuals who were observing them throughout the study. Despite efforts to facilitate the practice, such as a curtain between the investigators and participant, and a dimly lit, quiet room, it cannot be claimed that the meditation and relaxation conditions took place in the usual environment. Second, the current study does not take into account the potential influence of testing anxiety, that is, concern on the part of the meditators about "performing well." Although meditators did not voice any such concerns, and did not differ from the non-meditators in questionnaire measures of anxiety, this possibility cannot be ruled out. A third limitation is the fact that the meditation plus saline conditions did not appear to lower the heart rate substantially. This could again be due to the artificiality of the environment, thus preventing meditators from authentically engaging in their practice and displaying increased parasympathetic indices. However, retrospective ratings of the quality of the meditation practice indicated that meditators found their meditation practice during the infusions to be predominantly similar to their daily practice, reducing this as a possibility. The lack of heart rate reductions during saline could also be related to the type of meditation participants practiced. For example, some meditations are suggested to increase cardiovascular states whereas others are suggested to decrease them (Amihai and Kozhevnikov, 2014). Since most meditators were not instructed to use their meditation practice to reduce their heart rate (this would have been seen by the Vipassana practitioners as constraining and interfering with the meditation practice), it is possible that the mediators were performing the former type of meditation. Reports from the meditators about the nature and quality of their practice did not seem to indicate this as a possibility.

It is also possible that the lack of attenuated autonomic responses in the current study could be related to a non-ideal sample of meditators and/or a non-ideal meditation practice. For example, since the majority of the meditators came from a Vipassana background, their default meditation practice may not be intended to have effects on autonomic state. This argument is somewhat limited by the fact that

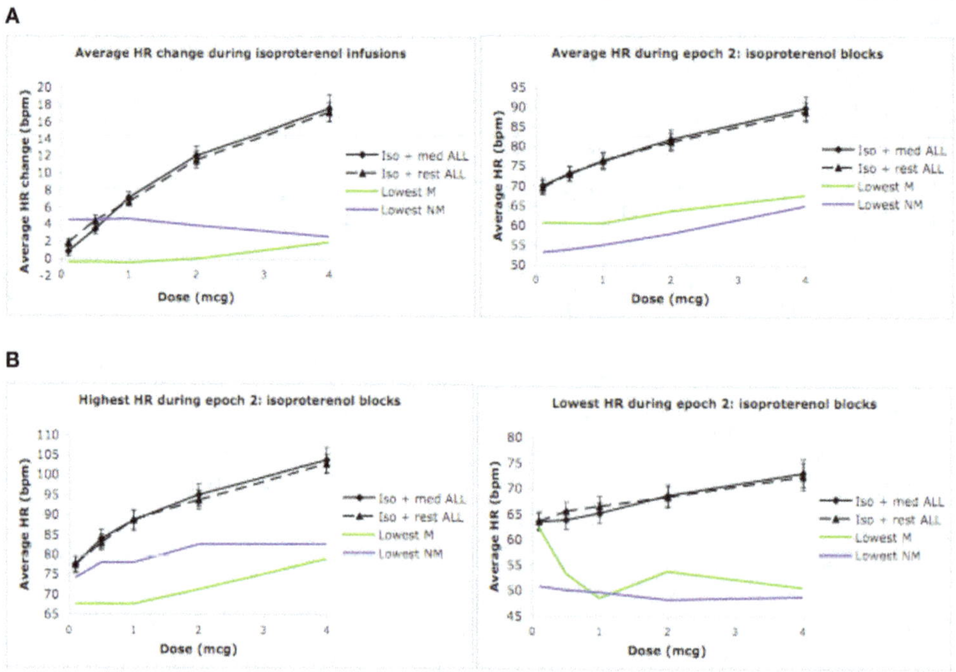

FIGURE 5 | Global and individual outlier heart rate changes during isoproterenol conditions. (A) Global mean heart rate change (epoch 2 minus epoch 1), and global mean heart rate observed during epoch 2 for all participants. **(B)** Global maximum heart rate change and global minimum heart rate change observed during epoch 2 for all participants. The green lines indicate the meditator who displayed the lowest response during the isoproterenol plus meditation condition. The blue lines indicate the non-meditators who displayed the lowest response during the isoproterenol plus relaxation condition. For purposes of clarity, the relaxation condition for the non-meditators is labeled as meditation. Error bars represent standard error of the mean.

Vipassana meditation techniques emphasize the cultivation of quiescent and tranquil states of mind, and show some evidence of altered sympatho-vagal balance (Krygier et al., 2013). Alternatively, it is theoretically possible that the default body scanning meditation practice practiced by the Vipassana meditators in this study could have been similar to a form of "open presence" rather than "focused attention" meditation. Since the former type of meditation has been associated with heightened arousal instead of relaxation, this might be expected to induce an increased autonomic response during meditation instead of a quiescent one. We observed a few marginal interactions between condition and group on heart rate with the meditators potentially exhibiting increased heart rates during meditation, that might potentially support this possibility. However, these non-significant effects were not consistently observed across the numerous heart rate measures employed. Further complicating this theoretical interpretation is evidence suggesting that open presence meditation can be associated with attenuated autonomic responses to startle challenges that probe sympathetic reactivity (Levenson et al., 2012). We conclude there is no direct evidence in the current study to support the notion of differentially decreased or increased autonomic indices in the meditators.

Another question about the current sample is whether the absence of effect could have been due to an inexperienced sample of meditators relative to the meditator in Dimsdale and Mills (2002), as the minimum requirement was only 2 years of meditative experience. Direct comparison between the studies is not possible, as the years of practice were not provided in that report. However, the notion that duration of practice is synonymous with meditative expertise is a matter of ongoing debate. Previous investigations of long term meditators have not always yielded evidence of increased ability, even for aspects of internal sensory experience that are routinely cultivated in Vipassana and Kundalini traditions (Khalsa et al., 2008; Daubenmier et al., 2013). A larger issue is the fact that the meditator described in that report was not formally trained in any particular tradition, making it impossible to select an appropriately representative group of meditators. In the current study we recruited formally trained meditators from multiple traditions, used extensive measurements of heart rate response, and examined individual outlier responses in both groups to determine whether there was specific evidence that meditation was associated with a lowered heart rate. Though some individuals displayed attenuated heart rate responses to isoproterenol, we found no evidence differentiating meditators from non-meditators.

A general point is that while these limitations warrant consideration, many of them would be similarly imposed by any empirical study of meditation and thus cannot be easily obviated. If further investigations of the current topic were continued, one helpful strategy would be to screen large samples of meditators to identify those individuals who can reliably demonstrate enhanced autonomic regulatory capabilities, and then perform detailed investigation into the cognitive and neurophysiological mechanisms underlying such abilities.

We feel it is also important to note that the current findings do not necessarily negate previously reported findings of decreased autonomic tone following the practice of meditation. Although our study appeared sufficiently powered to detect effects related to the isoproterenol vs. saline conditions, and the impact of different doses, there was lower power at the group level. While replicating this approach with a larger sample would help to address this limitation, the current results suggest that any effects, even if found reliable at the population level, would be small and would have limited consequence at the individual level. The current findings also have limited bearing on the effects of meditation on emotional experience or emotion regulation, effects consistently perceived at the individual level with training, as these constructs were not investigated. High levels of adrenergic hormones are often associated with acute anxiety, stress, anger, fear and hostility, and different forms of meditation are routinely practiced by some individuals to help ameliorate such experiences (Shirey, 2007; Robins et al., 2012; Serpa et al., 2014). Determining the biological mechanisms supporting those beneficial effects remains an important and active area of inquiry, regardless of whether they are derived from changes in autonomic reactivity or from changes in the central nervous system. Overall, these results simply suggest that the formal act of meditation may not be sufficient to physiologically counteract certain forms of elevated levels of peripheral cardiovascular adrenergic arousal or engender quicker recovery from such arousal.

Support

This research was supported by a grant from the National Center for Complementary and Alternative Medicine (NCCAM F31 AT003061-01A1), the National Center for Research Resources, General Clinical Research Centers Program (NIH M01-RR-59), a McDonnell Foundation Collaborative Award to DT (#220020387), and the William K. Warren Foundation.

Acknowledgments

We would like to thank Shinzen Young for assistance with participant recruitment, Erik St. Louis, Brian Olshansky, and Chirag Sandesara for assistance with participant screening, Sonia Schubert for nursing assistance with administration of the isoproterenol protocol, Becky Triplett and the IV ads pharmacy staff for isoproterenol preparation, and James Martins for safety monitoring.

References

Aftanas, L., and Golosheykin, S. (2005). Impact of regular meditation practice on EEG activity at rest and during evoked negative emotions. *Int. J. Neurosci.* 115, 893–909. doi: 10.1080/00207450590897969

Aftanas, L. I., and Golocheikine, S. A. (2002). Non-linear dynamic complexity of the human EEG during meditation. *Neurosci. Lett.* 330, 143–146. doi: 10.1016/S0304-3940(02)00745-0

Ahir, D. C. (1999). *Vipassana: A Universal Buddhist Meditation Technique*. Delhi: Sri Satguru Publications.

Amihai, I., and Kozhevnikov, M. (2014). Arousal vs. relaxation: a comparison of the neurophysiological and cognitive correlates of vajrayana and theravada meditative practices. *PLoS ONE* 9:e102990. doi: 10.1371/journal.pone.0102990

Astin, J. A., Shapiro, S. L., Eisenberg, D. M., and Forys, K. L. (2003). Mind-body medicine: state of the science, implications for practice. *J. Am. Board Fam. Pract.* 16, 131–147. doi: 10.3122/jabfm.16.2.131

Barnes, V. A., Davis, H. C., Murzynowski, J. B., and Treiber, F. A. (2004). Impact of meditation on resting and ambulatory blood pressure and heart rate in youth. *Psychosom. Med.* 66, 909–914. doi: 10.1097/01.psy.0000145902.91749.35

Barnes, V. A., Treiber, F. A., Turner, J. R., Davis, H., and Strong, W. B. (1999). Acute effects of transcendental meditation on hemodynamic functioning in middle-aged adults. *Psychosom. Med.* 61, 525–531. doi: 10.1097/00006842-199907000-00017

Beary, J. F., and Benson, H. (1974). A simple psychophysiologic technique which elicits the hypometabolic changes of the relaxation response. *Psychosom. Med.* 36, 115–120. doi: 10.1097/00006842-197403000-00003

Benson, H., Beary, J. F., and Carol, M. P. (1974). The relaxation response. *Psychiatry* 37, 37–46.

Bonadonna, R. (2003). Meditation's impact on chronic illness. *Holist. Nurs. Pract.* 17, 309–319. doi: 10.1097/00004650-200311000-00006

Brefczynski-Lewis, J. A., Lutz, A., Schaefer, H. S., Levinson, D. B., and Davidson, R. J. (2007). Neural correlates of attentional expertise in long-term meditation practitioners. *Proc. Natl. Acad. Sci. U.S.A.* 104, 11483–11488. doi: 10.1073/pnas.0606552104

Burley, M. (2000). *Hatha-yoga. Its Context, Theory and Practice*. New Delhi: Shri Jainendra Press.

Daubenmier, J., Sze, J., Kerr, C. E., Kemeny, M. E., and Mehling, W. (2013). Follow your breath: respiratory interoceptive accuracy in experienced meditators. *Psychophysiology* 50, 777–789. doi: 10.1111/psyp.12057

Davidson, R. J., Goleman, D. J., and Schwartz, G. E. (1976). Attentional and affective concomitants of meditation: a cross-sectional study. *J. Abnorm. Psychol.* 85, 235–238. doi: 10.1037/0021-843X.85.2.235

Dimsdale, J. E., and Mills, P. J. (2002). An unanticipated effect of meditation on cardiovascular pharmacology and physiology. *Am. J. Cardiol.* 90, 908–909. doi: 10.1016/S0002-9149(02)02726-1

Dimsdale, J., Ziegler, M., and Graham, R. (1988). The effect of hypertension, sodium, and race on isoproterenol sensitivity. *Clin. Exp. Hypertens. A* 10, 747–756.

Ekman, P., Davidson, R. J., Ricard, M., and Wallace, A. B. (2005). Buddhist and psychological perspectives on emotions and well-being. *Curr. Dir. Psychol. Sci.* 14, 59–63. doi: 10.1111/j.0963-7214.2005.00335.x

Farrow, J. T., and Hebert, J. R. (1982). Breath suspension during the transcendental meditation technique. *Psychosom. Med.* 44, 133–153. doi: 10.1097/00006842-198205000-00001

Fenz, W. D., and Plapp, J. M. (1970). Voluntary control of heart rate in a practitioner of yoga: negative findings. *Percept. Mot. Skills* 30, 493–494. doi: 10.2466/pms.1970.30.2.493

Harinath, K., Malhotra, A. S., Pal, K., Prasad, R., Kumar, R., Kain, T. C., et al. (2004). Effects of Hatha yoga and Omkar meditation on cardiorespiratory performance, psychologic profile, and melatonin secretion. *J. Altern. Complement. Med.* 10, 261–268. doi: 10.1089/107555304323062257

Hoffman, J. W., Benson, H., Arns, P. A., Stainbrook, G. L., Landsberg, G. L., Young, J. B., et al. (1982). Reduced sympathetic nervous system responsivity associated with the relaxation response. *Science* 215, 190–192. doi: 10.1126/science.7031901

Hughes, J. W., Fresco, D. M., Myerscough, R., van Dulmen, M. H., Carlson, L. E., and Josephson, R. (2013). Randomized controlled trial of mindfulness-based stress reduction for prehypertension. *Psychosom. Med.* 75, 721–728. doi: 10.1097/PSY.0b013e3182a3e4e5

Jevning, R., Wallace, R. K., and Beidebach, M. (1992). The physiology of meditation: a review. A wakeful hypometabolic integrated response. *Neurosci. Biobehav. Rev.* 16, 415–424. doi: 10.1016/S0149-7634(05)80210-6

Khalsa, S. S., Rudrauf, D., Damasio, A. R., Davidson, R. J., Lutz, A., and Tranel, D. (2008). Interoceptive awareness in experienced meditators. *Psychophysiology* 45, 671–677. doi: 10.1111/j.1469-8986.2008.00666.x

Kikushima, S., Kobayashi, Y., Nakagawa, H., and Katagiri, T. (1999). Triggering mechanism for neurally mediated syncope induced by head-up tilt test: role of catecholamines and response to propranolol. *J. Am. Coll. Cardiol.* 33, 350–357. doi: 10.1016/S0735-1097(98)00567-1

Kothari, L. K., Bardia, A., and Gupta, O. P. (1973). The yogic claim of voluntary control over the heart beat: an unusual demonstration. *Am. Heart J.* 86, 282–284. doi: 10.1016/0002-8703(73)90260-3

Krygier, J. R. J., Heathers, A. J., Shahrestani, S., Abbott, M., Gross, J. J., and Kemp, A. H. (2013). Mindfulness meditation, well-being, and heart rate variability: a preliminary investigation into the impact of intensive Vipassana meditation. *Int. J. Psychophysiol.* 89, 305–313. doi: 10.1016/j.ijpsycho.2013.06.017

Levenson, R. W., Ekman, P., and Ricard, M. (2012). Meditation and the startle response: a case study. *Emotion* 12, 650–658. doi: 10.1037/a0027472

Mills, P. J., Dimsdale, J. E., Ancoli-Israel, S., Clausen, J., and Loredo, J. S. (1998). The effects of hypoxia and sleep apnea on isoproterenol sensitivity. *Sleep* 21, 731–735.

Nyanaponika, T. (1969). *The Heart of Buddhist Meditation. A Handbook of Mental Training Based on the Buddha's Way of Mindfulness.* New York, NY: The Citadel Press.

Orme-Johnson, D. W. (1973). Autonomic stability and Transcendental Meditation. *Psychosom. Med.* 35, 341–349. doi: 10.1097/00006842-197307000-00008

Peng, C. K., Henry, I. C., Mietus, J. E., Hausdorff, J. M., Khalsa, G., Benson, H., et al. (2004). Heart rate dynamics during three forms of meditation. *Int. J. Cardiol.* 95, 19–27. doi: 10.1016/j.ijcard.2003.02.006

Robins, C. J., Keng, S. L., Ekblad, A. G., and Brantley, J. G. (2012). Effects of mindfulness-based stress reduction on emotional experience and expression: a randomized controlled trial. *J. Clin. Psychol.* 68, 117–131. doi: 10.1002/jclp.20857

Serpa, J. G., Taylor, S. L., and Tillisch, K. (2014). Mindfulness-based stress reduction (MBSR) reduces anxiety, depression, and suicidal ideation in veterans. *Med. Care* 52 (Suppl. 5), S19–S24. doi: 10.1097/MLR.0000000000000202

Shirey, M. R. (2007). An evidence-based solution for minimizing stress and anger in nursing students. *J. Nurs. Educ.* 46, 568–571.

Solberg, E. E., Ekeberg, O., Holen, A., Ingjer, F., Sandvik, L., Standal, P. A., et al. (2004). Hemodynamic changes during long meditation. *Appl. Psychophysiol. Biofeedback* 29, 213–221. doi: 10.1023/B:APBI.0000039059.20738.39

Taimni, I. K. (1961). *The Science of Yoga: The Yoga-sutras of Patanjali in Sanskit with Transliteration in Roman, Translation and Commentary in English.* Wheaton, IL: Theosophical Publication House.

Telles, S., Joshi, M., Dash, M., Raghuraj, P., Naveen, K. V., and Nagendra, H. R. (2004). An evaluation of the ability to voluntarily reduce the heart rate after a month of yoga practice. *Integr. Physiol. Behav. Sci.* 39, 119–125. doi: 10.1007/BF02734277

Telles, S., Nagarathna, R., and Nagendra, H. R. (1995). Autonomic changes during "OM" meditation. *Indian J. Physiol. Pharmacol.* 39, 418–420.

Travis, F., and Wallace, R. K. (1997). Autonomic patterns during respiratory suspensions: possible markers of Transcendental Consciousness. *Psychophysiology* 34, 39–46. doi: 10.1111/j.1469-8986.1997.tb02414.x

Wallace, R. K., Benson, H., and Wilson, A. F. (1971). A wakeful hypometabolic physiologic state. *Am. J. Physiol.* 221, 795–799.

Wenger, M. A., Bagchi, B. K., and Anand, B. K. (1961). Experiments in India on "voluntary" control of the heart and pulse. *Circulation* 24, 1319–1325. doi: 10.1161/01.CIR.24.6.1319

6

Respiratory Sensory Gating Measured by Respiratory-Related Evoked Potentials in Generalized Anxiety Disorder

Pei-Ying S. Chan [1,2], Chia-Hsiung Cheng [1,2], Shih-Chieh Hsu [3,4], Chia-Yih Liu [3,5], Paul W. Davenport [6] and Andreas von Leupoldt [7,8]*

[1] *Department of Occupational Therapy, College of Medicine, Chang Gung University, Taoyuan, Taiwan,* [2] *Healthy Ageing Research Center, Chang Gung University, Taoyuan, Taiwan,* [3] *Department of Psychiatry, Chang Gung Memorial Hospital, Taoyuan, Taiwan,* [4] *Department of Medicine, Chang Gung University, Taoyuan, Taiwan* [5] *Department of Traditional Chinese Medicine, Chang Gung University, Taoyuan, Taiwan,* [6] *Department of Physiological Sciences, College of Veterinary Medicine, University of Florida, Gainesville, FL, USA,* [7] *Research Group on Health Psychology, University of Leuven, Leuven, Belgium,* [8] *Department of Systems Neuroscience, University Medical Center Hamburg-Eppendorf, Hamburg, Germany*

Correspondence:
Pei-Ying S. Chan,
Department of Occupational Therapy, College of Medicine, Chang Gung University, No. 259, Wenhua 1st Road, Taoyuan 333, Taiwan
chanp@mail.cgu.edu.tw

The perception of respiratory sensations plays an important role both in respiratory diseases and in anxiety disorders. However, little is known about the neural processes underlying respiratory sensory perception, especially in patient groups. Therefore, the present study examined whether patients with generalized anxiety disorder (GAD) would demonstrate altered respiratory sensory gating compared to a healthy control group. Respiratory-related evoked potentials (RREP) were measured in a paired inspiratory occlusion paradigm presenting two brief occlusion stimuli (S1 and S2) within one inspiration. The results showed a significantly greater S2/S1 ratio for the N1 component of the RREP in the GAD group compared to the control group. Our findings suggest altered respiratory sensory processing in patients with GAD, which might contribute to altered perception of respiratory sensations in these patients.

Keywords: respiratory sensation, respiratory sensory gating, respiratory-related evoked potentials, generalized anxiety disorder

Introduction

The experience of respiratory sensations plays not only an important role in respiratory diseases such as asthma or chronic obstructive pulmonary disease (COPD), but also in anxiety disorders. For example, respiratory sensations such as breathlessness and chest tightness as well as ventilatory changes are diagnostic for anxiety disorders suggesting a close relationship between respiratory sensory processing and anxiety symptoms (American Psychiatric Association, 2013). Previous studies have demonstrated that negative affective states and traits including anxiety are related to over-perception of respiratory sensations irrespective of underlying ventilatory changes (Carr et al., 1994; Zaubler and Katon, 1996; Bogaerts et al., 2005; von Leupoldt and Dahme, 2007; von Leupoldt et al., 2013). However, few of the aforementioned studies used objective measures for respiratory perception, and the neural mechanisms that underlie the interrelationships between anxiety and increased perception of respiratory sensations are poorly understood (von Leupoldt et al., 2013).

The respiratory-related evoked potential (RREP) is a useful non-invasive electrophysiological method for studying the effects of anxiety on respiratory perception and its neural processing (von Leupoldt et al., 2013). With a single inspiratory occlusion paradigm (which is similar to the oddball paradigm used to test auditory sensory gating), a past report in healthy participants found that high anxious individuals on average presented increased RREP peak amplitudes for the long-latency peaks during negative emotional compared to neutral stimulation, which is the opposite pattern found in low anxious individuals (von Leupoldt et al., 2010, 2011). In addition, respiratory sensory gating, similar to other types of neural gating with exteroceptive stimuli such as sound and touch (Adler et al., 1998; Arnfred et al., 2001), was used to investigate central neural mechanisms of filtering repetitive respiratory stimuli within a short time window (Chan and Davenport, 2010a). A paired occlusion stimulation paradigm can be used to examine respiratory sensory gating by eliciting paired RREP waveforms, where the second stimulus (S2) results in a smaller N1 peak amplitude compared to the first stimulus (S1) in healthy individuals (Chan and Davenport, 2008). The N1 peak, primarily originating from the bilateral sensorimotor cortices, is thought to reflect both first and second order processing of respiratory sensory information in the cortex (Chan and Davenport, 2010a; von Leupoldt et al., 2013). A smaller RREP N1 peak S2/S1 ratio is indicative of an enhanced respiratory sensory gating function (i.e., filtering out more redundant sensory information).

There has been a robust amount of literature testing sensory neural gating in psychiatric disorders with exteroceptive, but not interoceptive stimuli (Ludewig et al., 2002; Holstein et al., 2010; Hunter et al., 2011; Markham and Koenig, 2011). These studies found that individuals with diseases including panic disorder and obsessive-compulsive disorder (OCD) had deficits in prepulse inhibition and sensorimotor gating (Ludewig et al., 2002; Kohl et al., 2013). It was reasoned that compromised sensory gating functions are associated with sensory overload or sensory "flooding" into the higher cortex, which may result in over-perception (Adler et al., 1998). Interoceptive respiratory stimuli have rarely been used to examine sensory gating of anxious individuals in previous studies. A recent report using RREPs has found that state anxiety modulated respiratory sensory gating, as demonstrated by a larger N1 peak S2/S1 ratio in higher compared to lower non-clinically anxious individuals (Chan et al., 2012). However, it remains unknown how clinical levels of anxiety, such as present in general anxiety disorder, are related to a similarly altered neural processing of respiratory sensations.

Therefore, the purpose of this study was to examine respiratory sensory gating elicited by a paired inspiratory occlusion paradigm in patients with generalized anxiety disorder (GAD). It was hypothesized that compared to a healthy control group, individuals with GAD would show a reduced respiratory sensory gating function as demonstrated by an elevated N1 peak S2/S1 ratio.

Materials and Methods

Participants

This study was approved by the Chang Gung Medical Foundation Institutional Review Board. A group of 20 patients with the diagnosis of GAD were recruited from the psychiatric outpatient clinic in a medical center located in northern Taiwan. All patients were interviewed by a psychiatrist with the mini-international neuropsychiatric interview (MINI), which is a structured diagnostic interview for DSM-IV diagnoses (Sheehan et al., 1998). Co-morbid diagnoses of substance abuse or psychosis served as exclusion criteria. A group of age-matched healthy controls (HCs) was recruited through public advertisements. All participants reportedly had no history of respiratory, cardiovascular, or neurological disease. All participants were required not to take any prescribed medication for at least 12 h before the experiment.

Experimental Procedure

After signing the informed consent, participants completed a standard pulmonary function test (PFT) with a spirometer (Cardinal Health Inc., Dublin, OH, USA). The PFT was conducted based on the guidelines of the American Thoracic Society and European Respiratory Society (Miller et al., 2005). In addition, participants were administered the Chinese-version questionnaires of the Beck Anxiety Inventory (BAI; Beck and Steer, 1990) and the Beck Depression Inventory-II (BDI-II; Beck et al., 1996).

During the experiment, participants were instructed to sit comfortably in an armed chair while breathing through a mouthpiece with their nose occluded. The mouthpiece was connected to a two-way non-rebreathing valve (Hans Rudolph Inc., Kansas City, MO, USA). The inspiratory port of the non-rebreathing valve was connected to a customized occlusion valve (Hans Rudolph Inc., Kansas City, MO, USA). A solenoid of a trigger box was connected to the occlusion valve and a pressure tank through reinforced tubing (Chan and Davenport, 2010a). The occlusion valve closure was manually controlled by the experimenter in the adjacent room. The participants' mouth pressure was monitored and recorded at the center of the non-rebreathing valve through a differential pressure transducer which connected to the pneumotachograph amplifier (1110 series, Hans Rudolph Inc., Kansas City, MO, USA) and a PowerLab signal recording unit (ADInstruments Inc., Bella Vista, NSW, Australia). For details of the respiratory apparatus setup, please refer to the review of Chan and Davenport (2010a).

The RREP method was previously described in a few comparable studies (Chan and Davenport, 2010b; Chan et al., 2014). In the current study, the electroencephalography (EEG) was sampled from cortical sites C3 and C4 at 1 kHz with a 40-channel EEG system (NuAmps, Compumedics Neuroscan Inc., Charlotte, NC, USA), bandpass filtered from DC to 50 Hz and referenced to the bilateral mastoids. Individual electrode impedance was set below 5 kΩ. The participants were provided with paired inspiratory occlusions of 150 ms each with a 500 ms inter-stimulus-interval. The paired stimuli were provided at the onset of inspiration randomly every 2–4 breaths. The onset of occlusion was identified as the start of mouth pressure change (Labchart V7, ADInstruments Inc., Bella Vista, NSW, Australia). At least 100 paired occlusions were collected for data analysis in every participant. The trigger box was set up to send parallel markers to the Neuroscan 4.5 recording software (Compumedics Neuroscan Inc., Charlotte, NC, USA). The participants were instructed to keep breathing normally, rather than to stop breathing, during inspiratory occlusions, and to count the number of breaths they felt obstructed during the experiment.

Data Analyses

For offline analysis, the EEG epoch was defined and extracted from 200 ms before to 1000 ms after the respiratory occlusion, i.e., for S1 and S2 separately. The first 200 ms served as the baseline. The signals were corrected for ocular movement using a built-in algorithm of the analysis software (BrainVision Analyzer 2, Brain Products GmbH., Gilching, Germany) and further filtered between 1 and 50 Hz (12 dB/octave roll-off). The artifacts were defined when greater than 100 and 60 μV, baseline to peak, for the four eye electrodes and all other electrodes, respectively. After the artifacts were extracted from the signals, the corresponding epochs were then averaged for S1 and S2 separately. The RREP N1 peak amplitudes for S1 and S2 were identified in a time window of 85–135 ms after occlusion onset at cortical sites C3 and C4 and the S2/S1 ratios were calculated.

Separate one-way analyses of variance (ANOVA) were performed to test for group differences in pulmonary function, non-respiratory parameters, and S2/S1 ratios. Based on our *a priori* hypothesis, the analyses for the N1 peaks were conducted with a one-tailed test. The critical p value was set at <0.05.

Results

Twenty patients with the diagnosis of GAD and 20 age-matched HC were recruited for this study. Data of two patients and two controls were excluded due to incomplete EEG data, which left the study with 18 patients (10 females and eight males) and 18 controls (nine females and nine males) for final analyses. The demographic data and the pulmonary function results of the two groups of participants are shown in **Table 1**. There was no statistical difference regarding age and pulmonary function between the two groups. The GAD group showed significantly higher scores than the HC group for the BDI-II [$F(1,35) = 10.31$, $p < 0.01$] and BAI [$F(1,35) = 21.12$, $p < 0.001$]. These scores (in the mid-range of the scales) indicated significantly higher levels of depression as well as anxiety in the GAD patients, respectively.

Figure 1 shows group averaged S1 and S2 RREP waveforms of the HC group (a) and the GAD group (b). The one-way ANOVA results revealed that the GAD group showed a significantly greater N1 S2/S1 ratio compared to the HCs at both electrode sites [**Figure 2**; C3: 1.06 ± 0.65 and 0.58 ± 0.27, $F(1,25) = 6.172$, $p < 0.05$; C4: 1.01 ± 0.38 and 0.67 ± 0.3, $F(1,32) = 8.28$, $p < 0.01$]. Further analyses on the S1 and S2 amplitudes for N1 showed that the HC group had significantly greater S1 amplitudes compared to the GAD group at the C3 electrode [$T(1,27) = 2.505$, $p < 0.05$]; however, the independent t-test did not show a significant difference for the S1 amplitude at C4 [$T(1,33) = 0.906$, $p > 0.05$]. Similarly, the analyses on the S2 amplitudes at electrodes C3 and C4 did not show significant differences between the two groups [$T(1,25) = -0.32$ and $T(1,32) = -0.686$, respectively, $p > 0.05$].

TABLE 1 | Baseline characteristics of study groups (group mean ± SD).

Variables	GAD patients	Healthy controls
N	18	18
Age (years)	47 ± 8.9	42 ± 9.3
Gender (female/male)	10/8	9/9
FEV1 (L)	2.7 ± 0.9	2.9 ± 0.6
FEV1 of predicted value (%)	86 ± 9.3	81.6 ± 10.7
BDI-II score	15.9 ± 11.1	6.3 ± 6.1*
BAI score	13 ± 7.6	4.2 ± 3*

*FEV1, forced expiratory volume in 1 sec; BDI-II, Beck Depression Inventory-II; BAI, Beck Anxiety Inventory. *Indicates a significant difference between the GAD group and the healthy control group ($p < 0.05$).*

Discussion

This study demonstrates that the paired inspiratory occlusion paradigm is a feasible measure for investigating respiratory sensory gating function in GAD patients. The results demonstrate that individuals with GAD, compared to HCs, show a higher RREP N1 S2/S1 ratio, which is suggestive of reduced respiratory sensory gating. This might be related to altered perception of respiratory sensations reported in the patient population.

The present findings are consistent with previous studies in healthy non-clinical samples, in which elevated levels of state anxiety and induced negative emotion were found to reduce respiratory sensory gating as evidenced by RREP methodology (Chan and Davenport, 2010b; Chan et al., 2012; Chenivesse et al., 2014). For example, using a similar paradigm of paired inspiratory occlusions in healthy individuals, Chenivesse et al. (2014) found reduced respiratory sensory gating during unpleasant compared to neutral emotional stimulation using affective picture series. Chan and Davenport (2010b) observed greater S2/S1 ratios for the RREP N1 peak during nicotine-withdrawal induced anxiety in college-aged smokers. In addition, Chan et al. (2012) tested the effect of state anxiety on the RREP in healthy college-aged students and found that higher anxious individuals demonstrated increased N1 S2/S1 ratios compared to lower anxious individuals. The present study suggests that reduced respiratory sensory gating is not only related to mild forms of state or trait anxiety, but might also be evident in more chronic, clinically relevant forms of anxiety such as in GAD. This increased afferent throughput of respiratory sensory signals might lead to an increased awareness of respiratory sensations and contribute to the altered perception of respiratory sensations in patients with anxiety disorders.

In the present study, the group difference in the N1 S2/S1 ratio was related to the fact that the HC group demonstrated higher S1 amplitudes compared to the GAD group for the N1 peak, at least for one out of two analyzed electrode sites. However, in a previous study in healthy participants, it was found that individuals with higher levels of anxiety had higher amplitudes of S2, while no group differences were observed for S1 (Chan et al., 2012). It can be speculated that the discrepancy between the present results and the previous report may be related to the counting task used in the present study, where participants were required to count the number of obstructed breaths during the experiment. Subsequently, the S1 N1 peak amplitude could have been modulated by focused attention generated by counting the stimulus, which needs to be explored further in future systematic studies. However, the observed general effect of the present study demonstrates a significantly smaller amplitude difference between

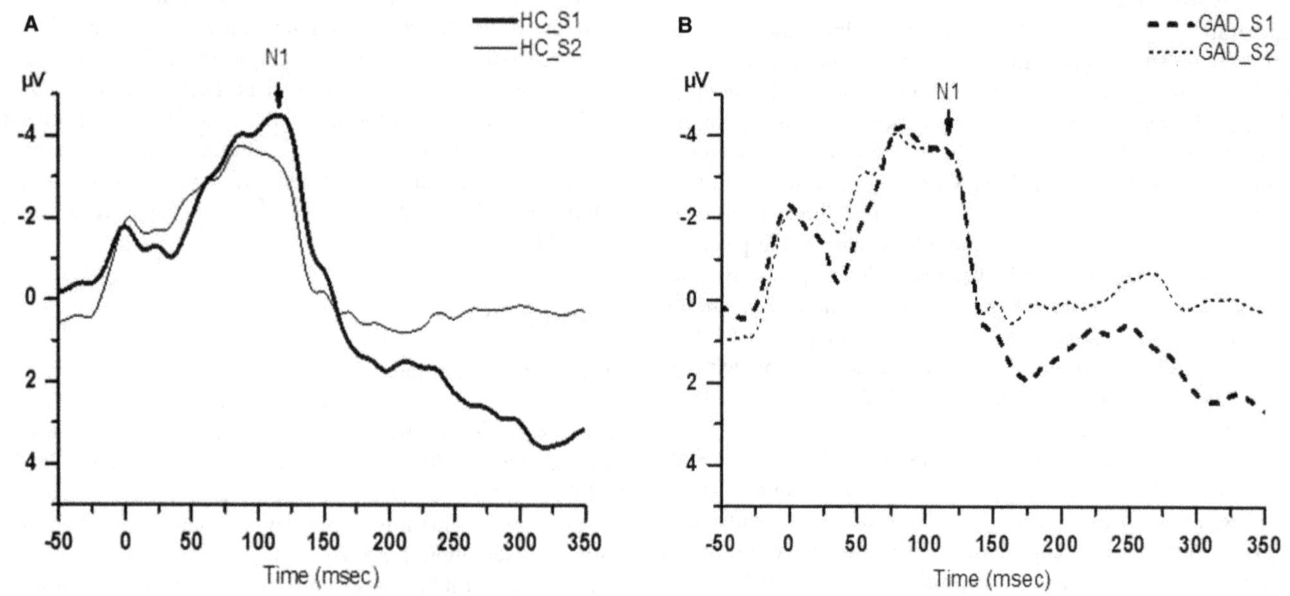

FIGURE 1 | Group averaged waveform from the C4 electrode. (A) The black solid lines represents the averaged S1 and S2 waveforms of the healthy control (HC) group ($N = 18$); **(B)** the black dashed lines represents the averaged S1 and S2 waveforms of the GAD group ($N = 18$).

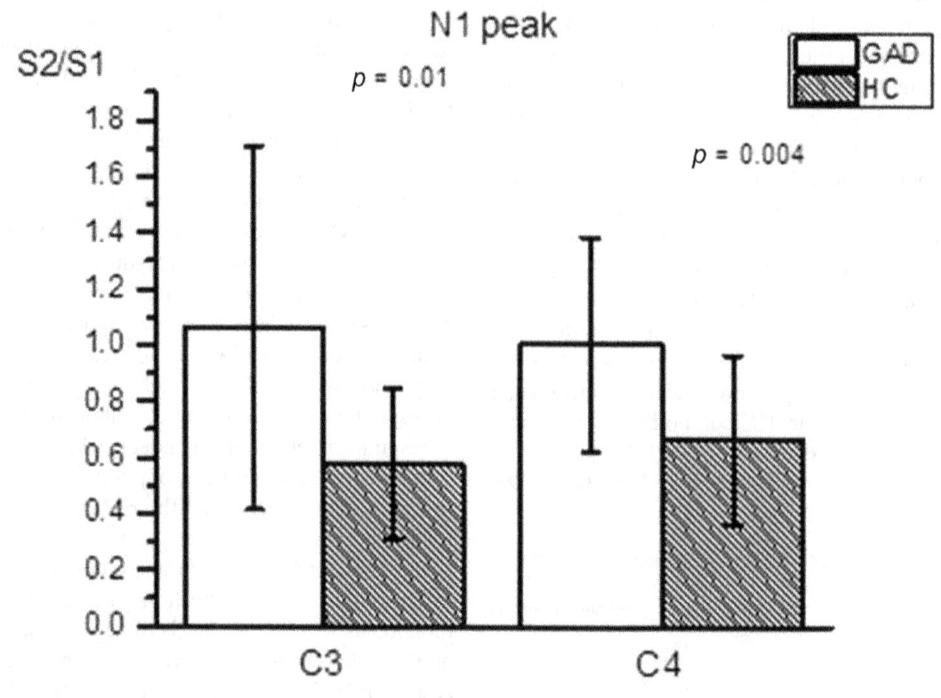

FIGURE 2 | Averaged RREP N1 peak S2/S1 ratio (mean ± SD) at the C3 and C4 electrodes. The shaded bars represent the averaged S2/S1 ratios of the HC group. The GAD group showed a significantly greater N1 S2/S1 ratio compared to the healthy controls at the C3 and C4 electrodes ($p = 0.01$ and 0.004, respectively).

S1 and S2 in GAD patients, which points toward higher sensory throughput. Thus, the present results converge with the findings of Chenivesse et al. (2014), who similarly found a reduction in respiratory sensory gating to be more closely associated with an amplitude modulation of S1 N1 and not S2 N1. The seemingly unaffected S2 amplitude in the present GAD group might be a function of the attentional processes for S1 acting on the neural throughput. However, we cannot fully exclude the alternative explanation that the observed increase in the RREP N1 S2/S1 ratio in the GAD group is unrelated to the concept of central neural gating and merely represent an altered neural response to the first of two paired stimuli in this population. This would be consistent with the notion of altered "gate-in ability" mentioned by Hu et al. (2012) where they found increased auditory S1 amplitude in healthy compared to patients with schizophrenia (Hu et al., 2012). Nevertheless, the extent to which a clinical anxiety disorder impacts respiratory gating at the throughput is unclear and requires further systematic investigation.

The present study is also in line with previous studies using exteroceptive stimuli, which have similarly demonstrated that several anxiety disorders are related to disrupted sensory gating in modalities other than the respiratory modality (Neylan et al., 1999; Ghisolfi et al., 2004; Hashimoto et al., 2008; Stewart and White, 2008; de Leeuw et al., 2010; Nanbu et al., 2010). For example, Hashimoto et al. (2008) tested auditory sensory gating with a paired click paradigm in patients with OCD and found a disrupted P50 S2/S1 ratio. Ghisolfi et al. (2004), using auditory stimulation, reported significantly greater S2/S1 ratios in post-traumatic stress disorder (PTSD) patients, compared to controls. Moreover, Nanbu et al. (2010) found an impaired P50 gating function in the OCD patients represented by elevated S2/S1 ratios with auditory stimuli compared to the HCs. Taken together, these past and the present findings suggest that impaired gating functions in anxiety disorders are not specific for one sensory modality, but can be observed across different interoceptive as well as exteroceptive modalities.

Some cautions need to be exercised when generalizing the results of the present study. Firstly, the study participants only self-reported to have no diagnosis of a respiratory disease or acute respiratory symptoms on the test day, such as experiencing a cold. Secondly, our study did not differentiate between patients with shorter disease duration and those with longer disease duration. Since it is sometimes observed by clinicians that patients with longer disease duration might also report rather reduced levels of respiratory sensations than those with shorter disease duration, future studies are encouraged to examine effect of disease duration on respiratory sensory gating. Finally, we cannot fully rule out residual effects of long-term anxiolytic, antidepressant, or antipsychotic medications in the GAD group. Previous studies have indeed demonstrated the potential modulating effects of medications on sensory gating in patients with anxiety disorders (Siegel et al., 2005; de Leeuw et al., 2010). Although our participants were asked to refrain from medication intake 12 h before the experiments, effects of long-term treatments remain unclear. Future studies are, therefore, encouraged to systematically examine the various potential effects of different treatments on respiratory sensory gating.

In summary, the present study suggests that patients with GAD show a larger S2/S1 ratio for the RREP N1 peak, which is suggestive of reduced respiratory sensory gating. Whether this pattern of neural processing of respiratory information varies between subgroups of GAD patients and the degree of anxiety requires more investigation. Future research is needed to clarify the effects of respiratory symptoms, medications, and duration of the disease on respiratory sensory gating in patients with GAD. Finally, GAD specific responses to the first of two paired respiratory stimuli and the respective impact on the subsequent gating ratio need further investigation.

Acknowledgments

The authors would like to thank for the financial support from Ministry of Science and Technology (103-2420-H-182-003-MY2) & Healthy Ageing Research Center at Chang Gung University (EMRPD1E1711 and CMRPD1B0332); and Ya-Jhih Ju for help during the data collection.

References

Adler, L. E., Olincy, A., Waldo, M., Harris, J. G., Griffith, J., Stevens, K., et al. (1998). Schizophrenia, sensory gating, and nicotinic receptors. *Schizophr. Bull.* 24, 189–202. doi: 10.1093/oxfordjournals.schbul.a033320

American Psychiatric Association. (2013). *Diagnostic and Statistical Manual of Mental Disorders: DSM-5*. Available at: http://dsm.psychiatryonline.org/book.aspx?bookid = 556 (accessed April 14, 2015).

Arnfred, S. M., Eder, D. N., Hemmingsen, R. P., Glenthoj, B. Y., and Chen, A. C. (2001). Gating of the vertex somatosensory and auditory evoked potential P50 and the correlation to skin conductance orienting response in healthy men. *Psychiatry Res.* 101, 221–235. doi: 10.1016/S0165-1781(01)00226-8

Beck, A. T., and Steer, R. A. (1990). *Beck Anxiety Inventory*. San Antonio, TX: Psychological Corp.

Beck, A. T., Steer, R. A., and Brown, G. K. (1996). *BDI-II, Beck Depression Inventory: Manual*. San Antonio, TX: Psychological Corp.

Bogaerts, K., Notebaert, K., Van Diest, I., Devriese, S., De Peuter, S., and Van den Bergh, O. (2005). Accuracy of respiratory symptom perception in different affective contexts. *J. Psychosom. Res.* 58, 537–543. doi: 10.1016/j.jpsychores.2004.12.005

Carr, R. E., Lehrer, P. M., Rausch, L. L., and Hochron, S. M. (1994). Anxiety sensitivity and panic attacks in an asthmatic population. *Behav. Res. Ther.* 32, 411–418. doi: 10.1016/0005-7967(94)90004-3

Chan, P. Y., and Davenport, P. W. (2008). Respiratory-related evoked potential measures of respiratory sensory gating. *J. Appl. Physiol.* 105, 1106–1113. doi: 10.1152/japplphysiol.90722.2008

Chan, P. Y., and Davenport, P. W. (2010a). Respiratory related evoked potential measures of cerebral cortical respiratory information processing. *Biol. Psychol.* 84, 4–12. doi: 10.1016/j.biopsycho.2010.02.009

Chan, P. Y., and Davenport, P. W. (2010b). The role of nicotine on respiratory sensory gating measured by respiratory-related evoked potentials. *J. Appl. Physiol.* 108, 662–669. doi: 10.1152/japplphysiol.00798.2009

Chan, P. Y., von Leupoldt, A., Bradley, M. M., Lang, P. J., and Davenport, P. W. (2012). The effect of anxiety on respiratory sensory gating measured by respiratory-related evoked potentials. *Biol. Psychol.* 91, 185–189. doi: 10.1016/j.biopsycho.2012.07.001

Chan, P. Y., von Leupoldt, A., Liu, C. Y., and Hsu, S. C. (2014). Respiratory perception measured by cortical neural activations in individuals with generalized anxiety disorder. *Respir. Physiol. Neurobiol.* 204, 36–40. doi: 10.1016/j.resp.2014.09.009

Chenivesse, C., Chan, P. Y., Tsai, H. W., Wheeler-Hegland, K., Silverman, E., von Leupoldt, A., et al. (2014). Negative emotional stimulation decreases respiratory sensory gating in healthy humans. *Respir. Physiol. Neurobiol.* 204, 50–57. doi: 10.1016/j.resp.2014.08.019

de Leeuw, A. S., Oranje, B., van Megen, H. J., Kemner, C., and Westenberg, H. G. (2010). Sensory gating and sensorimotor gating in medication-free obsessive-compulsive disorder patients. *Int. Clin. Psychopharmacol.* 25, 232–240. doi: 10.1097/YIC.0b013e328338c4f0

Ghisolfi, E. S., Margis, R., Becker, J., Zanardo, A. P., Strimitzer, I. M., and Lara, D. R. (2004). Impaired P50 sensory gating in post-traumatic stress disorder secondary to urban violence. *Int. J. Psychophysiol.* 51, 209–214. doi: 10.1016/j.ijpsycho.2003.09.002

Hashimoto, T., Shimizu, E., Koike, K., Orita, Y., Suzuki, T., Kanahara, N., et al. (2008). Deficits in auditory P50 inhibition in obsessive-compulsive disorder. *Prog. Neuropsychopharmacol. Biol. Psychiatry* 32, 288–296. doi: 10.1016/j.pnpbp.2007.08.021

Holstein, D. H., Vollenweider, F. X., Jancke, L., Schopper, C., and Csomor, P. A. (2010). P50 suppression, prepulse inhibition, and startle reactivity in the same patient cohort suffering from posttraumatic stress disorder. *J. Affect. Disord.* 126, 188–197. doi: 10.1016/j.jad.2010.02.122

Hu, L., Boutros, N. N., and Jansen, B. H. (2012). Sensory gating-out and gating-in in normal and schizophrenic participants. *Clin. EEG Neurosci.* 43, 23–31. doi: 10.1177/1550059411429524

Hunter, M., Villarreal, G., McHaffie, G. R., Jimenez, B., Smith, A. K., Calais, L. A., et al. (2011). Lateralized abnormalities in auditory M50 sensory gating and cortical thickness of the superior temporal gyrus in post-traumatic stress disorder: preliminary results. *Psychiatry Res.* 191, 138–144. doi: 10.1016/j.pscychresns.2010.09.012

Kohl, S., Heekeren, K., Klosterkotter, J., and Kuhn, J. (2013). Prepulse inhibition in psychiatric disorders—apart from schizophrenia. *J. Psychiatr. Res.* 47, 445–452. doi: 10.1016/j.jpsychires.2012.11.018

Ludewig, S., Ludewig, K., Geyer, M. A., Hell, D., and Vollenweider, F. X. (2002). Prepulse inhibition deficits in patients with panic disorder. *Depression Anxiety* 15, 55–60. doi: 10.1002/da.10026

Markham, J. A., and Koenig, J. I. (2011). Prenatal stress: role in psychotic and depressive diseases. *Psychopharmacology (Berl.)* 214, 89–106. doi: 10.1007/s00213-010-2035-0

Miller, M. R., Hankinson, J., Brusasco, V., Burgos, F., Casaburi, R., Coates, A., et al. (2005). Standardisation of spirometry. *Eur. Respir. J.* 26, 319–338. doi: 10.1183/09031936.05.00034805

Nanbu, M., Kurayama, T., Nakazawa, K., Matsuzawa, D., Komiya, Z., Haraguchi, T., et al. (2010). Impaired P50 suppression in fear extinction in obsessive-compulsive disorder. *Prog. Neuropsychopharmacol. Biol. Psychiatry* 34, 317–322. doi: 10.1016/j.pnpbp.2009.12.005

Neylan, T. C., Fletcher, D. J., Lenoci, M., McCallin, K., Weiss, D. S., Schoenfeld, F. B., et al. (1999). Sensory gating in chronic posttraumatic stress disorder: reduced auditory P50 suppression in combat veterans. *Biol. Psychiatry* 46, 1656–1664. doi: 10.1016/S0006-3223(99)00047-5

Sheehan, D. V., Lecrubier, Y., Sheehan, K. H., Amorim, P., Janavs, J., Weiller, E., et al. (1998). The Mini-International Neuropsychiatric Interview (M.I.N.I.): the development and validation of a structured diagnostic psychiatric interview for DSM-IV and ICD-10. *J. Clin. Psychiatry* 59(Suppl. 20), 22–33; quiz 34–57.

Siegel, S. J., Maxwell, C. R., Majumdar, S., Trief, D. F., Lerman, C., Gur, R. E., et al. (2005). Monoamine reuptake inhibition and nicotine receptor antagonism reduce amplitude and gating of auditory evoked potentials. *Neuroscience* 133, 729–738. doi: 10.1016/j.neuroscience.2005.03.027

Stewart, L. P., and White, P. M. (2008). Sensory filtering phenomenology in PTSD. *Depression Anxiety* 25, 38–45. doi: 10.1002/da.20255

von Leupoldt, A., Chan, P. Y., Bradley, M. M., Lang, P. J., and Davenport, P. W. (2011). The impact of anxiety on the neural processing of respiratory sensations. *Neuroimage* 55, 247–252. doi: 10.1016/j.neuroimage.2010.11.050

von Leupoldt, A., Chan, P. Y., Esser, R. W., and Davenport, P. W. (2013). Emotions and neural processing of respiratory sensations investigated with respiratory-related evoked potentials. *Psychosom. Med.* 75, 244–252. doi: 10.1097/PSY.0b013e31828251cf

von Leupoldt, A., and Dahme, B. (2007). Psychological aspects in the perception of dyspnea in obstructive pulmonary diseases. *Respir. Med.* 101, 411–422. doi: 10.1016/j.rmed.2006.06.011

von Leupoldt, A., Vovk, A., Bradley, M. M., Keil, A., Lang, P. J., and Davenport, P. W. (2010). The impact of emotion on respiratory-related evoked potentials. *Psychophysiology* 47, 579–586. doi: 10.1111/j.1469-8986.2009.00956.x

Zaubler, T. S., and Katon, W. (1996). Panic disorder and medical comorbidity: a review of the medical and psychiatric literature. *Bull. Menninger Clin.* 60(2 Suppl. A), A12–A38.

7

Neural Activity During Interoceptive Awareness and its Associations with Alexithymia—An fMRI Study in Major Depressive Disorder and Non-Psychiatric Controls

Christine Wiebking [1]* *and Georg Northoff* [2, 3, 4, 5, 6]

[1] Cluster of Excellence in Cognitive Sciences, Department of Sociology of Physical Activity and Health, University of Potsdam, Potsdam, Germany, [2] Mind, Brain Imaging and Neuroethics, Institute of Mental Health Research, University of Ottawa, Ottawa, ON, Canada, [3] Graduate Institute of Humanities in Medicine, Taipei Medical University, Taipei, Taiwan, [4] Taipei Medical University-Shuang Ho Hospital, Brain and Consciousness Research Center, New Taipei City, Taiwan, [5] Department of Psychology, National Chengchi University, Taipei, Taiwan, [6] Center for Cognition and Brain Disorders, Normal University Hangzhou, Hangzhou, China

***Correspondence:**
Christine Wiebking,
Cluster of Excellence in Cognitive Sciences, Department of Sociology of Physical Activity and Health, University of Potsdam, Am Neuen Palais 10 14469 Potsdam, Germany
wiebking@uni-potsdam.de;
christine.wiebking@gmx.de

Objective: Alexithymia relates to difficulties recognizing and describing emotions. It has been linked to subjectively increased interoceptive awareness (IA) and to psychiatric illnesses such as major depressive disorder (MDD) and somatization. MDD in turn is characterized by aberrant emotion processing and IA on the subjective as well as on the neural level. However, a link between neural activity in response to IA and alexithymic traits in health and depression remains unclear.

Methods: A well-established fMRI task was used to investigate neural activity during IA (heartbeat counting) and exteroceptive awareness (tone counting) in non-psychiatric controls (NC) and MDD. Firstly, comparing MDD and NC, a linear relationship between IA-related activity and scores of the Toronto Alexithymia Scale (TAS) was investigated through whole-brain regression. Secondly, NC were divided by median-split of TAS scores into groups showing low (NC-low) or high (NC-high) alexithymia. MDD and NC-high showed equally high TAS scores. Subsequently, IA-related neural activity was compared on a whole-brain level between the three independent samples (MDD, NC-low, NC-high).

Results: Whole-brain regressions between MDD and NC revealed neural differences during IA as a function of TAS-DD (subscale difficulty describing feelings) in the supragenual anterior cingulate cortex (sACC; BA 24/32), which were due to negative associations between TAS-DD and IA-related activity in NC. Contrasting NC subgroups after median-split on a whole-brain level, high TAS scores were associated with decreased neural activity during IA in the sACC and increased insula activity. Though having equally high alexithymia scores, NC-high showed increased insula activity during IA compared to MDD, whilst both groups showed decreased activity in the sACC.

Conclusions: Within the context of decreased sACC activity during IA in alexithymia (NC-high and MDD), increased insula activity might mirror a compensatory mechanism in NC-high, which is disrupted in MDD.

Keywords: major depressive disorder, alexithymia, interoceptive awareness, insula, sACC, interoception, fMRI, neuroimaging

Introduction

Alexithymia is a multifaceted personality construct characterized by individuals having difficulties identifying or describing feelings. The term literally means "no words for feelings" and was introduced by Sifneos (1973) based on observations of psychosomatic disorder patients. Subsequent investigations have identified a link between alexithymic traits and other mental illnesses, including affective disorders like major depression (Bankier et al., 2001; Honkalampi et al., 2001; Saarijärvi et al., 2001; Leweke et al., 2012) and somatoform disorder (Karvonen et al., 2005; Burba et al., 2006). The 20-item Toronto Alexithymia Scale (TAS-20, Bagby et al., 1994a,b; see De Gucht and Heiser, 2003 for review) is the most commonly used self-report measurement of alexithymia. In line with Sifneos' observations, TAS scores were found to be associated with increased levels of subjective body awareness (Nakao et al., 2002; Kano et al., 2007; Ernst et al., 2014) and depressive symptoms (Honkalampi et al., 2001; Saarijärvi et al., 2001). The described triangular interconnection between alexithymia, bodily awareness and depressive symptoms (see Harshaw, 2015 for review) is also evident in the general population (see Honkalampi et al., 2000 on depression and alexithymia; see Mattila et al., 2008 on alexithymia and somatization). This supports the view of alexithymia being a relatively stable personality trait (Luminet et al., 2001; Saarijarvi et al., 2006; Stingl et al., 2008) and emphasizes the need for more investigations in non-psychiatric participants. Despite these strong relationships between alexithymic traits and major depression, both of which include aberrant emotional and body awareness, the association between neural activity in response to body awareness and alexithymic traits in health and depression remains to be described.

A growing body of research in the field of neuroimaging has investigated body-related processes in the form of interoceptive stimuli. Neural activity during interoceptive awareness (IA)–the awareness of stimuli, such as the heartbeat, that originate within the body–is considered an important factor in the processing of emotions (Damasio, 1999; Lamm and Decety, 2008; Lamm and Singer, 2010; Gu et al., 2013). Alexithymia and IA are therefore linked by a similar feature, namely the perception of body-related and emotional stimuli. As proposed previously (e.g., by Gu et al., 2013; Kano and Fukudo, 2013), alexithymia might show differential associations with IA-related neural activity in brain regions implicated in subjective interoceptive and emotional experience. However, few neuroimaging studies have investigated neural activity during IA in affective disorders like major depressive disorder (MDD). These have shown aberrant neural activity during IA in MDD, particularly decreased neural activity in the insula (Wiebking et al., 2010, 2015; Avery et al., 2014). Although no investigations have directly targeted neural processes during IA in alexithymia, neuroimaging studies have shown altered neural activity in response to emotional stimuli in brain structures associated with IA such as the insula and the anterior cingulate cortex (ACC) (see Wingbermühle et al., 2012 for review). For example, Karlsson et al. (2008) used visual emotional stimuli in $H_2{}^{15}O$-PET (positron emission tomography) to investigate regional cerebral blood flow changes in healthy participants, which were recruited on the basis of their high and low TAS scores. During emotional processing, high alexithymic individuals showed increased cerebral blood flow in, amongst other brain regions, the insula. Less activation was reported in the anterior cingulate when comparing high vs. low alexithymic individuals. In another PET study comparing high and low non-psychiatric TAS scorers, Kano et al. (2007) investigated the effects of visceral stimulation. Alexithymic participants showed increased activity during colonic distension in the insula, but increased activity in the anterior cingulate during physical stimulation. In addition to PET, a growing number of functional magnetic resonance imaging (fMRI) studies have found that alexithymia is associated with aberrant task-evoked neural activity in response to emotional stimuli in the insula and ACC. In this context, increased neural activity in the insula, particularly in response to bodily stimuli, might be associated with hyperawareness of somatosensory signals in alexithymia, which has been described on the level of subjective emotional experience and behavior (e.g., De Gucht and Heiser, 2003; Nakao and Barsky, 2007; Wingbermühle et al., 2012; Kano and Fukudo, 2013).

The observation of enhanced insula activity seems to be a more consistent characteristic in alexithymia research compared to decreased ACC activity. Increased insula activity is frequently observed in response to awareness of interoceptive stimuli such as the heartbeat, breathing or bladder state (Simmons et al., 2006; Pollatos et al., 2007; Farb et al., 2013; Wiebking et al., 2014, 2015). In alexithymia, increased insula activation was shown in response to a variety of emotional tasks, such as viewing emotional pictures (negative or positive) (Deng et al., 2013), viewing facial expressions (happy or sad) (Lemche et al., 2013), viewing painful pictures (human hands and feet) (Moriguchi et al., 2007), a trauma script imagery task (Frewen et al., 2008) (posterior insula), or in response to an empathy for pain paradigm (Bird et al., 2010). However, simultaneously observed neural activity in regions of the ACC showed inconsistent patterns across these fMRI studies, ranging from increased activity (Deng et al., 2013; Lemche et al., 2013; see also Heinzel et al., 2010) to decreased responses (Moriguchi et al., 2007; Frewen et al., 2008; see also Berthoz et al., 2002; Leweke et al., 2004). Combining self-reported measures of body awareness and alexithymia with biochemical measurements, a recent MRS (magnetic resonance spectroscopy) study investigated the association between alexithymia (TAS), body awareness (Porges, 1993) and metabolite concentrations of GABA (gamma-aminobutyric acid) and glutamate in the insula and ACC (Ernst et al., 2014). Levels of glutamate, the primary excitatory neurotransmitter in the brain, were positively related to alexithymia and body awareness in the insula. The finding of increased glutamate-mediated excitatory transmission within the insula of alexithymic individuals supports the aforementioned studies of increased insula activity, although no direct neuro-biochemical relationship was demonstrated. Considering the methodological and interpretative shortcomings of MRS (Duncan et al., 2013), the results are further suggestive of an aberrant interoceptive signal-to-noise ratio in the insula in alexithymia. On the other hand, levels of GABA, the primary inhibitory neurotransmitter in the brain, were positively related to alexithymia in the ACC. Mediating a reduction of neural responses, an enhanced GABAergic transmission in the ACC may account for the decreased neural activity in alexithymia in this region (Ernst et al., 2014).

Despite the fact that certain brain regions (insula, ACC) and stimuli types (body-related interoceptive and emotional stimuli) play a role in alexithymia (see for reviews Wingbermühle et al., 2012; Moriguchi and Komaki, 2013; van der Velde et al., 2013), direct associations between alexithymic traits and neural activity during IA remain unclear. To clarify these questions, a well-established paradigm was used in the current fMRI study to investigate neural activity during internal (heartbeat counting) and external awareness (tone counting) in MDD and non-psychiatric controls (NC) (Wiebking et al., 2014, 2010, 2015). The 20-item TAS was used to objectify alexithymia. It is the most widely used self-report instrument to measure the degree of alexithymia and consist of three subscales: difficulties identifying feelings (DI), difficulties describing feelings (DD) and external oriented thinking (EO) (Bagby et al., 1994a,b). In the first step, a linear relationship between IA-related neural activity and TAS scores comparing MDD and NC was investigated on a whole-brain level, where TAS-DD showed significant effects. In the second step, the self-reported scores of alexithymia in the NC group, as assessed by TAS-DD, were dichotomized using a median-split. This led to two NC subgroups, those with high and those with low alexithymia scores. MDD and NC having high TAS-DD scores showed equally high alexithymia scores. Differences in fMRI response to IA between the three groups–high alexithymic NC, low alexithymic NC, and MDD–were then examined on the whole-brain level.

Based on previous neuroimaging findings in studies investigating neural activation of alexithymia or IA, it was hypothesized that differences would occur specifically during IA in the insula and medial-frontal cortex depending on alexithymia status. Though the literature depicts less consistent findings regarding medial prefrontal activity compared to insula responses in alexithymia, non-psychiatric alexithymic individuals were expected to show reduced neural activity during IA in the medial prefrontal regions, such as the ACC. This assumption was based on decreased task-evoked activity in the ACC in association with alexithymia (e.g., Berthoz et al., 2002; Leweke et al., 2004; Moriguchi et al., 2007; Frewen et al., 2008; Karlsson et al., 2008). Moreover, lower connectivity within medial frontal areas of the default mode network in alexithymia (Liemburg et al., 2012), lower ACC gray matter volumes in alexithymia (Borsci et al., 2009; Grabe et al., 2014) and a positive connection between alexithymia and GABA in the ACC (Ernst et al., 2014) support the assumption of decreased neural responses in alexithymia. According to consistent findings of increased insula activity in response to emotional and bodily stimuli in alexithymia (e.g., Moriguchi et al., 2007; Frewen et al., 2008; Deng et al., 2013; Lemche et al., 2013; see Wingbermühle et al., 2012; Kano and Fukudo, 2013 for reviews), increased neural activity during IA was expected in the insula in non-psychiatric alexithymic participants. In line with previous neuroimaging studies investigating neural activity during IA in MDD (Wiebking et al., 2010, 2015; Avery et al., 2014), the alexithymic MDD group was hypothesized to show reduced neural activity in the insula.

Methods

Participants

A group of 22 patients suffering from MDD and a group of 30 non-psychiatric controls (NC) underwent fMRI scanning. All NC completed self-report measurements of the TAS as well as fMRI scanning, but TAS scores for 6 MDD patients were missing. Hence, 30 NC (mean age 33.73 ± 11.62 years, range 22–60 years; mean years of education: 16.05 ± 2.42; 15 female participants) and 16 MDD patients (mean age 41.19 ± 11.78 years, range 23–58 years; mean years of education: 15.72 ± 2.88; 11 female participants) completed both study parts and were included in further analysis.

Patients with MDD were recruited in an acute state from the Department of Psychiatry (University of Magdeburg) or from the state hospital of Uchtspringe. Eligibility screening procedures included the 21-item Beck Depression Inventory (BDI, Beck et al., 1961) (n = 14 MDD: 30.57 ± 7.00; NC: n.a.) and the 20-item Beck Hopelessness Scale (BHS, Beck et al., 1974) (n = 14 MDD: 12.29 ± 4.60; NC: 4.60 ± 3.91) (please also refer to Supplementary Table 1). Diagnosis of MDD was made by the participants' treating psychiatrists according to DSM-IV standards (Diagnostic and Statistical Manual of Mental Disorders, 4th edition; American Psychiatric Association, 1994). Exclusion criteria included major medical illnesses, histories of seizures, metallic implants, a history of substance dependence, head trauma with loss of consciousness, pregnancy, and criteria for any psychiatric disorder other than MDD. NC were recruited from the local community and were questioned about psychiatric, neurological, or medical diseases using a custom-made semistructured clinical questionnaire. All participating individuals gave their written informed consent before participating in this study. The study was approved by the local ethics committee.

Psychometric Measures

The 20-item Toronto Alexithymia Scale (TAS-20, Bagby et al., 1994a) is the most widely used and validated self-report measure of alexithymia. The factors of the TAS-20 are replicable across cultures (Taylor et al., 2003) and the current study used the validated German version of the TAS-20 (Bach et al., 1996). Items are rated using a 5-point Likert scale, ranging from 1 (strongly disagree) to 5 (strongly agree). The 20-items are categorized in three dimensions: difficulties identifying feelings (TAS-DI, example item: "I am often confused about what emotion I am feeling"), difficulties describing feelings (TAS-DD, example item: "It is difficult for me to find the right words for my feelings") and externally oriented thinking (TAS-EO, example item: "I prefer talking to people about their daily activities rather than their feelings."). High TAS scores indicate high alexithymic traits, i.e., more difficulties describing or identifying feelings.

fMRI Paradigm

A well-established fMRI design for investigating interoceptive and exteroceptive awareness was used in this study (Supplementary Figure 2B). The basic concept of the paradigm was introduced by Critchley and Pollatos (Critchley et al., 2004; Pollatos et al., 2007) and further modified and applied in fMRI studies of non-psychiatric and depressed participants by Wiebking et al. (2010, 2014, 2015). Briefly, the paradigm consists of three independent conditions (Supplementary Figure 2B). Each condition was presented 48 times in total in a pseudo-randomized order for 9–13 s each. Participants were instructed to direct their awareness to the external or the internal environment and count corresponding stimuli such as externally applied tones or the own heartbeat. Alternatively, a condition without an active task required no counting and served as baseline activity.

In more detail, participants were made familiar with the fMRI task before the scanning session. All participants were instructed by the same researcher (CW) following a standardized protocol. Each participant received the same instructions and all had the possibility to practice the paradigm on a computer outside the MRI room. For practice and scanning sessions the software Presentation (Neurobehavioral Systems) was used. The fMRI paradigm used simple visual stimuli to indicate one of the three condition types. All visual stimuli were dark colored pictures centralized on the same light background and had the same picture dimensions. In the scanner, an LCD projector was used to project the visual stimuli onto a screen visible through a mirror mounted on the headcoil. To indicate an IA condition, the task type indicator–a dark colored picture of a stylized heart–was presented on the same screen (jittered between 9–13 s). During these conditions individuals were asked to concentrate on their body and silently count their own heartbeat. Any kind of manipulation, such as holding their breath or evaluating their pulse at the radial artery, was not allowed.

During exteroceptive awareness (EA), participants were asked to focus on externally applied tones. As long as the task-type indicator–a dark colored picture of a musical note–was visible (jittered between 9–13 s) on the screen, study participants counted the number of externally applied tones. Afterwards, the number of counted heartbeats or tones was indicated on a rating scale (4 s). The indicator on the scale was moved by the subject to the labeled position representing the number of beats that they counted. Left and right button presses were used to move the indicator to the left and right side on the scale. This feedback component allowed the monitoring of the participant's attendance to the task.

Auditory stimuli were presented via the scanner loudspeaker. Tones were presented throughout the scanning sessions at an individually adapted volume to match the difficulty of both counting tasks. To ensure equivalent difficulty of both tasks, participants were instructed to adjust the volume of the tone to the same level of perception difficulty as that of counting their own heartbeat. This was done at the beginning of each of the four scanning sessions, i.e., with the scanner acquiring images to also account for scanner noise. Similar to the implementation of the rating scale, participants used right and left button presses to move an indicator on a rating scale corresponding to increase or decrease the volume. To illustrate, where the heartbeat counting was more difficult, an individual would lower the volume of the external tone in order to make this aspect of the task equally difficult to the heartbeat counting. This was explained to the participants before the scan. Being a standard part of each scanning session, participants also practiced this procedure outside the MRI room on a computer including speakers. In addition, the presentation frequency of the tones was adapted to correspond to each participant's heart-rate. The heart-rate was recorded using the Siemens Physiological Monitoring Unit (PMU) as described previously (Wiebking et al., 2014). In order to control for habituation effects, the individual onset time of each tone was jittered by 200 ms. Conditions with no particular task (Shulman et al., 2009) were indicated by a dark cross (9–13 s). Participants were instructed to disengage, reduce any cognitive work during these periods and maintain an undirected awareness, i.e., focusing neither on internal nor external stimuli. The total experiment consisted of four scanning sessions of 9.6 min each.

MRI Data Acquisition and Pre-Processing

Functional echo planar images (EPI) were acquired using a 3-Tesla whole body MRI system (Siemens Trio, Erlangen, Germany). EPI with BOLD contrast were acquired using a body coil transmit and 8-channel receive headcoil. Thirty slices aligned at the AC-PC plane and covering the whole brain were acquired per volume. A total of 1160 volumes were collected over four scanning sessions per participant (FoV = $224 \times 224\,mm^2$; spatial resolution = $3.5 \times 3.5 \times 4\,mm^3$; $T_E = 30\,ms$; $T_R = 2000\,ms$; flip angle = 80°). High resolution T_1-weighted structural images were also acquired, using the following settings: MPRAGE; FoV = $256 \times 256\,mm^2$; spatial resolution = $1 \times 1 \times 2\,mm^3$; $T_E = 5\,ms$; $T_R = 1650\,ms$.

Functional data were processed using FSL (http://www.fmrib.ox.ac.uk/fsl/) (Smith et al., 2004; Woolrich et al., 2009). Functional images were corrected for head movement (MCFLIRT) (Jenkinson et al., 2002) and motion outliers, brain-extracted (BET) (Smith, 2002), high-pass filtered with a 100 s cut-off, and smoothed with a 5 mm FWHM Gaussian kernel. Structural data were processed according to the FSL-VBM pipeline (Douaud et al., 2007) (http://fsl.fmrib.ox.ac.uk/fsl/fslwiki/FSLVBM). First, structural images were brain-extracted and gray matter-segmented before being registered to the

MNI standard space using non-linear registration (Andersson et al., 2007). The resulting images were averaged and flipped along the x-axis to create a left-right symmetric, study-specific gray matter template. Next, all native gray matter images were non-linearly registered to this study-specific template and modulated to correct for local expansion (or contraction) due to the non-linear component of the spatial transformation. The modulated gray matter images were used to quantify the proportion of GM located in the regions of interest (see below).

Since structured noise still remains in the fMRI data after typical pre-processing steps, an independent component analysis (ICA) was applied to denoise the data and hence improve the sensitivity and specificity of the results. Using Probabilistic Independent Component Analysis (Beckmann and Smith, 2004) implemented in the MELODIC toolbox in FSL, a group ICA was performed on the pre-processed fMRI data. Components were visually inspected and classified as noise or signals of interest, according to a detailed description of an operationalized denoising procedure (Kelly et al., 2010). In particular, components were considered as noise when they showed a ring-like pattern in the periphery of the brain and tightly clustered areas in the frontal regions (McKeown et al., 1998), clusters with a location in the WM/cerebrospinal fluid or an association with blood vessels (Sui et al., 2009; Zou et al., 2009), spotted patterns diffusely spread over the brain, and time courses showing a saw-tooth pattern or spikes (McKeown et al., 1998). Individual timeseries from components that were identified as noise were removed from the original fMRI data through linear regression.

The statistical model for each participant involved the trial onsets and durations for the two conditions of interest (IA, EA) as well as parameters for motion outliers (FSL motion outliers) and movement (MCFLIRT). These were obtained during pre-processing steps and included as regressors to further minimize the effects of head movement. Specific condition effects for IA and EA (vs. implicit baseline) were tested by employing linear contrasts for each subject and different conditions using FEAT, Version 6.0 (Jenkinson et al., 2012). The resulting images were submitted to a second level random-effects analysis and subsequently two-sample unpaired *t*-tests were calculated on images obtained for each participant's volume set and different conditions using FLAME (details below).

Higher-Level Statistical Analyses (Regression, MANOVA, Group Comparisons)

In the first step, the inference of interest was whether a linear relationship between neural activity (IA vs. implicit baseline or EA vs. implicit baseline) and TAS scores differed between groups ($n = 16$ MDD and $n = 30$ NC). The result indicated that the group difference significantly varied during IA in medial frontal regions as a function of TAS-DD (Supplementary Figure 1A). In order to obtain more information about the underlying factors, the calculation of individual group slopes was required (see results for NC in **Figure 1A** and for MDD in Supplementary Figure 1B). Whole-brain results were corrected using Z (Gaussianized T/F) statistic images. These were thresholded using clusters determined by $Z > 2.3$ and a cluster significance threshold of $P < 0.01$ in case of NC (Worsley, 2001). In case of MDD patients, a less conservative threshold of $P < 0.05$ was used, as the higher threshold was without any result.

Next, three functional regions of interest (ROIs, **Figure 1B**) were defined according to significant correlations as a function of TAS-DD scores seen in **Figure 1A**. Mean BOLD responses within these regions were calculated for each participant and each condition using Featquery. Values were entered into SPSS 17 (SPSS inc., Chicago, IL). For visualization purposes only, mean IA-related activity within significant voxels of activation was plotted against TAS-DD scores in NC and MDD (**Figure 1B**, colors in correspondence to color of each ROI, values depicting residuals since corrected for the amount of GM in each ROI).

The NC group was then divided into two subgroups according to TAS-DD scores using a median-split (boxplots in **Figure 1C**). This resulted in a group with high TAS-DD scorers (NC-highDD, $n = 14$) and a group with low TAS-DD scorers (NC-lowDD, $n = 16$). Thus, the continuous TAS scores were turned into a categorical variable. This approach can be used to investigate differences (or similarities) of neural activity during IA within the neurotypical participant group (comparing IA-related neural activity between high and low alexithymic individuals not suffering from any psychiatric disorder). Similarly, neural IA-related activity of participants showing the same range of self-reported TAS scores can be compared (high alexithymic non-psychiatric participants and depressed patients). In addition, equal sample sizes across the three groups ensure stable results independent of possibly skewed subject numbers.

IA or EA-related BOLD responses within ROIs showed no extreme outliers (farther than three interquartile ranges away from the first or third quartile) within the MDD group ($n = 16$) or the NC subgroups of high ($n = 14$) and low TAS-DD scorers ($n = 16$). To test differences of neural activity between these groups, a multivariate analysis of variance (MANOVA) was performed (**Figure 1C**, right side). The three participant groups ($n = 16$ MDD, $n = 14$ NC-highDD, $n = 16$ NC-lowDD) were defined as the between-subjects factor. TAS scores (DD, difficulties describing feelings; DI, difficulties identifying feelings; EO, externally oriented thinking), BOLD responses for each condition (IA and EA) and age were entered as dependent within-subjects variables. Since gray matter volumes have been associated with alexithymia (Borsci et al., 2009; Grabe et al., 2014), the amount of GM within each ROI was also included. Bonferroni correction was used for *post-hoc* testing in order to reduce type I errors (**Table 1B** and **Figure 1C**, colors correspond to color of each ROI). Differences between BHS scores were assessed by calculating univariate analysis of variance, as two BHS values were missing in the MDD group, which would lead to listwise exclusion of these individuals in the MANOVA. To investigate an effect of region in the three different ROIs independent of group, IA-related BOLD responses were pooled together ($n = 46$). An ANOVA was performed accordingly (between-subjects factor: three regions, dependent

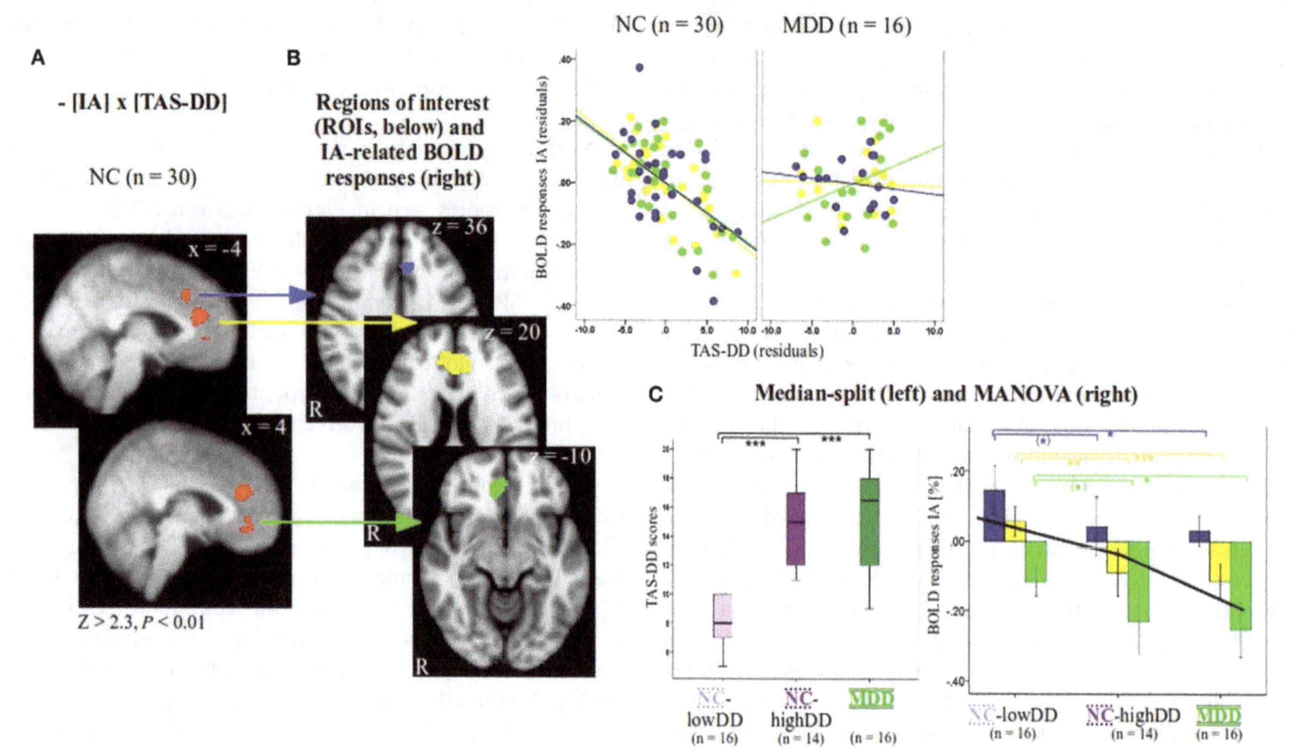

FIGURE 1 | (A) Individual group slope for non-psychiatric controls (NC, $n = 30$, $Z > 2.3$, $P < 0.01$, x-coordinates are given in MNI space) as a function of TAS-DD scores. Group differences of IA-related activity between NC and patients with major depressive disorder (MDD) as a function of TAS-DD scores and individual results investigating MDD ($n = 16$, $Z > 2.3$, $P < 0.05$) can be seen in Supplementary Figure 1A. To visualize the negative relationship between IA-related activity and TAS-DD scores in NC, signal changes of the three medial-frontal regions (see **B**) were plotted against TAS-DD scores (values depict residuals since corrected for the amount of gray matter). **(B)** Three regions of interest (blue, yellow, and green ROI) were defined according to significant results as a function of TAS-DD scores seen in **(A)** ($n = 30$ NC). Z-coordinates are given in MNI space. **(C)** The group of NC ($n = 30$) was divided on the basis of a median-split into two subgroups, those with high (NC-highDD, $n = 14$) and those with low TAS-DD scores (NC-lowDD, $n = 16$). The three groups (NC-lowDD, NC-highDD, MDD) were defined as between-subjects factor when calculating MANOVA. TAS scores, BOLD responses of interoceptive and exteroceptive awareness (IA, EA), the amount of gray matter within the ROIs and age were defined as dependent within-subjects variables. *Post-hoc* tests (Bonferroni corrected) revealed no differences of TAS-DD scores between NC-highDD and MDD. Both groups differed significantly compared to NC-lowDD [see boxplots on the left hand side of **(C)** and **Table 1B**]. IA-related BOLD responses show a similar pattern. No differences occurred between high-alexithymic participants (NC-highDD and MDD), whilst both groups differed significantly to NC-lowDD (see bar diagram on right hand side of **(C)** and **Table 1B** for details; values show mean ± 95% confidence intervals). Please note that the color code of bars corresponds to the color of each ROI and to the colors used in **Table 1B**. The black line in bar diagram **(C)** shows mean IA-related BOLD responses of all participants per region, ranging from the blue to the yellow and green ROI (independent of group, see also Supplementary Table 2). ANOVA (Supplementary Table 2) revealed a significant effect of region. *** indicates $P < 0.0001$, ** indicates $P < 0.001$, * indicates $P < 0.05$, (*) indicates $P < 0.1$.

variable: IA-related BOLD responses of $n = 46$ participants, see Supplementary Table 2).

The three groups (NC-highDD and NC-lowDD, which were defined according to self-reported TAS-DD scores, and MDD) revealed significant neural differences during IA in brain regions derived from negative correlations between TAS-DD and IA in a single group (NC). To further investigate the neural characteristics when directly comparing these *three* groups with each other, appropriate whole-brain comparisons were then performed (**Figure 2**, F-Test shown in Supplementary Figure 1B). In detail, IA-related activity was compared between non-psychiatric controls (NC) scoring high (NC-highDD) or low (NC-lowDD) on the TAS-DD scale (**Figure 2A**: NC-highDD vs. NC-lowDD, uncorrected, $P < 0.01$; NC-lowDD vs. NC-highDD, $Z > 2.8$, $P < 0.05$), between NC-highDD and patients with MDD (**Figure 2B**: NC-highDD vs. MDD, $Z > 3.4$, $P < 0.01$; no results for MDD vs. NC-highDD) and between NC-lowDD and MDD (**Figure 2C**: NC-lowDD vs. MDD, $Z > 3.4$, $P < 0.01$; no results for MDD vs. NC-lowDD). For visualization purposes, mean IA-related activity within significant voxels of activation was calculated and presented as bar diagrams (**Figure 2**, right side). Significant clusters of activation in **Figures 2B,C** that were located within insula/medial frontal cortex masks (as obtained from the Harvard-Oxford Cortical Structural Atlas included in FSL) were used for calculating mean IA and EA-related activity (see Supplementary Figures 2A,B).

TABLE 1 | (A) Using MANOVA, BOLD responses within the three regions of interest (defined according to Figures 1A,B) revealed a significant effect for group [*F*(hypothesis df: 26, error df: 62) = 3.76, *P* < 0.0001; Wilk's Lambda = 0.151, partial eta squared = 0.612]. (B) Significant between-subjects effects are highlighted by a box. *P*-values for pair-wise comparisons are based on Bonferroni corrections.

A	Multivariate outcome:	Wilk's Lambda	F (df)	*P*	Partial eta²	Observed Power[a]
		0.151	3.76 (26, 62)	< 0.0001	0.612	1

B

Box's M[†]: F (182, 4740.443) = 1.17, *P* > 0.05

Dependent variables	*P* (Levene's test of equality of error variances)	F (df) (Levene's test of equality of error variances)	Univariate outcome F(df)	*P* between-subjects	Partial eta²	Observed Power[a]	*P*-values (Bonferroni post-hoc) for multiple comparisons: NC-lowDD vs NC-highDD	NC-lowDD vs MDD	NC-highDD vs MDD
TAS-DD	0.001[b]	8.159 (2, 43)	28.086 (2, 43)	< 0.0001	0.566	1	**< 0.0001**	**< 0.0001**	1
TAS-DI	0.068	2.864 (2, 43)	23.135 (2, 43)	< 0.0001	0.518	1	0.210	**< 0.0001**	**0.0001**
TAS-EO	0.611	0.499 (2, 43)	3.336 (2, 43)	0.045	0.134	0.601	0.101	(0.088)	**1**
BOLD response IA (blue ROI in Fig. 1B)	0.149	1.994 (2, 43)	4.192 (2, 43)	0.022	0.163	0.707	(0.078)	**0.034**	1
BOLD response IA (yellow ROI in Fig. 1B)	0.233	1.508 (2, 43)	14.501 (2, 43)	< 0.0001	0.403	0.998	**< 0.001**	**< 0.0001**	1
BOLD response IA (green ROI in Fig. 1B)	0.007[c]	5.578 (2, 43)	4.731 (2, 43)	0.014	0.180	0.762	(0.075)	**0.018**	1
BOLD response EA (blue ROI in Fig. 1B)	0.019[c]	4.347 (2, 43)	1.695 (2, 43)	0.196	0.073	0.337	0.250	0.609	1
BOLD response EA (yellow ROI in Fig. 1B)	0.364	1.036 (2, 43)	2.323 (2, 43)	0.110	0.098	0.446	0.180	0.251	1
BOLD response EA (green ROI in Fig. 1B)	0.397	0.944 (2, 43)	2.452 (2, 43)	0.098	0.102	0.467	0.367	0.119	1
gray matter (blue ROI in Fig. 1B)	0.888	0.119 (2, 43)	1.010 (2, 43)	0.373	0.045	0.215	1	0.943	0.543
gray matter (yellow ROI in Fig. 1B)	0.638	0.454 (2, 43)	0.400 (2, 43)	0.673	0.018	0.110	1	1	1
gray matter (green ROI in Fig. 1B)	0.605	0.509 (2, 43)	0.120 (2, 43)	0.888	0.006	0.067	1	1	1
age	0.210	1.619 (2, 43)	2.307 (2, 43)	0.112	0.097	0.443	1	0.123	0.516
BHS[††]			19.951 (2, 41)	< 0.0001			0.132	**< 0.0001**	**< 0.001**

*Groups include non-psychiatric controls (NC) with low TAS-DD scores (lowDD, n = 16), NC with high TAS-DD scores (highDD, n = 14) and patients with major depressive disorder (MDD, n = 16). Dependent variables include TAS scores (DD, difficulties describing feelings; DI, difficulties identifying feelings; EO, externally oriented thinking), BOLD responses during interoceptive and exteroceptive awareness (IA and EA) in the three different regions of interest (ROIs, according to **Figure 1B**), the amount of gray matter within each ROI and age. Please note that the color code corresponds to the color of each ROI seen in **Figure 1B**. The regions were defined according to significant correlations (n = 30 NC) as a function of TAS-DD scores (see **Figure 1A** and Supplementary Figure 1A). Differences between BHS scores were assessed by calculating univariate analysis of variance, as two values were missing in the MDD group, which would lead to listwise exclusion of these individuals in the MANOVA calculations.*

[a] *alpha = 0.05*

[b] *Unadjusted one-way ANOVA [$F_{(2, 43)} = 6.069$, $P = 0.005$] confirms MANOVA result as well as adjusted by Brown–Forsythe and Welch's statistics [Brown-Forsythe: $F_{(2, 41.338)} = 6.013$, $P = 0.005$, Welch: $F_{(2, 28.090)} = 6.275$, $P = 0.006$]. For unequal variances the Games-Howell post-hoc results were checked with no differences compared to Bonferroni post-hoc results: NC-lowDD vs. NC-highDD: < 0.0001, NC-lowDD vs. MDD: < 0.0001, NC-highDD vs. MDD: 0.844.*

[c] *Unadjusted One-Way ANOVA shows no differences, Games-Howell post-hoc results reveal less strict results (for BOLD response IA: NC-lowDD vs. NC-highDD: 0.069. NC-lowDD vs. MDD: 0.010, NC-highDD vs. MDD: 0.915).*

[†] *Box's M is not significant, providing assurance that the assumption of equality of covariance matrices is not violated.*

[††] *Two missing values in MDD group. To avoid listwise exclusion of these data points in the MANOVA, the BHS results are based on univariate analysis of variance.*

Results

Comparing Patients Suffering from Major Depressive Disorder (MDD, *n* = 16) to Non-Psychiatric Controls (NC, *n* = 30) (Figure 1) and Determination of NC Subgroups

Firstly, we investigated whether a linear relationship between neural activity (IA vs. implicit baseline or EA vs. implicit baseline) and TAS scores differed between the *two* groups. 16 MDD patients were compared to the total group of 30 NC, i.e., this group was not classified according to their alexithymic state yet. Amongst the different TAS subscales, the relationship between IA-related activity and scores of the TAS-DD subscale was significantly different between these groups in medial-frontal regions (Supplementary Figure 1A). Individual group slopes were then calculated (see results for NC in **Figure 1A**

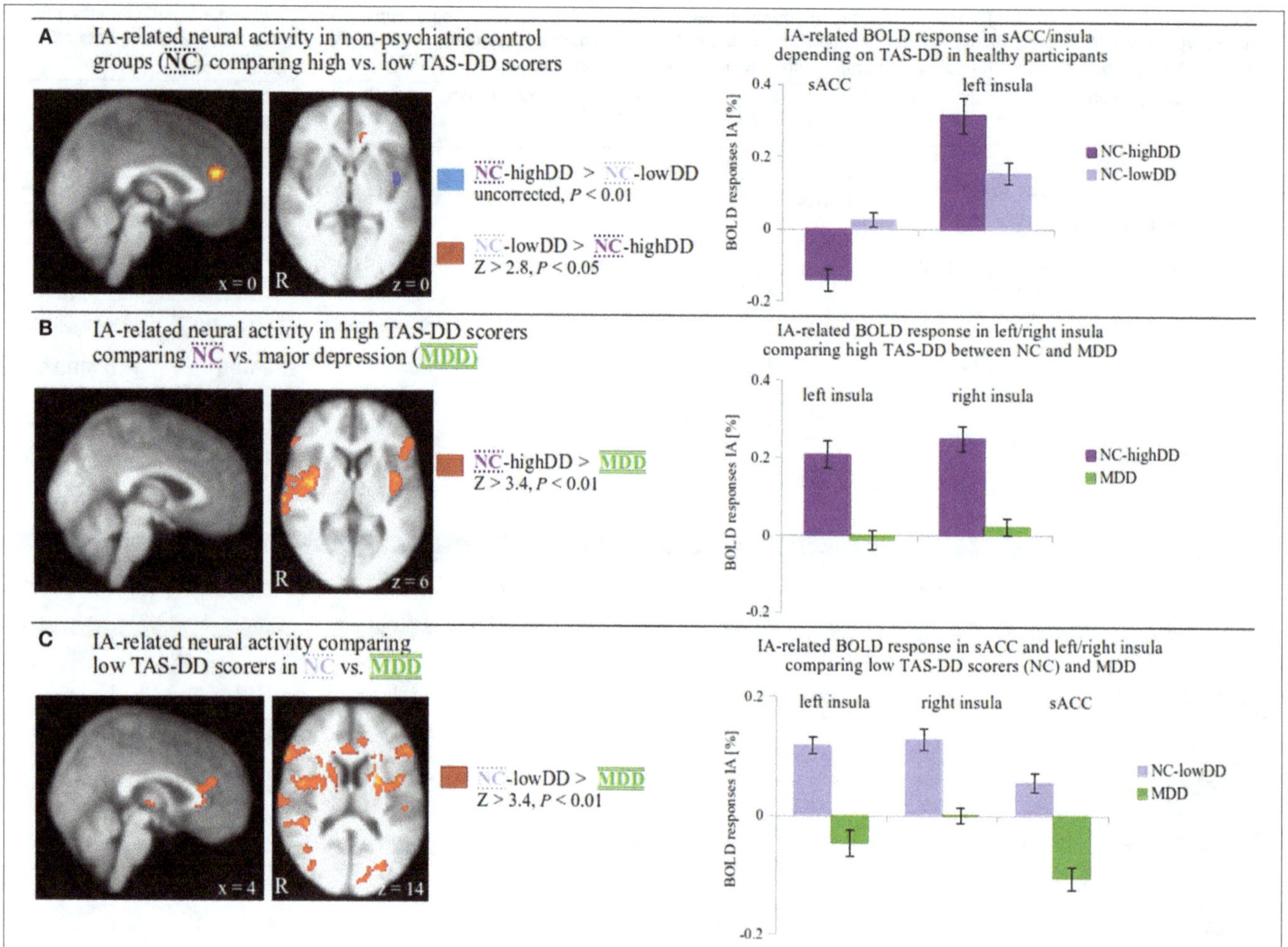

FIGURE 2 | (A) Comparing IA-related neural activity between non-psychiatric controls (NC) with low (lowDD) and high TAS-DD scores (highDD) on a whole-brain level, alexithymic participants (NC-highDD) show increased activity in the left insula whilst NC-lowDD show increased activity in the supragenual anterior cingulate cortex (sACC; BA 24/32, coordinates in MNI space). A closer look at underlying neural activity during IA (bar diagrams show mean signal change ± SEM)† reveal positive BOLD responses in both groups in the insula with greater activity in NC-highDD (dark purple bars). In the sACC, NC-lowDD (light purple bars) reveal small positive BOLD responses, whilst the alexithymic group shows negative BOLD responses. **(B)** IA-related neural activity is compared between high alexithymic groups, i.e., non-psychiatric controls scoring high on the TAS-DD scale (NC-highDD) and patients suffering from major depressive disorder (MDD, coordinates in MNI space.). TAS-DD scores show no differences between these groups. On the whole-brain level, NC-highDD show significantly greater neural activity during IA in the insula compared to MDD. IA-related BOLD responses within the left and right insula (bar diagrams show mean signal change ± SEM)† illustrate that NC-highDD show positive BOLD responses (dark purple bars), whereas MDD show small responses (green bars). **(C)** When considering TAS-DD scores and psychiatric status, the comparison between NC-lowDD and MDD seems the most distinct. On the whole-brain level, NC-lowDD show significantly greater neural activity during IA in the left/right insula and the sACC (coordinates given in MNI space.). Calculating BOLD responses of each region (bar diagrams show mean signal change ± SEM)†, MDD patients show negative BOLD responses particularly in the left insula and sACC (green bars). NC-lowDD reveal positive IA-related activity particularly in the bilateral insula (light purple bars).† Bar diagrams for illustration purposes only. No additional statistical tests were performed on these values in order to avoid circularity.

and results for MDD in Supplementary Figure 1B), which showed negative correlations in those regions between TAS-DD and IA-related activity in NC, but not in MDD. Hence, the differences seen comparing MDD and NC were due to significant negative associations between TAS-DD and IA-related activity in the NC group. The results underline that alexithymia is linearly associated with decreasing neural activity during IA in medial-frontal regions in non-psychiatric control participants. For visualization purposes, mean IA-related BOLD responses within correlating clusters (please refer to regions in **Figure 1B** and Supplementary Table 1 for mean values) were plotted against TAS-DD scores (scatterplot in **Figure 1B**, values controlled for the amount of gray volume in respective ROIs). Additional statistical tests were not performed on these values in order to avoid circularity (e.g., Kriegeskorte et al., 2009).

Secondly, examining the categorical effect of TAS-DD scores on neural activity, the NC group was divided in a group with high TAS-DD scores (mean TAS-DD in NC-highDD: 14.57 ± 2.93) and a group with low TAS-DD scores (mean TAS-DD in NC-lowDD: 8.25 ± 1.57) using median-split. The MDD group had a mean TAS-DD score of 15.25 (± 3.75). Hence, MDD and NC-highDD were matched according to their TAS score (please refer to **Table 1B** for group comparisons and Supplementary Table 1 for mean values of the remaining TAS scales). TAS scores, BOLD responses during IA and EA within each ROI, gray matter and age were submitted as within-subjects variables to a MANOVA, whilst the *three* subject groups (NC-lowDD, NC-highDD, MDD) were defined as between-subjects factor. The MANOVA revealed a significant effect for group [F(hypothesis df: 26, error df: 62) = 3.76, $P < 0.00001$; Wilk's Lambda = 0.151, partial eta squared = 0.612, please refer to **Table 1A**]. As highlighted in **Table 1B** by a box, significant group effects occurred for each of the TAS subscales. In addition, BOLD responses during IA showed significant group effects for each of the *three* ROIs, which were defined according to significant correlations in NC ($n = 30$) as a function of TAS-DD scores seen in **Figures 1A,B** (please note that the color of each ROI corresponds to the color code used in **Figure 1C** and **Table 1**). Neural activity during EA or the amount of gray matter within each ROI showed no significant group effect (mean values in Supplementary Table 1).

Bonferroni *post-hoc* tests (**Table 1B**) revealed no significant TAS-DD differences between NC-highDD and MDD, whilst both groups differed significantly to NC-lowDD (as detailed in boxplots of **Figure 1C**, left). High alexithymic groups (i.e., NC-highDD as well as MDD) were therefore matched for TAS-DD scores. A similar pattern of group differences was observed in regard to IA-related activity. No differences occurred between NC-highDD and MDD, whilst both groups differed significantly to NC-lowDD. This pattern holds true for each of the *three* ROIs (see bar diagram in **Figure 1C** and **Table 1B**).

Combining IA-related BOLD responses of all participants ($n = 46$) and comparing them between these *three* ROIs, an ANOVA showed a significant effect for region (Supplementary Table 2, $P < 0.0001$). Each region showed significantly different IA-related BOLD responses compared to the remaining *two* ROIs (Supplementary Table 2). In more detail, the most dorsal region (blue ROI in **Figure 1B**, $z = 36$) exhibited positive BOLD responses across all *three* subject groups (**Figure 1C**: blue bars represent BOLD responses by group, black line represents mean value across all $n = 46$ participants, see also Supplementary Table 2). Compared to these positive BOLD responses seen in the dorsal ROI in each group, the sACC region (yellow ROI in **Figure 1B**, $z = 20$, closely corresponding to regions in **Figures 2A,C**) showed relatively diminished neural activity in each group (**Figure 1C**: yellow bars represent BOLD responses by group, black line represents mean value across all $n = 46$ participants, see also Supplementary Table 2). Compared to IA-related BOLD responses seen in the sACC, the most ventral region (green ROI in **Figure 1B**, $z = -10$) showed negative BOLD responses across all groups (**Figure 1C**: green bars represent BOLD responses by group, black line represents mean value across all $n = 46$ participants, see also Supplementary Table 2). In summary, mean IA-related BOLD responses of all study participants differ significantly between the *three* ROIs, whilst the most dorsal region (blue ROI) showed positive IA-related BOLD responses, the sACC (yellow ROI) showed comparatively lower neural activity and the most ventral region (green ROI) showed negative IA-related BOLD responses.

Group Comparisons between NC with High-Alexithymic Traits (NC-highDD, *n* = 14), Low-Alexithymic Traits (NC-lowDD, *n* = 16) and MDD (*n* = 16) (Figure 2)

The MANOVA showed group differences during IA-related activity depending on TAS-DD scores. IA-related BOLD responses derived from regions that showed an association with TAS-DD scores in a group of non-psychiatric control participants, which was not subdivided according to their alexithymic trait. To investigate neural activity during IA between non-psychiatric control participants with high-alexithymic TAS-DD scores (NC-highDD), low-alexithymic TAS-DD scores (NC-lowDD) and depressed patients (having equally high TAS-DD scores), whole-brain comparisons between these three groups were performed (F-Test shown in Supplementary Figure 1B).

In a first step, IA-related activity on the whole-brain level was compared between the two non-psychiatric groups scoring high or low on the TAS-DD (**Figure 2A**, $Z > 2.8$, $P < 0.05$, red region). Non-psychiatric controls (NC) with low TAS-DD scores (NC-lowDD) showed increased IA-related activity in the supragenual anterior cingulate cortex (sACC; BA 24/32) when compared to NC with high TAS-DD scores (NC-highDD). Having a closer look at the underlying BOLD responses, NC-lowDD (light purple bars) revealed only small positive BOLD responses (bar diagrams in **Figure 2A**). The difference seen in the sACC was attributed to negative BOLD responses in the alexithymic group (dark purple bars, mean values in Supplementary Table 1). In contrast, IA-related activity was increased in NC-highDD in the left insula when compared to NC-lowDD (**Figure 2A**, uncorrected, $P < 0.01$, blue region). Plotting IA-related BOLD responses of this region depicted robust positive BOLD responses in both groups; however, alexithymic individuals showed the highest BOLD response in the left insula compared to low-alexithymic participants.

In a second step, IA-related activity on the whole-brain level was compared between high alexithymic groups, i.e., non-psychiatric controls scoring high on the TAS-DD scale and patients with MDD (**Figure 2B**, $Z > 3.4$, $P < 0.01$). Although TAS-DD scores did not differ between these groups (**Figure 1C** left, **Table 1B**), NC-highDD showed increased IA-related activity in the bilateral insula. Calculating BOLD responses within the left and right insula (bar diagrams in **Figure 2B**) showed low neural activity during IA in MDD (green bars), whereas NC-highDD showed again robust positive BOLD responses (dark purple bars, mean values in Supplementary Table 1). High alexithymic groups showed no differences in the sACC due to comparable negative BOLD responses in both groups [please compare negative BOLD response in the sACC in MDD (**Figure 2C**, green bar) to negative BOLD response in the sACC in NC-highDD (**Figure 2A**, dark purple bar)].

In a final step (**Figure 2C**), the two most distinct groups considering both TAS-DD scores and psychiatric status (NC-lowDD vs. MDD) were compared. On the whole-brain level,

NC-lowDD showed greater neural activity during IA in the bilateral insula and the sACC ($Z > 3.4$, $P < 0.01$). Again, IA-related BOLD responses were calculated for each region and plotted by region (bar diagrams in **Figure 2C**). Whilst NC-lowDD showed positive BOLD responses particularly in the bilateral insula (light purple bars), MDD showed small and reduced neural activity particularly in the sACC (green bars, mean values in Supplementary Table 1). Please note that calculated signal changes seen in bar diagrams in **Figures 2 A–C** serve visualization purposes; no additional statistical tests were performed on these values in order to avoid circularity (Kriegeskorte et al., 2009).

Discussion

This study forges a link between neural activity during IA and alexithymia in non-psychiatric controls and depressed individuals. To investigate IA, a well-established fMRI paradigm was used. Alexithymia was assessed using the TAS. In non-psychiatric controls (NC), TAS scores were negatively associated with IA-related activity in medial-frontal regions such as the supragenual anterior cingulate cortex (sACC; BA 24/32). Following a dimensional approach, where alexithymia is seen on a continuum and individuals may exhibit higher or lower degrees of alexithymia, high alexithymia in NC was associated with decreased activity in the sACC and increased activity in the insula when compared to low-alexithymic NC. Though having equally high TAS scores, high-alexithymic NC showed increased activity in the bilateral insula compared to MDD. Neural IA-activity in the sACC was similar between both high-alexithymic groups (i.e., MDD and NC-highDD).

sACC: Decreased Neural Activity in Response to IA in High-Alexithymic Controls and Major Depressive Disorder

When comparing neural activity during IA as a function of TAS-DD scores between non-psychiatric controls (NC, group consisting of $n = 30$ participants independent of TAS scores) and depressed participants (MDD, $n = 16$), group differences occurred in the medial-frontal cortex and specifically the sACC. In NC there is a clear negative relationship between IA-related neural activity and TAS-DD scores (**Figure 1A**), i.e., high alexithymia scores are associated with decreased neural activity in medial-frontal regions. Depressed patients showed no associations between TAS scores and neural activity in these areas.

The finding of decreased neural activity in non-psychiatric alexithymia is a well-documented neural response seen in many fMRI-studies (Berthoz et al., 2002; Moriguchi et al., 2007; Wingbermühle et al., 2012; Terasawa et al., 2013). Prior studies, however, used affective stimuli, such as emotional pictures of faces or body parts, to investigate neural correlates of alexithymia, implicating arousal and valence effects (e.g., van der Velde et al., 2013). The current study extends the existing literature by showing similar neural responses in non-psychiatric participants performing an IA task, i.e., independent of externally induced (visual-emotional or physical) stimulation. Participants merely shifted their awareness toward their own body (heartbeat counting) or toward the external environment (tone counting). Through this rather natural design of awareness switching it is possible to investigate the link between alexithymic traits on the subjective level and body awareness at the neural level. Both measures are positively associated on the subjective level of self-reports (e.g., Nyboe Jacobsen et al., 2006; Moriguchi and Komaki, 2013; Ernst et al., 2014). However, this is the first study linking alexithymia to neural activity during IA. Our first finding supports the assumption of decreasing neural activity in medial-frontal brain areas, as a function of alexithymic traits (see also bar diagram in **Figure 1C**).

Further, we investigated BOLD differences/similarities between groups showing different/similar alexithymic status. For this, the subjective TAS-DD scores were used to subdivide the non-psychiatric group into high-alexithymic (NC-highDD) and low-alexithymic individuals (NC-lowDD). Instead of correlating self-reported alexithymia scores linearly with neural activity, as done previously, this step served as a dimensional approach to study different degrees of alexithymia. Thus, the continuous TAS scores were transformed into categorical variables (dichotomized using a median-split) leading to NC subgroups with high or low alexithymia scores. The advantage of this approach is twofold, as it can be used to investigate differences (or similarities) of IA-related neural activity (a) within a non-psychiatric control group comparing high/low alexithymic individuals (NC-highDD vs. NC-lowDD) and (b) between MDD and non-psychiatric individuals having the same range of TAS scores (NC-highDD vs. MDD). In addition, equal sample sizes across the three groups ensure stable results independent of potentially skewed subject numbers. When comparing healthy high and low TAS scorers on the whole-brain level, the results underline negative IA-related BOLD responses in the sACC in high-alexithymic compared to low-alexithymic NC participants (**Figure 2A**). This pattern may represent a common association between alexithymia and IA-related activity in the sACC, given the fact that both participant groups include individuals without psychopathological symptoms. This finding is accompanied by concomitantly increased IA-related BOLD responses in the insula, when comparing high vs. low-alexithymic NC participants, suggesting again this pattern (in combination with decreased sACC activity) may be a characteristic property of IA-related activity in alexithymia. The described neural differences in the sACC of healthy participants are in line with findings on the biochemical level, which show a positive relationship between levels of the inhibitory neurotransmitter GABA in the ACC and alexithymia. Though biochemical findings need to be interpreted with caution (Duncan et al., 2013), the features of GABA-mediated reductions in neural activity may account for decreased IA-related neural activity in alexithymia in this region (Ernst et al., 2014). When comparing IA-related activity between groups showing no differences in self-reported TAS scores (MDD and NC-highDD), decreased activity in the sACC in response to IA is similar between groups (please compare regional negative BOLD response in the sACC in MDD in **Figure 2C** to negative BOLD response in the sACC in NC-highDD in **Figure 2A**). This finding is suggestive of decreased sACC activity in high-alexithymic individuals independent of depression. Whole-brain results of neural activity during IA in the sACC are summarized in **Figure 3A**.

Furthermore, IA-related activity in the sACC appears to follow a regional dorsal-to-ventral gradient. As indicated by a black line in bar diagram **Figure 1C** (representing IA-related BOLD responses of all participants per region, ranging from the dorsal to the ventral ROI, independent of group, see Supplementary Table 2), IA-related activity decreased linearly from dorsal to ventral regions. This finding may shed light on the inconsistencies regarding ACC activity in alexithymia, since other studies have shown increasing ACC activity in alexithymia in response to emotional picture stimuli (Kano et al., 2007; Heinzel et al., 2010; Deng et al., 2013; Lemche et al., 2013; van der Velde et al., 2013). Besides possible differential effects of task type, task/arousal-load and valence-dependency in alexithymia (McRae et al., 2008; van der Velde et al., 2013), future studies should consider regional-specific activity patterns of the employed functional tasks. This could be realized by comparing BOLD responses along a dorsal-ventral gradient (or posterior-anterior, if applicable).

Briefly, within a sample representing the general population, high alexithymia was associated with increased insula and decreased sACC activity in response to IA compared to low

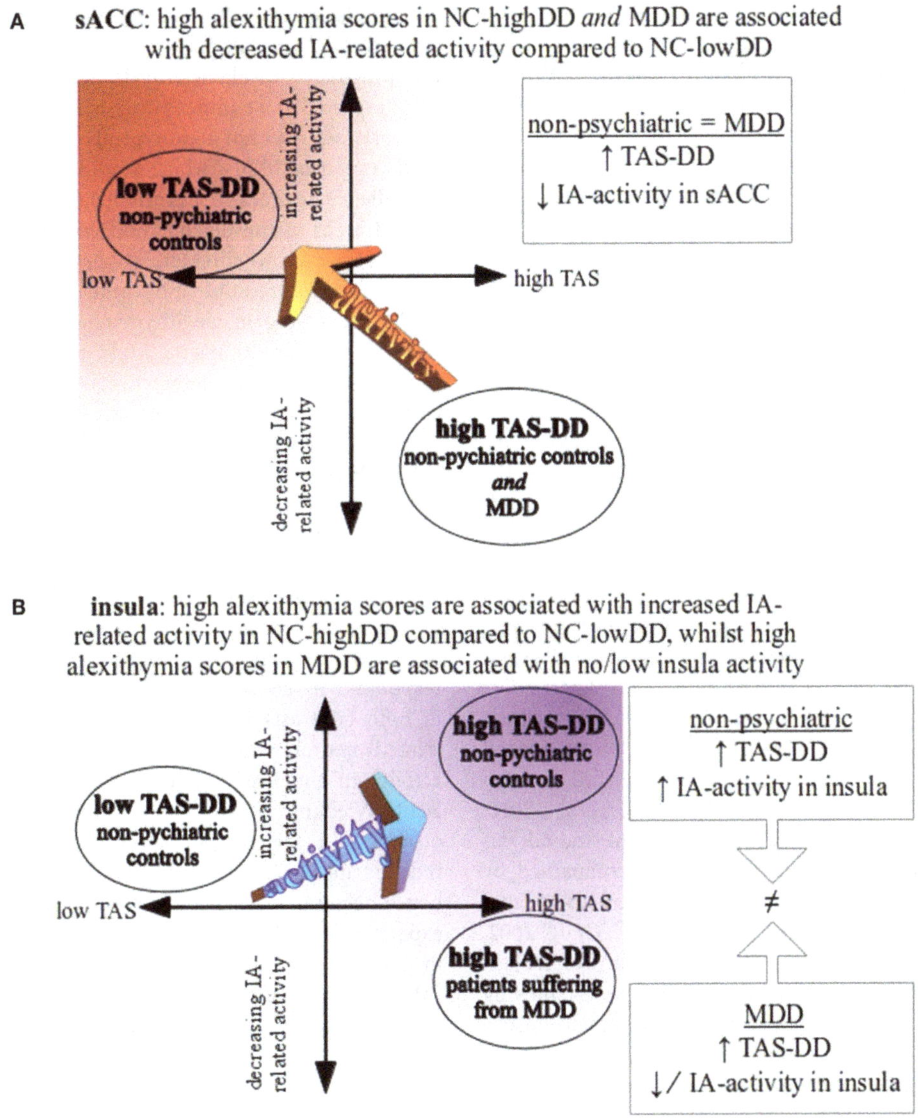

FIGURE 3 | Schematic overview of study-specific results. (A) In the sACC, both high-alexithymic groups (NC-highDD and MDD) show a comparable pattern of negative IA-related activity in the supragenual anterior cingulate cortex (sACC, see also **Figures 2A,C**). Low TAS-DD scorers, however, show positive IA-related activity in the sACC (see also **Figure 2C**). **(B)** Comparing NC-highDD and NC-lowDD, participants with high alexithymia show greater positive BOLD responses in the insula during IA. As both groups consist of participants without psychopathological symptoms, this pattern may represent the typical association between alexithymia and IA-related activity in the insula. When comparing groups that show similar TAS-DD scores (NC-highDD and MDD), however, depressed patients show significantly lower IA-related activity in the insula (see also **Figure 2B**).

alexithymia. Considering high-alexithymic depressed patients in comparison to high-alexithymic healthy participants, both groups showed comparable IA-related deactivation in the same sACC region. In addition, NC-highDD showed increased insula activity. Assuming a balance between sACC and insula activity as exemplified by the low-alexithymic group (low sACC activity and relatively higher positive insula responses during IA), the observed pattern in the high-alexithymic healthy group (decreased sACC and increased insula activity) could be interpreted as compensatory mechanisms to maintain a balanced system of IA-related neural activity between insula and sACC. In MDD, aberrant IA-related insula activity might account for alexithymic traits rather than sACC activity, as the latter shows no differentiation between high-alexithymic groups (MDD and NC-highDD).

Insula: Increased Neural Activity in Response to IA in High-alexithymic Controls, but Aberrant in Major Depressive Disorder

In concordance with fMRI studies investigating IA in the insula using a heartbeat counting task (Critchley et al., 2004; Wiebking et al., 2014), non-psychiatric participant groups (NC-highDD and NC-lowDD) showed robust positive IA-related activity in the insula (**Figures 2A–C**). Non-psychiatric alexithymic participants (NC-highDD) showed significantly higher neural activity in response to IA, when compared to NC-lowDD. As proposed by Kano and Moriguchi (Kano and Fukudo, 2013; Moriguchi and Komaki, 2013), heightened neural insula activity suggests amplified neural responses in alexithymic individuals in response to body-related stimuli. However, the present study is the first to show this neural pattern holds true during IA (as asked for by Kano and Fukudo, 2013), i.e., without externally applied emotional or physical stimulation. The current finding of increased IA-related activity in the insula of alexithymic individuals is supportive of theories suggesting increased awareness of somatosensory signals in alexithymia (e.g., De Gucht and Heiser, 2003; Kano et al., 2007; Ernst et al., 2014). However, heightened awareness of somatosensory signals in alexithymia does not necessarily correlate with objectified measures of sensitivity (Kano et al., 2007), specifically when dealing with psychiatric populations (Wiebking et al., 2010; Terhaar et al., 2012; Avery et al., 2014). The current study draws a link between alexithymia (that is typically accompanied by increased somatic awareness) and increased insula activity (that is related to somatic and emotional awareness, e.g., Craig, 2002, 2004; Critchley, 2005; Singer et al., 2009) in response to an IA task in non-psychiatric controls compared to MDD patients (showing the same degree of alexithymia). A direct link to somatosensory sensitivity, however, was not investigated in the current study.

Although speculative, increased IA-related insula activity in non-psychiatric alexithymia (NC-highDD) might mirror a compensatory mechanism within the scope of decreased sACC activity. In more detail, impaired emotional processing capacities, inherent in alexithymia (and mirrored by decreased activity in the sACC in MDD and NC-highDD), require a higher degree of body-related, interoceptive signals (mirrored by increased activity in the insula in healthy NC-highDD, but not in MDD) in order to sufficiently process emotional stimuli. This assumption is supported by close connections between insula and ACC on the anatomical level (Mesulam and Mufson, 1982; Nieuwenhuys et al., 2007; Moisset et al., 2010; in primates), as well as on the functional level showing connectivity (Taylor et al., 2009; Horn et al., 2010) and co-activation in response to various emotional paradigms (e.g., Singer et al., 2004; Bartels and Zeki, 2004). The insula can be described as a region integrating multimodal signals (Kelly et al., 2012; Farb et al., 2013) and both insula and ACC are crucial nodes of the salience network. Particularly the insula is seen as the integral hub of this network, which is assumed to mediate information flow across other neural networks, evaluate the most homeostatically relevant internal and external stimuli and, consequentially, guide behavioral responses to salient stimuli (Seeley et al., 2007; Craig, 2009; Menon and Uddin, 2010). As decreased sACC activity in response to IA is accompanied by upregulated insula activity in healthy alexithymics, it could be interpreted as a compensatory mechanism within the salience network found within neurotypical brain structures. This process is disrupted in MDD, which would mirror altered integrity of the salience network in alexithymia (as demonstrated for medial frontal areas of the default mode network by Liemburg et al., 2012). However, whether alterations within the salience network during IA are predominantly influenced by aberrant insula or sACC processing in alexithymia needs clarification in future analyses using additional methods, such as connectivity analyses or psychophysiological interactions.

In a second comparison, neural activity during IA was compared between high TAS-DD scorers (NC-highDD and MDD, **Figure 2B**). TAS-DD scores between both groups did not differ, but non-psychiatric controls showed increased IA-related activity in the bilateral insula compared to the MDD group. A similar pattern can be seen when comparing NC-lowDD to MDD (**Figure 2C**). Again, the non-psychiatric group (NC-lowDD) showed robust positive BOLD responses, but depressed participants revealed only low or negative BOLD responses. In contrast to the sACC, the insula was affected in both contrasts involving MDD. Research investigating IA-related neural activity in MDD observed impaired neural activity in the insula in depressed individuals (Avery et al., 2014; Wiebking et al., 2015), suggesting that the differences observed in the current study might be related to overall reduced insula activity independent of alexithymic traits, and hence account for general deficits of physiological processing capacities in MDD. Reduced functional connectivity between insula and sACC in psychopathology within the salience network (Menon, 2011) supports this assumption. Whole-brain results of IA-related neural activity in the insula are summarized in **Figure 3B**.

Taken together, the current study is the first investigating neural activity in the insula during IA (heartbeat awareness) in non-psychiatric participants scoring high or low on the TAS-DD subscale and patients suffering from MDD scoring equally high on the TAS-DD scale. Higher positive BOLD-responses in response to IA were seen in non-psychiatric alexithymic individuals (NC-highDD), but not in the MDD group with equally high TAS-DD scores. This finding suggests an upregulated insula activity in response to interoceptive, body-related awareness in non-psychiatric alexithymia and may be seen as a compensatory mechanism within the scope of decreased sACC activity. As depressed patients show similarly decreased BOLD responses in the sACC, but unbalanced insula activity, these processes seem to be disrupted in depression.

Limitations

Patients with MDD participating in the current study were under medication. Thus, a medication effect in comparison to the non-psychiatric group cannot be fully excluded. Future studies should control for these influences more rigorously and include unmedicated patients as well. Moreover, it must be pointed out that the investigation of participants' sensitivity to internal processes, i.e., the distinction between good and poor heartbeat perceivers (Pollatos and Schandry, 2004; Herbert et al., 2011) and subsequent whole-brain comparisons, was not targeted in this study. Whilst the performance of heartbeat perception on the behavioral level is inconclusive in depression (Dunn et al., 2007; Terhaar et al., 2012), possible differences in accuracy might be related to cognitive impairments and/or reduced heartbeat evoked potentials in MDD (Terhaar et al., 2012) and thus resemble confounding factors for such analyses. Specifically the impaired heartbeat evoked potentials suggest that the neural activity underlying the interoceptive stimulus response might be altered in MDD *per se* (Avery et al., 2014). The central interest of the current study was to investigate neural activity during awareness directed toward the internal (heartbeat) or external environment (tones), independent of individual task performance or sensitivity (please refer to Garfinkel et al., 2014 distinguishing interoceptive accuracy from interoceptive awareness). Further research carefully considering those factors and including long counting periods without auditory stimulation is needed to clarify the exact relationship between heartbeat counting performance/accuracy and neural activity during heartbeat counting in health and depression. Finally, methodological concerns could be raised with respect to the subjective measure used (TAS-20). Results of this self-report questionnaire may be biased by self-presentation concerns and social desirability. Controlling for these factors, future studies investigating alexithymia should additionally use observer-based scales, such as the Toronto Structured Interview for Alexithymia (TSIA, Bagby et al., 2006), which show high correlations with TAS-20.

Conclusions

In contrast to neuroimaging studies on emotional processes of alexithymia, often done through emotional picture viewing, the current study focuses on the relationship between neural activity during awareness of body-related, interoceptive stimuli (heartbeat) in combination with subjective scores of alexithymia in depressed and non-psychiatric participants. The results of the current fMRI study show that alexithymia is associated with altered IA-related activity in the sACC (significantly decreased activity in NC-highDD and MDD) and insula (significantly increased activity in NC-highDD, but not MDD). In the context of decreased interoceptive activity in the sACC, increased insula activity in response to interoceptive signals could be a compensatory mechanism within the salience network in non-psychiatric alexithymia, which is decoupled in MDD.

Author Contributions

CW and GN conceptualized research; CW developed paradigm, acquired data, analyzed data and wrote the paper.

Acknowledgments

CW would like to thank NW Duncan, MV Rector and NF Wagner for their helpful comments and suggestions on the manuscript. Thanks also to the staff from the Department of Neurology (OvG University Magdeburg) for their technical support, to the staff from the state hospital of Uchtspringe, specifically A Bauer, for assisting with patient recruitment, to M de Greck for technical support and mentoring and to all study participants for their cooperation. CW acknowledges the support of the German Research Foundation (Deutsche Forschungsgemeinschaft, DFG) and the Open Access Publication Fund of the University of Potsdam. GN acknowledges: Hope of Depression Research Foundation (HDRF), Canadian Institutes of Health Research (CIHR), EJLB-Michael Smith Foundation (CIHR–EJLB).

References

American Psychiatric Association (1994). *Diagnostic and Statistical Manual of Mental Disorders (DSM IV)*. Washington, DC: American Psychiatric Press.

Andersson, J. L. R., Jenkinson, M., and Smith, S. (2007). *Non-linear Registration aka Spatial Normalisation. FMRIB Technial Report TR07JA2*, Oxford. Available online at: www.fmrib.ox.ac.uk/analysis/techrep

Avery, J. A., Drevets, W. C., Moseman, S. E., Bodurka, J., Barcalow, J. C., and Simmons, W. K. (2014). Major depressive disorder is associated with abnormal interoceptive activity and functional connectivity in the Insula. *Biol. Psychiatry* 76, 258–266. doi: 10.1016/j.biopsych.2013.11.027

Bach, M., Bach, D., de Zwaan, M., Serim, M., and Böhmer, F. (1996). [Validation of the German version of the 20-item Toronto Alexithymia Scale in normal persons and psychiatric patients]. *Psychother. Psychosom. Med. Psychol.* 46, 23–28.

Bagby, R. M., Parker, J. D. A., and Taylor, G. J. (1994a). The twenty-item Toronto Alexithymia Scale-I. Item selection and cross-validation of the factor structure. *J. Psychosom. Res.* 38, 23–32. doi: 10.1016/0022-3999(94)90005-1

Bagby, R. M., Taylor, G. J., and Parker, J. D. (1994b). The Twenty-item Toronto Alexithymia Scale-II. Convergent, discriminant, and concurrent validity. *J. Psychosom. Res.* 38, 33–40. doi: 10.1016/0022-3999(94)90006-X

Bagby, R. M., Taylor, G. J., Parker, J. D. A., and Dickens, S. E. (2006). The development of the Toronto Structured Interview for Alexithymia: item selection, factor structure, reliability and concurrent validity. *Psychother. Psychosom.* 75, 25–39. doi: 10.1159/000089224

Bankier, B., Aigner, M., and Bach, M. (2001). Alexithymia in DSM-IV disorder: comparative evaluation of somatoform disorder, panic disorder, obsessive-compulsive disorder, and depression. *Psychosomatics* 42, 235–240. doi: 10.1176/appi.psy.42.3.235

Bartels, A., and Zeki, S. (2004). The neural correlates of maternal and romantic love. *Neuroimage* 21, 1155–1166. doi: 10.1016/j.neuroimage.2003.11.003

Beck, A. T., Ward, C. H., Mendelson, M., Mock, J., and Erbaugh, J. (1961). An inventory for measuring depression. *Arch. Gen. Psychiatry* 4, 561–571. doi: 10.1001/archpsyc.1961.01710120031004

Beck, A. T., Weissman, A., Lester, D., and Trexler, L. (1974). The measurement of pessimism: the hopelessness scale. *J. Consult. Clin. Psychol.* 42, 861–865. doi: 10.1037/h0037562

Beckmann, C. F., and Smith, S. M. (2004). Probabilistic independent component analysis for functional magnetic resonance imaging. *IEEE Trans. Med. Imaging* 23, 137–152. doi: 10.1109/TMI.2003.822821

Berthoz, S., Artiges, E., Van De Moortele, P.-F., Poline, J.-B., Rouquette, S., Consoli, S. M., et al. (2002). Effect of impaired recognition and expression of emotions on frontocingulate cortices: an fMRI study of men with alexithymia. *Am. J. Psychiatry* 159, 961–967. doi: 10.1176/appi.ajp.159.6.961

Bird, G., Silani, G., Brindley, R., White, S., Frith, U., and Singer, T. (2010). Empathic brain responses in insula are modulated by levels of alexithymia but not autism. *Brain* 133, 1515–1525. doi: 10.1093/brain/awq060

Borsci, G., Boccardi, M., Rossi, R., Rossi, G., Perez, J., Bonetti, M., et al. (2009). Alexithymia in healthy women: a brain morphology study. *J. Affect. Disord.* 114, 208–215. doi: 10.1016/j.jad.2008.07.013

Burba, B., Oswald, R., Grigaliunien, V., Neverauskiene, S., Jankuviene, O., and Chue, P. (2006). A controlled study of alexithymia in adolescent patients with persistent somatoform pain disorder. *Can. J. Psychiatry* 51, 468–471.

Craig, A. D. (2002). How do you feel? Interoception: the sense of the physiological condition of the body. *Nat. Rev. Neurosci.* 3, 655–666. doi: 10.1038/nrn894

Craig, A. D. (2004). Human feelings: why are some more aware than others? *Trends Cogn. Sci.* 8, 239–241. doi: 10.1016/j.tics.2004.04.004

Craig, A. D. (2009). How do you feel–now? The anterior insula and human awareness. *Nat. Rev. Neurosci.* 10, 59–70. doi: 10.1038/nrn2555

Critchley, H. D. (2005). Neural mechanisms of autonomic, affective, and cognitive integration. *J. Comp. Neurol.* 493, 154–166. doi: 10.1002/cne.20749

Critchley, H. D., Wiens, S., Rotshtein, P., Ohman, A., and Dolan, R. J. (2004). Neural systems supporting interoceptive awareness. *Nat. Neurosci.* 7, 189–195. doi: 10.1038/nn1176

Damasio, A. R. (1999). *The Feeling of What Happens?: Body and Emotion in the Making of Consciousness 1st Edn.*, New York, NY: Harcourt Brace.

De Gucht, V., and Heiser, W. (2003). Alexithymia and somatisation. *J. Psychosom. Res.* 54, 425–434. doi: 10.1016/S0022-3999(02)00467-1

Deng, Y., Ma, X., and Tang, Q. (2013). Brain response during visual emotional processing: an fMRI study of alexithymia. *Psychiatry Res.* 213, 225–229. doi: 10.1016/j.pscychresns.2013.03.007

Douaud, G., Smith, S., Jenkinson, M., Behrens, T., Johansen-Berg, H., Vickers, J., et al. (2007). Anatomically related grey and white matter abnormalities in adolescent-onset schizophrenia. *Brain* 130, 2375–2386. doi: 10.1093/brain/awm184

Duncan, N. W., Wiebking, C., Munoz-Torres, Z., and Northoff, G. (2013). How to investigate neuro-biochemical relationships on a regional level in humans? Methodological considerations for combining functional with biochemical imaging. *J. Neurosci. Methods* 221, 183–188. doi: 10.1016/j.jneumeth.2013.10.011

Dunn, B. D., Dalgleish, T., Ogilvie, A. D., and Lawrence, A. D. (2007). Heartbeat perception in depression. *Behav. Res. Ther.* 45, 1921–1930. doi: 10.1016/j.brat.2006.09.008

Ernst, J., Böker, H., Hättenschwiler, J., Schüpbach, D., Northoff, G., Seifritz, E., et al. (2014). The association of interoceptive awareness and alexithymia with neurotransmitter concentrations in insula and anterior cingulate. *Soc. Cogn. Affect. Neurosci.* 9, 857–863. doi: 10.1093/scan/nst058

Farb, N. A. S., Segal, Z. V, and Anderson, A. K. (2013). Attentional modulation of primary interoceptive and exteroceptive cortices. *Cereb. Cortex* 23, 114–126. doi: 10.1093/cercor/bhr385

Frewen, P. A., Lanius, R. A., Dozois, D. J. A., Neufeld, R. W. J., Pain, C., Hopper, J. W., et al. (2008). Clinical and neural correlates of alexithymia in posttraumatic stress disorder. *J. Abnorm. Psychol.* 117, 171–181. doi: 10.1037/0021-843X.117.1.171

Garfinkel, S. N., Seth, A. K., Barrett, A. B., Suzuki, K., and Critchley, H. D. (2014). Knowing your own heart: distinguishing interoceptive accuracy from interoceptive awareness. *Biol. Psychol.* 104, 65–74. doi: 10.1016/j.biopsycho.2014.11.004

Grabe, H. J., Wittfeld, K., Hegenscheid, K., Hosten, N., Lotze, M., Janowitz, D., et al. (2014). Alexithymia and brain gray matter volumes in a general population sample. *Hum. Brain Mapp.* 35, 5932–5945. doi: 10.1002/hbm.22595

Gu, X., Hof, P. R., Friston, K. J., and Fan, J. (2013). Anterior insular cortex and emotional awareness. *J. Comp. Neurol.* 521, 3371–3388. doi: 10.1002/cne.23368

Harshaw, C. (2015). Interoceptive dysfunction?: toward an integrated framework for understanding somatic and affective disturbance in depression. *Psychol. Bull.* 141, 311–363. doi: 10.1037/a0038101

Heinzel, A., Schäfer, R., Müller, H.-W., Schieffer, A., Ingenhag, A., Eickhoff, S. B., et al. (2010). Increased activation of the supragenual anterior cingulate cortex during visual emotional processing in male subjects with high degrees of alexithymia: an event-related fMRI study. *Psychother. Psychosom.* 79, 363–370. doi: 10.1159/000320121

Herbert, B. M., Herbert, C., and Pollatos, O. (2011). On the relationship between interoceptive awareness and alexithymia: is interoceptive awareness related to emotional awareness? *J. Pers.* 79, 1149–1175. doi: 10.1111/j.1467-6494.2011.00717.x

Honkalampi, K., Hintikka, J., Antikainen, R., Lehtonen, J., and Viinamäki, H. (2001). Alexithymia in patients with major depressive disorder and comorbid cluster C personality disorders: a 6-month follow-up study. *J. Pers. Disord.* 15, 245–254. doi: 10.1521/pedi.15.3.245.19211

Honkalampi, K., Hintikka, J., Tanskanen, A., Lehtonen, J., and Viinamäki, H. (2000). Depression is strongly associated with alexithymia in the general population. *J. Psychosom. Res.* 48, 99–104. doi: 10.1016/S0022-3999(99)00083-5

Horn, D. I., Yu, C., Steiner, J., Buchmann, J., Kaufmann, J., Osoba, A., et al. (2010). Glutamatergic and resting-state functional connectivity correlates of severity in major depression - the role of pregenual anterior cingulate cortex and anterior insula. *Front. Syst. Neurosci.* 4:33. doi: 10.3389/fnsys.2010.00033

Jenkinson, M., Bannister, P., Brady, M., and Smith, S. (2002). Improved optimization for the robust and accurate linear registration and motion correction of brain images. *Neuroimage* 17, 825–841. doi: 10.1006/nimg.2002.1132

Jenkinson, M., Beckmann, C. F., Behrens, T. E. J., Woolrich, M. W., and Smith, S. M. (2012). FSL. *Neuroimage* 62, 782–790. doi: 10.1016/j.neuroimage.2011.09.015

Kano, M., and Fukudo, S. (2013). The alexithymic brain: the neural pathways linking alexithymia to physical disorders. *Biopsychosoc. Med.* 7:1. doi: 10.1186/1751-0759-7-1

Kano, M., Hamaguchi, T., Itoh, M., Yanai, K., and Fukudo, S. (2007). Correlation between alexithymia and hypersensitivity to visceral stimulation in human. *Pain* 132, 252–263. doi: 10.1016/j.pain.2007.01.032

Karlsson, H., Näätänen, P., and Stenman, H. (2008). Cortical activation in alexithymia as a response to emotional stimuli. *Br. J. psychiatry* 192, 32–38. doi: 10.1192/bjp.bp.106.034728

Karvonen, J. T., Veijola, J., Kokkonen, P., Läksy, K., Miettunen, J., and Joukamaa, M. (2005). Somatization and alexithymia in young adult Finnish population. *Gen. Hosp. Psychiatry* 27, 244–249. doi: 10.1016/j.genhosppsych.2005.04.005

Kelly, C., Toro, R., Di Martino, A., Cox, C. L., Bellec, P., Castellanos, F. X., et al. (2012). A convergent functional architecture of the insula emerges across imaging modalities. *Neuroimage* 61, 1129–1142. doi: 10.1016/j.neuroimage.2012.03.021

Kelly, R. E. Jr., Alexopoulos, G. S., Wang, Z., Gunning, F. M., Murphy, C. F., Morimoto, S. S., et al. (2010). Visual inspection of independent components: defining a procedure for artifact removal from fMRI data. *J. Neurosci. Methods* 189, 233–245. doi: 10.1016/j.jneumeth.2010.03.028

Kriegeskorte, N., Simmons, W. K., Bellgowan, P. S., and Baker, C. I. (2009). Circular analysis in systems neuroscience: the dangers of double dipping. *Nat. Neurosci.* 12, 535–540. doi: 10.1038/nn.2303

Lamm, C., and Decety, J. (2008). Is the extrastriate body area (EBA) sensitive to the perception of pain in others? *Cereb. Cortex* 18, 2369–2373. doi: 10.1093/cercor/bhn006

Lamm, C., and Singer, T. (2010). The role of anterior insular cortex in social emotions. *Brain Struct. Funct.* 214, 579–591. doi: 10.1007/s00429-010-0251-3

Lemche, E., Brammer, M. J., David, A. S., Surguladze, S., a, Phillips, M. L., Sierra, M., et al. (2013). Interoceptive-reflective regions differentiate alexithymia traits in depersonalization disorder. *Psychiatry Res.* 214, 66–72. doi: 10.1016/j.pscychresns.2013.05.006

Leweke, F., Leichsenring, F., Kruse, J., and Hermes, S. (2012). Is alexithymia associated with specific mental disorders? *Psychopathology* 45, 22–28. doi: 10.1159/000325170

Leweke, F., Stark, R., Milch, W., Kurth, R., Schienle, A., Kirsch, P., et al. (2004). [Patterns of neuronal activity related to emotional stimulation in alexithymia]. *Psychother. Psychosom. Med. Psychol.* 54, 437–444. doi: 10.1055/s-2004-828350

Liemburg, E. J., Swart, M., Bruggeman, R., Kortekaas, R., Knegtering, H., Curciæ-Blake, B., et al. (2012). Altered resting state connectivity of the default mode network in alexithymia. *Soc. Cogn. Affect. Neurosci.* 7, 660–666. doi: 10.1093/scan/nss048

Luminet, O., Bagby, R. M., and Taylor, G. J. (2001). An evaluation of the absolute and relative stability of alexithymia in patients with major depression. *Psychother. Psychosom.* 70, 254–260. doi: 10.1159/000056263

Mattila, A. K., Kronholm, E., Jula, A., Salminen, J. K., Koivisto, A.-M., Mielonen, R.-L., et al. (2008). Alexithymia and somatization in general population. *Psychosom. Med.* 70, 716–722. doi: 10.1097/PSY.0b013e31816ffc39

McKeown, M. J., Makeig, S., Brown, G. G., Jung, T. P., Kindermann, S. S., Bell, A. J., et al. (1998). Analysis of fMRI data by blind separation into independent spatial components. *Hum. Brain Mapp.* 6, 160–188. doi:10.1002/(SICI)1097-0193(1998)6:3<160::AID-HBM5>3.0.CO;2-1

McRae, K., Reiman, E. M., Fort, C. L., Chen, K., and Lane, R. D. (2008). Association between trait emotional awareness and dorsal anterior cingulate activity during emotion is arousal-dependent. *Neuroimage* 41, 648–655. doi: 10.1016/j.neuroimage.2008.02.030

Menon, V. (2011). Large-scale brain networks and psychopathology: a unifying triple network model. *Trends Cogn. Sci.* 15, 483–506. doi: 10.1016/j.tics.2011.08.003

Menon, V., and Uddin, L. Q. (2010). Saliency, switching, attention and control: a network model of insula function. *Brain Struct. Funct.* 214, 655–667. doi: 10.1007/s00429-010-0262-0

Mesulam, M. M., and Mufson, E. J. (1982). Insula of the old world monkey. III: efferent cortical output and comments on function. *J. Comp. Neurol.* 212, 38–52. doi: 10.1002/cne.902120104

Moisset, X., Bouhassira, D., Denis, D., Dominique, G., Benoit, C., and Sabaté, J.-M. (2010). Anatomical connections between brain areas activated during rectal distension in healthy volunteers: a visceral pain network. *Eur. J. Pain* 14, 142–148. doi: 10.1016/j.ejpain.2009.04.011

Moriguchi, Y., Decety, J., Ohnishi, T., Maeda, M., Mori, T., Nemoto, K., et al. (2007). Empathy and judging other's pain: an fMRI study of alexithymia. *Cereb. Cortex* 17, 2223–2234. doi: 10.1093/cercor/bhl130

Moriguchi, Y., and Komaki, G. (2013). Neuroimaging studies of alexithymia: physical, affective, and social perspectives. *Biopsychosoc. Med.* 7:8. doi: 10.1186/1751-0759-7-8

Nakao, M., and Barsky, A. J. (2007). Clinical application of somatosensory amplification in psychosomatic medicine. *Biopsychosoc. Med.* 1:17. doi: 10.1186/1751-0759-1-17

Nakao, M., Barsky, A. J., Kumano, H., and Kuboki, T. (2002). Relationship between somatosensory amplification and alexithymia in a Japanese psychosomatic clinic. *Psychosomatics* 43, 55–60. doi: 10.1176/appi.psy.43.1.55

Nieuwenhuys, R., Voogd, J., and Van Huijzen, C. (2007). *The Human Central Nervous System: A Synopsis and Atlas, 4th Edn.* Berlin Heidelberg: Springer.

Nyboe Jacobsen, L., Smith Lassen, I., Friis, P., Videbech, P., and Wentzer Licht, R. (2006). Bodily symptoms in moderate and severe depression. *Nord. J. Psychiatry* 60, 294–298. doi: 10.1080/08039480600790358

Pollatos, O., and Schandry, R. (2004). Accuracy of heartbeat perception is reflected in the amplitude of the heartbeat-evoked brain potential. *Psychophysiology* 41, 476–482. doi: 10.1111/1469-8986.2004.00170.x

Pollatos, O., Schandry, R., Auer, D. P., and Kaufmann, C. (2007). Brain structures mediating cardiovascular arousal and interoceptive awareness. *Brain Res.* 1141, 178–187. doi: 10.1016/j.brainres.2007.01.026

Porges, S. W. (1993). *Body Perception Questionnaire (German version).* Laboratory of Developmental Assessment, University of Maryland.

Saarijarvi, S., Salminen, J. K., and Toikka, T. (2006). Temporal stability of alexithymia over a five-year period in outpatients with major depression. *Psychother. Psychosom.* 75, 107–112. doi: 10.1159/000090895

Saarijärvi, S., Salminen, J. K., and Toikka, T. B. (2001). Alexithymia and depression: a 1-year follow-up study in outpatients with major depression. *J. Psychosom. Res.* 51, 729–733. doi: 10.1016/S0022-3999(01)00257-4

Seeley, W. W., Menon, V., Schatzberg, A. F., Keller, J., Glover, G. H., Kenna, H., et al. (2007). Dissociable intrinsic connectivity networks for salience processing and executive control. *J. Neurosci.* 27, 2349–2356. doi: 10.1523/JNEUROSCI.5587-06.2007

Shulman, R. G., Hyder, F., and Rothman, D. L. (2009). Baseline brain energy supports the state of consciousness. *Proc. Natl. Acad. Sci. U.S.A.* 106, 11096–11101. doi: 10.1073/pnas.0903941106

Sifneos, P. E (1973). The prevalence of "alexihtymic" characteristic mechanisms in psychosomatic patients. *Psychother. Psychosom.* 21, 133–136.

Simmons, A., Strigo, I., Matthews, S. C., Paulus, M. P., and Stein, M. B. (2006). Anticipation of aversive visual stimuli is associated with increased insula activation in anxiety-prone subjects. *Biol. Psychiatry* 60, 402–409. doi: 10.1016/j.biopsych.2006.04.038

Singer, T., Critchley, H. D., and Preuschoff, K. (2009). A common role of insula in feelings, empathy and uncertainty. *Trends Cogn. Sci.* 13, 334–340. doi: 10.1016/j.tics.2009.05.001

Singer, T., Seymour, B., Doherty, J. O., Kaube, H., Dolan, R. J., and Frith, C. D. (2004). Empathy for pain involves the affective but not sensory components of pain. *Science* 303, 1157–1163. doi: 10.1126/science.1093535

Smith, S. M. (2002). Fast robust automated brain extraction. *Hum. Brain Mapp.* 17, 143–155. doi: 10.1002/hbm.10062

Smith, S. M., Jenkinson, M., Woolrich, M. W., Beckmann, C. F., Behrens, T. E. J., Johansen-Berg, H., et al. (2004). Advances in functional and structural MR image analysis and implementation as FSL. *Neuroimage* 23, S208–S219. doi: 10.1016/j.neuroimage.2004.07.051

Stingl, M., Bausch, S., Walter, B., Kagerer, S., Leichsenring, F., and Leweke, F. (2008). Effects of inpatient psychotherapy on the stability of alexithymia characteristics. *J. Psychosom. Res.* 65, 173–180. doi: 10.1016/j.jpsychores.2008.01.010

Sui, J., Adali, T., Pearlson, G. D., and Calhoun, V. D. (2009). An ICA-based method for the identification of optimal FMRI features and components using combined group-discriminative techniques. *Neuroimage* 46, 73–86. doi: 10.1016/j.neuroimage.2009.01.026

Taylor, G. J., Bagby, R. M., and Parker, J. D. (2003). The 20-Item Toronto Alexithymia Scale IV. Reliability and factorial validity in different languages and cultures. *J. Psychosom. Res.* 55, 277–283. doi: 10.1016/S0022-3999(02)00601-3

Taylor, K. S., Seminowicz, D. A., and Davis, K. D. (2009). Two systems of resting state connectivity between the insula and cingulate cortex. *Hum. Brain Mapp.* 30, 2731–2745. doi: 10.1002/hbm.20705

Terasawa, Y., Shibata, M., Moriguchi, Y., and Umeda, S. (2013). Anterior insular cortex mediates bodily sensibility and social anxiety. *Soc. Cogn. Affect Neurosci.* 8, 259–266. doi: 10.1093/scan/nss108

Terhaar, J., Viola, F. C., Bär, K.-J., and Debener, S. (2012). Heartbeat evoked potentials mirror altered body perception in depressed patients. *Clin. Neurophysiol.* 123, 1950–1957. doi: 10.1016/j.clinph.2012.02.086

van der Velde, J., Servaas, M. N., Goerlich, K. S., Bruggeman, R., Horton, P., Costafreda, S. G., et al. (2013). Neural correlates of alexithymia: a meta-analysis of emotion processing studies. *Neurosci. Biobehav. Rev.* 37, 1774–1785. doi: 10.1016/j.neubiorev.2013.07.008

Wiebking, C., Bauer, A., de Greck, M., Duncan, N. W., Tempelmann, C., and Northoff, G. (2010). Abnormal body perception and neural activity in the insula in depression: an fMRI study of the depressed "material me." *World J. Biol. Psychiatry* 11, 538–549. doi: 10.3109/15622970903563794

Wiebking, C., de Greck, M., Duncan, N. W., Tempelmann, C., Bajbouj, M., and Northoff, G. (2015). Interoception in insula subregions as a possible state marker for depression - an exploratory fMRI study investigating healthy, depressed and remitted participants. 9:82. *Front. Behav. Neurosci.* doi: 10.3389/fnbeh.2015.00082

Wiebking, C., Duncan, N. W., Tiret, B., Hayes, D. J., Marjanska, M., Doyon, J., et al. (2014). GABA in the insula - a predictor of the neural response to interoceptive awareness. *Neuroimage* 86, 10–18. doi: 10.1016/j.neuroimage.2013.04.042

Wingbermühle, E., Theunissen, H., Verhoeven, W. M. A., Kessels, R. P. C., and Egger, J. I. M. (2012). The neurocognition of alexithymia: evidence from neuropsychological and neuroimaging studies. *Acta Neuropsychiatr.* 24, 67–80. doi: 10.1111/j.1601-5215.2011.00613.x

Woolrich, M. W., Jbabdi, S., Patenaude, B., Chappell, M., Makni, S., Behrens, T., et al. (2009). Bayesian analysis of neuroimaging data in FSL. *Neuroimage* 45, S173–S186. doi: 10.1016/j.neuroimage.2008.10.055

Worsley, K. J. (2001). "Statistical analysis of activation images," in *Functional MRI: An Introduction to Methods*, eds P. Jezzard, P. Matthews and S. Smith (Oxford: Oxford University Press), 1–23.

Zou, Q., Wu, C. W., Stein, E. A., Zang, Y., and Yang, Y. (2009). Static and dynamic characteristics of cerebral blood flow during the resting state. *Neuroimage* 48, 515–524. doi: 10.1016/j.neuroimage.2009.07.006

8

Psychophysiology of Duration Estimation in Experienced Mindfulness Meditators and Matched Controls

*Simone Otten[1], Eva Schötz[1], Marc Wittmann[2], Niko Kohls[3], Stefan Schmidt[4,5] and Karin Meissner[1]**

[1] Meissner Lab, Institute of Medical Psychology, Ludwig-Maximilians-University, Munich, Germany, [2] Institute for Frontier Areas of Psychology and Mental Health, Freiburg, Germany, [3] Division Integrative Health Promotion, University of Applied Sciences and Arts, Coburg, Germany, [4] Department of Psychosomatic Medicine, University Medical Center Freiburg, Freiburg, Germany, [5] Institute for Transcultural Health Studies, European University Viadrina, Frankfurt (Oder), Germany

***Correspondence:**
Karin Meissner,
Institute of Medical Psychology,
Goethestr. 31, 80336 Munich,
Germany
karin.meissner@med.lmu.de

Recent research suggests that bodily signals and interoception are strongly related to our sense of time. Mindfulness meditators train to be aware of their body states and therefore could be more accurate at interval timing. In this study, $n = 22$ experienced mindfulness meditators and $n = 22$ matched controls performed both, an acoustic and a visual duration reproduction task of 8, 14, and 20 s intervals, while heart rate and skin conductance were continuously assessed. In addition, participants accomplished a heart beat perception task and two selective attention tasks. Results revealed no differences between meditators and controls with respect to performance in duration reproduction or attentional capacities. Additionally no group difference in heart beat perception scores was found. Across all subjects, correlational analyses revealed several associations between performance in the duration reproduction tasks and psychophysiological changes, the latter being also related to heart beat perception scores. Furthermore, former findings of linearly increasing cardiac periods and decreasing skin conductance levels during the auditory duration estimation task (Meissner and Wittmann, 2011) could be replicated, and these changes could also be observed during a visual duration reproduction task. In contrast to our earlier findings, the heart beat perception test was not related with timing performance. Overall, although experienced meditators did not differ from matched controls with respect to duration reproduction and interoceptive awareness, this study adds significantly to the emerging view that time perception is related to autonomic regulation and awareness of body states.

Keywords: mindfulness meditation, time perception, interoceptive awareness, autonomic regulation, attention

Introduction

The diversity of psychological and neurophysiological models of time perception is indicative of the fact that the neural substrates and the processes accounting for the experience of time are still unknown (Wittmann and Paulus, 2009; Merchant et al., 2013). Recent conceptualizations that are based on empirical findings hint on the relationship between the experience of time and self-and body processes, that is, on interoceptive states (Craig, 2009, 2015; Wittmann, 2009, 2013):

Recordings in functional magnetic resonance imaging (fMRI) indicated that activity within the insular cortex increased continuously during the perception of duration (Wittmann et al., 2010, 2011). Moreover, the experience of temporal delay between external acoustic stimulation and the heart beat is related to anterior insula activation (Critchley et al., 2004). Given the close connection between the insular cortex and interoceptive processes, the integration and perception of ascending bodily signals would substantially contribute to the experience of time.

Recordings of physiological signals (heart periods, skin conductance levels, respiratory periods) during a duration reproduction task, with intervals of 8, 14, and 20 s, showed not only a linear increase of cardiac periods during the encoding of temporal intervals, but also revealed a positive relationship between the slope of this increase in cardiac periods and duration reproduction accuracy (Meissner and Wittmann, 2011). A recent study also reported a positive correlation between a greater control of the vagal tone and duration reproduction accuracy for intervals from 0.5 to 40 s (Pollatos et al., 2014). That is, the increase in cardiac (parasympathetic) tone, as registered in the brain, could represent a bodily marker for the estimation of duration. It is important to mention that it is not the mean heart rate during a target interval that correlated with timing behavior (as assessed by Schwarz et al., 2013) but the linear increase of cardiac periods that showed this relationship. In addition, results of Meissner and Wittmann (2011) indicated that duration reproduction accuracy correlated positively with the subjects' ability to accurately perceive their own heart beats. A possible explanation would be that individuals, who have a greater awareness of their body states and thus a better access to visceral feedback, are more accurate at interval timing. Alternatively, the correlations between duration reproduction accuracy and both, the slope of cardiac slowing and interoceptive awareness could be due to increased attentional capacities of good performers in time estimation tasks, who then simply would be able to focus more adequately on time perception. In this scenario, the steeper slope of cardiac periods in good performers would reflect their higher attentional load, and the relationship with interoceptive awareness could be due to the recently reported association between interoceptive awareness and attention to external acoustic or visual stimuli that are used for presentation of temporal intervals (Matthias et al., 2009).

In this study, we investigated the role of body signals, interoceptive awareness, and attentional capacities for interval timing by focusing on a group of participants who regularly train their attentional capacities to internal and external stimuli, namely experts in mindfulness meditation. Mindfulness is understood as bringing awareness to each present moment in time with an accepting and non-judgmental attitude. It can be developed and cultivated through introspective training such as meditation (Kabat-Zinn, 2005; Sauer et al., 2013). Mindfulness meditators during a typical meditation session focus on their breathing-in and breathing-out or perform a "body scan" where attention is systematically directed to each area of the body from the toes to the top of the head (Morone et al., 2008). Effects of mindfulness meditation are manifold but basically comprise increased body awareness, attention regulation, and emotion regulation (Hölzel et al., 2011; Tang et al., 2015) all of which are mental factors that are involved in time perception (Wittmann and Schmidt, 2014). Accordingly, regular mindfulness meditation has been shown to alter the cortical representation of interoceptive attention (Farb et al., 2013). Moreover, there is evidence indicating that mindfulness meditators after intensive practice have increased attentional capacities to external stimuli (Jha et al., 2007; Zeidan et al., 2010; Moore et al., 2012). These findings suggest that mindfulness meditators may process time differently than matched controls. Indeed, mindfulness meditators report that subjective time slows down during mindfulness-oriented meditation and in daily life (Kabat-Zinn, 2005).

In this study, we aimed to compare the role of bodily signals and attentional capacities to internal and external stimuli for time estimation in $n = 22$ experienced mindfulness meditators and $n = 22$ matched controls. Participants performed two duration reproduction tasks, while heart rate and skin conductance were monitored. In addition, they accomplished a heart beat perception task and two selective attention tasks. We hypothesized that in comparison to non-meditating controls, the mindfulness meditation experts would reproduce time intervals more accurately. We also expected that mindfulness meditators would show higher attentional capacities and would score higher in interoceptive awareness. Finally, we investigated whether the expected differences between meditators and controls would go along with alterations of bodily signals during the duration estimation tasks (Meissner and Wittmann, 2011; Wittmann and Schmidt, 2014).

A secondary aim of the study was to investigate whether we could replicate our former findings of linearly increasing cardiac periods and decreasing skin conductance levels during the auditory duration estimation task (Meissner and Wittmann, 2011), and whether these changes could also be observed during an otherwise similar visual duration reproduction task.

Materials and Methods

Design

We conducted a cross-sectional study comparing $n = 22$ experienced mindfulness meditators with $n = 22$ matched controls. The study was conducted at the Institute of Medical Psychology, University of Munich, as part of a larger two-center study, which compared 42 mindfulness meditators with 42 matched controls in their ability of time perception using a broad range of behavioral tasks (Wittmann et al., 2015). The study was approved by the ethical committees of the partaking Universities (Munich and Freiburg).

Participants

Twenty-two experienced mindfulness meditation practitioners with at least three years of continuous practice and at least 2 h of practice a week over the last 8 weeks were recruited, as well as $n = 22$ matched controls without any meditation experience. The matching criteria were sex, age (± 3 years) and education (± 1 level of 5). Participants were recruited by advertisements in meditation centers, on online platforms of the universities, and by word of mouth. Meditators were included when reporting to regularly practice a form of meditation, which had a dominant orientation toward awareness of the present moment. Therefore, we included individuals if they participated in forms of mindfulness meditation, Vipassana meditation, or Soto Zen. Control subjects were required to have no experience with any form of meditation including Yoga or Tai-Chi. Age was restricted to the range between 21 and 50 years in order to constrain age-related effects in the psychophysical tasks and psychophysiological measurements. Participants had to be fluent in German and were additionally required to report good health and no known medical or psychological problems as assessed with a detailed screening questionnaire. All participants signed an informed consent and received a moderate financial compensation (€20) for participation.

Instruments

Heart Beat Perception Task

The heart beat perception task consists of four heart beat counting intervals (35, 25, 45, and 60 s), during which participants are asked to attend to their own heart beats and count them silently (Pollatos et al., 2007). The beginning and end of each counting interval is signaled by a start and stop tone. A heart beat perception score is calculated for each participant across the four trials according to the following equation where high scores (maximum = 1) indicate accurate heart beat perception:

$$1 - 1/4\Sigma((|\text{recorded heart beats—counted heart beats}|)/\text{recorded heart beats})$$

Freiburg Mindfulness Inventory (FMI)

The FMI (Walach et al., 2006; Kohls et al., 2009) contains 14 items which evaluate self-reported mindfulness on the basis of a two-dimensional structure utilizing a 4-point item scale format. These dimensions are "presence" as ability to attend to the present moment ("I am open to the experience of the present moment") and "acceptance" as non-judgmental attitude ("I am patient with myself when things go wrong").

The Attention Network Test (ANT)

The ANT (Fan et al., 2002) assesses the processing efficiency of the three attention networks of (1) alerting, (2) orienting, and (3) executive attention (conflict effect). They are quantified by means of computerized reaction time measures for differently cued and un-cued stimulus conditions. Participants have to respond to either the left or right arrow key on the computer keyboard depending on stimulus configuration. An overall reaction time score and an index of accuracy are also calculated.

Divided Attention

Divided attention was assessed by a subtest of the "Test Battery for Attentional Performance" [TAP, Version 1.7;Psytest: Herzogenrath (Zimmermann and Fimm, 1997, 2002)]. The subtest comprised a dual task paradigm in which the participants were required to simultaneously monitor visual and acoustic stimuli. In the visual task participants had to detect whether crosses that appeared in a random configuration in a 4 × 4 matrix form the corners of a square. The acoustical task included a regular sequence of high and low beeps, whereas participants had to detect an irregularity in the sequence of these beeps.

Visual and Auditory Duration Reproduction

In two separate computer tasks running on Psychtoolbox for Matlab subjects had to reproduce the duration of (1) a visual and (2) an auditory stimulus with intervals of 8, 14, and 20 s duration. In each trial, a green square (resp. a sinus tone of 440 Hz) was first presented for one of the three durations. After the pause of either 4.5, 5, or 5.5 s duration a second yellow square (resp. a sinus tone of 500 Hz) appeared. It had to be stopped by pressing the space bar when participants felt that this second stimulus had reached the duration of the first. The reproduction task contained six presentations of each of the three durations summing up to 18 trials. At the beginning and the end of each stimulus, an onset resp. offset marker signal was sent from the Matlab program to the physiological acquisition software. Participants were requested not to use mental strategies such as inner counting but to rely on their subjective feeling of elapsed time. To further prevent mental counting, participants were additionally given a secondary working-memory task. Before each trial four numbers were presented. At the end of each trial one number appeared and the subjects responded whether the number was one of the four previously presented by pressing the left or right arrow button, respectively for "yes" and "no." The accuracy of duration reproduction can be calculated from the temporal reproduction performance. In the following, the mean of the six reproduced time intervals of each of the three durations is referred to as "reproduced duration." A duration reproduction score (referred to as "duration reproduction accuracy") was calculated for each participant and each of the three durations from six trials according to the following equation:

$$1 - 1/6\Sigma((|\text{presented duration—reproduced duration}|)/\text{presented duration})$$

Similar to the heart beat perception score high scores (maximum 1) indicate accurate duration reproduction.

Physiological Recordings

Heart rate and electrodermal activity were continuously recorded using a BIOPAC MP 150 device (BIOPAC Systems Inc., Goleta, CA, USA) with AcqKnowledge 4.1 software for data acquisition. The sampling rate was 500 Hz for the ECG signal and 15.625 Hz for the electrodermal signal.

Heart rate was measured using three disposable Ag/AgCl electrodes which were positioned in an Einthoven Lead I configuration and connected to the BIOPAC amplifier module ECG100C. Skin conductance was measured using two disposable Ag/AgCl electrodes which were attached to the thenar and hypothenar portions of the left hand and connected to the BIOPAC amplifier module GSR100C.

Data Reduction and Statistical Analysis

Intervals between successive R peaks (cardiac periods) were extracted from the electrocardiogram signal using the peak-detection function implemented in AcqKnowledge 4.2. Cardiac periods were examined and screened for artifacts based on the procedure developed by Porges and Byrne (1992). Cardiac periods were re-sampled at 15.625 Hz by using cubic interpolation. Cardiac periods and skin conductance levels were averaged for every second across the six presentations of each of the three durations (8, 14, 20 s). Skin conductance levels were log-transformed prior to statistical analysis to obtain normal distributions.

Differences between study groups where analyzed by Student-*t*-tests for normally distributed data, Mann–Whitney *U*-tests for non-parametric data, and Chi-Quadrat tests for categorial data. Group differences for performance in the auditory and visual duration reproduction tasks were explored by mixed-design ANOVAs, with the between-subjects factor "group" (mindfulness meditators, controls) and the within-subjects factor "interval" (8, 14, 20 s). To evaluate changes over time in cardiac periods and skin conductance levels separate mixed-design ANOVAs were performed for the encoding and reproduction phase of each duration interval. Due to individual differences in the length of the individually reproduced intervals, the analyses for physiological changes during the reproduction intervals for auditory and visual reproduction tasks were restricted to seconds that were available for all subjects. In the auditory reproduction task we analyzed the first 6, 8, and 12 s of the 8, 14, and 20 s intervals in the reproduction phase of the task, respectively. In the visual reproduction tasks the analyses were restricted to the first 6, 8, and 15 s for cardiac periods and 6, 8, 12 s for skin conductance levels of the 8, 14, and 20 s intervals in the reproduction phase of the task, respectively. Greenhouse-Geisser corrections for repeated measures ANOVAs were applied where appropriate.

Individual linear slopes of cardiac periods and skin conductance levels for each of the 18 encoding intervals were determined using linear regression. In order to normalize the data, individual time series were converted into z-scores before performing the linear regression analyses. Individual slopes were then averaged for each participant across the six presentations for each of the three durations. Group differences between slopes were explored by mixed-design ANOVAs with "interval" as the within-subjects factor and "group" as between-subjects factor.

Spearman-Rho's correlations were used to explore relationships between behavioral results in the duration reproduction tasks, associated psychophysiological changes, heart beat perception scores, and results in the attention tests.

For statistical tests the significance level was set to $p \leq 0.05$. Whenever multiple tests were conducted, the significance level was Bonferroni-corrected.

Results

Study Sample

On average, the mindfulness meditators had a meditation experience of 10.4 years (± 7.5 SD) and had practiced 7 h per week (± 5 SD). Meditation and control groups did not differ with respect to age, education, body mass index, physical activity, or current stress levels. As expected, meditators showed significantly higher values of self-reported mindfulness in the "Freiburg Mindfulness Inventory" (FMI) in both subscales of "presence" and "acceptance," as well as in the sum score (**Table 1**).

Heart Beat Perception Scores

The mean heart beat perception scores were 0.79 (± 0.18 SD) in meditators and 0.78 (± 0.15 SD) in controls, with no significant difference in-between (Mann–Whitney *U*-test, $p = 0.751$; **Figure 1**).

TABLE 1 | Characteristics of the two study groups.

Variable	Mindfulness meditators (*n* =22)	Matched controls (*n* = 22)	*p*-value[a]
Age (mean ± SD)	39.7 ± 7.9	39.5 ± 8.0	0.953[b]
Sex (female (%))	12 (55)	12 (55)	0.619[c]
Body mass index (mean ± SD)	21.6 ± 2.3	21.9 ± 2.7	0.675
Educational level			0.794[c]
Primary school; *Hauptschule*, n (%)	0	1 (5)	
Secondary modern school; *Realschule*, n (%)	1 (5)	1 (5)	
Grammer school; *Gymnasium*, n (%)	4 (18)	4 (18)	
University or college degree; *Studium*, n (%)	17 (77)	16 (73)	
Physical Activity (ln IPAQ, mean ± SD)	7.84 ± 0.65	7.76 ± 1.10	0.778
Current stress level (SQCB, mean ± SD)	12.73 ± 3.7	12.45 ± 4.5	0.822
Mindfulness (FMI)			
Presence (mean ± SD)	25.2 ± 2.8	20.3 ± 3.9	**<0.001**
Acceptance (mean ± SD)	19.3 ± 2.1	16.4 ± 37.5	**0.001**
Sum score (mean ± SD)	44.7 ± 4.2	37.5 ± 5.8	**<0.001**

IPAQ, International Physical Activity Questionnaire; SQCB, Short Questionnaire on Current Burden; FMI, Freiburg Mindfulness Inventory.
Bold values signify p < 0.05.
[a] *t-Test if not otherwise indicated.*
[b] *Mann–Whitney U-Test.*
[c] *Chi-Square Test.*

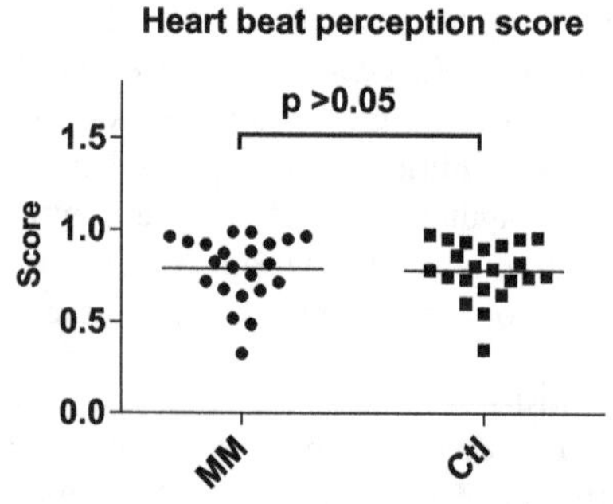

FIGURE 1 | Statistical dot plot with means (black bars) showing heart beat perception scores in mindfulness meditators (MM) and matched controls (Ctl). Mann–Whitney *U*-Tests were used to evaluate group differences.

Attentional Capacities

Meditators and controls did not differ in any of the variables of the Attention Network Test (**Figure 2**; Supplementary Table 1) or in the mean accuracy for the TAP subtest "divided attention" (Supplementary Table 1).

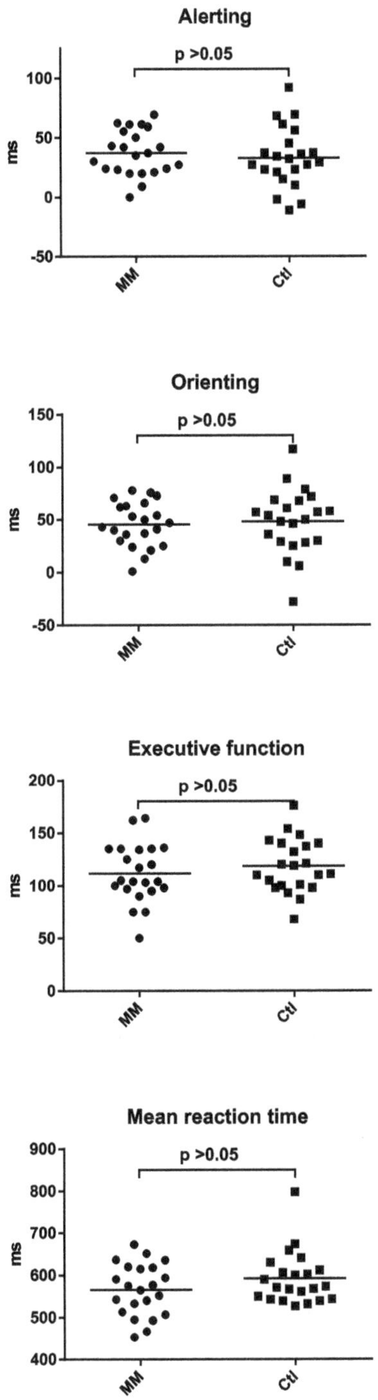

FIGURE 2 | Statistical dot plots with means (black bars) showing the results of the Attention Network Test (ANT) in mindfulness meditators (MM) and matched controls (Ctl). *T*-Tests were used to evaluate group differences.

Duration Reproduction Tasks

Neither in the auditory nor in the visual duration reproduction tasks reproduced duration differed significantly between mindfulness meditators and matched controls (**Figure 3**; Supplementary Table 2). The respective analyses for the accuracy of duration reproduction also did not reveal any significant group differences (Supplementary Table 2).

Psychophysiological Changes in Duration Reproduction Tasks

Cardiac Periods

In both the auditory and visual reproduction tasks mean cardiac periods increased over time during the encoding intervals of 8, 14, and 20 s duration and decreased during the first 4 s after the tone had stopped (**Figure 4**). Mixed-design ANOVAs showed significant main effects for time for each of the three encoding intervals of 8, 14, and 20 s duration during both the auditory and the visual tasks, and linear trends fitted the data reasonably well; no significant main effects for group or interaction effects for time by group were revealed (**Table 2**).

Compared with the respective encoding intervals, cardiac periods showed similar patterns during the reproduction intervals of the respective tasks (**Figure 4**). Mixed-design ANOVAs confirmed some of the changes to be significant, but the results were more ambiguous than during the encoding intervals (**Table 2**).

Skin Conductance Levels

Both in the auditory and visual reproduction tasks, mean skin conductance levels decreased over time during the encoding intervals of 8, 14, and 20 s duration and increased during the first 4 s after the tone had stopped (**Figure 5**). Mixed-design ANOVAs showed significant main effects for time for each of the three encoding intervals of 8, 14, and 20 s duration in the auditory task, and linear trends fitted the data reasonably well (**Table 3**). In the visual duration reproduction task main effects for time were significant for the encoding intervals of 14 and 20 s duration. No significant main effects for group were found in neither task. A significant interaction effect group by time was revealed during the 8 s-encoding interval of the visual reproduction task [$F = 4.90$ (1.62, 63.23); $p = 0.0163$], in that meditators showed a sharper decline of skin conductance levels than controls (**Figure 5**; **Table 3**).

Skin conductance levels showed similar patterns during the reproduction intervals when compared with the respective encoding intervals (**Figure 5**). Mixed-design ANOVAs confirmed some of the main effects of time to be significant, but again, results were much more ambiguous than during the encoding intervals (**Table 3**).

Slopes of Cardiac Periods during the Encoding Intervals of the Auditory and Visual Reproduction Tasks

Across groups, cardiac periods increased by 0.09 z-scores per second, 0.05 z-scores per second, and 0.04 z-scores per second during the 8, 14, and 20 s intervals of the auditory reproduction task, respectively. The slopes of cardiac periods did not differ

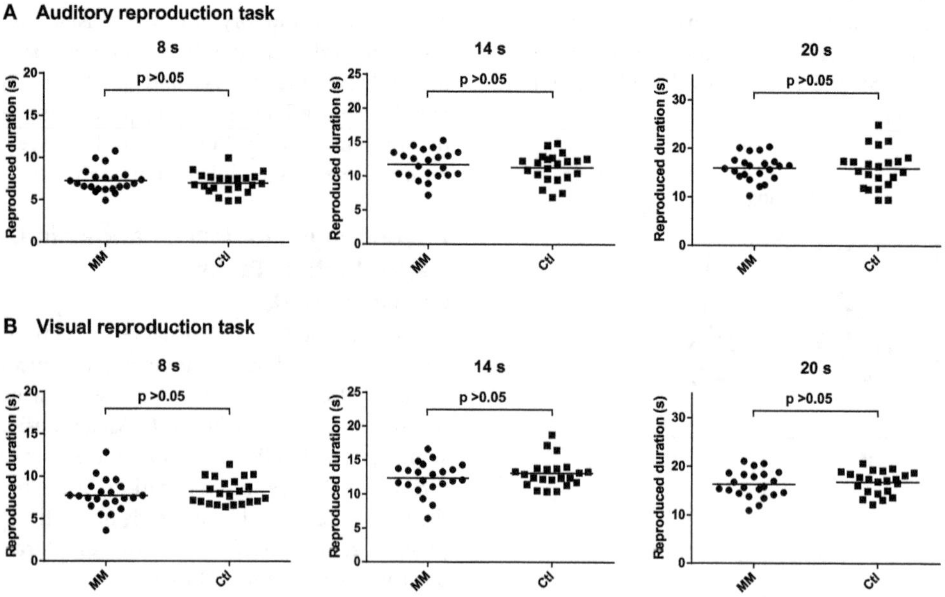

FIGURE 3 | Statistical dot plots with means (black bars) showing reproduced duration for the three intervals of the auditory (A) and visual (B) duration reproduction tasks in mindfulness meditators (MM) and matched controls (Ctl). Mixed analyses of variance revealed no significant differences between groups [main effect of group, auditory task: $F_{(1,\ 42)} = 0.4$, $p = 0.514$; visual task: $F_{(1,\ 42)} = 0.9$, $p = 0.345$]. *P*-values correspond to the results of simple effects *T*-tests.

significantly between meditators and controls (Supplementary Table 3).

Cardiac periods increased by 0.08 z-scores per second, 0.04 z-scores per second, and 0.04 z-scores per second during the 8, 14, and 20 s encoding intervals of the visual reproduction task, respectively. Again, the slopes of meditators and controls did not differ (Supplementary Table 3).

Slopes of Skin Conductance Levels during the Encoding Intervals of the Auditory and Visual Reproduction Tasks

Across groups, skin conductance levels decreased by 0.012 z-scores per second, 0.011 z-scores per second and 0.011 z-scores per second during the 8, 14, and 20 s intervals of the auditory reproduction task, respectively. A significant difference between groups was revealed for the 8-s interval, with a steeper decline in the group of meditators (-0.020 ± 0.026 vs. -0.004 ± 0.020, $p = 0.005$; **Figure 6**, Supplementary Table 3).

During the visual reproduction task, skin conductance levels decreased across all subjects by 0.004, 0.008 and 0.010 z-scores per second during the 8, 14, and 20 s intervals, respectively. Meditators consistently showed a steeper decline of skin conductance levels than controls (**Figure 6**, Supplementary Table 3).

Correlational Analyses

Since no differences between groups were revealed the correlational analyses were performed for the whole study sample ($n = 44$).

Attentional Capacities

Correct answers in the TAP subtest for divided attention correlated positively with heart beat perception scores across all subjects ($r = 0.381$, $p = 0.011$), while the three components of the ANT alerting, orienting, and executive function and ANT reaction times showed no significant association (results not shown). Correct answers in the TAP subtest for divided attention also correlated positively with reproduced duration during the 8 s-encoding interval of the auditory reproduction task (Spearman's $r = 0.396$, $p = 0.008$), and with reproduced duration and reproduction accuracy during the 20 s encoding interval of the visual reproduction task (Spearman's $r = 0.396$, $p = 0.009$; and $r = 0.390$, $p = 0.010$, respectively). Alerting, orienting and executive function in the ANT as well as reaction times did not show such correlations (results not shown).

Heart Beat Perception Scores

Heart beat perception scores were not associated with any of the performance variables in the auditory and visual duration estimation tasks, or with the slope of cardiac periods but correlated negatively with the slope of skin conductance levels during the 14 s-encoding interval of the auditory task (Spearman's $r = -0.404$, $p = 0.009$) and with the slope of skin conductance levels during the 8 s-encoding interval of the visual task (Spearman's $r = -0.451$, $p = 0.003$) (Supplementary Table 4).

Reproduced Duration and Duration Reproduction Accuracy

During the 8 s-interval of the auditory task, the slopes of cardiac periods correlated positively with reproduced duration (**Figure 7**;

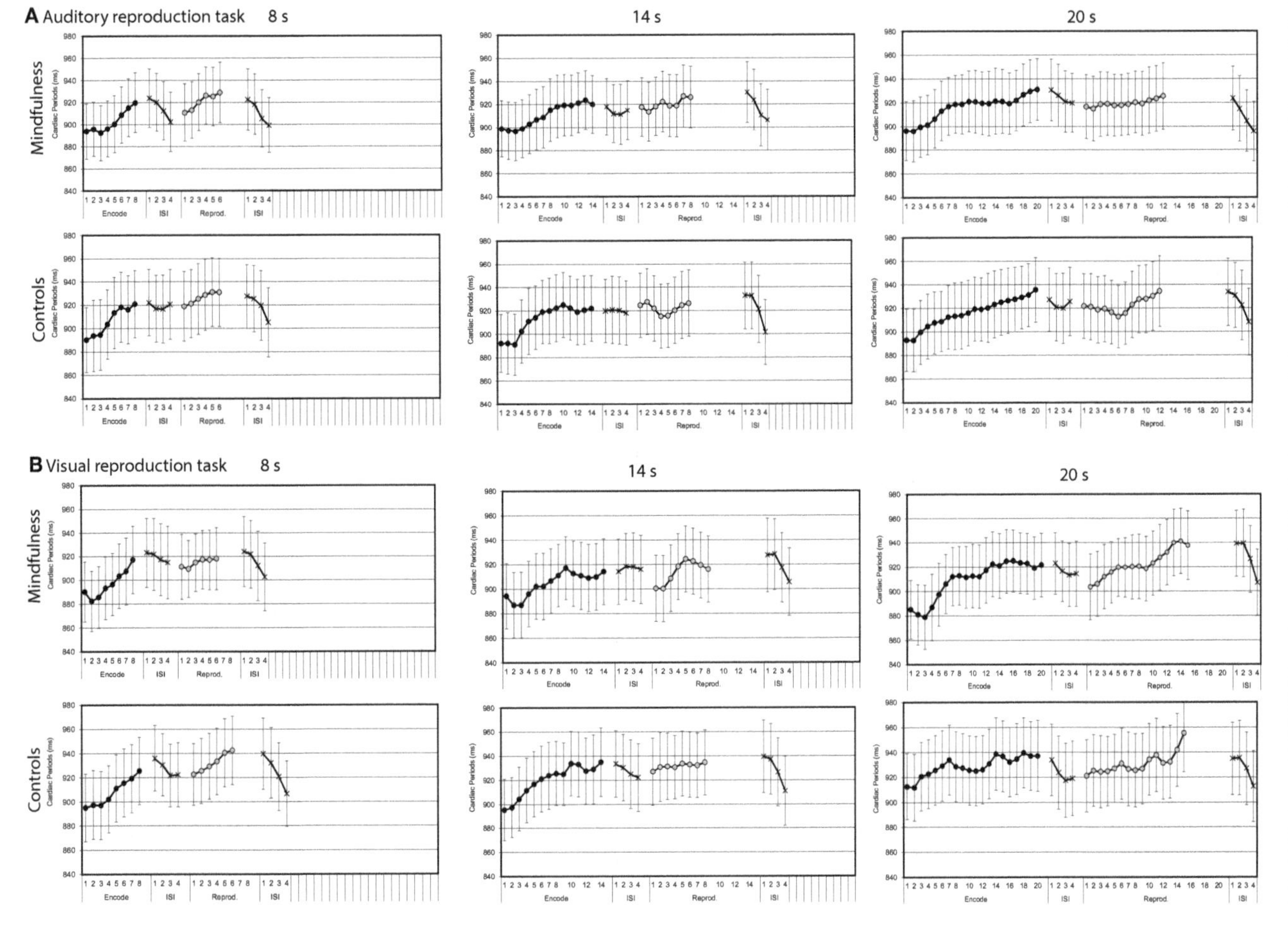

FIGURE 4 | Cardiac periods by groups (mindfulness meditators, controls). Mean second-to-second changes of cardiac periods (ms) during the encoding intervals, the reproduction intervals, and first 4 s of the subsequent inter-stimulus intervals (ISI) for tones of 8, 14, and 20 s duration of both groups (for statistical results, see **Table 2**). **(A)** Auditory reproduction task. **(B)** Visual reproduction task. Error bars represent standard error of the mean.

Supplementary Table 5), while the slopes of skin conductance levels during the auditory 8-s intervals showed a negative correlation with duration reproduction accuracy (Spearman's $r = -0.433, p = 0.005$; Supplementary Table 6).

Discussion

We studied the role of attentional capacities, cardiac awareness, and bodily signals for performance in duration reproduction tasks in $n = 22$ experienced mindfulness meditators and $n = 22$ age-, sex-, and education matched controls. Contrary to our hypotheses, results revealed no differences between meditators and controls with respect to performance in duration reproduction abilities, heart beat perception scores, or attentional capacities. Similarly, the slope of cardiac periods did not differ between groups. Meditators, however, showed a steeper decline of skin conductance levels during the 8 s-encoding intervals of both, the visual and auditory duration reproduction task (**Figure 5**, **Table 3**), which could be due to higher relaxation abilities in experienced meditators (Steinhubl et al., 2015). However, since we did not find any other group differences, we do not want to over-emphasize this finding. Across measures, we observed large inter-individual differences in both meditators and controls, which should be considered in future research.

Our negative findings have to be interpreted in the way that experienced meditators did not perform more accurately in time perception in the seconds' range. Related, psychophysiological parameters, although in general to some degree associated with time perception in the seconds range (see below), did not differ between the two groups studied. Although mindfulness meditators learn to focus on their bodily sensations, they did not differ from carefully matched controls with regard to interoceptive awareness in the heart beat perception task. Two recent studies have also failed to show that experienced meditators are more accurate in the heart beat perception task (Khalsa et al., 2008; Melloni et al., 2013). Our findings are in contrast to studies showing how self-reported aspects of interoception in daily life are enhanced after extended mindfulness meditation (Bornemann et al., 2014). Moreover, objective indices of judgments related to body sensitivity such as tactile discrimination reveal a better performance after extended body-centered meditation (Fox et al., 2012; Mirams

TABLE 2 | Results of the mixed-design ANOVAs for cardiac periods during the three encoding and reproduction intervals of the auditory and visual duration reproduction tasks.

	Interval	Main effect for time		Linear trend		Interaction effect (time by group)		Main effect for group	
		F[df, error(df)]	*p*	*F*[df, error(df)]	*p*	*F*[df, error(df)]	*p*	*F*[df, error(df)]	*p*
Auditory Task	**AUDITORY ENCODING INTERVAL**								
	8 s	16.38 (2.46, 103.28)	**<0.001****	32.15 (1, 42)	**<0.001****	1.15 (2.46, 103.28)	0.328	0.009 (1, 42)	0.926
	14 s	15.96 (4.74, 199.10)	**<0.001****	74.21 (1, 42)	**<0.001****	0.99 (4.74, 199.10)	0.423	0.002 (1, 42)	0.961
	20 s	13.25 (5.78, 242.93)	**<0.001****	59.33 (1, 42)	**<0.001****	0.40 (5.78, 242.93)	0.875	0.000 (1, 42)	0.999
	AUDITORY REPRODUCTION INTERVAL								
	8 s	4.83 (1.93, 80.92)	**0.011****	6.66 (1, 42)	**0.013****	0.22 (1.926, 80.912)	0.797	0.018 (1, 42)	0.893
	14 s	1.43 (3.57, 150.10)	0.233	1.06 (1, 42)	0.307	1.49 (3.57, 150.10)	0.212	0.002 (1, 42)	0.962
	20 s	2.44 (4.22, 177.05)	**0.045***	5.40 (1, 42)	**0.025***	0.65 (4.22, 177.05)	0.633	0.006 (1, 42)	0.939
Visual Task	**VISUAL ENCODING INTERVAL**								
	8 s	15.07 (2.523, 100.92)	**<0.001****	26.28 (1, 40)	**<0.001****	0.33 (2.523, 100.92)	0.770	0.08 (1, 40)	0.779
	14 s	14.73 (4.48, 179.34)	**<0.001****	63.07 (1, 40)	**<0.001****	0.99 (4.48, 179.34)	0.421	0.17 (1, 40)	0.680
	20 s	14.23 (5.22, 208.75)	**<0.001****	41.03 (1, 40)	**<0.001****	2.50 (5.22, 208.75)	**0.029***	0.26 (1, 40)	0.614
	VISUAL REPRODUCTION INTERVAL								
	8 s	4.91 (2.33, 93.26)	**0.007****	8.25 (1, 40)	**0.006****	1.05 (2.33, 93.26)	0.363	0.22 (1, 40)	0.639
	14 s	3.66 (2.88, 115.14)	0.146	5.56 (1, 40)	0.023*	1.84 (2.88, 115.14)	0.146	0.26 (1, 40)	0.638
	20 s	8.44 (4.89, 195.49)	**<0.001****	38.88 (1, 40)	**<0.001****	1.24 (4.89, 195.49)	0.292	0.05 (1, 40)	0.828

$^{*}p < 0.05$.
$^{**}p < 0.0167$ (Bonferroni-corrected level of significance).
Bold values signify $p < 0.05$.

et al., 2013). Expert mindfulness meditators have in general shown to be more sensitive to inner urges and impulses which are important for decision-making and movement initiation (Jo et al., 2014, 2015). The discrepancies with our results may be explained by a dissociation between interoceptive sensibility (i.e., self-attributed interoception) and interoceptive accuracy as assessed in an objective heart beat counting task (Garfinkel et al., 2015): Only self-attributed interoception but not interoceptive accuracy might be enhanced in mindfulness meditators as compared to non-meditating controls. In this context it is important to note that most findings of a relationship between mindfulness meditation and a specific mental function were collected in longitudinal studies. That is, individuals who attended a meditation course or retreat lasting several days or weeks improved performance in attention (Maclean et al., 2010) and timing tasks (Droit-Volet et al., 2015). Most likely, in cross-sectional studies as ours large inter-individual differences not controlled for can mask effects of meditation experience.

Heart beat perception scores were positively related with correct answers in the TAP subtest of divided attention, which confirms previous findings of a relationship between attention to external and internal stimuli (Matthias et al., 2009). Moreover, we found several correlations between heart beat perception scores and performance variables in the duration reproduction tasks, but none of them passed the threshold of significance (Supplementary Table 4). Heart beat perception scores, however, correlated negatively with the slopes of skin conductance levels during the 8 s- and 14 s- interval of the visual and auditory tasks, respectively (Supplementary Table 4). These results suggest that better heart beat perceivers tended to show a steeper decrease of both heart rate and skin conductance levels during the shorter encoding intervals of our tasks. Our results are only partially consistent with results in our previous cohort of younger subjects, who were not selected with respect to meditation experience (Meissner and Wittmann, 2011): In that study, heart beat perception scores were found to correlate positively with duration reproduction accuracy for intervals of 8 and 14 s duration in an identical auditory reproduction task. It should be noted that heart beat perception scores in our present cohort were higher than before [0.78 ± 0.16 SD vs. 0.62 ± 0.18 SD in (Meissner and Wittmann, 2011)]. A ceiling effect may thus explain why we could not reproduce these correlations of our former study. A recent study showed that better heart beat perceivers could better synchronize their heart cycle with the start and stop responses in a time reproduction task with significant correlation for a short 2 s time interval (Pollatos et al., 2014)—a further hint suggesting that interoceptive awareness is indeed associated with better performance in time perception. Again, the heart beat perception score of 0.65 ± 0.19 SD in their study was lower than that in our present cohort.

Further, results revealed that reproduced duration correlated positively with the slope of cardiac periods during the 8-s interval of the auditory task, although only moderately, while duration reproduction accuracy showed a negative correlation with the slopes of skin conductance levels. These results provide further evidence that performance in duration reproduction tasks is linked to changes in autonomic tone during the encoding of duration, namely a shift from sympathetic to parasympathetic dominance (Meissner and Wittmann, 2011; Pollatos et al., 2014). However, in our previous study we found this relationship only

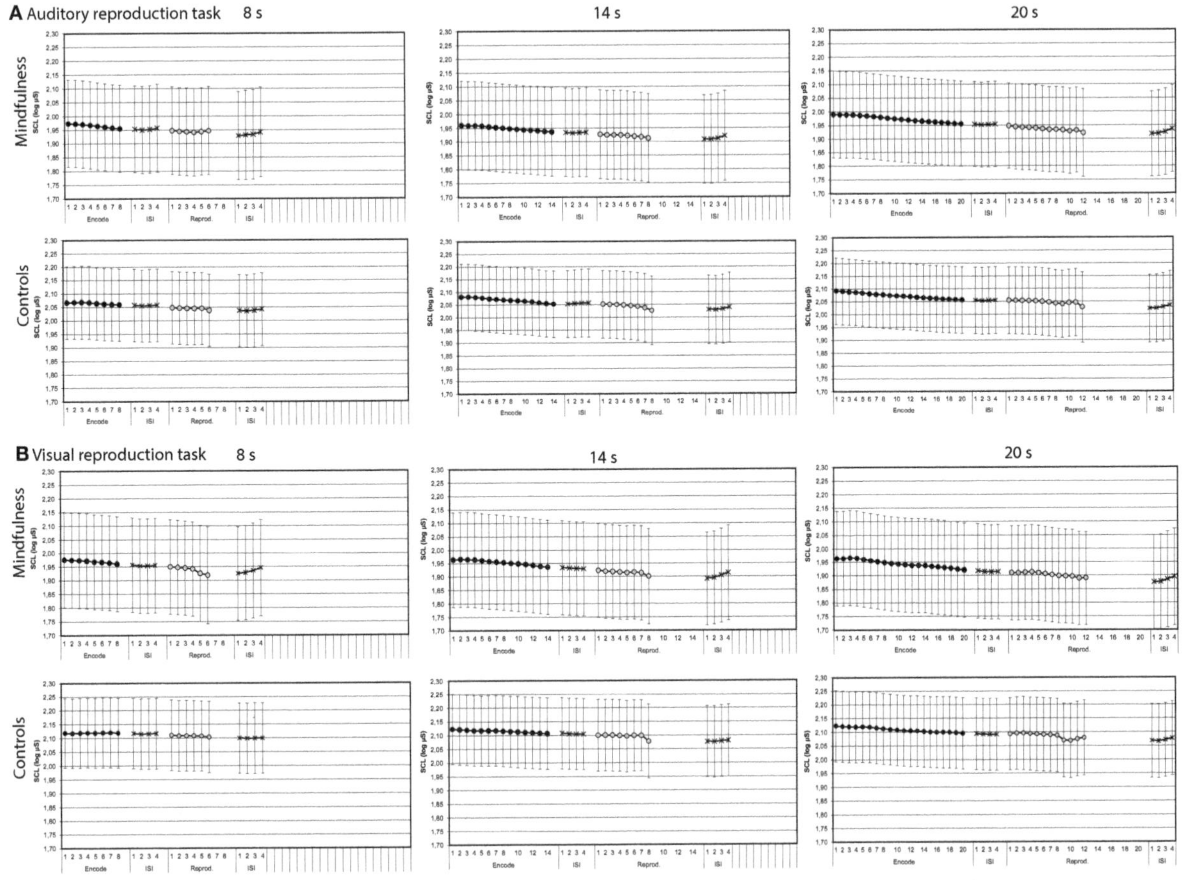

FIGURE 5 | Skin conductance levels by groups (mindfulness meditators, controls). Mean second-to-second changes of skin conductance levels (log μS) during the encoding interval, the reproduction interval, and first 4 s of the subsequent inter-stimulus intervals (ISI) for tones of 8, 14, and 20 s duration of both groups (for statistical results, see **Table 3**). **(A)** Auditory reproduction task. **(B)** Visual reproduction task. Error bars represent standard error of the mean.

TABLE 3 | Results of the mixed-design ANOVAs for skin conductance levels during the three encoding intervals of the auditory and visual duration reproduction tasks.

	Interval	Main effect for time		Linear trend		Interaction effect (time by group)		Main effect for group	
		F[df, error(df)]	*p*	*F*[df, error(df)]	*p*	*F*[df, error(df)]	*p*	*F*[df, error(df)]	*p*
Auditory Task	**AUDITORY ENCODING INTERVAL**								
	8 s	22.01 (1.62, 63.08)	**<0.001****	27.92 (1, 39)	**<0.001****	2.46 (1.62, 63.08)	0.104	0.26 (1, 39)	0.638
	14 s	42.39 (1.55, 60.52)	**<0.001****	54.42 (1, 39)	**<0.001****	0.27 (1.55, 60.52)	0.706	0.33 (1, 39)	0.571
	20 s	29.38 (1.38, 53.75)	**<0.001****	35.69 (1, 39)	**<0.001****	0.15 (1.38, 53.75)	0.780	0.24 (1, 39)	0.631
	AUDITORY REPRODUCTION INTERVAL								
	8 s	0.29 (1.19, 46.38)	0.632	0.39 (1, 39)	0.536	0.42 (1.19, 46.34)	0.557	0.23 (1, 39)	0.634
	14 s	4.40 (1.14, 44.57)	**0.037***	7.02 (1, 39)	**0.012****	0.39 (1.14, 44.57)	0.561	0.34 (1, 39)	0.565
	20 s	1.80 (1.19, 46.30)	0.186	5.11 (1, 39)	**0.029***	0.08 (1.19, 46.30)	0.826	0.29 (1, 39)	0.594
Visual task	**VISUAL ENCODING INTERVAL**								
	8 s	3.56 (1.62, 63.23)	**0.043***	4.27 (1, 39)	**0.045***	4.90 (1.62, 63.23)	**0.016****	0.477 (1, 39)	0.494
	14 s	22.45 (1.72, 67.11)	**<0.001****	29.75 (1, 39)	**<0.001****	2.68 (1.72, 67.11)	0.084	0.543 (1, 39)	0.446
	20 s	35.12 (2.33, 91.00)	**<0.001****	62.40 (1, 39)	**<0.001****	2.18 (2.33, 91.00)	0.110	0.567 (1, 39)	0.456
	VISUAL REPRODUCTION INTERVAL								
	8 s	3.65 (1.54, 59.87)	**0.043***	7.22 (1, 39)	**0.011****	1.94 (1.54, 59.87)	0.162	0.608 (1, 39)	0.440
	14 s	3.69 (1.34, 52.22)	**0.048***	8.50 (1, 39)	**0.006****	0.16 (1.34, 52.22)	0.763	0.678 (1, 39)	0.415
	20 s	3.78 (1.61, 62.91)	**0.037***	4.08 (1, 39)	0.030*	0.40 (1.61, 62.91)	0.626	0.686 (1, 39)	0.413

*$p < 0.05$.

**$p < 0.0167$ (Bonferroni-corrected level of significance).

Bold values signify $p < 0.05$.

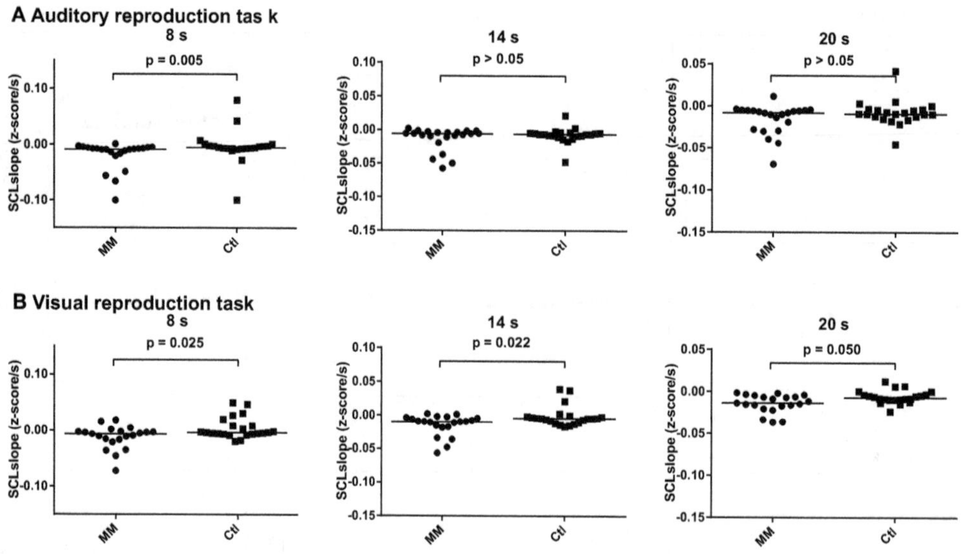

FIGURE 6 | Statistical dot plots with median (black bars) showing skin conductance slopes during the three encoding intervals of the auditory (A) and visual (B) duration reproduction tasks. Mann–Whitney *U*-Tests were used to evaluate group differences.

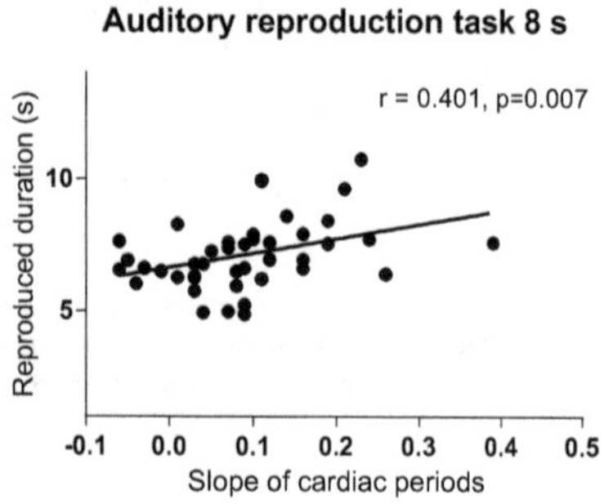

FIGURE 7 | Relationship between the slope of cardiac intervals and reproduced duration of the 8-s interval of the auditory reproduction task. *R*, Spearman's correlation coefficient.

for the slopes of cardiac periods and not for the slopes of skin conductance levels, and furthermore for intervals of 14 s and 20 s duration (Meissner and Wittmann, 2011). These discrepancies may be linked to differences in the study sample with respect to age, education, and interoceptive awareness.

In contrast to our previous study, participants performed not only an auditory duration reproduction task but also an otherwise similar duration reproduction task in the visual domain. Remarkably, the analyses of psychophysiological changes confirmed not only our previous finding that cardiac periods increased continuously during the encoding and reproduction intervals of the auditory task (Meissner and Wittmann, 2011), but extend these findings in showing a likewise increase of cardiac periods during the encoding and reproduction intervals of the visual task (**Figure 4**, **Table 2**). Similarly, the continuous decrease of skin conductance levels described earlier for the auditory task (Meissner and Wittmann, 2011) could be replicated and was also found during the visual task (**Figures 4, 5, Tables 2, 3**). As in our former study, linear trends fitted the continuous changes in cardiac periods and skin conductance levels during the encoding of auditory time intervals reasonably well, and the same was true with respect to the visual task (**Tables 2, 3**). These results suggest that the psychophysiological changes during the encoding of duration are very similar for both auditory and visual stimuli and thus are not modality dependent. That is, an interoceptive mechanism related to time perception may be operating. In both cases, the encoding of duration appears to be accompanied by linear decreases of heart rate and skin conductance levels, indicating a shift from sympathetic to parasympathetic dominance. Furthermore, several relationships of these autonomic changes with both performance and interoceptive awareness scores could be shown. These associations could not be attributed to a stronger attentional load, since attentional capacities, if at all, were found to correlate negatively with the slopes of cardiac periods during the encoding of duration. Thus, even though we found some relationships between attentional capacities and performance in the duration estimation tasks, there was no evidence that the characteristic changes of cardiac periods and skin conductance levels during the encoding of time are related to attentional capacities. Rather, a steady change in autonomic tone, as registered in the brain, could represent a bodily marker for the estimation of duration.

Overall, our results are further indicative of an at least moderate relationship between the experience of time and the heart beat, although we did not find differences between meditators and controls. Recent research has shown how strongly intertwined emotions and feeling states of the body are with the processing of duration (Droit-Volet and Meck, 2007; Dirnberger et al., 2012; Droit-Volet et al., 2013; Wackermann et al., 2014).

The association between duration reproduction performance and increasing neural activation in the insular cortex on the one hand and body functions and body experience on the other hand point to the notion that subjective time is embodied (Meissner and Wittmann, 2011; Wittmann, 2013). Body signals may function as an internal reference for subjective time when judging the duration of external events. With this study, we add to findings that may eventually lead to a theory of how we as humans perceive time.

However, it is only fair to state that relative to the overall number of potential correlations across cells (defined by duration, modality, and psychophysiological measures) the found number of correlations cannot be judged as high. That is, one cannot conclude that the specific psychophysiological measures employed (that is, assessing body functions of heart and skin conductance) are sufficient indicators of subjective time. Rather these physiological measures may contribute to the feeling of passing time in a similar way as they are selective indicators of the present condition of the body (Craig, 2015). Subjective time may result or be mediated through the integration of all bodily signals of which we only were able to assess a selection.

Acknowledgments

We gratefully acknowledge that this study was funded by the Bial foundation Porto, grant no. 52/12 to MW, KM, and SS: "The embodied experience of time: modulations of mindfulness meditation". KM received additional support by the Schweizer-Arau Foundation (Germany) and the Theophrastus Foundation (Germany).

References

Bornemann, B., Herbert, B. M., Mehling, W. E., and Singer, T. (2014). Differential changes in self-reported aspects of interoceptive awareness through 3 months of contemplative training. *Front. Psychol.* 5:1504. doi: 10.3389/fpsyg.2014.01504

Craig, A. D. (2009). Emotional moments across time: a possible neural basis for time perception in the anterior insula. *Philos. Trans. R. Soc. Lond. B Biol. Sci.* 364, 1933–1942. doi: 10.1098/rstb.2009.0008

Craig, A. D. (2015). *How do you Feel? An Interoceptive Moment with your Neurobiological Self.* Princeton: Princeton University Press.

Critchley, H. D., Wiens, S., Rotshtein, P., Ohman, A., and Dolan, R. J. (2004). Neural systems supporting interoceptive awareness. *Nat. Neurosci.* 7, 189–195. doi: 10.1038/nn1176

Dirnberger, G., Hesselmann, G., Roiser, J. P., Preminger, S., Jahanshahi, M., and Paz, R. (2012). Give it time: neural evidence for distorted time perception and enhanced memory encoding in emotional situations. *Neuroimage* 63, 591–599. doi: 10.1016/j.neuroimage.2012.06.041

Droit-Volet, S., Fanget, M., and Dambrun, M. (2015). Mindfulness meditation and relaxation training increases time sensitivity. *Conscious. Cogn.* 31, 86–97. doi: 10.1016/j.concog.2014.10.007

Droit-Volet, S., Fayolle, S., Lamotte, M., and Gil, S. (2013). Time, emotion and the embodiment of timing. *Timing Time Percept.* 1, 99–126. doi: 10.1163/22134468-00002004

Droit-Volet, S., and Meck, W. H. (2007). How emotions colour our perception of time. *Trends Cogn. Sci.* 11, 504–513. doi: 10.1016/j.tics.2007.09.008

Fan, J., Mccandliss, B. D., Sommer, T., Raz, A., and Posner, M. I. (2002). Testing the efficiency and independence of attentional networks. *J. Cogn. Neurosci.* 14, 340–347. doi: 10.1162/089892902317361886

Farb, N. A., Segal, Z. V., and Anderson, A. K. (2013). Mindfulness meditation training alters cortical representations of interoceptive attention. *Soc. Cogn. Affect. Neurosci.* 8, 15–26. doi: 10.1093/scan/nss066

Fox, K. C., Zakarauskas, P., Dixon, M., Ellamil, M., Thompson, E., and Christoff, K. (2012). Meditation experience predicts introspective accuracy. *PLoS ONE* 7:e45370. doi: 10.1371/journal.pone.0045370

Garfinkel, S. N., Seth, A. K., Barrett, A. B., Suzuki, K., and Critchley, H. D. (2015). Knowing your own heart: distinguishing interoceptive accuracy from interoceptive awareness. *Biol. Psychol.* 104, 65–74. doi: 10.1016/j.biopsycho.2014.11.004

Hölzel, B. K., Lazar, S. W., Gard, T., Schuman-Olivier, Z., Vago, D. R., and Ott, U. (2011). How does mindfulness meditation work? Proposing mechanisms of action from a conceptual and neural perspective. *Perspect. Psychol. Sci.* 6, 537–559. doi: 10.1177/1745691611419671

Jha, A. P., Krompinger, J., and Baime, M. J. (2007). Mindfulness training modifies subsystems of attention. *Cogn. Affect. Behav. Neurosci.* 7, 109–119. doi: 10.3758/CABN.7.2.109

Jo, H. G., Hinterberger, T., Wittmann, M., and Schmidt, S. (2015). Do meditators have higher awareness of their intentions to act? *Cortex* 65c, 149–158. doi: 10.1016/j.cortex.2014.12.015

Jo, H. G., Wittmann, M., Borghardt, T. L., Hinterberger, T., and Schmidt, S. (2014). First-person approaches in neuroscience of consciousness: brain dynamics correlate with the intention to act. *Conscious. Cogn.* 26, 105–116. doi: 10.1016/j.concog.2014.03.004

Kabat-Zinn, J. (2005). *Coming to Our Senses.* New York, NY: Hyperion

Khalsa, S. S., Rudrauf, D., Damasio, A. R., Davidson, R. J., Lutz, A., and Tranel, D. (2008). Interoceptive awareness in experienced meditators. *Psychophysiology* 45, 671–677. doi: 10.1111/j.1469-8986.2008.00666.x

Kohls, N., Sauer, S., and Walach, H. (2009). Facets of mindfulness–results of an online study investigating the freiburg mindfulness inventory. *Pers. Individ. Diff.* 46, 224–230. doi: 10.1016/j.paid.2008.10.009

Maclean, K. A., Ferrer, E., Aichele, S. R., Bridwell, D. A., Zanesco, A. P., Jacobs, T. L., et al. (2010). Intensive meditation training improves perceptual discrimination and sustained attention. *Psychol. Sci.* 21, 829–839. doi: 10.1177/0956797610371339

Matthias, E., Schandry, R., Duschek, S., and Pollatos, O. (2009). On the relationship between interoceptive awareness and the attentional processing of visual stimuli. *Int. J. Psychophysiol.* 72, 154–159. doi: 10.1016/j.ijpsycho.2008.12.001

Meissner, K., and Wittmann, M. (2011). Body signals, cardiac awareness, and the perception of time. *Biol. Psychol.* 86, 289–297. doi: 10.1016/j.biopsycho.2011.01.001

Melloni, M., Sedeno, L., Couto, B., Reynoso, M., Gelormini, C., Favaloro, R., et al. (2013). Preliminary evidence about the effects of meditation on interoceptive sensitivity and social cognition. *Behav. Brain Funct.* 9:47. doi: 10.1186/1744-9081-9-47

Merchant, H., Harrington, D. L., and Meck, W. H. (2013). Neural basis of the perception and estimation of time. *Annu. Rev. Neurosci.* 36, 313–336. doi: 10.1146/annurev-neuro-062012-170349

Mirams, L., Poliakoff, E., Brown, R. J., and Lloyd, D. M. (2013). Brief body-scan meditation practice improves somatosensory perceptual decision making. *Conscious. Cogn.* 22, 348–359. doi: 10.1016/j.concog.2012.07.009

Moore, A., Gruber, T., Derose, J., and Malinowski, P. (2012). Regular, brief mindfulness meditation practice improves electrophysiological markers of attentional control. *Front. Hum. Neurosci.* 6:18. doi: 10.3389/fnhum.2012.00018

Morone, N. E., Greco, C. M., and Weiner, D. K. (2008). Mindfulness meditation for the treatment of chronic low back pain in older adults: a randomized controlled pilot study. *Pain* 134, 310–319. doi: 10.1016/j.pain.2007.04.038

Pollatos, O., Herbert, B. M., Kaufmann, C., Auer, D. P., and Schandry, R. (2007). Interoceptive awareness, anxiety and cardiovascular reactivity to isometric exercise. *Int. J. Psychophysiol.* 65, 167–173. doi: 10.1016/j.ijpsycho.2007.03.005

Pollatos, O., Yeldesbay, A., Pikovsky, A., and Rosenblum, M. (2014). How much time has passed? Ask your heart. *Front Neurorobot* 8:15. doi: 10.3389/fnbot.2014.00015

Porges, S. W., and Byrne, E. A. (1992). Research methods for measurement of heart rate and respiration. *Biol. Psychol.* 34, 93–130. doi: 10.1016/0301-0511(92)90012-J

Sauer, S., Walach, H., Schmidt, S., Hinterberger, T., Lynch, S., Büssing, A., et al. (2013). Assessment of mindfulness: review on state of the art. *Mindfulness* 4, 3–17. doi: 10.1007/s12671-012-0122-5

Schwarz, M. A., Winkler, I., and Sedlmeier, P. (2013). The heart beat does not make us tick: the impacts of heart rate and arousal on time perception. *Atten. Percept. Psychophys.* 75, 182–193. doi: 10.3758/s13414-012-0387-8

Steinhubl, S. R., Wineinger, N. E., Patel, S., Boeldt, D. L., Mackellar, G., Porter, V., et al. (2015). Cardiovascular and nervous system changes during meditation. *Front. Hum. Neurosci.* 9:145. doi: 10.3389/fnhum.2015.00145

Tang, Y.-Y., Hölzel, B. K., and Posner, M. I. (2015). The neuroscience of mindfulness meditation. *Nat. Rev. Neurosci.* 16, 213–225. doi: 10.1038/nrn3916

Wackermann, J., Meissner, K., Tankersley, D., and Wittmann, M. (2014). Effects of emotional valence and arousal on acoustic duration reproduction assessed via the "dual klepsydra model". *Front. Neurorobot.* 8:11. doi: 10.3389/fnbot.2014.00011

Walach, H., Buchheld, N., Buttenmüller, V., Kleinknecht, N., and Schmidt, S. (2006). Measuring mindfulness—the Freiburg mindfulness inventory (FMI). *Pers. Individ. Diff.* 40, 1543–1555. doi: 10.1016/j.paid.2005.11.025

Wittmann, M. (2009). The inner experience of time. *Philos. Trans. R. Soc. Lond. B Biol. Sci.* 364, 1955–1967. doi: 10.1098/rstb.2009.0003

Wittmann, M. (2013). The inner sense of time: how the brain creates a representation of duration. *Nat. Rev. Neurosci.* 14, 217–223. doi: 10.1038/nrn3452

Wittmann, M., Otten, S., Schotz, E., Sarikaya, A., Lehnen, H., Jo, H. G., et al. (2015). Subjective expansion of extended time-spans in experienced meditators. *Front. Psychol.* 5:1586. doi: 10.3389/fpsyg.2014.01586

Wittmann, M., and Paulus, M. P. (2009). Temporal horizons in decision making. *J. Neurosci. Psychol. Econ.* 2, 1–11. doi: 10.1037/a0015460

Wittmann, M., and Schmidt, S. (2014). "Mindfulness meditation and the experience of time," in *Meditation – Neuroscientific Approaches and Philosophical Implications. Studies in Neuroscience, Consciousness and Spirituality 2,* eds S. Schmidt and H. Walach (New York, NY: Springer-Verlag), 199-210.

Wittmann, M., Simmons, A. N., Aron, J. L., and Paulus, M. P. (2010). Accumulation of neural activity in the posterior insula encodes the passage of time. *Neuropsychologia* 48, 3110–3120. doi: 10.1016/j.neuropsychologia.2010.06.023

Wittmann, M., Simmons, A. N., Flagan, T., Lane, S. D., Wackermann, J., and Paulus, M. P. (2011). Neural substrates of time perception and impulsivity. *Brain Res.* 1406, 43–58. doi: 10.1016/j.brainres.2011.06.048

Zeidan, F., Johnson, S. K., Diamond, B. J., David, Z., and Goolkasian, P. (2010). Mindfulness meditation improves cognition: evidence of brief mental training. *Conscious. Cogn.* 19, 597–605. doi: 10.1016/j.concog.2010.03.014

Zimmermann, P., and Fimm, B. (1997). *Test for Assessing Attentional Performance (TAP). Version 1.5.* Herzogenrath: Psychologische Testsysteme.

Zimmermann, P., and Fimm, B. (2002). "A test battery for attentional performance," in *Applied Neuropsychology of Attention. Theory, Diagnosis and Rehabilitation,* eds M. Leclerq and P. Zimmermann (London: Psychology Press), 110–151.

9

The Relationships between Interoception and Alexithymic Trait: The Self-Awareness Questionnaire in Healthy Subjects

Mariachiara Longarzo [1], Francesca D'Olimpio [1], Angela Chiavazzo [1], Gabriella Santangelo [1,2], Luigi Trojano [1] and Dario Grossi [1]**

[1] *Laboratory of Neuropsychology, Department of Psychology, Second University of Naples, Caserta, Italy,* [2] *Hermitage Capodimonte, Napoli, Italy*

Correspondence:
Mariachiara Longarzo and Dario Grossi, Laboratory of Neuropsychology, Department of Psychology, Second University of Naples, Viale Ellittico 31, 81100 Caserta, Italy
mariachiara.longarzo@unina2.it; dario.grossi@unina2.it

Interoception is the basic process enabling evaluation of one's own bodily states. Several previous studies suggested that altered interoception might be related to disorders in the ability to perceive and express emotions, i.e., alexithymia, and to defects in perceiving and describing one's own health status, i.e., hypochondriasis. The main aim of the present study was to investigate the relationships between alexithymic trait and interoceptive abilities evaluated by the "Self-Awareness Questionnaire" (SAQ), a novel self-report tool for assessing interoceptive awareness. Two hundred and fifty healthy subjects completed the SAQ, the Toronto Alexithymia Scale-20 items (TAS-20), and a questionnaire to assess hypochondriasis, the Illness Attitude Scale (IAS). The SAQ showed a two-factor structure, with good internal consistency (Cronbach's alpha = 0.88). We observed significant direct correlations between SAQ, TAS-20 and two of its subscales, and the IAS. Regression analysis confirmed that the difficulty in identifying and expressing emotions is significantly related with awareness for one's own interoceptive feelings and with a tendency to misinterpret and amplify bodily sensations. From a clinical point of view, the assessment of interoceptive awareness by the SAQ could be pivotal in evaluating several psychopathological conditions, such as the somatoform disorders.

Keywords: interoceptive awareness, emotion, alexithymia, hypochondriasis, health, insular cortex

Introduction

Interoception is the basic process collecting information coming from one's own body, such as heartbeat, hunger, thirst, breathing, or visceral sensations. Such information is less distinct than that provided by "exteroceptive" somatosensory systems (e.g., touch, or skin temperature), but contributes to maintain homeostasis (Craig, 2003), and allows the brain to build a sense of one's own physical condition and to answer questions such as "how do you feel now?" (Craig, 2009).

The relationships between interoception and emotion are still under debate. Classic and modern theories of emotion maintain that bodily states contribute to or are essential for emotional experience (James, 1884; Lange, 1885; Bard, 1928; Cannon, 1932; Lane and Schwartz, 1987). Several studies using heartbeat perception tasks demonstrated a positive association between interoception and intensity of emotional experience (e.g., Schandry, 1981; Jones, 1994; Wiens et al., 2000; Barrett et al., 2004; Critchley et al., 2004; Wiens, 2005; Herbert et al., 2007; Pollatos et al., 2007a,b). Thus, according to Craig (2004, 2009), individual differences in emotional awareness may be directly related to individual differences in interoception. Similarly, Pollatos et al. (2005) and Dunn et al. (2010) reported that individuals with high sensitivity to interoceptive signals showed higher arousal in response to emotional visual stimuli.

The close relationships between interoception and emotional awareness have recently been supported by a neurofunctional study. Zaki et al. (2012) demonstrated that the anterior insular cortex, which is considered essential for both interoception monitoring and emotional processing (Craig, 2004, 2009; Critchley et al., 2004; Pollatos et al., 2007a), was activated during tasks in which healthy subjects were required to monitor their own heartbeat or to assess their own emotional experience after viewing videos of people recounting emotional stories. On this basis, the anterior insular cortex has been regarded as "a convergence zone" of interoceptive and emotional awareness (Zaki et al., 2012; see also Lamm and Singer, 2010). Enhanced activation in a network including the anterior insular cortex, the anterior cingulate cortex, ventromedial prefrontal cortex and somatosensory cortex has been reported in subjects with high interoceptive awareness (Critchley et al., 2004).

Emotional awareness and processing of interoceptive signals are thought to be impaired in alexithymia (Lane et al., 1998, 2000; Craig, 2004), defined by Sifneos (1973) as a reduced ability to identify and describe one's own emotions in patients with psychosomatic diseases. Alexithymia has been hypothesized to be associated with altered activation and morphology in the anterior cingulate cortex and the anterior insular cortex (Lane et al., 1998, 2000; Berthoz et al., 2002; Kano et al., 2003; Borsci et al., 2009). Consistent with such findings, Herbert et al. (2011) demonstrated that the ability to process interoceptive signals (assessed on a heartbeat perception task) is negatively associated with all aspects of alexithymia.

Available studies on the relationships between alexithymia and interoception, however, did not take into account the distinct facets of this complex construct. In this respect, Terasawa et al. (2013) underscored that tasks assessing heartbeat detection (e.g., Schandry, 1981) provide a measure of interoceptive "sensitivity," i.e., of the accuracy in objective tests of bodily sensations, but are not suitable to investigate other aspects of interoception, such as interoceptive "sensibility" and interoceptive "awareness." Interoceptive "sensibility" would express one's own tendency to be focused on internal states, whereas interoceptive "awareness" (IA) would imply cognitive appreciation of interoceptive sensations, and the ability of expressing bodily feelings. Interoceptive "sensibility" and IA are best addressed by means of self-report questionnaires (Terasawa et al., 2013). The distinction of the three aspects of interoception can help explaining the reason why studies using self-report questionnaires or "objective" heartbeat detection tasks provide divergent findings about the relationships between interoception and alexithymic trait. In fact, contrary to findings reported by Herbert et al. (2011), Ernst et al. (2014) observed a positive correlation between IA (assessed by a self-report tool, the Body Perception Questionnaire; Porges, 1993) and alexithymic trait in a small sample of healthy subjects in a study on neurotransmitter concentration in the insula and in the anterior cingulate cortex.

A positive correlation between IA and alexithymic trait could be consistent with the view that close attention toward interoceptive sensations could hamper interpretation of one's own emotional feelings (Biondi, 1991). Accordingly, in their review Kano and Fukudo (2013) suggested that persons with alexithymia tend to over-report physical symptoms and interpret even low-intensity emotion-related sensations as signs of illness. This hypothesis is consistent with the results of a PET study showing stronger awareness of visceral sensations in alexithymics, with greater activity in the posterior insula and in the rostral anterior cingulate cortex (Kano et al., 2007). Moreover, Tominaga et al. (2014) found a strong correlation between alexithymia (assessed by Toronto Alexithymia Scale-20 items; Bagby et al., 1994) and hypochondriasis and somatoform disorders (assessed by Somatosensory Amplification Scale; Barsky et al., 1988; Nakao and Barsky, 2007). Taken together, such observations would show a significant direct correlation between measures of IA and alexithymic trait, when self-report assessment tools are used.

The aim of the present study was to investigate the hypothesis of a direct association between IA and alexithymic trait in a large sample of healthy subjects. In a systematic review of the available self-report measures assessing body-awareness, Mehling et al. (2009) concluded that no instruments assess all aspects of interoception, but mainly explore anxiety or emotions, without sufficient details on physical sensations and with a few items on body awareness. For these reasons, Mehling et al. (2012) developed the Multidimensional Assessment of Interoceptive Awareness (MAIA) to evaluate dimensions of body awareness such as quality of body sensations, attention regulation for body sensations, emotional awareness for physiological signs of emotion or tendency to evaluate one's own body as safe and trustworthy. However, to the best of our knowledge, there is no tool evaluating perception of a wide range of bodily sensations. For this purpose, we devised a new self-report measure of IA to be used in a large non-clinical sample. This questionnaire, termed "Self-Awareness Questionnaire" (SAQ), is short and easy to administer and refers to commonly felt bodily sensations. This novel self-report tool is based on the "How do you feel questionnaire" (Grossi et al., 2014), which allowed to demonstrate that poor interoceptive awareness is associated with insular damage in stroke patients, and includes further items derived from other available questionnaires (e.g., Barsky et al., 1988; Porges, 1993).

Materials and Methods

Participants

We recruited 250 healthy subjects among students and academic staff of the Departments of Psychology and Political Science at the Second University of Naples. To be included in the present study, subjects had to meet the following criteria: (1) lack of current or past history of alcohol or drug abuse, (2) lack of current or past history of major psychiatric diseases, (3) lack of history of brain injury, stroke, or any other major clinical condition, (4) lack of past or current use of psychoactive medications. The eligibility criteria were assessed by means of a brief semi-structured clinical interview.

All individuals were naïve to the scopes and purposes of the study and gave their written informed consent to participate without any reward.

Materials and Methods

All participants completed three questionnaires assessing: (i) IA, (ii) the ability to identify and describe emotions, and (iii) attitudes associated with hypochondriasis. To assess IA, and to specifically investigate how and how frequently subjects feel signals arising from their own body, we used an extended version of the "How do you feel questionnaire" (Grossi et al., 2014). The questionnaire included 35 items (Appendix 1) to be rated on a 5-point Likert scale (0 = never; 1 = sometimes; 2 = often; 3 = very often; 4 = always). The total score ranges 0–140, with higher scores meaning higher IA. In a preliminary study on an independent sample of 50 healthy students, we required participants to rate whether each item was clearly comprehensible and whether it assessed common physical sensations. All items were considered simple to comprehend, and suitable to address bodily sensations; all 35 items were thus included in the SAQ.

To assess the ability to identify and describe emotions, we used the Toronto Alexithymia Scale-20 items (TAS-20; Bagby et al., 1994), the most widely used self-report tool to assess the Alexithymia construct. The 20 items explore three factors reflecting the main aspects of the alexithymia: difficulty in identifying feelings; difficulty in describing feelings; externally oriented thinking. Each item has to be rated on a 5-point Likert scale (from 1 = "completely agree" to 5 = "strongly disagree"). The total score ranges 20–100, with higher scores indicating higher levels of alexithymia. The Italian version of TAS-20 has been demonstrated to show good test-retest reliability (0.86) and adequate internal consistency (Cronbach's alpha: 0.75) in a wide sample of healthy adults and of medical and psychiatric outpatients (Bressi et al., 1996).

The Illness Attitude Scale (IAS; Kellner, 1987) investigates attitude, fear and beliefs associated with hypochondriac behavior, and includes 27 items rated on a 5-point Likert scale (from 0 = "no" to 4 = "most of the time"). The total score ranges 0–108, with higher scores indicating more severe hypochondriac symptoms. The IAS is a reliable instrument, distinguishing hypochondriac patients from psychiatric patients and healthy individuals (Kellner, 1987). The IAS has been translated in several languages and its psychometric properties are well established (Sirri et al., 2008); an Italian version of the scale has been used in studies on clinical samples (e.g., Fava et al., 2000), but its psychometric properties have not been assessed specifically.

The study was approved by the Local Ethics Committees.

Statistical Analysis

Quality of data for SAQ questionnaire was evaluated by computing percentage of missing or invalid items; a percentage <5% is considered as an index of acceptable data quality. Moreover, data quality was assessed using mean, median, skewness, kurtosis and extent of ceiling and floor effects. Floor and ceiling effects <15% were defined as optimal (Cronbach, 1951). Internal consistency of the SAQ and of its dimensions was evaluated by Cronbach's alpha (McHorney and Tarlov, 1995); a value $\geq$0.70 was considered acceptable (Scientific Advisory Committee of the Medical Outcomes Trust, 2002). Scaling assumptions referring to the correct grouping of items were checked using alpha correction if item is removed (item is removed if the alpha value increases).

Furthermore, we run an Exploratory Factor Analysis, by Principal Axis Factor analysis (PAF), to explore the latent structure of the scale and to perform data reduction; we extracted the components explaining an amount of variance greater than a single item do (i.e., with Eigenvalues > 1; Kaiser, 1960). We computed factor loadings after oblimin rotation, allowing factors to correlate.

Thirdly, we assessed correlations of SAQ with TAS-20 (total and subscales scores for both measures) and IAS total score by computing Pearson's correlation coefficients (Bonferroni correction for multiple comparisons was adopted to reduce type-I errors).

Last, to investigate whether IA explained a significant portion of variance of alexithymia, a regression analysis was performed on TAS total score, using the SAQ total score as a predictor. The specific contribution of SAQ in predicting TAS scores was also assessed by means of hierarchical multiple regression analysis performed on TAS total scores, in which we first entered demographic data (age and gender), then IAS total score, and last SAQ total score as independent variables.

Results

Two hundred and fifty healthy subjects (175 females, 75 males) participated in the present study. **Table 1** shows the participants' demographic features and data about the psychometric variables.

Data collected for each item of the SAQ were computable and there was no missing value. There was no floor or ceiling effect; skewness was 0.78 and kurtosis was 0.30.

Descriptive analyses on normality distribution and communality indexes from PAF, suggested removing 7 of the 35 items (Items: 1, 2, 5, 8, 20, 22, 29). The mean score on the resulting 28-item SAQ "awareness index" (max value = 112) for the present sample was 23.33 (*SD* = 11.67). PAF extracted three factors with an eigenvalue higher than 1. The screen test indicated a two-factor solution, which accounted for about 29% of the total variance. Factor loadings obtained after oblimin rotation are shown in **Table 2**.

TABLE 1 | Participants' demographic and psychometric variables.

	Mean	Standard deviation	Range
Age (years)	27.9	9.4	18–57
Education (years)	14.3	2.3	5–18
IAS	33.6	15.4	4–81
TAS-20	34.0	9.8	9–69
SAQ	27.4	13.5	2–78

TABLE 2 | Factor analysis of SAQ.

Items	F1	F2
3. I feel my heart beat in my ears	**0.41**	0.21
4. I feel very hot in comparison to others	−0.16	**0.62**
6. I feel pain extremely	**0.32**	0.02
7. I feel my stomach tightening	**0.49**	0.18
9. I feel a sudden hunger pang	0.01	**0.47**
10. I feel my back ache	0.10	**0.32**
11. I feel pins and needles	0.19	**0.32**
12. I feel that I can't get enough air into my lungs	**0.72**	−0.09
13. I have an extra-strong heartbeat	**0.60**	0.13
14. I feel full and bloated after eating	0.21	**0.43**
15. I have the sudden urge to urinate	0.17	**0.36**
16. I feel as if I am on fire	0.23	**0.51**
17. I feel a burning sensation in my stomach	**0.32**	0.13
18. I feel a pain in my stomach	0.24	**0.30**
19. I feel very cold in comparison to others	**0.64**	0.08
21. I feel like I have to throw up	**0.44**	0.17
23. I feel chilled	**0.60**	−0.10
24. I feel my legs heavy	0.25	**0.41**
25. I feel my throat dry	0.21	**0.47**
26. I have heavy feeling in my chest	**0.69**	−0.06
27. I feel my heart thudding	**0.71**	−0.10
28. I feel sudden thirst pangs	−0.10	**0.65**
30. I feel breathless without engaging in any type of exertion and effort	**0.32**	**0.31**
31. I feel my ears burning	0.14	**0.30**
32. I feel a lump in my throat	**0.63**	−0.05
33. I feel faint	**0.37**	0.07
34. I feel my palms sweaty	−0.14	**0.51**
35. I have difficulty in swallowing	**0.30**	0.07

For each item the highest factor loading is printed in bold.

The first factor (F1) was the most relevant (eigenvalue after rotation = 5.81), and mainly (but not exclusively) included items related to visceral sensations (items 3, 6, 7, 12, 13, 17, 19, 21, 23, 26, 27, 30, 32, 33, 35). The second factor (F2, eigenvalue = 4.50) mainly included items referring to somatosensory sensations (items 4, 9, 10, 11, 14, 15, 16, 18, 24, 25, 28, 30, 31, 34). The SAQ awareness index significantly correlated with both F1 ($r = 0.89$, $p < 0.01$) and F2 factors ($r = 0.86$, $p < 0.01$); the two factors significantly correlated with each other ($r = 0.55$, $p < 0.01$). The SAQ awareness index and its two factors showed good internal consistency (F1: alpha = 0.85; F2: alpha = 0.81; total alpha = 0.88). Correlation analyses (**Table 3**) revealed that the SAQ awareness index (and its two factors F1 and F2) significantly correlated with the TAS-20 total score and its two subscales "difficulty identifying feelings" and "difficulty describing feelings," but not with the TAS-20 subscale assessing "externally oriented thinking." In addition, the SAQ awareness index and its factors F1 and F2 significantly correlated with the IAS total score. All such correlations were positive, meaning that high scores on the SAQ and on its factors were associated to high scores on the questionnaires assessing two specific facets of alexithymia and hypochondriasis. By the same token, high IAS total scores were associated with high scores on TAS-20 and with high scores on two components of TAS-20, "difficulty identifying feelings" and "difficulty describing feelings."

The results from the regression analysis showed that Interoception was a moderately significant predictor of Alexithymia, explaining 13% of the variance (beta = 0.37, $p < 0.001$). The results from the hierarchical regression models showed that: age and gender were not significant predictors of total TAS score [step# 1: $R2 = 0.002$; $F_{(2, 247)} = 1.28$; $p = 0.28$; beta for age = 0.08, beta for gender = 0.06]; hypochondria was a significant predictor of alexithymia [step# 2: $R2 = 0.12$; $F_{(3, 246)} = 11.96$; $p < 0.001$; beta = 0.35]; IA, as evaluated by the SAQ, was still a significant predictor of TAS scores accounting for a further 8% of the variance [step# 3: $R2 = 0.20$; $F_{(4, 245)} = 16.21$; $p < 0.001$; beta for gender = 0.18, beta for IAS = 0.21, beta for SAQ = 0.33].

Discussion

The aim of the present study was to investigate the relationships between bodily awareness, i.e., "interoceptive awareness" after Terasawa et al. (2013), and the ability to identify and to describe emotions. For this purpose we used a self-report questionnaire (SAQ) specifically assessing how and how frequently participants felt signals from their own body. The SAQ demonstrated a good internal consistency; items clustered into two factors, the first mainly related to visceral feelings, and the second mainly related to somatosensory feelings.

In exploring the relationships between IA and emotion processing, we observed significant positive relationships of SAQ awareness index and of its two factors with TAS-20. In particular, the two subscales of the TAS-20 investigating "difficulty in identifying feelings" and "difficulty in describing feelings" showed strong positive relationships with the SAQ awareness index and its two factors. It is important to underscore that the correlation of IA with alexithymic trait was positive in the present study, consistent with findings reported by Ernst et al. (2014), who observed that a high IA, as assessed by a self-report index, correlated with high alexithymic trait. Although studies assessing interoceptive "sensitivity" (e.g., via the heartbeat detection task) reported an inverse correlation (Herbert et al., 2011), our and Ernst et al.'s findings clearly support the idea that IA is directly correlated with alexithymic trait.

The association between hypochondriasis and interoception might suggest that high IA is related to a strong concern for one's own bodily sensations. Along these lines, Salkovskis and Warwick (1986) maintained that cognitive processes concerning body, health and illness might increase attention toward bodily signals. Barsky et al. (1990) subsumed this hypothesis in the Somatosensory Amplification model, according to which hypochondriac subjects tend to focus on their somatic sensations and experience them as intense and disturbing. This group of subjects are hypervigilant about their own body, concentrate on their own physical sensations and consider them as dangerous.

TABLE 3 | Correlation analyses between SAQ with IAS and TAS-20.

	TAS-20				IAS
	Difficulty identifying feelings	**Difficulty describing feelings**	**Externally oriented thinking**	**Total score**	**Total score**
SAQ – Awareness index	0.47**	0.29**	−0.02	0.34**	0.32**
SAQ – F1 factor	0.40**	0.20*	−0.08	0.25**	0.31**
SAQ – F2 factor	0.45**	0.33**	0.04	0.37**	0.26**
IAS – Total score	0.36**	0.27**	0.10	0.33**	–

Bonferroni corrected p-values: $^{*}p < 0.05$; $^{**}p < 0.01$.

The significant positive correlations among the SAQ awareness index (and its two factors), TAS-20 scores and IAS total score suggest that focusing one's own attention on bodily feelings can hamper interpretation of one's own emotional feelings, and of bodily sensations that are direct expressions of such feelings. In other words, these results might be compatible with the idea that IA, emotion processing and hypochondriasis trait are different facets of a condition in which individuals focus on their own body signals and do not recognize the emotional nature of such signals. These individuals might be prone to interpret their physiological modifications as a cue that something life threatening is going on. Increased attention to one's own somatic signals might lead to a corresponding inattention to one's own emotions. These findings seem to confirm the interconnection between these conditions. However, from the present study we cannot infer the causal directionality of such correlations, and it remains entirely plausible that high scores of TAS-20 factors may contribute to the development of somatoform disorders through attention, amplification and misinterpretation of somatic sensations with emotional arousal (Kano and Fukudo, 2013).

It is important to take into account that our conclusions are based on observations gathered from a sample of relatively young and educated healthy adults, and this might limit generalization of our findings. However, most available scales, including the Italian version of the TAS-20 used in this study, have been developed and tested on similar samples. Moreover, in our study we did not assess validity and reliability of the SAQ questionnaire and did not investigate its relationships with objective measures of interoception (e.g., the heartbeat detection task). We also acknowledge that the inclusion of clinical samples would have added further informative values to the present study, allowing ascertaining whether the same relationships among IA, alexithymia and hypochondriasis also apply in individuals with somatoform disorders. These potential limitations might trigger further extensive investigation, in which the different aspects of interoception are assessed concurrently, and levels of anxiety and of depressive symptoms are controlled, since they could modulate alexithymia, and particularly the "difficulty identifying emotion" and "difficulty describing emotion" factors of TAS-20 (Tominaga et al., 2014).

Despite the above limitations, our study demonstrated relevant interactions between IA, alexithymia and hypochondriasis. IA is necessary for maintaining homeostasis, but excessive attention to the body can interfere with the ability to interpret bodily signals correctly. From a clinical point of view, the assessment of IA could be pivotal in evaluating several psychopathological conditions. Misinterpretation of physical sensations, rather than physical sensations in themselves, can determine concerns in individuals with a high level of attention to their own body; this could result in a vicious circle involving metacognition and evaluation of one's own thinking and symptoms (Marcus et al., 2007). These observations suggest that IA could contribute in maintaining such syndromes. The SAQ might reveal useful in comprehending the role of IA in clinical samples of patients with somatoform disorders.

References

Bagby, R. M., Parker, J. D., and Taylor, G. J. (1994). The twenty-item Toronto Alexithymia Scale-I. Item selection and cross-validation of the factor structure. *J. Psychosom. Res.* 38, 23–32. doi: 10.1016/0022-3999(94)90005-1

Bard, P. (1928). A diencephalic mechanism for the expression of rage with special reference to the sympathetic nervous system. *Am. J. Physiol.* 84, 490–515.

Barrett, L. F., Quigley, K. S., Bliss-Moreau, E., and Aronson, K. R. (2004). Interoceptive sensitivity and self-reports of emotional experience. *J. Pers. Soc. Psychol.* 87, 684–697. doi: 10.1037/0022-3514.87.5.684

Barsky, A. J., Goodson, J. D., Lane, R. S., and Cleary, P. D. (1988). The amplification of somatic symptoms. *Psychosom. Med.* 50, 510–519. doi: 10.1097/00006842-198809000-00007

Barsky, A. J., Wyshak, G., and Klerman, G. L. (1990). The somatosensory amplification scale and its relationship to hypochondriasis. *J. Psychiatr. Res.* 24, 323–334. doi: 10.1016/0022-3956(90)90004-A

Berthoz, S. M., Artiges, E., Van der Moortele, P. F., Poline, J., Consoli, S. M., and Martinot, J. L. (2002). Effect of impaired recognition and expression of emotions on fronto-cingulate cortices: an fMRI study of men with alexithymia. *Am. J. Psychiatry* 159, 961–967. doi: 10.1176/appi.ajp.159.6.961

Biondi, M. (1991). *La Psicosomatica Nella Pratica Clinica*. Roma: Pensiero Scientifico Editore.

Borsci, G., Boccardi, M., Rossi, R., Rossi, G., Perez, J., Bonetti, M., et al. (2009). Alexithymia in healthy women: a brain morphology study. *J. Affect. Disord.* 114, 208–215. doi: 10.1016/j.jad.2008.07.013

Bressi, C., Taylor, G., Parker, J., Bressi, S., Brambilla, V., Aguglia, E., et al. (1996). Cross validation of the factor structure of the 20-items Toronto Alexithymia Scale: an Italian multicenter study. *J. Psychosom. Res.* 41, 551–559. doi: 10.1016/S0022-3999(96)00228-0

Cannon, W. B. (1932). *The Wisdom of the Body*. New York, NY: W.W. Norton and Co.

Craig, A. D. (2003). Interoception: the sense of the physiological condition of the body. *Curr. Opin. Neurobiol.* 13, 500–505. doi: 10.1016/S0959-4388(03)00090-4

Craig, A. D. (2004). Human feelings: why are some more aware than others? *Trends Cogn. Sci.* 8, 39–241. doi: 10.1016/j.tics.2004.04.004

Craig, A. D. (2009). How do you feel now? The anterior insula and human awareness. *Nat. Rev. Neurosci.* 10, 59–70. doi: 10.1038/nrn2555

Critchley, H. D., Wiens, S., Rothstein, P., Ohlman, A., and Dolan, R. J. (2004). Neural system supporting interoceptive awareness. *Nat. Neurosci.* 7, 189–195. doi: 10.1038/nn1176

Cronbach, L. J. (1951). Coefficient alpha and the internal structure of tests. *Psychometrika* 16, 297–334. doi: 10.1007/BF02310555

Dunn, B. D., Galton, H. C., Morgan, R., Evans, D., Oliver, C., Meyer, M., et al. (2010). Listening to your heart: how interoception shapes emotion experience and intuitive decision making. *Psychol. Sci.* 21, 1835–1844. doi: 10.1177/0956797610389191

Ernst, J., Böker, H., Hättenschwiler, J., Schüpbach, D., Northoff, G., Seifritz, E., et al. (2014). The association of interocpetive awareness and alexithymia with neurotransmitter concentrations in insula and anterior cingulate. *Soc. Cogn. Affect. Neurosci.* 9, 857–863. doi: 10.1093/scan/nst058

Fava, G. A., Grandi, S., Rafanelli, C., Fabbri, S., and Cazzaro, M. (2000). Explanatory therapy in hypochondriasis. *J. Clin. Psychiatry* 61, 317–322. doi: 10.4088/JCP.v61n0414

Grossi, D., Di Vita, A., Palermo, L., Sabatini, U., Trojano, L., and Guariglia, C. (2014). The brain network for self-feeling: a symptom-lesion mapping study. *Neuropsychologia* 63, 92–98. doi: 10.1016/j.neuropsychologia.2014.08.004

Herbert, B. M., Herbert, C., and Pollatos, O. (2011). On the relationship between interoceptive awareness and alexithymia: is interoceptive awareness related to emotional awareness? *J. Pers.* 79, 1149–1175. doi: 10.1111/j.1467-6494.2011.00717.x

Herbert, B. M., Pollatos, O., and Schandry, R. (2007). Interoceptive sensitivity and emotion processing: an EEG study. *Int. J. Psychophysiol.* 65, 214–227. doi: 10.1016/j.ijpsycho.2007.04.007

James, W. (1884). What is an emotion? *Mind* IX, 188–205. doi: 10.1093/mind/os-IX.34.188

Jones, G. E. (1994). "Perception of visceral sensation: a review of recent findings, methodologies, and future directions," in *Advance Psychophysiology*, Vol. 5, eds J. R. Jennings and P. K. Ackles (London: Jessica Kingsley publisher), 55–192.

Kaiser, H. F. (1960). The application of electronic computers to factor analysis. *Educ. Psychol. Meas.* 20, 141–151. doi: 10.1177/001316446002000116

Kano, M., and Fukudo, S. (2013). The alexithymic brain: the neural pathways linking alexithymia to physical disorders. *Biopsychosoc. Med.* 7:1. doi: 10.1186/1751-0759-7-1

Kano, M., Fukudo, S., Gyoba, J., Kamachi, M., Tagawa, M., Mochizuki, H., et al. (2003). Specific brain processing of facial expressions in people with alexithymia: an H2 15O-PET study. *Brain* 126(Pt 6), 1474–1484. doi: 10.1093/brain/awg131

Kano, M., Hamaguchi, T., Itoh, M., Yanai, K., and Fukudo, S. (2007). Correlation between alexithymia and hypersensitivity to visceral stimulation in human. *Pain* 132, 252–263. doi: 10.1016/j.pain.2007.01.032

Kellner, R. (1987). *Abridged Manual of the Illness Attitude Scale (Mimeographed).* Albuquerque, NM: Department of Psychiatry, University of New Mexico.

Lamm, C., and Singer, T. (2010). The role of anterior insular cortex in social emotions. *Brain Struct. Funct.* 214, 579–591. doi: 10.1007/s00429-010-0251-3

Lane, R. D., Reimann, E. M., Axelrod, B., Yun, L. S., Holmes, A., and Schwartz, G. E. (1998). Neural correlates of levels of emotional awareness: evidence of an interaction between emotion and attention in the anterior cingulated cortex. *J. Cogn. Neurosci.* 10, 525–535. doi: 10.1162/089892998562924

Lane, R. D., and Schwartz, G. E. (1987). Levels of emotional awareness: a cognitive-developmental theory and its application to psychopathology. *Am. J. Psychiatry* 144, 133–143.

Lane, R. D., Sechrest, L., Riedel, R., Shapiro, D. E., and Kaszniak, A. W. (2000). Pervasive emotion recognition deficit common to alexithymia and the repressive coping style. *Psychosom. Med.* 62, 492–501. doi: 10.1097/00006842-200007000-00007

Lange, C. G. (1885). "The mechanism of the emotions," in *The Classical Psychologists*, ed B. Rand (Boston, MA: Houghton Mifflin), 672–684.

Marcus, D. K., Gurley, J. R., Marchi, M. M., and Bauer, C. (2007). Cognitive and perceptual variables in hypochondriasis and health anxiety: a systematic review. *Clin. Psychol. Rev.* 27, 127–139. doi: 10.1016/j.cpr.2006.09.003

McHorney, C. A., and Tarlov, A. R. (1995). Individual-patient-monitoring in clinical practice: are available health surveys adequate? *Qual. Life Res.* 4, 293–307. doi: 10.1007/BF01593882

Mehling, W. E., Gopisetty, V., Daubenmier, J., Price, C. J., Hecht, F. M., and Stewart, A. (2009). Body awareness: construct and self-report measures. *PLoS ONE* 4:e5614. doi: 10.1371/journal.pone.0005614

Mehling, W. E., Price, C., Daubenmier, J. J., Acree, M., Bartmess, E., and Stewart, A. (2012). The Multidimensional Assessment of Interoceptive Awareness (MAIA). *PLoS ONE* 7:e48230. doi: 10.1371/journal.pone.0048230

Nakao, M., and Barsky, A. J. (2007). Clinical application of somatosensory amplification in psychosomatic medicine. *Biopsychosoc. Med.* 1:17. doi: 10.1186/1751-0759-1-17

Pollatos, O., Gramann, K., and Schandry, R. (2007a). Neural systems connecting interoceptive awareness and feelings. *Hum. Brain Mapp.* 28, 9–18. doi: 10.1002/hbm.20258

Pollatos, O., Herbert, B. M., Matthias, E., and Schandry, R. (2007b). Heart rate response after emotional presentation is modulated by interoceptive awareness. *Int. J. Psychophysiol.* 63, 117–124. doi: 10.1016/j.ijpsycho.2006.09.003

Pollatos, O., Kirsch, W., and Schandry, R. (2005). Brain structures involved in interoceptive awareness and cardioafferent signal processing: a dipole source localization study. *Hum. Brain Mapp.* 26, 54–64. doi: 10.1002/hbm.20121

Porges, S. (1993). *Body Perception Questionnaire.* Laboratory of Developmental Assessment, University of Maryland. Available online at: http://stephenporges.com/index.php/publicationss/21-body-perception-questionnaires

Salkovskis, P. M., and Warwick, H. (1986). Morbid preoccupation, health anxiety and reassurance: a cognitive-behavioral approach to hypochondriasis. *Behav. Res. Ther.* 24, 597–602. doi: 10.1016/0005-7967(86)90041-0

Schandry, R. (1981). Heart beat perception and emotional experience. *Psychophysiology* 18, 483–488. doi: 10.1111/j.1469-8986.1981.tb02486.x

Scientific Advisory Committee of the Medical Outcomes Trust. (2002). Assessing health status and quality-of-life instruments: attributes and review criteria. *Qual. Life Res.* 11, 193–205. doi: 10.1023/A:1015291021312

Sifneos, P. E. (1973). The prevalence of "alexithymic" characteristic mechanisms in psychosomatic patients. *Psychother. Psychosom.* 22, 255–262. doi: 10.1159/000286529

Sirri, R., Grandi, S., and Fava, G. A. (2008). The illness attitude scale. A clinicmetric index for assessing hypochondriacal fears and beliefs. *Psychother. Psychosom.* 77, 337–350. doi: 10.1159/000151387

Terasawa, Y., Shibata, M., Moriguchi, Y., and Umeda, S. (2013). Anterior insular cortex mediates bodily sensibility and social anxiety. *Soc. Cogn. Affect. Neurosci.* 8, 259–266. doi: 10.1093/scan/nss108

Tominaga, T., Choi, H., Nagoshi, Y., Wada, Y., and Fukui, K. (2014). Relationship between alexithymia and coping strategies in patients with somatoform disorders. *Neuropsychiatr. Dis. Treat.* 10, 55–62. doi: 10.2147/NDT.S55956

Wiens, S. (2005). Interoception in emotional experience. *Curr. Opin. Neurol.* 18, 442–447. doi: 10.1097/01.wco.0000168079.92106.99

Wiens, S., Mezzacappa, E. S., and Katkin, E. (2000). Heartbeat detection and the experience of emotions. *Cogn. Emot.* 14, 417–427. doi: 10.1080/026999300378905

Zaki, J., Davis, J. I., and Ochsner, K. N. (2012). Overlapping activity in anterior insula during interoception and emotional experience. *Neuroimage* 62, 493–499. doi: 10.1016/j.neuroimage.2012.05.012

10

Interoception and Symptom Reporting: Disentangling Accuracy and Bias

Sibylle Petersen[1,2], Ken Van Staeyen[1], Claus Vögele[2], Andreas von Leupoldt[1] and Omer Van den Bergh[1]*

[1] *Research Group on Health Psychology, KU Leuven, Leuven, Belgium,* [2] *Institute for Health and Behaviour, University of Luxembourg, Luxembourg, Luxembourg*

Correspondence:
Sibylle Petersen,
Research Group on Health Psychology, KU Leuven,
Tiensestraat 102, 3000 Leuven,
Belgium
sibylle.petersen@ppw.kuleuven.be

Anxiety and anxiety sensitivity are positively related to accuracy in the perception of bodily sensations. At the same time, research consistently reports that these traits are positively related to bias, resulting in the report of more and more intense symptoms that poorly correspond with physiological dysfunction. The aim of this study was to test the relationship of accuracy and bias in interoception. Furthermore, we tested the impact of individual differences in negative affect and symptom report in daily life on interoceptive accuracy and bias. Individuals higher in symptom report in daily life and negative affect were marginally more accurate in an interoceptive classification task in which participants were asked to identify different respiratory stimuli (inducing breathing effort) as belonging to a high or low intensity category. At the same time, bias in overestimating intensity of stimuli was significantly increased in participants higher in symptom report and negative affect, but only for more ambiguous stimuli. Results illustrate that interoceptive accuracy and bias need to be considered independently to understand their interaction with psychological factors and to disentangle (mis)perception of bodily sensations from liberal or conservative perceptual decision strategies.

Keywords: classification, interoception, bias, accuracy, decision strategies, sensitivity, negative affect, symptom report

Introduction

Nothing is closer to us than our own body but few things seem so elusive than the perception of bodily sensations. The brain is not passively 'measuring' signals from the body, but these signals interact with emotions and influence decision-making, behavior, and attention which in turn can change (perception of) bodily signals (Damasio, 1994; Craig, 2004). It is thus little surprising that correlations between self-reported bodily sensations and physiological changes are usually low in healthy individuals and in patients with somatic disease (e.g., Banzett et al., 2000; Petersen et al., 2011). Research consistently reports that negative affect, anxiety, and anxiety sensitivity are related to stronger nocebo effects (Van den Bergh et al., 1997, 2004) and higher levels of symptom report (Watson and Pennebaker, 1989; Petersen et al., 2011). Furthermore, in anxiety disorders, elevated self-report of somatic sensations is combined with a tendency for catastrophizing interpretations (e.g., Barlow, 1988; Barsky et al., 1994; Carleton et al., 2014).

In contrast to these low correlations between self-report and physiology particularly in individuals high in negative affect, a meta-analysis reports medium to strong effect sizes for a positive relationship between accuracy in heartbeat detection and psychological variables such as anxiety sensitivity and anxiety (Domschke et al., 2010). Higher anxiety sensitivity is also related to lower detection thresholds for respiratory stimuli (Petersen and Ritz, 2011). Furthermore, a study measuring respiratory-related evoked potentials (RREPs) in participants watching negative emotional pictures found increased later components of RREPs in individuals higher compared to lower in anxiety (von Leupoldt et al., 2011). The authors interpreted these increases as higher motivated attention toward respiratory signals in more anxious individuals in negative affective contexts.

Thus, while individuals higher in negative affective states and traits seem to be more biased in interoception, they seem at the same time to be more accurate. Attempts to reconcile these seemingly paradox findings are faced with a methodological challenge. Research on interoceptive accuracy uses mostly heartbeat detection tasks (e.g., Domschke et al., 2010). In one type of heartbeat detection task, participants are asked to decide whether a sound signal is matching their heartbeat (e.g., Asmundson et al., 1993; Harver et al., 1993). Data collected in such a task could be used to differentiate bias and accuracy in a signal detection approach. Unfortunately, performance levels are around chance in the majority of participants, raising questions about the usefulness of this paradigm (Jones, 1994; Domschke et al., 2010). Most studies on interoception use another type of heartbeat detection task and follow a mental tracking paradigm in which participants are instructed to count their heartbeats during intervals of different length (Schandry, 1981). This paradigm, however, cannot differentiate sensitivity and bias.

Confounding sensitivity and bias means to lose crucial information. At the same level of sensitivity, individuals can apply liberal or conservative decision strategies which may result in very different forms of coping behavior following either a liberal "better safe than sorry" or a conservative "wait and see" approach (Macmillan and Creelman, 2004; Lynn and Barrett, 2014).

Making perceptual decisions and classifying bodily sensations, for example, as symptom or as benign sensation, is an inherent part of interoception (Petersen et al., 2014) and may present a challenge (Carleton et al., 2014). A headache may be painful, but *perhaps* not painful enough to take medication; breathlessness may be strong, but *perhaps* not strong enough to signal that we should stop exercise; heartbeat may be elevated, but *probably* not a sign of a heart disease. Although these sensations are clearly above detection threshold they are ambiguous regarding their category. We cannot be 100% sure about classification of ambiguous sensations at the border of two interoceptive categories, but we can optimize decision strategies by weighting the risk of missing a symptom against the risk of false alarms, that is, the risk of classifying a benign sensation as pathological (Griffiths et al., 2008). These processes of forming probabilistic beliefs about bodily sensations have been suggested to underlie interoception at every stage of information processing and may not always be deliberate or even consciously accessible (Edwards et al., 2012).

In this study, we tested the relationship between accuracy and bias in an interoceptive classification task in which participants were asked to correctly categorize different respiratory stimuli (respiratory loads increasing breathing effort) as belonging to either a low or a high intensity category and to indicate their place within these categories by labeling them as A1, A2, A3, A4, B1, B2, B3, and B4, with increasing numbers being related to increasing stimulus intensity (Petersen et al., 2014). A novel feature of this study is that we tested classification of a number of stimuli varying in intensity and assigned to two categories. Most signal detection paradigms test only one (usually low intensity) stimulus representing the signal against signal absent trials (for exceptions, see Kepecs et al., 2008; Yang et al., 2014). Interoceptive categories, however, usually subsume a range of different signals (e.g., signs of airway obstruction in asthma may come in different degrees of intensity and a variety of symptoms on multiple dimensions may fall in the category 'cold symptoms') and an important task is to classify ambiguous signals as belonging to one of various sensation classes.

We tested accuracy in the classification of multiple stimuli as belonging to one of two intensity categories. Furthermore, we tested how bias changes across the different stimuli within interoceptive categories. We expected that it would be more challenging to classify sensations closer to the border of two interoceptive categories than for more prototypical/central sensations, for example, a stimulus which is not clearly high or low, but moderate. It is important to note that also if non-categorization information would be given, an A1 stimulus would be more easily classified as low than an A4 stimulus. We tested whether interoceptive bias would be higher for more ambiguous sensations and whether individual differences in negative affect and symptom report in daily life would affect this bias. Higher increase in bias toward the shared category border could be interpreted a strategy to rather risk false alarms than to miss signals which indicated potential harm (Lynn and Barrett, 2014). Lower increase in bias toward the category border could be interpreted as the category border being incorporated into interoceptive decision-making, serving as a red flag and reducing misclassification particularly for stimuli at this category border.

We expected that higher bias for more ambiguous stimuli (better safe than sorry approach ignoring the category border) would lead to more accurate results in the classification task (as predicted by Lynn and Barrett, 2014), explaining the seeming paradox of higher bias and higher accuracy in individuals with negative affective expectations toward symptoms or high in general negative affect.

Materials and Methods

Participants

Participants were 54 women (mean age 21.04 years, SD = 1.8) without known chronic or acute disease. They were selected based on prescreens for high and low habitual report of bodily symptoms in daily life using the Checklist for Symptoms in Daily Life (Wientjes and Grossman, 1994). Participants completed this symptom checklist consisting of 39 bodily sensations (e.g., tingling in arms and hands, back pain, etc.) with regard to how often they had experienced these sensations within the last year (scale 1 *never* −5 *very often*). The questionnaire has good reliability (Chronbach's alpha 0.70–0.90, Wientjes and Grossman, 1994). The pre-screen was filled in by 355 individuals and mean score was 91.95 (SD = 18.59). We invited participants with values in the symptom checklist of either higher than 80 or lower than 60 consecutively from this prescreen list.

Participants completed the same checklist also after the laboratory appointment. After the task, four participants from the high symptom reporter group did no longer reach high values on the symptom checklist and were excluded from the analysis. Thirteen of the participants in the low symptom report group reached values of 61–75 on the symptom checklist after the task (which induced symptoms), and were included in the low symptom report group. This resulted in $n = 25$ for the low symptom report group and $n = 25$ for the high symptom report group. Participants signed informed consent prior to participation. They received course credit or reimbursement for participation. The study was approved by the local ethics committee.

Instruments

We induced feelings of breathing effort by presenting eight different respiratory loads using the instrument Powerbreathe K5 (POWERbreathe International Ltd., Southam, UK). The instrument allows gradually increasing breathing resistance at inhalation by reducing the diameter of the breathing port. We used the software Breathelink to program presentation of loads. Exhalation was unrestricted. Resistances of the eight respiratory loads were chosen to ensure that differences between adjacent breathing loads could be distinguished by healthy volunteers, but were similar enough to leave an uncertainty margin for classification (7, 9, 11, 14, 18, 23, 28, and 37 cmH_2O).

Participants completed the Checklist for Symptoms in Daily Life (Wientjes and Grossman, 1994) twice during the experimental session. Firstly, they were asked to indicate which of the symptoms they had perceived in the last year to test whether results of the pre-screen were reliable. Secondly, we asked them to indicate which symptoms they had experienced during the actual task. We used the Positive and Negative Affect Schedule (Watson et al., 1988), a five point rating scale assessing negative and positive affect within the last 4 weeks with ten mood related adjectives as items for each subscale.

Protocol

At the start of the protocol, participants signed an informed consent form giving information about the study. In the first block of the interoceptive classification task, we presented the eight loads in random order, each load four times for two breaths. During presentation, the label of the load was presented on a computer screen (label A1 together with the 6 cmH_2O load, A2 - 9 cmH_2O, A3 - 11 cmH_2O, A4 - 14 cmH_2O, label B1 - 18 cmH_2O, B2 - 23 cmH_2O, B3 - 28 cmH_2O, and B4 - 37 cmH_2O). Participants were instructed that in the following block, they would have to solve a categorization task. They were asked to memorize the sensation and the label so that they would be able in a second block, when loads were presented without label, to indicate the correct category (A or B) and the location of the stimulus within its category (1–4). In previous studies, we found that participants are able to distinguish and label eight different loads assigned to two categories after this training procedure above chance level (Petersen et al., 2014). Furthermore, in a study comparing a group of participants receiving category information and a control group receiving the information that stimuli were labeled with numbers only (increasing consecutively from lowest to highest stimulus), we found that categorization and an arbitrary category boundary indeed change perception of interoceptive stimuli. Differentiation between stimuli falling into the same categories was reduced and discrimination between categories was more pronounced in the experimental compared to the control group (Petersen et al., 2014, Study 1). Similar results were found in studies on visual perception using arbitrary category boundaries in a similar paradigm (Tajfel and Wilkes, 1963; Corneille et al., 2002), and in a study using an odor classification paradigm where mice were trained to classify six odor stimuli as belonging to one of two similar odor categories (Kepecs et al., 2008).

In Block 2, loads were presented again in pseudo-randomized order with 18 presentations per load, that is, 144 load presentations overall, with two breaths per load presentation. We presented the loads in a way that each load was preceded at least once by every other load to reduce the impact of order effects. Participants were asked to classify each load by assigning the correct labels (A1–B4). They did not receive feedback on their performance. After Block 2, participants completed the PANAS, the symptom checklist for symptoms last year, and the symptom checklist for symptoms directly after the task. Finally, we asked them to answer demographic questions on age, height, weight, and chronic or acute disease.

Data Analysis

We used SPSS 20 for all analyses and used the SPSS syntax for *c* and *d'* indices proposed by Stanislaw and Todorov (1999). We calculated d'_{class} measures reflecting how accurate participants were in distinguishing categories A and B. In contrast to signal detection, classification is a choice between two types of signals and not between signal and noise. Still, the formula is identical, only that A or B is treated as 'noise' and the other category as 'signal.' We calculated one mean d'_{class} index as mean of a d'_{Aclass} (treating A as signal and B as noise) and d'_{Bclass} (treating B as signal and A as noise).

Furthermore, we used signal detection analysis to calculate classification criteria c_{class}. We calculated $c_{class} = -(z(H) + z(F))/2$, with z(H) as z-transformed hit rates of rating A if a load of Category A was presented, and z(F) as z-transformed false alarm rates of reporting A1 if a load from B was presented. This resulted in four c_{class} indices per category ($c_{classA1}$, $c_{classA2}$, $c_{classA3}$, $c_{classA4}$, $c_{classB1}$, $c_{classB2}$, $c_{classB3}$, and $c_{classB4}$). Thus, c_{class} reflects bias, taking into account classification errors and correct classification with c indices below zero indicating a liberal bias and c indices above zero indicating a conservative bias.

In a classification paradigm, the meaning of the terms liberal and conservative depends on the standard used. A tendency to require little evidence to classify a stimulus as B is liberal in a "better safe than sorry" approach, if missing B is regarded to be more costly than missing A. We calculated c_{class} indices in a way that the term 'liberal' refers to a decision strategy that favors classifying A as B for stimuli actually belonging to A, or classifying B as A for stimuli actually belonging to B. In other words, a liberal tendency in our analysis is a strategy that favors misclassification. We expect this liberal tendency to be strongest for more ambiguous stimuli and to become weaker for clearly high or clearly low stimuli.

We tested in a univariate analysis of covariance (ANCOVA) whether d'_{class} would differ between groups of high and low habitual symptom reporters, controlling for negative affect and symptoms experienced during the task as covariates. In further repeated-measures ANCOVAs we tested whether c_{class} values would be significantly reduced as functions of closeness to the shared category border between A and B. The within-individual factor was Classification with four levels of the within-individual factor Classification per category ($c_{classA1}$, $c_{classA2}$, $c_{classA3}$, $c_{classA4}/c_{classB1}$, $c_{classB2}$, $c_{classB3}$, $c_{classB4}$). We included Symptom Report (high/low) as between-individual variable to test whether individuals higher in habitual symptom report would show more liberal tendencies to assign stimuli to the B category. Again, we included negative affect and symptoms experienced during the task within the ANCOVA model as covariates.

As alternative way of analysis, we performed an individual regression slopes analysis, following the procedure described by Pfister et al. (2013). In this analysis, we tested whether the individual regression slopes of participants indicating a decline toward a more liberal bias from A1 to A4 and an increase to more conservative bias from B1 to B4 would be steeper for participants in the High Symptom Report group. We compared the single slopes in a one way ANCOVA with negative affect as covariate and symptoms reported directly after the task included as control variable. Furthermore, we calculated the absolute value for the slopes across Categories A and B and compared, whether steepness of slopes would differ between A and B in a repeated-measures ANCOVA with negative affect as covariate, controlling for symptoms experienced during the task. Furthermore, in the supplemental material we present data on accuracy of differentiation within categories.

Results

Table 1 summarizes mean values across groups of high and low symptom reporters for self-report scales. We tested group differences in a *t*-test for independent groups.

Accuracy of Classification

High symptom reporters were marginally more accurate classifying loads as A or B, main effect for Symptom Report $F(1,49) = 3.60$, $p = 0.064$, $\eta^2_p = 0.073$ (**Figure 1**). Effects for negative affect and symptoms reported during the task were not significant, $Fs < 1$.

Misclassifications and Response Bias

Classification bias c_{class} per load became less conservative toward the shared category border between A and B in both groups, but this decrease was significant only for Category A $F(3,44) = 18.27$, $p < 0.001$, $\eta^2_p = 0.555$, and not for Category B $F(3,44) = 2.06$, $p = 0.119$, $\eta^2_p = 0.123$ (**Figure 2**, statistics are reported with the covariate Negative Affect). Please note that the lack of

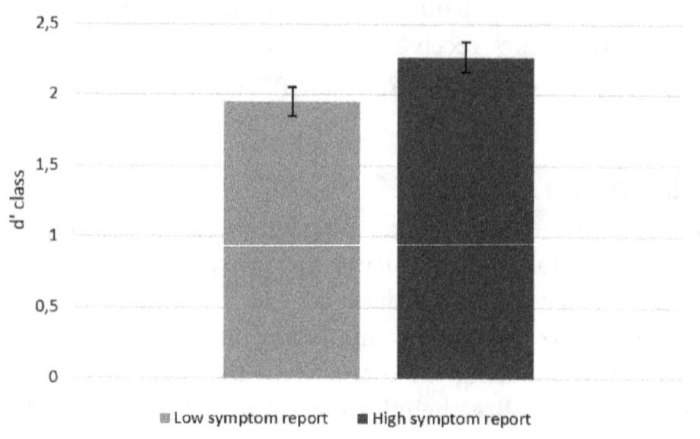

FIGURE 1 | Accuracy in classifying loads as belonging to A or B. Error bars represent standard errors of the mean.

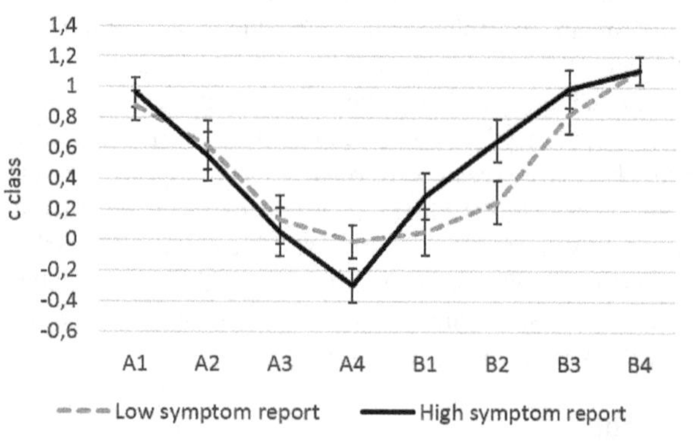

FIGURE 2 | C_{class} values calculated from z-transformed correct identification rates of a load as A or B and z-transformed false alarm rates. Error bars represent standard error of the mean.

TABLE 1 | Participants' characteristics (standard deviation in parentheses).

	Symptom experienced last year	Negative affect	Symptoms experienced during the task
Low symptom reporters	60.12* (SD = 7.05, range: 42–75)	16.84* (SD = 7.10, range: 10–34)	57.04* (SD = 9.24, range: 41–75)
High symptom reporters	99.88 (SD = 14.06, range: 81–138)	24.48 (SD = 7.84, range: 11–28)	78.68 (SD = 16.47, range: 41–107)

*$p < 0.001$.

significant results for Category B does not necessarily imply that the effect for Category A was larger than for Category B. In the next section, we report a comparison of effects with individual slope analysis. The decrease in Category A, was significantly stronger in high symptom reporters, interaction Classification X Symptom Report $F(3,44) = 3.97$, $p = 0.014$, $\eta_p^2 = 0.213$. This interaction was non-significant for category B, $F(3,44) = 1.65$, $p = 0.192$, $\eta_p^2 = 0.101$. In other words, across Category A, participants high in habitual symptom report showed a more pronounced change from a conservative to a liberal criterion than participants low in habitual symptom report. Also after exclusion of the covariate Negative Affect from this model, the interaction effect of Classification X Symptom Report remained significant, $F(3,44) = 3.18, p = 0.033$, $\eta_p^2 = 0.175$.

For Category A, the interaction of the within-individual factor Classification with the covariate Negative Affect was marginally significant, $F(3,44) = 2.29$, $p = 0.092$, $\eta_p^2 = 0.135$, but not for Category B, $F(3,46) = 2.06$, $p = 0.120$, $\eta_p^2 = 0.123$. None of the other effects was significant (all $Fs < 1.18$).

Post hoc tests revealed that none of the single c_{class} indices differed significantly between groups regardless whether negative affect was included in the model or not (all $ps > 0.103$). We performed explorative *post hoc* tests with t-tests for independent groups to test whether the c_{class} index for A4 indicated indeed a *liberal* bias or only a *less conservative* bias. This explorative t-test testing differences from zero, suggests that only for high habitual symptom reporters, c_{class} for A4 was significantly smaller than zero, that is, only in this group bias changed significantly from conservative to liberal $t(24) = -2.327$, $p = 0.027$.

Individual Slopes Analysis

Individual slopes analysis revealed that the shift toward a liberal bias (as expressed in a negative slope for Category A) was stronger in the High Symptom Report group (mean: −0.386, SD = 0.143) compared to the Low Symptom Report group (mean: −0.357, SD = 0.126), $F(1,47) = 4.59$, $p = 0.037$, $\eta_p^2 = 0.089$. The effect of negative affect was non-significant and we report statistics without including negative affect as covariate. For Category B, the shift to a more conservative bias from B1 to B4 did not differ significantly between groups of Low (mean: 0.366, SD = 0.159) and High Symptom Report (mean: 0.293, SD = 0.184), $F(1,47) = 2.34$, $p = 0.128$, $\eta_p^2 = 0.049$. Comparing the absolute value of single slopes (ignoring the direction of slope) in a repeated-measures ANCOVA revealed that the steepness of slope was marginally higher for Category A (mean: 0.375, SD = 0.125) than for Category B (mean: 0.339, SD = 0.155), main within-individual effect $F(1,46) = 3.27$, $p = 0.077$, $\eta_p^2 = 0.065$. The interaction with the between-individual factor Symptom Report was significant, $F(1,47) = 4.64$, $p = 0.036$, $\eta_p^2 = 0.090$. While for participants from the Low Symptom Report group the absolute amount of slopes for Category A (mean: 0.357, SD = 0.126) and B (mean: 0.366, SD = 0.158) were not significantly different ($p = 0.828$), the absolute amount of slopes differed significantly for the High Symptom Report group between Category A (mean: 0.392, SD = 0.125) and B (mean: 0.311, SD = 0.150) with a steeper slope for Category A. In other words, change in bias across categories was stronger for category A than for Category B, but only in the High Symptom Report group. Again negative affect had no significant effect and including it did not change patterns of significance and we report results without including this covariate.

Correlations between Bias and Accuracy

Correlations (Pearson's correlation coefficient, two-tailed, **Figure 3**) between d'_{class} and c_{class} were negative for category A, but significant for stimulus A4 only. In other words, higher accuracy was related to a more liberal decision criterion for the stimulus at the category border of A. This 'better safe than sorry' approach for this particularly ambiguous stimulus was related to overall better classification results. Furthermore, d'_{class} was significantly positively related to c_{class} for stimulus B1 and B2, that is, higher accuracy was related to a more conservative decision criterion (being reluctant to decide that stimuli B1 and B2 belonged to A) for stimuli at the border of category B. None of the other correlations was significant, $p > 0.053$.

Discussion

Participants higher in symptom report in daily life scored higher on negative affect and reported more symptoms during the experimental task. They were also marginally more accurate in the interoceptive classification task. This is consistent with results of prior research showing a relationship between interoceptive accuracy/awareness and negative affective traits and states (e.g., Domschke et al., 2010; Petersen et al., 2011; von Leupoldt et al., 2011). A novel finding of this study is that classification strategies changed across categories. Bias became more liberal with increasing closeness to the border between categories. This change from conservative to more liberal decision strategies was significantly stronger in high symptom reporters and positively related to classification sensitivity d'_{class}. Furthermore, for Category A (but not for B), the decline of individual slopes was steeper in the High compared

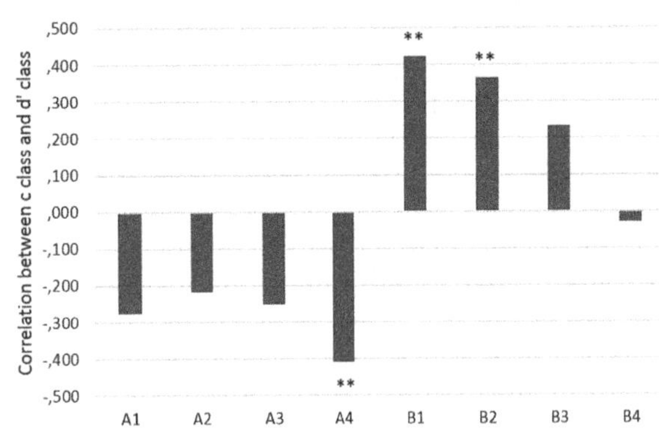

FIGURE 3 | Correlation coefficients between c_{class} indices and d'_{class} ($p < 0.01$).**

to the Low Symptom Report group. Thus, the information that there was a category border between A4 and B1 was effectively reducing bias toward the category border, but only in participants who did not have strong affective expectations about stimuli.

Results confirm hypotheses proposed recently in signal detection research. Lynn and Barrett (2014) have suggested that increased uncertainty about visual stimuli will lead to more extreme decision strategies in signal detection tasks. These more extreme strategies for more ambiguous stimuli are suggested to *optimize* decision-making and behavior (Lynn et al., 2012). Lynn and Barrett (2014) give the example of a person who walks more carefully in a dimly lit room compared to a brightly lit room (i.e., under higher or lower ambiguity of visual signals). Optimizing speed relative to sight allows avoiding injury or breaking objects. They suggest that "extreme bias may reflect not an impairment, but a normal adaptive mechanism that offsets the single impairment, poor sensitivity" (Lynn and Barrett, 2014, p. 1670). Calibrating bias to increased uncertainty for ambiguous sensations (closer to category borders) may be highly adaptive. More extreme decision strategies under higher ambiguity (closeness to category borders) were more successful in the present study. Higher classification accuracy (d'_{class}) was related to *more* bias at the borders of categories and increased bias in this laboratory task was related to bias in symptom report in daily life.

It is important to note that the border of the two neutrally labeled categories A and B was not intrinsically meaningful, as it would be the case for a border that marks a transition between a sensation and a symptom category (e.g., increased airway resistance which is either benign or indicates an oncoming asthma attack). It could be questioned whether arbitrary category boundaries as such affected perception. Prior research using a design with an experimental group receiving category information and a control group for which stimuli were numbered consecutively from lowest to highest without categories has found that arbitrary category boundaries between categories A and B affect interoception (Petersen et al., 2014, Study 1). Differentiation between categories was increased compared to differentiation within categories in the experimental group (receiving category information) compared to differentiation between identical stimuli in control groups (not receiving category information). Similar effects have also been found for visual perception (e.g., Tajfel and Wilkes, 1963; Corneille et al., 2002). Furthermore, a study testing classification decisions in mice which were trained to classify six odor stimuli as belonging to two categories (which shared a boundary and for which similarity between stimulus A3 and B1 was the same as between A3 and A2) found that detection of higher ambiguity for stimuli closer to category boundaries does not require meta-cognition and that ambiguity determines speed of decisions, overall accuracy, and is correlated with prefrontal cortex activity (Kepecs et al., 2008). For meaningful category labels and borders, or a paradigm which would trigger meta-cognition in participants about the costs of misclassifying stimuli, results might be even stronger than observed in this minimal paradigm using neutral labels.

It is tempting to speculate that the observed increasingly liberal bias for ambiguous stimuli at category borders in high symptom reporters may be a first step toward interoceptive threat-generalization. If a stimulus at a category border is consistently misclassified as belonging to a more intense category, this may lead in the long run to establishing a new and lower category border. This process may continue until more and more stimuli, which were initially in a low (or "safe sensation") category, are subsumed in a higher (or "dangerous symptom") category. Following this lead, fear-generalization in anxiety disorders (Lissek et al., 2008) may be interpreted as increasingly more liberal strategies spreading from category borders to more and more stimuli within a category.

Ambiguity of sensations because of their location close to a category border and their no longer clearly low or clearly high magnitude may be clinically as relevant as ambiguity related to a poor signal to noise ratio of sensations (i.e., minimal stimulation such as heartbeat at rest). Research on the role of uncertainty in panic disorder found that intolerance of uncertainty was substantially and significantly related to symptom report in panic disorder even after controlling for anxiety sensitivity (Carleton et al., 2014). Results from this correlation study suggest that patients seemed to find uncertainty about the decision whether a clearly detectable sensation is a sign of pathology (or not) at least as aversive as the sensation as such. Probing deeper into interoceptive classification strategies in anxiety disorder, future research should test whether the relationship of negative affective states and traits and interoceptive accuracy and bias is mediated by increased feelings of aversiveness related to uncertainty about classification of sensations as pathological or benign. Trait constructs such as intolerance for ambiguity which are closely related to anxiety may be interesting in that regard (Birrell et al., 2011).

Negative affect was increased in individuals high in habitual symptom report. Negative affect has been suggested to be related to a general lack of inhibition and not specifically to interoception (Bogaerts et al., 2015), but again, our results suggest that this effect is not a general effect across all types of stimuli within a category, but higher for more ambiguous stimuli. The relationship between negative affect and perceptual bias has been confirmed in a large number of studies (e.g., Bar-Haim et al., 2007; Cisler and Koster, 2010), but testing decision strategies under uncertainty may help to shed light on the processes underlying this relationship.

A limitation for generalization of the present results is that categories were distinct and defined only by one dimension (inspiratory resistance). Interoceptive categories are multidimensional and may overlap on some dimensions and be distinct on others. Sensations experienced during an asthma attack, for example, and sensations related to panic attacks belong to two distinct diagnostic categories, but are partly overlapping in how they are experienced by patients (Lehrer et al., 2002). Multidimensionality and overlap of categories in interoception may increase ambiguity of sensations, which in turn may lead to more extreme forms of bias for ambiguous sensations.

A further important limitation of this study is that we included only young female participants. Attention to interoceptive stimuli as well as expression of negative affect and fear may be higher in women than in men. Future research needs to address gender and age differences in interoceptive classification.

Conclusion

The relationship between sensitivity and bias is not uniform across interoceptive categories. To understand the relationship between interoception and individual difference variables, paradigms are needed which do not confound sensitivity and bias and vary the degree of ambiguity stimuli have regarding their classification.

Author Contributions

SP and OVdB contributed equally to the development of the study concept and study design. AvL contributed conceptually to the study design and data analysis. Testing, data collection, and data analysis were performed by SP, and KvS in collaboration with OVdB as senior researcher. All authors, SP, OVdB, KvS, AvL, and CV drafted the paper. All authors approved the final version of the paper for submission.

References

Asmundson, G. J., Sandler, L. S., Wilson, K. G., and Norton, G. R. (1993). Panic attacks and interoceptive acuity for cardiac sensations. *Behav. Res. Ther.* 31, 193–197. doi: 10.1016/0005-7967(93)90071-2

Banzett, R. B., Dempsey, J. A., O'Donnell, D. E., and Wamboldt, M. Z. (2000). Symptom perception and respiratory sensation in asthma. *Am. J. Respir. Crit. Care Med.* 162, 1178–1182. doi: 10.1164/ajrccm.162.3.9909112

Bar-Haim, Y., Lamy, D., Pergamin, L., Bakermans-Kranenburg, M. J., and van IJzendoorn, M. H. (2007). Threat-related attentional bias in anxious and non-anxious individuals: a meta-analytic study. *Psychol. Bull.* 133, 1–24. doi: 10.1037/0033-2909.133.1.1

Barlow, D. (1988). *Anxiety and its Disorders: The Nature and Treatment of Anxiety and Panic.* New York, NY: Guilford Press.

Barsky, A. J., Cleary, P. D., Sarnie, M. K., and Ruskin, J. N. (1994). Panic disorder, palpitations, and the awareness of cardiac activity. *J. Nerv. Ment. Dis.* 182, 63–71. doi: 10.1097/00005053-199402000-00001

Birrell, J., Meares, K., Wilkinson, A., and Freeston, M. (2011). Toward a definition of intolerance of uncertainty: a review of factor analytical studies of the Intolerance of Uncertainty Scale. *Clin. Psychol. Rev.* 31, 1198–1208. doi: 10.1016/j.cpr.2011.07.009

Bogaerts, K., Rayen, L., Lavrysen, A., Van Diest, I., Janssens, T., Schruers, K., et al. (2015). Unraveling the relationship between trait negative affectivity and habitual symptom reporting. *PLoS ONE* 10:e0115748. doi: 10.1371/journal.pone.0115748

Carleton, R. N., Duranceau, S., Freeston, M. H., Boelen, P. A., McCabe, R. E., and Antony, M. M. (2014). "But it might be a heart attack": intolerance of uncertainty and panic disorder symptoms. *J. Anxiety Disord.* 28, 463–470. doi: 10.1016/j.janxdis.2014.04.006

Cisler, J. M., and Koster, E. H. (2010). Mechanisms of attentional biases towards threat in anxiety disorders: an integrative review. *Clin. Psychol. Rev.* 30, 203–216. doi: 10.1016/j.cpr.2009.11.003

Corneille, O., Klein, O., Lambert, S., and Judd, C. M. (2002). On the role of familiarity with units of measurement in categorical accentuation: Tajfel and Wilkes (1963) revisited and replicated. *Psychol. Sci.* 13, 380–383. doi: 10.1111/j.0956-7976.2002.00468.x

Craig, A. D. (2004). Human feelings: why are some more aware than others? *Trends Cogn. Sci.* 8, 239–241. doi: 10.1016/j.tics.2004.04.004

Damasio, A. R. (1994). The somatic marker hypothesis and the possible functions of the prefrontal cortex. *Philos. Trans. R. Soc. Lond. B Biol. Sci.* 351, 1413–1420. doi: 10.1098/rstb.1996.0125

Domschke, K., Stevens, S., Pfleiderer, B., and Gerlach, A. L. (2010). Interoceptive sensitivity in anxiety and anxiety disorders: an overview and integration of neurobiological findings. *Clin. Psychol. Rev.* 30, 1–11. doi: 10.1016/j.cpr.2009.08.008

Edwards, M. J., Adams, R. A., Brown, H., Pareés, I., and Friston, K. J. (2012). A Bayesian account of 'hysteria'. *Brain* 135, 3495–3512. doi: 10.1093/brain/aws129

Griffiths, T. L., Kemp, C., and Tenenbaum, J. B. (2008). "Bayesian models of cognition," in *Cambridge Handbook of Computational Cognitive Modeling*, ed. R. Sun (Cambridge, MA: Cambridge University Press), 59–100.

Harver, A., Katkin, E. S., and Bloch, E. (1993). Signal-detection outcomes on heartbeat and respiratory resistance detection tasks in male and female subjects. *Psychophysiology* 30, 223–230. doi: 10.1111/j.1469-8986.1993.tb03347.x

Jones, G. E. (1994). "Perception of visceral sensations: a review of recent findings, methodologies, and future directions," in *Advances in Psychophysiology*, eds J. R. Jennings, P. K. Ackles, and M. G. H. Cole (London: Jessica Kingsley Publishers), 55–191.

Kepecs, A., Uchida, N., Zariwala, H. A., and Mainen, Z. F. (2008). Neural correlates, computation and behavioural impact of decision confidence. *Nature* 455, 227–231. doi: 10.1038/nature07200

Lehrer, P., Feldman, J., Giardino, N., Song, H. S., and Schmaling, K. (2002). Psychological aspects of asthma. *J. Consult. Clin. Psychol.* 70, 691–711. doi: 10.1037/0022-006X.70.3.691

Lissek, S., Biggs, A. L., Rabin, S. J., Cornwell, B. R., Alvarez, R. P., Pine, D. S., et al. (2008). Generalization of conditioned fear-potentiated startle in humans: experimental validation and clinical relevance. *Behav. Res. Ther.* 46, 678–687. doi: 10.1016/j.brat.2008.02.005

Lynn, S. K., and Barrett, L. F. (2014). "Utilizing" signal detection theory. *Psychol. Sci.* 25, 1663–1673. doi: 10.1177/0956797614541991

Lynn, S. K., Zhang, X., and Barrett, L. F. (2012). Affective state influences perception by affecting decision parameters underlying bias and sensitivity. *Emotion* 12, 726–736. doi: 10.1037/a0026765

Macmillan, N. A., and Creelman, C. D. (2004). *Detection Theory: A User's Guide.* Hove: Psychology press.

Petersen, S., and Ritz, T. (2011). Situational self-awareness and report bias in the detection of respiratory loads in healthy individuals: over-report is linked to external focus. *Br. J. Health Psychol.* 15, 401–416. doi: 10.1348/135910709X466676

Petersen, S., Schroijen, M., Moelders, C., Zenker, S., and Van den Bergh, O. (2014). Interoceptive categorization: the role of perceptual organization on processing sensations from inside. *Psychol. Sci.* 25, 1059–1066. doi: 10.1177/0956797613519110

Petersen, S., van den Berg, R., Janssens, T., and Van den Bergh, O. (2011). Illness and symptom perception: a theoretical approach towards an integrative measurement model. *Clin. Psychol. Rev.* 31, 428–439. doi: 10.1016/j.cpr.2010.11.002

Pfister, R., Schwarz, K. A., Carson, R., and Janczyk, M. (2013). Easy methods for extracting individual regression slopes: comparing SPSS, R, and Excel. *Tutor. Quant. Methods Psychol.* 9, 72–78.

Schandry, R. (1981). Heartbeat perception and emotional experience. *Psychophysiology* 18, 483–488. doi: 10.1111/j.1469-8986.1981.tb02486.x

Stanislaw, H., and Todorov, N. (1999). Calculation of signal detection theory measures. *Behav. Res. Methods Instrum. Comput.* 31, 137–149. doi: 10.3758/BF03207704

Tajfel, H., and Wilkes, A. L. (1963). Classification and quantitative judgement. *Br. J. Psychol.* 54, 101–114. doi: 10.1111/j.2044-8295.1963.tb00865.x

Van den Bergh, O., Stegen, K., and Van de Woestijne, K. P. (1997). Learning to have psychosomatic complaints: conditioning of respiratory behavior and somatic complaints in psychosomatic patients. *Psychosom. Med.* 59, 13–23.

Van den Bergh, O., Winters, W., Devriese, S., Van Diest, I., Vos, G., and De Peuter, S. (2004). Accuracy of respiratory symptom perception in persons with high and low negative affectivity. *Psychol. Health* 19, 213–222. doi: 10.1080/08870440410001675627

von Leupoldt, A., Chan, P. Y. S., Bradley, M. M., Lang, P. J., and Davenport, P. W. (2011). The impact of anxiety on the neural processing of respiratory sensations. *Neuroimage* 55, 247–252. doi: 10.1016/j.neuroimage.2010.11.050

Watson, D., Clark, L. A., and Tellegen, A. (1988). Development and validation of a brief measure of positive and negative affect. *J. Pers. Soc. Psychol.* 54, 1063–1070. doi: 10.1037/0022-3514.54.6.1063

Watson, D., and Pennebaker, J. (1989). Health complaints, stress, and distress: exploring the central role of negative affectivity. *Psychol. Rev.* 96, 234–254. doi: 10.1037/0033-295X.96.2.234

Wientjes, C. J., and Grossman, P. (1994). Over-reactivity of the psyche or the soma? Interindividual associations between psychosomatic symptoms, anxiety, heart rate, and end-tidal partial carbon dioxide pressure. *Psychosom. Med.* 56, 533–540. doi: 10.1097/00006842-199411000-00009

Yang, F., Wu, Q., and Li, S. (2014). Learning-induced uncertainty reduction in perceptual decisions is task-dependent. *Front. Hum. Neurosci.* 8:282. doi: 10.3389/fnhum.2014.00282

11

Interoceptive Fear Learning to Mild Breathlessness as a Laboratory Model for Unexpected Panic Attacks

Meike Pappens, Evelien Vandenbossche, Omer Van den Bergh and Ilse Van Diest*

Health Psychology, University of Leuven, Leuven, Belgium

***Correspondence:**
Meike Pappens,
Health Psychology, University of Leuven, Tiensestraat 102, B-3000 Leuven, Belgium
meike.pappens@ppw.kuleuven.be

Fear learning is thought to play an important role in panic disorder. Benign interoceptive sensations can become predictors (conditioned stimuli – CSs) of massive fear when experienced in the context of an initial panic attack (unconditioned stimulus – US). The mere encounter of these CSs on a later moment can induce anxiety and fear, and precipitate a new panic attack. It has been suggested that fear learning to interoceptive cues would result in unpredictable panic. The present study aimed to investigate whether fear learning to an interoceptive CS is possible without declarative knowledge of the CS–US contingency. The CS consisted of mild breathlessness (or: dyspnea), the US was a suffocation experience. During acquisition, the experimental group received six presentations of mild breathlessness immediately followed by suffocation; for the control group both experiences were always separated by an intertrial interval. In the subsequent extinction phase, participants received six unreinforced presentations of the CS. Expectancy of the US was rated continuously and startle eyeblink electromyographic, skin conductance, and respiration were measured. Declarative knowledge of the CS–US relationship was also assessed with a post-experimental questionnaire. At the end of acquisition, both groups displayed the same levels of US expectancy and skin conductance in response to the CS, but the experimental group showed a fear potentiated startle eyeblink and a different respiratory response to the CS compared to the control group. Further analyses on a subgroup of CS–US unaware participants confirmed the presence of startle eyeblink conditioning in the experimental group but not in the control group. Our findings suggest that interoceptive fear learning is not dependent on declarative knowledge of the CS–US relationship. The present interoceptive fear conditioning paradigm may serve as an ecologically valid laboratory model for unexpected panic attacks.

Keywords: fear conditioning, panic, respiration, interoceptive fear, CS–US contingency, dual process theory

Introduction

Fear learning to initially harmless cues is considered one of the mechanisms in the etiology of panic disorder (Bouton et al., 2001; Kessler et al., 2006; Acheson et al., 2012; De Cort et al., 2012; Pappens et al., 2012). Low-level interoceptive bodily sensations (such as mild sensations of dyspnea; a slight increase in heart rate; etc.) are thought to acquire fear-evoking properties because they were once experienced in the context of an isolated panic attack (unconditioned stimulus – US). A panic-predictive connotation is attributed to the cues (conditioned stimuli – CSs), causing them to trigger anxiety and/or fear on a later encounter. The successful application of exposure techniques in the treatment of panic disorder demonstrates the relevance of learning theory for this disorder. For example, interoceptive exposure to mild bodily sensations significantly reduces long-term anxiety in panic patients (McHugh et al., 2009).

An important feature of panic attacks is that they can occur either 'expectedly' or 'unexpectedly' (DSM-V). It has been suggested that especially in the latter case fear conditioning to an interoceptive sensation is responsible for triggering a new culmination of fear and fear-related arousal (Vickers and McNally, 2005; Johnson et al., 2014).

Up until today interoceptive fear conditioning remains a poorly studied phenomenon, despite the fact that approximately 40% of panic attacks are defined as unexpected (Shulman et al., 1994). Over the past years, a few studies have investigated fear conditioning to an interoceptive CS in the laboratory. For example, Acheson et al. (2007) used 5 s applications of 20% CO_2-enriched air as an interoceptive CS and 15 s of the same substance as US in a panic-relevant interoceptive fear conditioning design. In a follow-up study (Acheson et al., 2012) they investigated discriminatory learning in interoceptive fear conditioning. Earlier, we successfully established and extinguished fear to a mild sensation of breathlessness by combining it with a distressing breathing obstruction and investigated the differences between exteroceptive versus interoceptive fear conditioning and the influence of vagal tone on interoceptive fear conditioning processes (Pappens et al., 2012, 2013, 2014). Together these studies demonstrated that fear can be acquired to benign interoceptive sensations and they provided valid experimental paradigms to study the characteristics of interoceptive fear conditioning in the laboratory.

An unstudied yet fundamental issue remains whether interoceptive fear conditioning can happen outside of awareness as is suggested by its presumable role in unpredictable panic. For decades, there has been an ongoing discussion in the scientific community on the necessity of propositional knowledge of the CS–US contingency as a prerequisite for associative fear learning. Proponents of a single-process theory of fear learning consider high-level cognitive processes as the foundation of all associative learning. As a consequence fear learning is based on propositional knowledge of the CS–US contingency (e.g., Purkis and Lipp, 2001; Lovibond and Shanks, 2002; Mitchell et al., 2009; Sebastiani et al., 2011; Newell and Shanks, 2014).

In contrast, a dual-process theory of fear learning describes one type of learning that leads to declarative knowledge and another type that activates the fear system without the person being necessarily aware of the CS–US contingency (e.g., Hamm and Vaitl, 1996; LeDoux, 1996; Clark and Squire, 1998; Öhman and Soares, 1998; Öhman and Mineka, 2001; Knight et al., 2003; Beaver et al., 2005; Smith et al., 2005; Tabbert et al., 2006, 2011; Weike et al., 2007; Kindt et al., 2009; Raes et al., 2010; Schultz and Helmstetter, 2010; Soeter and Kindt, 2010). In the past years, a great number of studies have sought to provide evidence for the existence of two distinct fear learning pathways. For example, Beaver et al. (2005) demonstrated fear learning to a subliminally administered – and thus not consciously perceived-danger cue (CS+). Also, fear learning to a CS+ was present in skin conductance responses (SCRs) although it was hardly discriminable from a CS- (Schultz and Helmstetter, 2010), but these results could not be replicated (Sevenster et al., 2014). Clark and Squire (1998) examined startle eyeblink responses in a classical aversive conditioning study in patients with deficits in declarative memory as a result of severe bilateral hippocampal lesions. As expected, the patients were not capable of reporting upon the CS–US relationship. However, they did display fear potentiated startle eyeblink responses to the danger cue, which were probably triggered through activation of the amygdala (LeDoux, 2000). Also, the intake of propranolol, a substance blocking noradrenergic release (Kindt et al., 2009) resulted in a dissociation in fear learning outcomes (self-reported fear and the startle eyeblink response on the one hand, versus skin conductance and US-expectancy ratings on the other hand).

In a previous study in healthy participants we also demonstrated a divergence between different fear measures in a panic-relevant interoceptive fear conditioning paradigm (Pappens et al., 2013). The experimental group received six paired administrations of a CS (a mild feeling of breathlessness) and a US (a strong feeling of breathlessness) during acquisition. The control group received the same amount of CSs and USs in acquisition, but in an unpaired fashion. Extinction was the same for both groups – it consisted of six CS-only administrations. Our measures included startle eyeblink responses electromyographic (EMG), respiratory measures and self-reported fear. While fear conditioning emerged in startle eyeblink EMG and respiration in the experimental group only, no group differences occurred in self-reported fear. Thus, although the experimental group showed a potentiated startle eyeblink response and an increased ventilation during the CS, they did not to report more fear to the CS compared to the control group. These findings suggest that our interoceptive paradigm is able to install fear without propositional knowledge of its presence. However, a direct measure of declarative knowledge about the CS–US contingency was lacking in the study nor was it possible to compare participants that were either aware or not of the CS–US relationship. This limited the conclusions that could be drawn with respect to the necessity of propositional knowledge to install interoceptive fear learning or existence of fear learning.

With the current study we aimed to replicate and extend our previous study. We replaced self-reported fear by a continuous measure of self-reported US-expectancy (DIAL) and we included SCRs. Declarative knowledge of the CS–US relationship was also evaluated post-experimentally by assessing the participants' knowledge of the experienced contingencies.

Our hypotheses were the following:

We expected to replicate the findings of Pappens et al. (2013) during acquisition: (a) higher startle eyeblink EMG responses during the CS than during ITI in the experimental but not in the control group; (b) less reduction in respiratory frequency and in tidal volume during the CS load in the experimental compared to the control group; and we predicted: (c) US expectancy during the CS load not to be greater for the experimental than for the control group. (d) since SCR was not included in Pappens et al. (2013) and given the mixed results in literature (e.g., Kindt et al., 2009; Schultz and Helmstetter, 2010; Sevenster et al., 2014) we held no specific predictions for this outcome measure.

Given the hypothesized role of interoceptive fear learning in unexpected panic, we expected to observe similar differences during acquisition in the experimental subgroup of CS–US unaware participants versus those in the control group: (a) higher startle eyeblink EMG responses during the CS than during ITI in the experimental but not in the control group; (b) less of a reduction in respiratory frequency and in tidal volume during the CS load in the experimental compared to the control group; (c) no differences in US expectancy in the experimental versus control group; (d) again, we held no specific predictions for SCR.

Materials and Methods

Participants

Fifty-six psychology students participated in this study (51 women, $M = 19.16$ years, range 18–25 years) in return for a course credit. Persons with a current or past history of cardiovascular disease, chronic, or acute respiratory disease, pregnancy, current, or past history of drug or alcohol abuse or dependence, psychotropic drug use and any current or past psychiatric disorder including panic and anxiety disorder were excluded from the study. The study protocol was approved by the Medical Ethics Committee of the University of Leuven in accordance with the Declaration of Helsinki. All subjects signed an informed consent form stating – amongst other information – that participation was voluntary and that they could withdraw from the study at any moment.

Stimuli and Apparatus

Stimuli

The CS consisted of a non-aversive linear resistive load of 0.98 kPa*s/l, whereas an aversive linear resistive load of 3.91 kPa*s/l served as the US. Resistive loads were administered during both the inspiratory and expiratory phases of the respiratory cycle. The CS was presented for 8 s, the US for 30 s (Pappens et al., 2013).

Breathing Apparatus

A mouthpiece was mounted onto a bacterial filter that was fitted on a pneumotachograph (Fleisch No. 2, Epalinges, Switzerland). The pneumotachograph was connected to a non-rebreathing valve of which the inspiratory and expiratory port were installed on a three-way *Y*-valve (stopcock type) using a vinyl tube (inner diameter: 3.5 cm; length 100 cm). This set-up enabled easy switching between room air and loaded breathing. The signal from the pneumotachograph was amplified using a pressure transducer (Sine Wave Carrier Demodulator CD15, Valydine EngineeringTM) and was calibrated daily with a 1 l syringe. Fractional end-tidal CO_2 ($FetCO_2$) was measured using an infrared capnograph (POET II, Criticare, USA) that sampled expired air from the breathing circuit close to the mouthpiece. The capnograph was calibrated daily using a calibration gas containing 7.5% CO_2. Air flow and CO_2 waveforms were digitized at 20 Hz.

Skin Conductance

Electrodermal activity was recorded with Fukuda standard Ag/AgCl electrodes (1 cm diameter) filled with a Unibase electrolyte and attached to the hypothenar palm of the non-dominant hand, which was cleaned with tap water before the start of the procedure. The inter-electrode distance was 2.5 cm. A Coulbourn skin conductance coupler (LabLinc v71–23) provided a constant 0.5 V across electrodes. The signal was digitized at 100 Hz.

Startle Eyeblink Response

Orbicularis oculi electromyographic activity (EMG) was recorded as an index of the eyeblink component of the startle response with three Ag/AgCl Sensormedics electrodes (0.25 cm diameter) filled with electrolyte gel. After cleaning the skin to reduce inter-electrode resistance, electrodes were placed on the left side of the face (Blumenthal et al., 2005). The raw signal was amplified by a Coulbourn isolated bioamplifier with bandpass filter (LabLinc v75–04). The recording bandwidth of the EMG signal was between 13 Hz and 10 kHz. The signal was rectified online and smoothed by a Coulbourn multifunction integrator (LabLinc v76–23 A) with a time constant of 50 ms. The EMG signal was digitized at 1000 Hz from 500 ms before the onset of the auditory startle probe (a 95 dB burst of white noise with a rise time <1 ms presented binaurally for 50 ms through headphones) until 1000 ms after probe onset.

US Expectancy Dial

Participants were asked to continuously rate their expectancy of the US with a custom built dial (Pappens et al., 2012). The scale on which they had to rate US-expectancy, ranged from zero ("I am certain that the heavy breathing resistance is not coming now") to one hundred ("I am certain that the heavy breathing resistance is coming now"). The dial produced an analog signal, which was digitized and stored at 10 Hz (Pappens et al., 2012).

Post-Experimental Contingency Assessment

After the experimental phase, participants were questioned orally about their knowledge of the CS–US relationship with the following open question: "can you tell us whether you detected a certain order in the administration of the different types of stimuli?" Participants of the experimental group were categorized as 'aware' when correctly identifying that a light breathing resistance (CS) immediately preceded a strong breathing resistance (US) and as 'unaware' if they could not verbalize this relationship. In the control group a person was labeled as 'aware' when he/she correctly identified that CS and US were separated by an intertrial interval (ITI) and as 'unaware' if they did not.

Software

All devices were connected to a PC through a National Instruments PCI-6221 16-Bit acquisition card (National Instruments, Austin, TX, USA). Affect 4.0 software (Spruyt et al., 2010) was used for stimulus presentation and data acquisition. Physiological signals were treated off-line with PSPHA (De Clerck et al., 2006), a modular script-based program to generate and apply calibration factors and to extract parameters from each of the recorded signals.

Procedure

Upon their arrival, the experimenter told the participants that psychophysiological and subjective responses would be measured during three different types of breathing trials: normal breathing, mildly restricted breathing and heavily restricted breathing. She showed the participants how to continuously indicate their US expectancy with the online dial. Then she attached the electrodes and explained that brief bursts of noise would be administered through the headphones but that these could be ignored. Following this, participants took in the mouthpiece and put on the nose-clip and the headphones.

After the administration of 10 acoustic startle probes (30 s in between probes) to habituate the startle eyeblink response the experimental phase was started. Participants received three CS pre-exposure trials, followed by six acquisition trials and six extinction trials. Half of the participants were randomly assigned to the experimental (paired) group, the other half to the control (unpaired) group.

A pre-exposure trial consisted of a 20 s baseline, a CS (8 s) presentation and an ITI of 30 s without stimulus. For the experimental group, acquisition trials consisted of baseline (20 s), CS (8 s), US (30 s), and ITI (variable between 25 and 35 s). The control group received the following sequence during the acquisition trials: baseline (20 s), CS (8 s), ITI (25–35 s), and US (30 s). Extinction trials consisted of baseline (20 s), CS (8 s), and ITI (55–65 s) for both groups. Startle probes were applied in every trial at random times between 6.5–7.5 s after CS onset, between 21–23 s after US onset, and between 21–23 s after ITI onset (Pappens et al., 2013). After the experimental phase of the study participants were questioned about their knowledge of the CS–US relationship.

Before leaving, participants were fully debriefed.

Response Definition, Scoring, and Statistical Analysis

Startle eyeblink EMG data, SCRs and respiratory signals were processed with PSychoPHysiological Analysis (PSPHA; De Clerck et al., 2006).

The EMG response was calculated by subtracting the mean baseline value (0–20 ms after probe onset) from the peak value (between 21–175 ms after probe onset). Each startle eyeblink was visually checked on artifacts. Distorted blinks or trials with spontaneous blinks during baseline were removed from the data. Rejected blinks ($n = 11$) were replaced by the means of the previous and following trial. EMG data were standardized and T-transformed within persons (Blumenthal et al., 2005).

Respiratory rate (RR, in cycles per minute, cpm) and tidal volume (V_T, in ml) were extracted on a breath-by-breath basis and then averaged across the CS episode, and across the 20 s baseline episode preceding the CS. Baseline values were subtracted from CS values for statistical analysis.

Skin conductance responses were calculated by subtracting the mean skin conductance level (SCL) during baseline (2 s before the CS onset) from the maximum SCR during the subsequent 8 s CS period. All SCRs below 0.05 were coded as a non-response by setting them to zero and values were Log10 (SCR + 1) transformed.

The mean of the online US-expectancy ratings (DIAL) were calculated for the first and the last second of the CS presentation per trial.

All data (EMG, RR, V_T, SCR, DIAL) were averaged across three pre-exposure trials and across every two subsequent trials in acquisition and extinction, resulting in one pre-exposure block, three acquisition blocks, and three extinction blocks.

Due to technical difficulties, the data of five participants (two of the experimental group and three of the control group) were excluded from all analyses, bringing the total number of participants on 51. For DIAL and respiration, data of one additional participant of the experimental group was lost due to technical difficulties.

Potential pre-existing group differences were tested with a first set of analyses on the pre-exposure data. Respiratory parameters and SCR data were entered in a repeated measures ANOVA (RM ANOVA) design with group (experimental/control) as a between subject variable. Probe (CS/ITI) and Time (second 1, second 8) were, respectively, added as within subject variables for startle eyeblink EMG and DIAL.

The *a priori* hypotheses that served to examine whether we could replicate the results of Pappens et al. (2013) were analyzed in mixed ANOVA designs. Startle eyeblink EMG data were entered in a group (experimental/control) × probe (CS/ITI) × block (1, 2, 3) design. Respiratory parameters and SCR data were entered in a design with group (experimental/control) and block (1, 2, 3) as a between subject and a within subject variable respectively. DIAL data were entered in a group (experimental/control) × time (second 1, second 8) × block (1, 2, 3) design. Only acquisition data were entered in the analyses – extinction data are depicted in the figures for illustrative purposes only.

Participants were assigned to the 'aware' or 'unaware' group based on the post-experimental assessment of awareness. We opted to use this CS–US contingency awareness measure and not the CS–US expectancy dial because it produced a discrete 'YES' or 'NO' awareness criterion. (Data of the CS–US expectancy dial are more difficult to unambiguously categorize as aware/not aware in a between-subject design and less clearly result in a dichotomy. Former studies who did use online dial data for categorization were within-subject paradigms; e.g., Schultz and Helmstetter, 2010).

Using the post-experimental CS–US contingency awareness data, 35 of 51 included participants were labeled as 'unaware'; 16 in the experimental group and 19 in the control group. For DIAL and respiration, 34 'unaware' and 16 'aware' persons were identified in the 50 included participants.

Hypotheses were tested also in the 'unaware' subgroup by entering the data in a mixed ANOVA design with group (experimental/control) as a between-subject variable and block (1, 2, 3) as a within-subject variable. Probe (CS/ITI) and Time (Second 1, Second 8) were added as a within-subject variable for the EMG and US-expectancy analyses, respectively.

Alpha was set at 0.05. Greenhouse–Geisser corrections were applied where appropriate. Uncorrected degrees of freedom and corrected *p*s are reported, together with η_p^2. Additional testing of significant results were analyzed with two-tailed planned comparisons. Statistical analyses for all measures were accomplished with Statistical Version 12.

Results

Pre-Exposure

US-Expectancy

No effects involving the group factor were observed in US expectation. Main effect of group: $F(1,48) = 0.54$, $p = 0.47$, $\eta_p^2 = 0.01$; group × time: $F(1,48) = 0.54$, $p = 0.47$, $\eta_p^2 = 0.01$.

US-expectancy increased from second 1 to second 8; main effect of time: $F(1,48) = 22.88$, $p < 0.001$, $\eta_p^2 = 0.32$, see **Figure 1**.

Startle Eyeblink EMG

No *a priori* group differences were found in startle eyeblink data during pre-exposure: group: $F(1,49) = 0.94$, $p = 0.34$, $\eta_p^2 = 0.02$; group × probe: $F(1,49) = 0.88$, n.s., $\eta_p^2 = 0.02$, see **Figure 2**.

Respiratory Rate

As expected, the effect of group was not significant: $F(1,48) = 0.74$, $p = 0.40$, $\eta_p^2 = 0.015$, see **Figure 3**.

Tidal Volume (V_T)

We observed no differences between the experimental and the control group in V_T during pre-exposure: $F(1,48) = 0.14$, $p = 0.71$, $\eta_p^2 = 0.003$. See **Figure 4**.

Skin Conductance Response

Although **Figure 5** suggests that the experimental group displayed higher SCRs during pre-exposure, such effect of group was not significant: $F(1,49) = 2.75$, $p = 0.10$, $\eta_p^2 = 0.05$.

Replication of Pappens et al. (2013)

US-Expectancy

No significant effects involving the group factor were observed in US expectancy. Effect of group: $F(1,48) = 2.69$, $p = 0.11$, $\eta_p^2 = 0.05$; group × block: $F(2,96) = 0.17$, $p = 0.85$, $\eta_p^2 = 0.003$; group × time: $F(1,48) = 0.12$, $p = 0.73$, $\eta_p^2 = 0.003$. The critical group × block × time interaction was also not significant: $F(2,96) = 0.62$, $p = 0.54$, $\eta_p^2 = 0.01$.

Participants' US expectancy increased from the first to the last second of the CS; main effect of time: $F(1,48) = 20.99$, $p < 001$, $\eta_p^2 = 0.30$. Ratings also increased over blocks; main effect of block: $F(2,96) = 10.24$, $p < 0.001$, $\eta_p^2 = 0.18$. See **Figure 1**.

Startle Eyeblink EMG

Further testing of the significant probe x group interaction [$F(1,49) = 7.99$, $p < 0.01$, $\eta_p^2 = 0.14$] indicated that CS probes evoked larger responses than ITI probes in the experimental group, $F(1,49) = 23.01$, $p < 0.001$, but not in the control group, $F(1,49) = 0.73$, $p = 0.40$.

The main effect of group was not significant, $F(1,49) = 0.01$, $p = 0.91$, $\eta_p^2 = 0.00$; nor was the block × group interaction, $F(2,98) = 1.07$, $p = 0.35$, $\eta_p^2 = 0.02$, or the probe × block × group interaction, $F(2,98) = 0.18$, $p = 0.83$, $\eta_p^2 = 0.00$.

We also observed a significant effect of probe: $F(1,49) = 16.18$, $p<0.01$, $\eta_p^2 = 0.25$. See **Figure 2**.

Respiratory Rate

Both the main effect of group and the group × block interaction were not significant; $F(1,48) = 0.89$, $p = 0.35$, $\eta_p^2 = 0.02$, and $F(2,96) = 1.13$, $p = 0.33$, $\eta_p^2 = 0.02$, respectively.

However, in line with results of Pappens et al. (2013) and as suggested by **Figure 3**, two-tailed planned comparisons did indicate that a linear decrease in RR in response to the CS (change scores from baseline to CS) from the first to the last block of acquisition was not significant for the experimental group, $F(1,48) = 0.10$, $p = 0.76$, while it was for the control group, $F(1,48) = 5.18$, $p = 0.013$ (see **Figure 3**). This effect was mainly driven by an increase in RR in the control group during baseline (data not presented).

Tidal Volume (V_T)

No significant effects involving the group factor were observed: group, $F(1,48) = 0.58$, $p = 0.45$, $\eta_p^2 = 0.01$; group × block interaction, $F(2,96) = 2.09$, $p = 0.13$, $\eta_p^2 = 0.04$. See **Figure 4**.

Skin Conductance Response

No significant effects involving the group factor emerged for SCR; group: $F(1,49) = 1.29$, $p = 0.26$, $\eta_p^2 = 0.03$; group × block interaction: $F(2,98) = 0.11$, $p = 0.90$, $\eta_p^2 = 0.002$, see **Figure 5**.

Analysis on Subgroup of 'Unaware' Participants

US Expectancy

No differences in US expectancy were observed between the experimental and the control group during acquisition. Effect of group: $F(1,32) = 0.94$, $p = 0.34$, $\eta_p^2 = 0.03$; group × time:, $F(1,32) = 0.25$, $p = 0.62$, $\eta_p^2 = 0.008$; group × block:, $F(2,64) = 0.40$, $p = 0.67$, $\eta_p^2 = 0.01$; group × block × time: $F(2,64) = 1.27$, $p = 0.29$, $\eta_p^2 = 0.04$.

A significant main effect of time was present, $F(1,32) = 22.88$, $p < 0.001$, $\eta_p^2 = 0.42$, with an increase in US expectancy during the CS from second 1 to second 8; and also of block, $F(2,64) = 7.73$, $p < 0.001$, $\eta_p^2 = 0.19$, indicating that US expectancy augmented across acquisition blocks. See **Figure 8**.

Startle Eyeblink EMG

Across acquisition, a decrease in startle responses during the CS probe was observed for the control group, $F(1,33) = 17.12$, $p < 0.001$, but not for the experimental group, $F(1,33) = 0.75$, $p = 0.39$; significant group × probe × block interaction: $F(2,66) = 3.26$, $p = 0.04$, $\eta_p^2 = 0.09$.

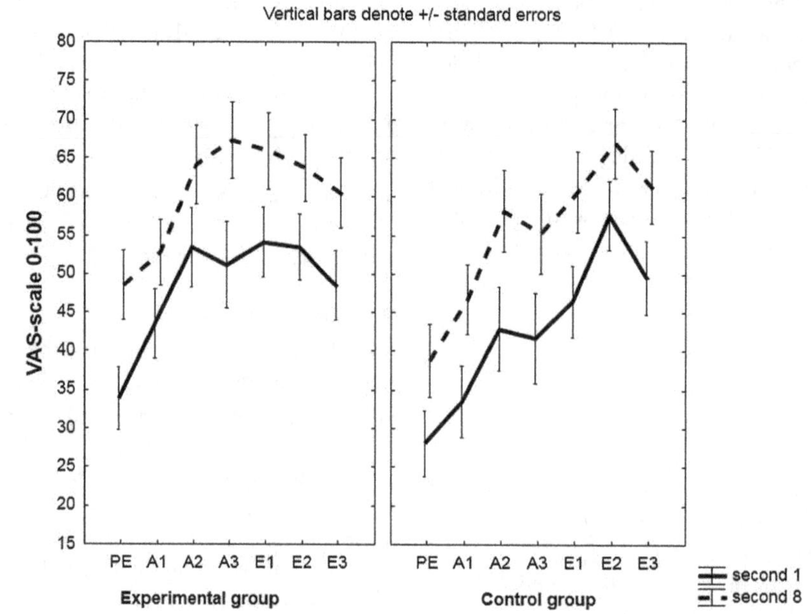

FIGURE 1 | Unconditioned stimulus (US)-expectancy during the conditioned stimulus (CS)-period. Mean US-expectancy ratings of second 1 and second 8 during the conditioned stimuli (CSs) on a VAS-scale ranging from 0 (certainly no breathing resistance) to 100 (certainly breathing resistance) for the experimental and the control group. Responses were averaged across three pre-exposure trials (PE), two acquisition trials (A1, A2, A3) and two extinction trials (E1, E2, E3).

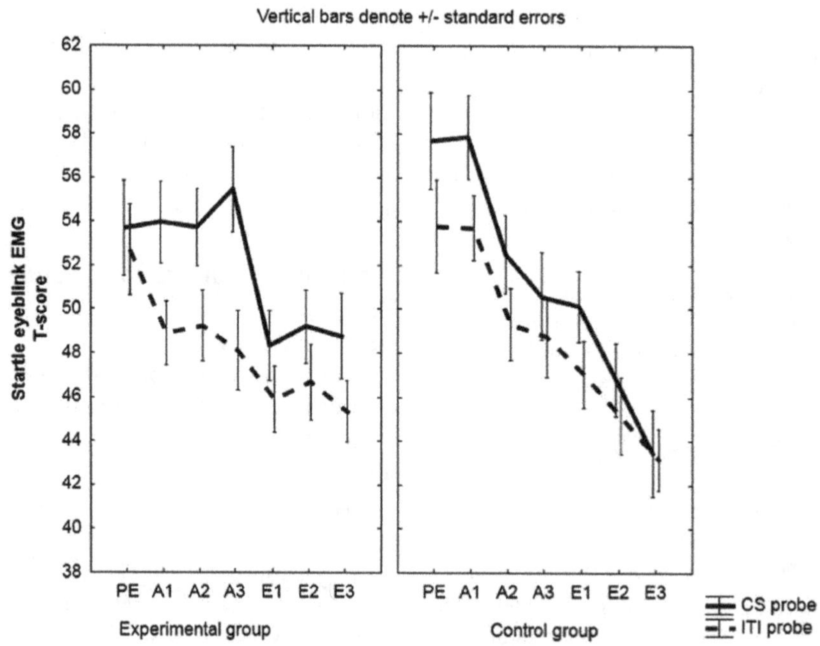

FIGURE 2 | Startle eyeblink responses electromyographic (EMG). Mean EMG-responses (T-scores) during the CS and during ITI for the experimental and the control group. Startle eyeblink responses were averaged across three pre-exposure trials (PE), two acquisition trials (A1, A2, A3) and two extinction trials (E1, E2, E3).

Also the two-way interactions were significant; probe × group interaction: $F(1,33) = 4.69$, $p = 0.04$, $\eta_p^2 = 0.12$, and block × group interaction: $F(2,66) = 5.56$, $p < 0.01$, $\eta_p^2 = 0.14$.

Finally, probes administered during the CS evoked stronger startle responses than those during ITI; main effect of probe: $F(1,33) = 10.7$, $p < 0.01$, $\eta_p^2 = 0.24$). See **Figure 6**.

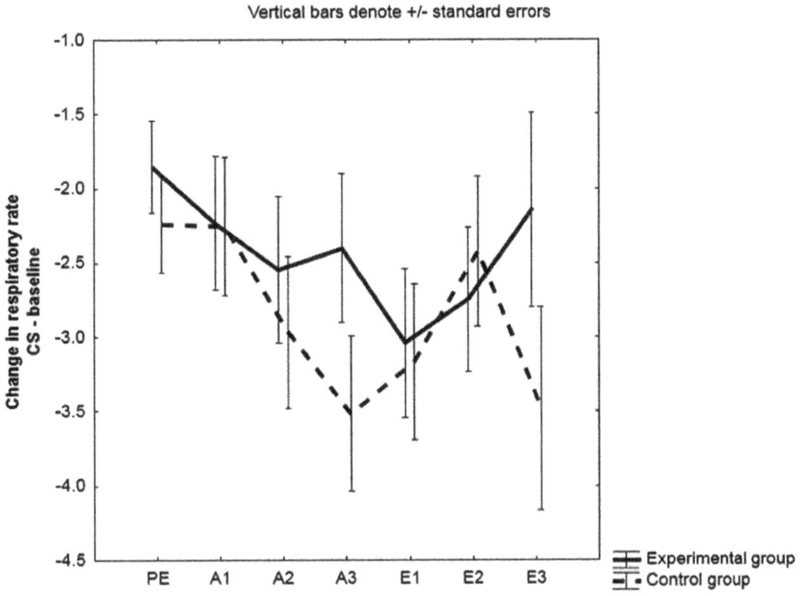

FIGURE 3 | Change in respiratory rate (RR). Mean changes in RR (in cycles per minute, cpm) during the CS relative to baseline (CS minus baseline) for the experimental and the control group. Respiratory responses were averaged across three pre-exposure trials (PE), two acquisition trials (A1, A2, A3) and two extinction trials (E1, E2, E3).

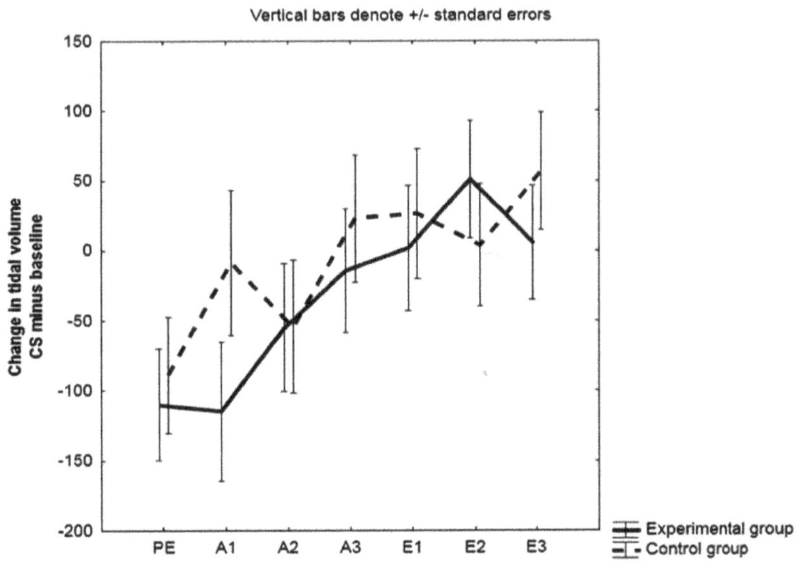

FIGURE 4 | Change in Tidal Volume (V_T). Mean changes in V_T (in ml) during the CS relative to baseline (CS minus baseline) for the experimental and the control group. Respiratory responses were averaged across three pre-exposure trials (PE), two acquisition trials (A1, A2, A3), and two extinction trials (E1, E2, E3).

Respiratory Rate

We observed no differences in RR between the experimental and the control group. Effect of Group: $F(1,32) = 1.91$, $p = 0.18$, $\eta_p^2 = 0.06$. Group × Block interaction: $F(2,64) = 1.26$, $p = 0.29$, $\eta_p^2 = 0.04$.

However, in line with our results in the total group sample, we did observe a linear decrease in RR across acquisition blocks in the control group: $F(1,32) = 5.83$, $p = 0.02$, but not in the experimental group: $F(1,32) = 0.01$, $p = 0.91$, see **Figure 7**. Again, this effect was driven by an increase in RR in the control group during baseline (data not presented).

Tidal Volume (V_T)

No significant effects involving the group factor were observed in V_T.

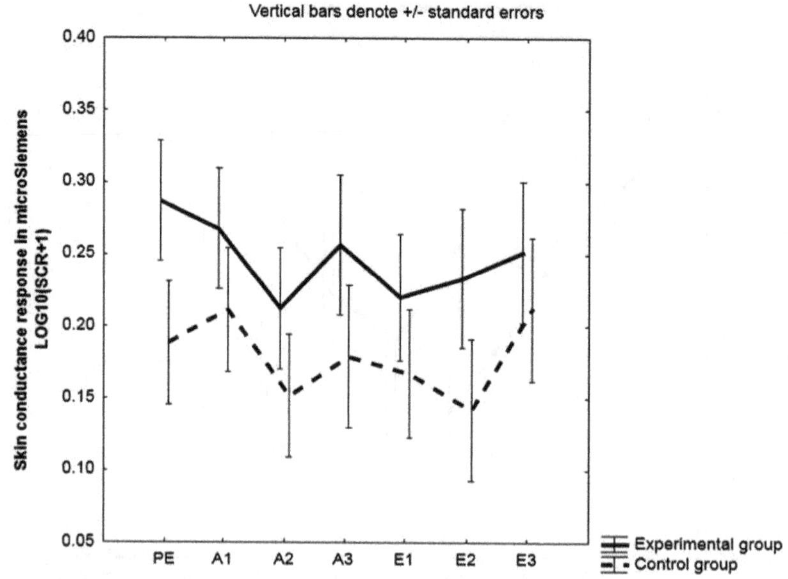

FIGURE 5 | Skin conductance responses (SCR) during the CS-period. Mean SCRs [LOG10 (1 + SCR)] in microSiemens during the CSs. SCRs were averaged across three pre-exposure trials (PE), two acquisition trials (A1, A2, A3), and two extinction trials (E1, E2, E3).

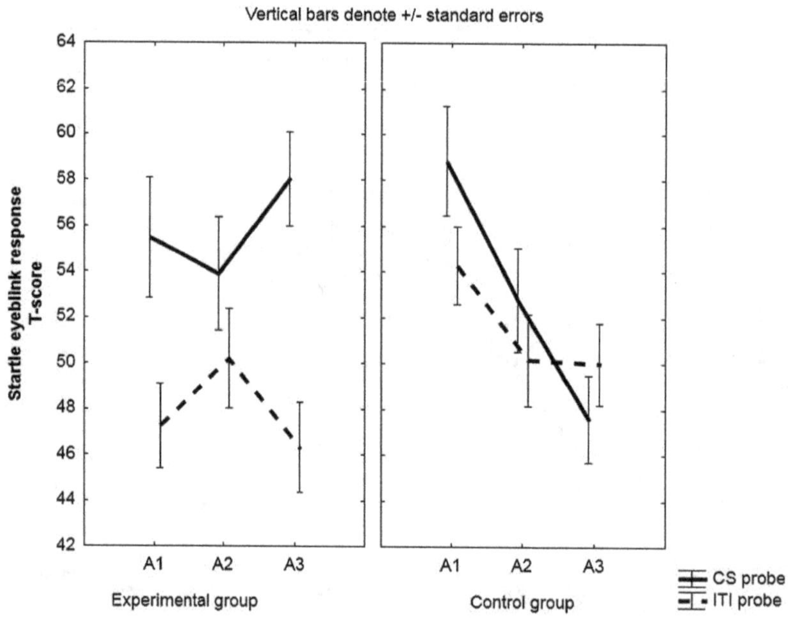

FIGURE 6 | Startle eyeblink responses in unaware participants (EMG). Mean EMG-responses (T-scores) during the CS and during ITI for the experimental and the control group. Startle eyeblink responses were averaged across two acquisition trials (A1, A2, A3).

Effect of Group: $F(1,32) = 0.45$, $p = 0.50$, $\eta_p^2 = 0.01$. Group × Block interaction: $F(2,64) = 0.9$, $p = 0.41$, $\eta_p^2 = 0.03$.

Skin Conductance Response

No significant effects involving the group factor were observed during acquisition: main effect of Group, $F(1,33) = 1.3$, $p = 0.26$, $\eta_p^2 = 0.04$; Group × Block: $F(2,66) = 0.46$, $p = 0.63$, $\eta_p^2 = 0.01$.

Discussion

The aim of the current study was to examine the relationship between interoceptive fear conditioning and CS–US contingency awareness. To this end we first sought to replicate the results of a previous study (Pappens et al., 2013) that suggested fear learning had occurred without declarative knowledge of the CS–US relationship. Second, we aimed to examine fear learning in the

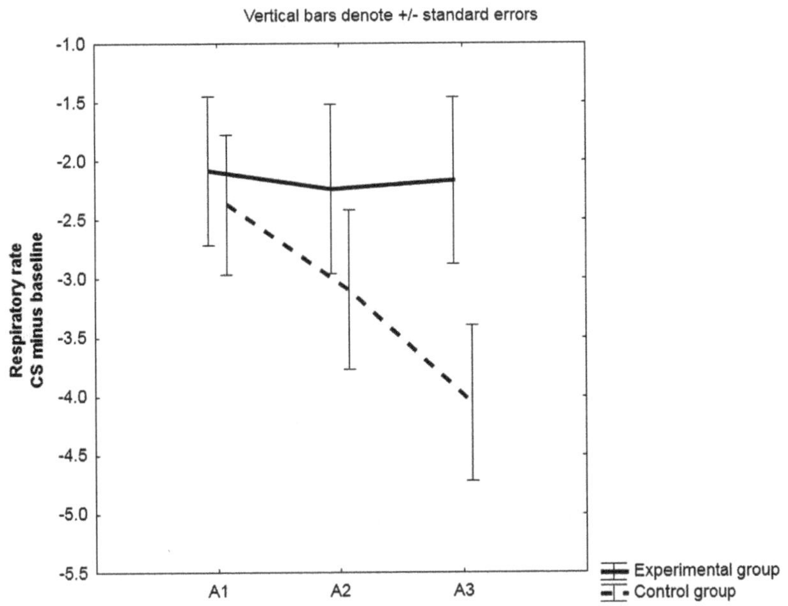

FIGURE 7 | Change in RR in unaware participants. Mean changes in RR (in cycles per minute, cpm) during the CS relative to baseline (CS minus baseline) for the experimental and the control group. Respiratory responses were averaged across two acquisition trials (A1, A2, A3).

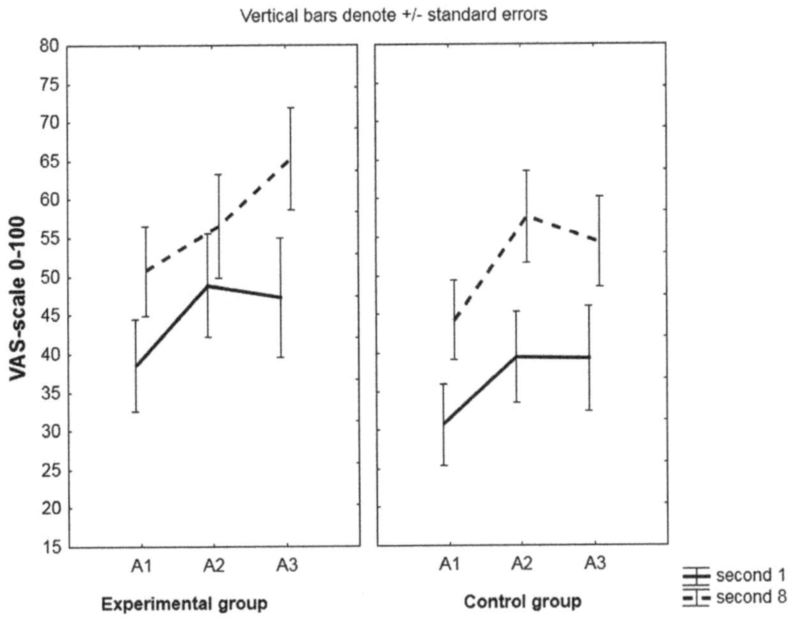

FIGURE 8 | US-expectancy during the CS-period in the subgroup of unaware participants. Mean US-expectancy ratings of second 1 and second 8 during the CS on a VAS-scale ranging from 0 (certainly no heavy breathing resistance) to 100 (certainly heavy breathing resistance) for the experimental and the control group during the three acquisition blocks (A1, A2, A3).

subgroup of CS–US unaware participants. Our design consisted of the administration of six explicitly paired presentations of mild breathlessness (CS) followed by strong dyspnea (US) in the experimental group while the control group received the same number of CSs and USs in an explicitly unpaired fashion. Our measures included startle eyeblink EMG, respiratory parameters, skin conductance, a continuous US expectancy measure and a discrete post-experimental CS–US contingency assessment. Based on Pappens et al. (2013), we predicted to observe a dissociation in fear outcomes, that is, fear learning (or: group) effects in startle potentiation and respiration, but not in self-reported US-expectancy. We expected the same pattern to occur in a subgroup of participants who failed to verbalize the CS–US relationship at the end of the experiment.

Our findings confirm that a mild respiratory sensation can acquire fear-evoking properties because of its predictive relationship with an aversive respiratory event. These results corroborate earlier findings demonstrating interoceptive fear conditioning (Acheson et al., 2007, 2012; Pappens et al., 2012, 2014) and they successfully replicate the data of Pappens et al. (2013). Most importantly, we replicated the expected dissociation of the dependent measures: during acquisition experimental and control group differed in fear responding both in startle eyeblink EMG and respiratory measures, but not in US expectancy.

In the second part of this study we examined the relationship between fear learning and declarative knowledge of the CS-US relationship in a more direct way by selecting a subgroup of participants who could not correctly report this relationship despite 100% explicit contingent presentations of CS and US. Despite the absence of propositional knowledge and the lack of fear conditioning effects in US-expectancy, they did display fear learning in startle eyeblink responses and – to a lesser extent – in RR. These results demonstrate fear acquisition without declarative knowledge of the CS–US relationship.

In contrast to Pappens et al. (2013), the current study did include a SCR measurement, but we failed to observe conditioning effects in SCR. Several authors have argued that fear conditioned changes in SCR, just as US-expectancy, may primarily reflect propositional learning (declarative knowledge of the CS–US contingency), whereas startle eyeblink potentiation more directly reflects subcortical emotional learning that can dissociate from propositional learning (e.g., Hamm and Vaitl, 1996; Öhman and Mineka, 2001; Weike et al., 2007; Soeter and Kindt, 2010). However, previous experiments generated mixed results as to whether conditioning of SCR requires 'conscious' knowledge of the CS–US relationship (Hamm and Vaitl, 1996; Purkis and Lipp, 2001; Lovibond and Shanks, 2002; Hamm and Weike, 2005; Weike et al., 2007; Sevenster et al., 2014) or whether 'unconscious' fear learning also results in conditioned SCRs (Bechara et al., 1995; Knight et al., 2003; Schultz and Helmstetter, 2010). Our results rather support the former hypothesis: we did not observe learning effects in SCR in a group that failed to learn declarative knowledge of the CS–US relationship.

For several reasons, the present interoceptive fear conditioning paradigm seems especially apt for the experimental study of panic. First, 40% of panic attacks are described as coming 'out of the blue' (Shulman et al., 1994) although the efficacy of exposure therapy suggests that panic attacks are conditioned events triggered by a cue (McHugh et al., 2009; Meuret et al., 2011). The dissociation observed in our data can offer a nice framework for understanding this apparent paradox. Despite being unable to verbalize the CS–US relationship, participants did display conditioned defensive fear responses. For this reason, our interoceptive fear conditioning paradigm could represent a strong tool for the further study of unexpected panic.

Second, clinically, the dissociation we observed in fear measures could offer an explanation for the high relapse rate observed in anxiety patients after exposure therapy (Vervliet et al., 2013): although verbal assessment indicated their fear had disappeared, residual fear might still have been present at the end of therapy, precipitating the recovery of fear at a later instance. The dissociation observed in our data and the failure of total fear extinction observed in startle eyeblinks in the experimental group (Pappens et al., 2013) shows that our paradigm could serve to study the phenomenon of return of fear.

Third, the fact that we installed fear learning to a fear relevant *respiratory* sensation is of special interest for the study of panic. In the phenomenology of panic, next to cardiac sensations, respiratory symptoms are prevalent (e.g., Ley, 1989; Briggs et al., 1993; Klein, 1993; Kircanski et al., 2009; Meuret et al., 2011) and seem vulnerable for fear conditioning processes. This may be related to the fact that both response systems produce strong internal sensations that may enter awareness and that both are referring to vital functions, the failure of which may represent a sudden life threatening experience causing fear of imminent death. Last, this study represents a successful replication of an earlier experiment (Pappens et al., 2013) corroborating the usefulness of this paradigm as a source of stable and reliable results.

Most studies that reported unconscious fear learning manipulated the level of awareness experimentally by adding a distractor task (e.g., Smith et al., 2005; Tabbert et al., 2006, 2011; Weike et al., 2007), through subliminal conditioning (Öhman and Soares, 1998; Raes et al., 2010) or by obscuring discriminability between the danger and the safety cue (Knight et al., 2003; Schultz and Helmstetter, 2010). Proponents of a single process theory refuted these results based on the lack of a consensus concerning the criteria used to determine the effectiveness of blocking a stimulus from awareness (Sebastiani et al., 2011). A strength of our study is that it created a dissociation without the inclusion of an awareness manipulation. It would, however, be presumptuous to interpret our data as strong evidence for a dual process theory of fear learning since it cannot be excluded that one measure (e.g., the startle eyeblink response) might be more sensitive than another one (e.g., SCR or US expectancy), yet still reflect the same, single learning process. Our data do suggest that different response systems can at least have different activation thresholds in the learning of the contingency between CS and US.

A first limitation of our study is that the effects obtained are rather small. Interoceptive fear conditioning seems overall difficult to install with a homoreflexive paradigm, i.e., a design in which CS and US share the same response system and have similar initial sensory properties (Dworkin, 2000). Future studies should thus focus on the identification of individuals that are more prone to develop fear for harmless interoceptive sensations. For example, we demonstrated that persons scoring high on Fear of Suffocation (a subscale of the Claustrophobia Questionnaire) are more vulnerable for maladaptive breathing during obstructed breathing (Pappens et al., 2013). Also, Melzig et al. (2011) demonstrated that people who score high on Anxiety Sensitivity, are more sensitive to respiratory cues.

Second, since an increase of US expectancy was already present during pre-exposure it is possible that later fear learning effects in US expectancy were masked due to ceiling effects.

The increase during pre-exposure is in line with previous results (Pappens et al., 2013) and is likely caused by the inherent relatedness of the mild resistive load (applied as CS) and the breathing occlusion (applied as US). Based on the findings of another study, we have recently argued that pre-existing expectations about the CS–US relationship may indeed hamper detecting the true contingencies (Schroijen et al., 2015).

In summary, this study successfully installed interoceptive fear to mild dyspnea. Fear learning was observed in startle blink EMG and respiratory measures but not in US expectancy and SCR. The same pattern of results was observed in subgroup of participants who failed to acquire declarative knowledge of the CS–US relationship. We propose that our interoceptive fear learning paradigm might serve as an ecologically valid laboratory model for unexpected panic.

Acknowledgments

The present study was supported by a KU Leuven grant (PF/10/005) to the Centre for Excellence of Generalization Research in Ill Health and Psychopathology and a post-doctoral fellowship of FWO Vlaanderen to MP.

References

Acheson, D. T., Forsyth, J. P., and Moses, E. (2012). Interoceptive fear conditioning and panic disorder: the role of conditioned stimulus–unconditioned stimulus predictability. *Behav. Ther.* 6, 174–189. doi: 10.1016/j.beth.2011.06.001

Acheson, D. T., Forsyth, J. P., Prenoveau, J. M., and Bouton, M. E. (2007). Interoceptive fear conditioning as a learning model of panic disorder: an experimental evaluation using 20% CO2-enriched air in a non- clinical sample. *Behav. Res. Ther.* 45, 2280–2294. doi: 10.1016/j.brat.2007.04.008

Beaver, J. D., Mogg, K., and Bradley, B. P. (2005). Emotional conditioning to masked stimuli and modulation of visuospatial attention. *Emotion* 5, 67–79.

Bechara, A., Tranel, D., Damasio, H., Adolphs, R., Rockland, C., and Damasio, A. R. (1995). Double dissociation of conditioning and declarative knowledge relative to the amygdala and hippocampus in humans. *Science* 269, 1115–1118.

Blumenthal, T. D., Cuthbert, B. N., Filion, D. L., Hackley, S., Lipp, O. V., and Van Boxtel, A. (2005). Committee report: guidelines for human startle eyeblink electromyographic studies. *Psychophysiology* 42, 1–15. doi: 10.1111/j.1469 8986.2005.00271.x

Bouton, M. E., Mineka, S., and Barlow, D. H. (2001). A modern learning theory perspective on the etiology of panic disorder. *Psychol. Rev.* 108, 4–32.

Briggs, A. C., Stretch, D. D., and Brandon, S. (1993). Subtyping of panic disorder by symptom profile. *Br. J. Psychiatry* 163, 201–209. doi: 10.1192/bjp.163.2.201

Clark, R. E., and Squire, L. R. (1998). Classical conditioning and brain systems: the role of awareness. *Science* 280, 77–81.

De Clerck, A., Verschuere, B., Crombez, G., and De Vlieger, P. (2006). Psychophysiological analysis (PSPHA): a modular script based program for analyzing psychophysiological data. *Behav. Res. Methods* 38, 504–510. doi: 10.3758/BF03192805

De Cort, K., Griez, E., Büchler, M., and Schruers, K. (2012). The role of "interoceptive" fear conditioning in the development of panic disorder. *Behav. Ther.* 1, 203–215. doi: 10.1016/j.beth.2011.06.005

Dworkin, B. R. (2000). "Interoception," in *Handbook of Psychophysiology*, 2nd Edn, eds J. T. Cacioppo, L. G. Tassinary, and G. G. Berntson (New York: Cambridge University Press), 482–506.

Hamm, A. O., and Vaitl, D. (1996). Affective learning: awareness and aversion. *Psychophysiology* 33, 698–710.

Hamm, A. O., and Weike, A. I. (2005). The neuropsychology of fear learning and fear regulation. *Int. J. Psychophysiol.* 57, 5–14. doi: 10.1016/j.ijpsycho.2005.01.006

Johnson, P. L., Federici, L. M., and Shekhar, A. (2014). Etiology, triggers and neurochemical circuits associated with unexpected, expected, and laboratory-induced panic attacks. *Neurosci. Biobehav. Rev.* 46, 429–454. doi: 10.1016/j.neubiorev.2014.07.027

Kessler, R. C., Chiu, W. T., Jin, R., Ruscio, A. M., Shear, K., and Walters, E. E. (2006). The epidemiology of panic attacks, panic disorder, and agoraphobia in the National Comorbidity Survey Replication. *Arch. Gen. Psychiat.* 63, 415–424.

Kindt, M., Soeter, M., and Vervliet, B. (2009). Beyond extinction: erasing human fear responses and preventing the return of fear. *Nat. Neurosci.* 12, 256–258. doi: 10.1038/nn.2271

Kircanski, K., Phil, C., Craske, M. G., Epstein, A. M., and Wittchen, H.-U. (2009). Subtypes of panic attacks: a critical review of the empirical literature. *Depress. Anxiety* 26, 878–887. doi: 10.1002/da.20603

Klein, D. F. (1993). False suffocation alarms, spontaneous panics, and related conditions: an integrative hypothesis. *Arch. Gen. Psychiatry* 50, 306–317. doi: 10.1001/archpsyc.1993.01820160076009

Knight, D., Nguyen, H., and Bandettini, P. (2003). Expression of conditional fear with and without awareness. *Proc. Natl. Acad. Sci. U.S.A.* 100, 15280–15283.

LeDoux, J. E. (1996). *The Emotional Brain.* New York: Simon & Schuster.

LeDoux, J. E. (2000). Emotion circuits in the brain. *Annu. Rev. Neurosci.* 23, 155–184. doi: 10.1146/annurev.neuro.23.1.155

Ley, R. (1989). Dyspneic-fear and catastrophic cognitions in hyperventilatory panic attacks. *Behav. Res. Ther.* 27, 549–554. doi: 10.1016/0005-7967(89)90089-2

Lovibond, P. F., and Shanks, D. R. (2002). The role of awareness in Pavlovian conditioning: empirical evidence and theoretical implications. *J. Exp. Psychol.* 1, 3–26.

McHugh, R. K., Smits, J. A., and Otto, M. W. (2009). Empirically supported treatments for panic disorder. *Psychiat. Clin. N. Am.* 32, 593–610. doi: 10.1016/j.psc.2009.05.005

Melzig, C. A., Holtz, K., Michalowski, J. M., and Hamm, A. O. (2011). Interoceptive threat leads to defensive mobilization in highly anxiety sensitive persons. *Psychophysiology* 48, 745–754. doi: 10.1111/j.1469-8986.2010.01150

Meuret, A. E., Rosenfield, D., Wilhelm, F. H., Zhou, E., Conrad, A., Ritz, T., et al. (2011). Do unexpected panic attacks occur spontaneously? *Biol. Psychiatry* 70, 985–991. doi: 10.1016/j.biopsych.2011.05.027

Mitchell, C. J., De Houwer, J., and Lovibond, P. F. (2009). The propositional nature of human associative learning. *Behav. Brain Sci.* 32, 183–198.

Newell, B. R., and Shanks, D. R. (2014). Unconscious influences on decision making: a critical review. *Behav. Brain Sci.* 37, 1–19. doi: 10.1017/S0140525X12003214

Öhman, A., and Mineka, S. (2001). Fears, phobias, and preparedness: toward an evolved module of fear and fear learning. *Psychol. Rev.* 108, 483–522.

Öhman, A., and Soares, J. J. (1998). Emotional conditioning to masked stimuli: expectancies for aversive outcomes following nonrecognized fear relevant stimuli. *J. Abnorm. Psychol.* 127, 69–82.

Pappens, M., Schroijen, M., Sütterlin, S., Smets, E., Van den Bergh, O., Thayer, J., et al. (2014). Resting heart rate variability predicts safety learning and fear extinction in an interoceptive fear conditioning paradigm. *PLoS ONE* 9:e105054. doi: 10.1371/journal.pone.0105054

Pappens, M., Smets, E., Vansteenwegen, D., Van den Bergh, O., and Van Diest, I. (2012). Learning to fear suffocation: a new paradigm for interoceptive fear conditioning. *Psychophysiology* 6, 821–828.

Pappens, M., Van den Bergh, O., Vansteenwegen, D., Ceunen, E., De Peuter, S., and Van Diest, I. (2013). Learning to fear obstructed breathing: comparing interoceptive and exteroceptive cues. *Biol. Psychol.* 92, 36–42.

Purkis, H. M., and Lipp, O. V. (2001). Does affective learning exist in the absence of contingency awareness? *Learn. Motiv.* 32, 84–99.

Raes, A. K., Koster, E. H., Van Damme, S., Fias, W., and De Raedt, R. (2010). Aversive conditioning under conditions of restricted awareness: effects on spatial cueing. *Q. J. Exp. Psychol.* 1, 23.

Schroijen, M., Pappens, M., Schruers, K., Van den Bergh, O., Vervliet, B., and Van Diest, I. (2015). Generalization of fear to respiratory sensations. *Behav. Ther.* (in press). doi: 10.1016/j.beth.2015.004

Schultz, D. H., and Helmstetter, F. J. (2010). Classical conditioning of autonomic fear responses is independent of contingency awareness. *J. Exp. Psychol. Anim. Behav. Process.* 36, 495–500. doi: 10.1037/a0020263

Sebastiani, L., Castellani, E., and D'Alessandro, L. (2011). Emotion processing without awareness: features detection or significance evaluation? *Int. J. Psychophysiol.* 80, 150–156. doi: 10.1016/j.ijpsycho.2011.02.019

Sevenster, D., Beckers, T., and Kindt, M. (2014). Fear conditioning of SCR but not the startle reflex requires conscious discrimination of threat and safety. *Front. Behav. Neurosci.* 8, 32. doi: 10.3389/fnbeh.2014.00032

Shulman, I. D., Cox, B. J., Swinson, R. P., Kuch, K., and Reichman, J. T. (1994). Precipitating events, locations and reactions associated with initial unexpected panic attacks. *Behav. Res. Ther.* 32, 17–20.

Smith, C., Clark, R., Manns, J., and Squire, L. (2005). Acquisition of differential delay eyeblink conditioning is independent of awareness. *Behav. Neurosci.* 119, 78–86.

Soeter, M., and Kindt, M. (2010). Dissociating response systems: erasing fear from memory. *Neurobiol. Learn. Mem.* 94, 30–41. doi: 10.1016/j.nlm.2010.03.004

Spruyt, A., Clarysse, J., Vansteenwegen, D., Baeyens, F., and Hermans, D. (2010). Affect 4.0: A free software package for implementing psychological and psychophysiological experiments. *Exp. Psychol.* 57, 36–45. doi: 10.1027/1618-3169/a000005

Tabbert, K., Merz, C. J., Klucken, T., Schweckendiek, J., Vaitl, D., Wolf, O. T., et al. (2011). Influence of contingency awareness on neural, electrodermal and evaluative responses during fear conditioning. *Soc. Cogn. Affect. Neurosci.* 6, 495–506. doi: 10.1093/scan/nsq070

Tabbert, K., Stark, R., Kirsch, P., and Vaitl, D. (2006). Dissociation of neural responses and skin conductance reactions during fear conditioning with and without awareness of stimulus contingencies. *Neuroimage* 32, 761–770. doi: 10.1016/j.neuroimage.2006.03.038

Vervliet, B., Craske, M. G., and Hermans, D. (2013). Fear extinction and relapse: state of the art. *Annu. Rev. Clin. Psychol.* 9, 215–248. doi: 10.1146/annurev-clinpsy-050212-185542

Vickers, K., and McNally, R. J. (2005). Respiratory symptoms and panic in the National Comorbidity Survey: a test of Klein's suffocation false alarm theory. *Behav. Res. Ther.* 43, 1011–1018.

Weike, A. I., Schupp, H. T., and Hamm, A. O. (2007). Fear acquisition requires awareness in trace but not delay conditioning. *Psychophysiology* 44, 170–180. doi: 10.1111/j.1469-8986.2006.00469.x

12

When Interoception Helps to Overcome Negative Feelings Caused by Social Exclusion

Olga Pollatos [1]*, *Ellen Matthias* [1] *and Johannes Keller* [2]

[1] Health Psychology, Institute of Psychology and Education, University of Ulm, Ulm, Germany, [2] Social Psychology, Institute of Psychology and Education, University of Ulm, Ulm, Germany

***Correspondence:**
Olga Pollatos,
Health Psychology, Institute
of Psychology and Education,
University of Ulm, Albert-Einstein-Allee
47, 89069 Ulm, Germany
olga.pollatos@uni-ulm.de

Social exclusion affects mental and physical health. The ability to regulate emotional responses to social exclusion is therefore essential for our well-being. As individual differences in detecting bodily signals (interoceptive sensitivity, IS) have been associated with the ability of emotion regulation, we aimed at exploring whether IS fosters coping with social exclusion and flexibility in emotion regulation. The first study investigated subjective feelings and behavioral affiliation tendencies in response to ostracism using a cyberball paradigm. Sixty-nine participants were assessed who differed with respect to IS. The second study examined habitual emotion regulation processes focusing on suppression and reappraisal as well as IS in 116 participants. Main results were that the effect of ostracism on distress and behavioral affiliation tendencies were qualified by IS—being ostracized had less impact on participants with stronger IS. Furthermore, Study 2 revealed that IS was associated with habitually stronger emotion regulation strategies. We conclude that having access to bodily signals helps (IS) reducing aversive states provoked by social exclusion, probably due to the fact that IS is associated with emotion regulation strategies.

Keywords: interoception, interoceptive sensitivity, ostracism, emotion regulation, embodied cognition

Introduction

Social Exclusion and Health

Social exclusion is associated with adverse effects for mental and physical health (see e.g., Zöller et al., 2010). Social exclusion, i.e., ostracism, is a ubiquitous phenomenon across the lifespan that threatens the fundamental human need to belong to a group. According to Baumeister and Leary (1995) the need to belong is a powerful, fundamental, and extremely pervasive motivation. Being ostracized elicits social pain, loneliness, anxiety and sadness (e.g., Hawkley et al., 2011). Experimental approaches to studying ostracism use behavioral manipulations that induce being excluded or ignored, and one of the most frequent methods employed is cyberball (Williams, 2006). Cyberball is a virtual ball toss game that participants play using a laptop or computer. Being socially rejected in cyberball is associated with reduction in belongingness, self-esteem, and control as well as negative affect such as anger and sadness (Sebastian et al., 2010).

While the cyberball paradigm was used in numerous studies, data on personality factors that moderate feelings of exclusion caused by cyberball are sparse: Onoda et al. (2010) recently demonstrated that participants with low self-esteem experienced increased social pain as compared to individuals with higher trait self-esteem. Other moderator variables were not reported, especially no effect of introversion-extraversion, individualism-collectivism, need for belonging, and loneliness

(Williams, 2006). A possible moderator candidate stems from research on emotion processing and emotion regulation, as being socially excluded causes a variety of unpleasant feelings the ostracized person has to deal with. A salient cue signaling how comfortable people feel in social interactions is the space between them, the "interpersonal distance" (see e.g., Perry et al., 2013, 2015). As cyberball causes social pain and negative affect, the measure of interpersonal distance can be used to monitor its effect on the level later social interactions.

Emotion Regulation and Interoception

An important human ability is to regulate such negative emotions caused by being socially excluded. There exist different strategies to regulate one's emotions (Gross and John, 2003; Ochsner et al., 2004; Goldin et al., 2008; Pollatos and Gramann, 2012), and individuals differ in their use of emotion regulation strategies with implications for their well-being and social functioning (Gross and John, 2003; Goldin et al., 2008). Füstös et al. (2013) state that awareness of one's emotional state is an essential variable for emotion regulation, following that this process might also be linked to the awareness of one's bodily state. The perception of bodily changes (interoception) is a central concept in several theories of emotions (James, 1884; Damasio, 1994; Craig, 2004) that postulate a relationship between interoceptive processes and emotions.

Having these results in mind one could assume that a higher emotional arousal in emotion induction would also hamper the ability to regulate emotions. But a recent study by Füstös et al. (2013) demonstrated that the ability to perceive bodily signals (interoceptive sensitivity, IS) facilitated the downregulation of affect-related arousal when participants were instructed to use reappraisal, a common emotion regulation strategy (Gross and John, 2003; Goldin et al., 2008). Also Weiss et al. (2014) reported a positive relationship between emotion regulation abilities as assessed by questionnaire and IS.

Social Exclusion and Interoception

Füstös et al. (2013) assume that a greater sensitivity to one's bodily state facilitates the regulation of emotional responses. They suggested that the detection of ongoing bodily changes is easier or more accurate, and this might in turn facilitate the discrimination and regulation of different emotional states (Füstös et al., 2013). Whether these mechanisms might also facilitate the regulation of unpleasant affect elicited as response to social rejection is unclear till now. A study by Werner et al. (2013) supports this assumption: Participants took part in a discussion round and after a certain time they were excluded from the discussion. IS modulated positive and negative affect and perceived acceptance respectively rejection during exclusion. Whether the change in subjective feelings caused by exclusion leads to motivational engagement to overcome the situation and whether such a behavior is also modulated by interoceptive processes was not addressed in the former study. Ferri et al. (2013) hypothesized that IS might contribute to interindividual differences concerning social attitudes and interpersonal space; the social situation they used in the study involved an experimenter who performed movements at different distances from the participant's hand. This setup involves no direct social interaction, but the role of interoceptive signals might be much more important in social relevant situation as implemented in the cyberball paradigm.

The idea that perceiving internal signals more precisely facilitates processes of comparisons of different internal states related to emotions and their regulation in social relevant situations would be supported when there is further evidence that interoception interacts with emotion regulation in everyday life. Emotion regulation strategies as assessed by questionnaire might therefore an interesting tool to evaluate regulation capacities in general. The idea that IS supports emotion regulation is in accordance with data from Feldman Barrett et al. (2001) who demonstrated that persons with highly differentiated emotion experience could better regulate their emotions in everyday situations.

Interoceptive Sensitivity—Measurement

A common method to assess interoception is the ability to perceive one's heartbeats accurately (Schandry, 1981; Critchley et al., 2004; Dunn et al., 2007; Pollatos et al., 2008). This ability can be measured by using validated and reliable heartbeat perception tasks (Whitehead and Drescher, 1980; Schandry, 1981), in which participants are instructed to perceive their own heartbeats without feeling for their pulse. There is convincing evidence that higher IS is associated with more intense feelings and higher activation of underlying brain structures during emotional stimulation (Wiens, 2005; Pollatos et al., 2007a; Dunn et al., 2010; Füstös et al., 2013). IS was also associated with cognitive functions like decision-making, selective attention or self-regulation during physical exercise (Pollatos et al., 2007b; Werner et al., 2009b; Dunn et al., 2010; Lenggenhager et al., 2013).

Aim of the Studies

To further elucidate whether IS interacts with feelings of social exclusion and emotion regulation in general, we conducted a study on healthy participants. We hypothesized that IS is related to (1) better coping with social exclusion, (2) less motivation to engage in behavior serving to overcome these feelings of exclusion, and (3) better emotion regulation capacity and more flexibility in general. To experimentally vary social exclusion we employed a standardized paradigm (cyberball) known to affect well-being. Additionally, as ostracism causes a threat to fundamental needs (belonging, self-esteem, control, and meaningful existence, see e.g., Jamieson et al., 2010) and therefore individuals are motivated to fortify these needs, we used preferred interpersonal distance (see Perry et al., 2013) as one measure to examine behavioral tendencies to cope with the threat to the need for affiliation in the ostracism paradigm. The second study builds on the first by examining emotion regulation in general (using an established self-report measure).

Materials and Methods—Study 1

Participants were screened for health status using an anamnestic questionnaire. Exclusion criteria were any history of any axis 1 disorders, in particular anxiety disorders or depression according to the Diagnostic and Statistical Manual of Mental Disorders as well as drug use (except of contraceptives). All participants gave their written informed consent. They received an amount of 10€ for their participation. Sixty-nine female participants (mean age 23.6, SD 3.7) were included in the main experiment. The experiment was conducted in accordance with the Declaration of Helsinki, the local ethics committee approved the study.

Procedure

First, IS was assessed. ECG electrodes were placed to the right mid-clavicle and lower left rib cage. We used four heartbeat counting phases (varying in length) in accordance with the Mental Tracking Method suggested by Schandry (Schandry, 1981). Participants were asked to count their own heartbeats silently and to report the number of counted heartbeats at the end of the counting phase. IS was calculated according to the following transformation:

$$\tfrac{1}{4}\Sigma(1 - (|\text{recorded heartbeats} - \text{counted heartbeats}|)/\text{recorded heartbeats})$$

The mean score was 0.70 (SD 0.14).

Then, electrodes were detached and the cyberball paradigm started. As we also wanted to assess the effect of cyberball on interpersonal relations, we slightly varied the paradigm in the following manner: There were always two experimental supervisors present up to this stage of the experiment. One of them then left before the other experimental supervisor explained the cyberball paradigm to participants. Participants were randomly assigned to one of the three cyberball conditions: inclusion, social exclusion and social exclusion due to pretended technical failure. Technical failure condition was introduced to assess effects of being implicitly excluded. Participants were told that they take part in a mental visualization exercise in which they toss a ball over the internet with two other players (see Hawkley et al., 2011). Importantly, one of the other two players was the one experimental supervisor that the participants had met before. His/her photo was also depicted on the screen while the other player was unfamiliar to the participant. Participants were told that all persons involved in the procedure were connected via internet. We also took a photo of the participant in the beginning so that his/her photo was then used in the cyberball setting. Then one of three conditions took place: In the inclusion condition, participants receive the ball one-third of the time. In the social exclusion and technical failure conditions participants receive the ball only at the beginning and are then ignored. At the end of the failure condition a screen appears with an error message in which it is explained that due to connection problems of the internet the participant was no longer connected with the other players.

Subjective mood was assessed as one outcome variable; this was carried out immediately prior and directly after cyberball using the German version of the Profile of Mood States (Dalbert, 1991). Nineteen items are rated on a seven-point Likert scale comprising four aspects of negative (fatigue, depression, anger, sadness, 13 items) and positive mood (six items), translated into a negativity index (range 19–133) with higher scores reflecting greater negativity.

Feelings of exclusion were evaluated as described by (Bolling et al., 2011) using a 10-item questionnaire given to participants immediately after playing cyberball. The items were taken from the Needs Threat Scale that checks for distress following exclusion (Eisenberger et al., 2003; Hawkley et al., 2011) as adapted by Bolling et al. (2011). The Needs Threat Scale included statements about feelings of control, belongingness, and self-esteem on a Likert scale from 1 = "not at all" to 5 = "extremely." Example items were: "During the game, I felt ignored."; "I felt rejected"; "I felt like an outsider." Consistent with previous research, we computed as mean needs with higher scores reflect greater needs threat (see also Hawkley et al., 2011).

We also assessed preferred interpersonal distance as suggested by Perry et al. (2013), operationalized as distance chosen between the participant and the one cyberball player who was also present in the laboratory prior to cyberball and whose photograph had been shown in the cyberball paradigm before. For the assessment of interpersonal distance, the participant and the experimental supervisor were placed directly facing each other with a start distance of three meters, and then the experimenter approached the participant until s/he said "stop" to signal a distance s/he evaluated as appropriate. This procedure was chosen very similar to the one suggested by Perry et al. (2013) in virtual reality to assess one's preferred interpersonal distance. Our exclusion took also place on the screen with photos of the protagonist, while the later assessment of the preferred interpersonal distance was carried out in real space. The detailed instruction was to signal the distance the participant felt comfortable with in accordance to Perry et al. (2013). This preferred interpersonal distance was noted in meters.

Study 2—Materials and Methods

Participants were screened with the procedure as applied in Study 1. They received an amount of 5€ for their participation. 116 participants (27 male) took part, their mean age was 25.6 (SD 3.2). The local ethics committee approved the experiment.

Procedure

First, all participants filled in the German version of the Emotion Regulation Questionnaire (Abler and Kessler, 2009) developed by Gross and John (2003). The ERQ is one of the first validated instruments for the investigation of emotion regulation processes. The questionnaire tests two common regulation strategies: suppression (example item: I control my emotions by not expressing them) and reappraisal (example item: I control my emotions by changing the way I think about the situation I'm in). The German version of the ERQ consists of 10 items assessing habitual reappraisal and suppression on a seven point scale.

Afterward, the heartbeat detection task as described in Study 1 was conducted. The mean heartbeat perception score was 0.71 (SD 0.16).

Data analyses

Study 1: We calculated three regression analyses using the following dependent variables:

a. feelings of exclusion as measured by the needs index,
b. negative feelings as measured by the negativity index, and
c. preferred interpersonal distance.

The variables a-c. as were z-standardized and then regressed on the following variables:

d. z-standardized interoceptive sensitivity,
e. dummy 1 (codes: 0 = inclusion; 1 = technical failure; 0 = exclusion),
f. dummy 2 (codes: 0 = inclusion; 0 = technical failure; 1 = exclusion),
g. the interaction involving standardized interoceptive sensitivity and dummy 1, and hour the interaction involving standardized interoceptive sensitivity and dummy 2.

In accordance to Aiken and West (1991) simple slopes analyses were applied with one SD above and below the mean IS score in order to examine possible differences between participants with high versus low IS.

Furthermore, Pearson's correlation analyses were performed between the needs index and interpersonal distance.

Study 2: One regression analysis (forward selection) with reappraisal and suppression as predictors and interoceptive sensitivity as criterion was carried out.

Results

Study 1

Table 1 summarizes the needs index, the negativity index and preferred interpersonal distance evoked by the different cyberball conditions.

The first regression analysis with a. the *needs index* as criterion revealed that the criterion was explained by both dummies as well as both interaction terms [$F(5,63) = 83.77, p < 0.001, R = 0.93, R^2$ adjusted = 0.86]. A significant effect of condition was reflected in significant effects of dummy 1 ($T = 15.93$, $\beta = 0.86$, $p < 0.001$) and dummy 2 ($T = 17.35$, $\beta = 0.93$, $p < 0.001$). Crucially, both interaction terms between IS and dummy 1 ($T = -2.90$, $\beta = -0.20, p < 0.01$) as well as dummy 2 ($T = -3.79, \beta = -0.26$, $p < 0.001$) were significant.

Figure 1 summarizes theses effects. Both social exclusion conditions caused greater needs index scores as compared to the mean scores in the social inclusion condition. Importantly, IS qualified the effect of ostracism. High IS (one SD above mean) was associated with less pronounced needs scores in both social exclusion conditions as compared to low IS (one SD below mean).

With respect to the b. *negativity index*, the criterion was explained by condition only [dummy 1: $T = 4.87$, $\beta = 0.55$, $p < 0.001$; dummy 2: $T = 4.80, \beta = 0.65, p < 0.001; F(5,68) = 9.28$, $p < 0.001, R = 0.65, R^2$ adjusted = 0.38]. IS and both interaction terms were not significant. These effects indicate an increase in negative feelings for both social exclusion conditions (SE: mean before 57.8, mean after 77.5; SET: mean before 56.8, mean after 74.6), while no change was observed after social inclusion (mean before: 57.1, mean after 59.0).

We also obtained a significant positive correlation between the needs index and the negativity index after cyberball ($r = 0.57$, $p < 0.001$): The more participants felt excluded, the more negative feelings increased and vice versa.

The third regression analysis with c. *interpersonal distance* as criterion showed that the criterion was explained by dummy 2 and the interaction term between IS and dummy 2 [$F(5,63) = 6.57$, $p < 0.001$, $R = 0.58$, R^2 adjusted = 0.29]. Interpersonal distance increased after exclusion only as reflected in the significant effect of dummy 2 ($T = -5.31, \beta = -0.64, p < 0.001$), while dummy 1 was only marginally significant ($T = -1.85, \beta = -0.22, p = 0.07$). The interaction term between IS and dummy 2 was significant ($T = 2.09$, $\beta = 0.34$, $p < 0.05$). These effects are depicted in **Figure 2**.

TABLE 1 | Needs index, negativity index (before and after cyberball) and preferred interpersonal distance contrasting the different conditions ($N = 69$ total; $N = 24$; SE, social exclusion; $N = 23$; SET, social exclusion, technical failure; $N = 22$; SI, social inclusion).

	Mean(SD)		
	SE	**SET**	**SI**
Needs index	3.9 (0.7)	3.8 (0.7)	1.4 (0.3)
Negativity index			
Before	57.8 (8.0)	56.8 (6.0)	57.1 (8.3)
After	77.5 (8.9)	74.6 (15.8)	59.0 (8.4)
Interpersonal distance (in m)	0.69 (0.17)	0.88 (0.25)	0.97 (0.14)

Preferred interpersonal distance was smaller after exclusion (mean distance 0.69 meters) as compared to both exclusion with technical failure explanation (mean 0.88 meters) and inclusion (mean 0.97 meters). Higher IS (one SD above mean) was associated with an decrease of interpersonal distance after social exclusion, which was more pronounced for the group with low IS (one SD below mean) and depicts the significant interaction effect.

In a last step we obtained a significant inverse correlation between the needs index and interpersonal distance of $r = -0.40$ ($p < 0.01$) indicating that a greater threat of needs was associated with a smaller interpersonal distance.

Study 2

ERQ Emotion Regulation and IS

Mean ERQ scores were 28.2 (SD 6.2) for reappraisal and 13.2 (SD 4.5) for suppression. The consecutive regression analysis with IS as criterion and reappraisal as well as suppression as predictors revealed significant effects of reappraisal ($T = 3.14$, $\beta = 0.27, p < 0.01$) and suppression [$T = 3.22, \beta = 0.28, p < 0.01$; $F(2,113) = 9.71$, $p < 0.001$, $R = 0.38$, $R^2 = 0.15$]. Higher IS was associated with both higher reappraisal and higher suppression. These effects are depicted in **Figure 3** (z-standardized scores for all variables depicted).

Discussion

Our data provide new evidence for the relevance of the perception of bodily signals regarding ostracism and emotion regulation. Being socially rejected using the cyberball paradigm is typically associated with a threat to elementary needs such as belongingness, self-esteem or control, as well as an increase in negative affect as demonstrated in former studies (Eisenberger et al., 2003; Williams et al., 2006; Sebastian et al., 2010; Hawkley et al., 2011). Importantly, IS moderated this effect: Higher IS was found to be associated lower levels of (a) distress and (b)

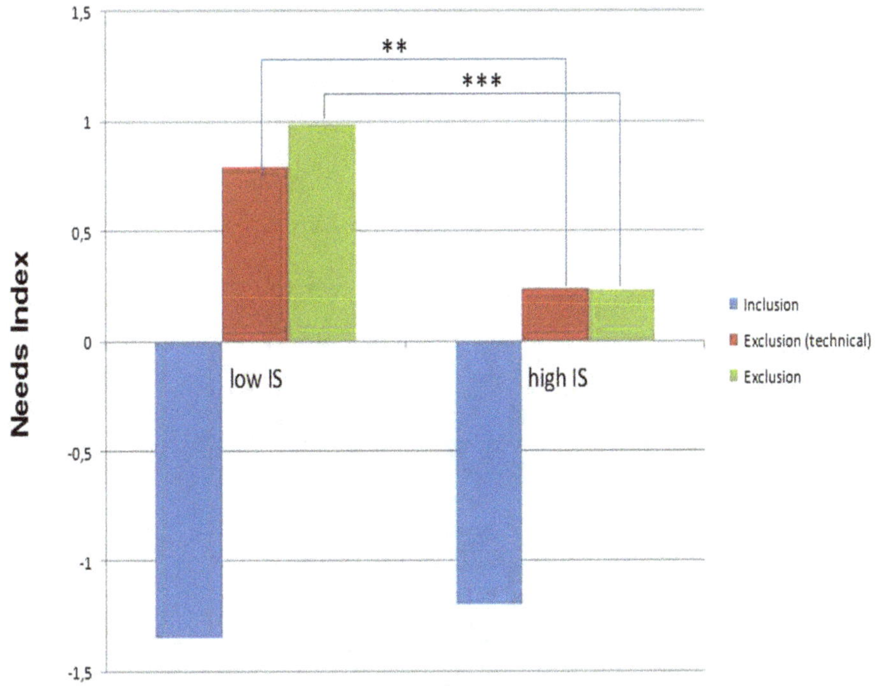

FIGURE 1 | Needs index after cyberball contrasting participants with high and low IS in the three experimental conditions ($N = 69$; ***$p < 0.001$; **$p < 0.01$).

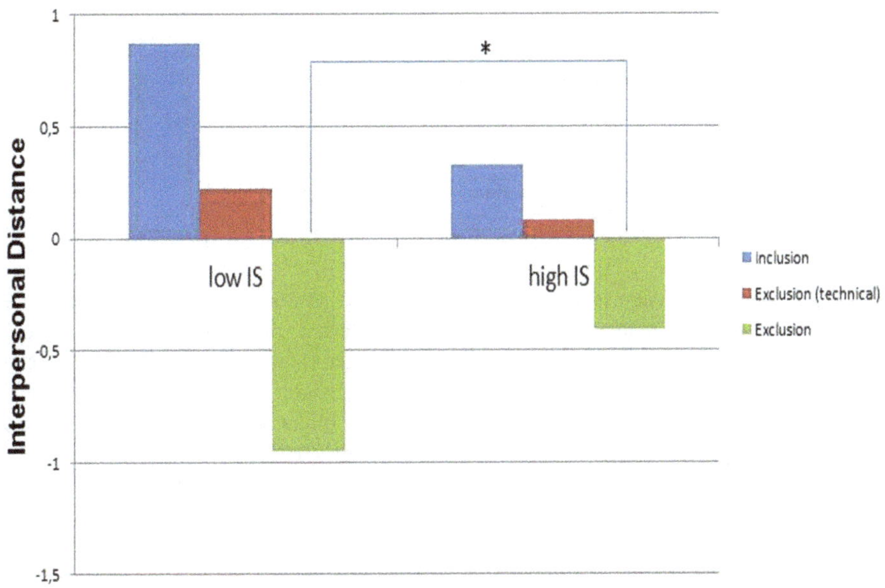

FIGURE 2 | Interpersonal distance after cyberball contrasting participants with high and low IS in the three experimental conditions ($N = 69$; *$p < 0.05$).

behavioral affiliation tendencies following social exclusion. In accordance with these results IS was correlated with higher scores on the emotion regulation questionnaire.

These results are in accordance with Füstös et al. (2013) who demonstrated that IS facilitates the downregulation of affect-related arousal and corresponding neural activation when applying reappraisal as emotion regulation strategy. Our results are also in accordance to data on social exclusion in a discussion as provided by Werner et al. (2013). They suggested that individuals with high IS reduce aversive states by using somatic information for self-regulation to a greater extent. One possible explanation builds on data from Feldman Barrett and co-authors (Feldman Barrett et al., 2001): Here, emotionally differentiated participants reported a wide range of emotion

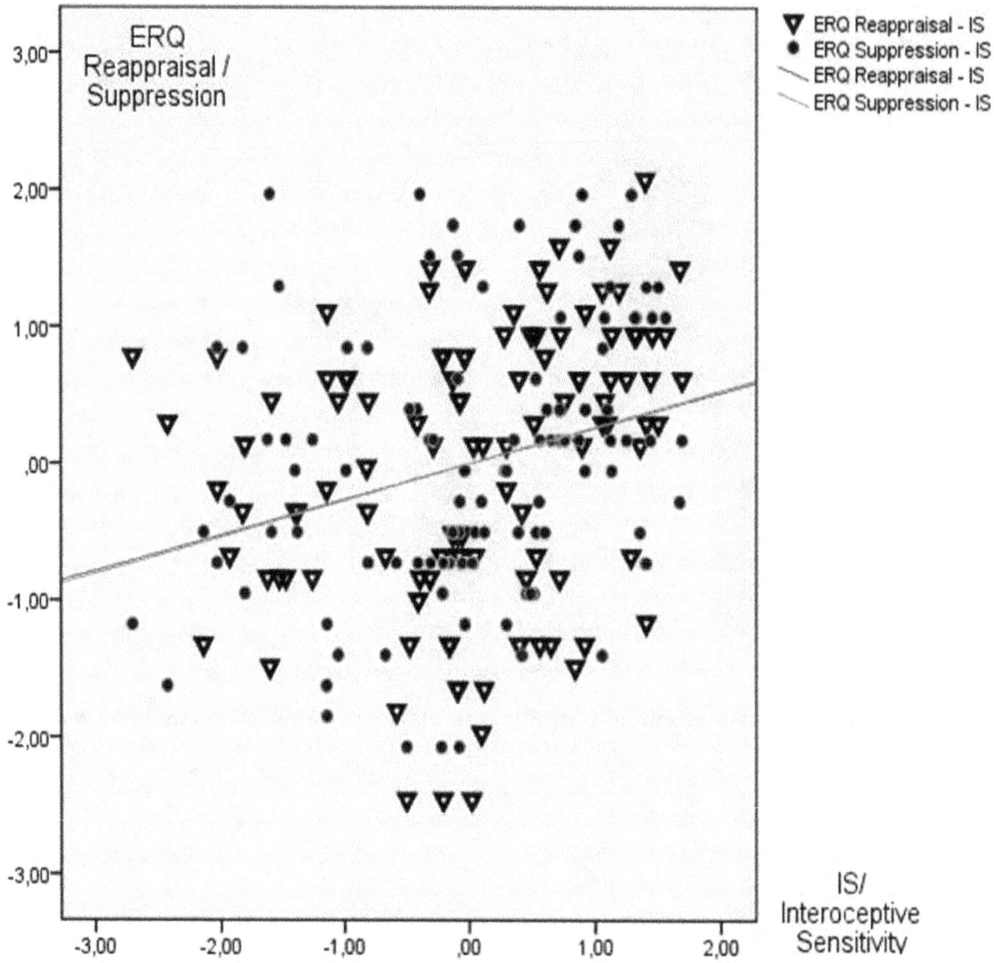

FIGURE 3 | Scatterplot between IS and both reappraisal and suppression (N = 116; z-scores depicted).

regulation strategies used in daily life. We assume that IS facilitates the efficiency of different emotion regulation strategies by providing a more fine-tuned feedback of the actual emotional state. This is also the case in situations characterized by social exclusion: Both the feelings of exclusion as well as the need to affiliate as reflected by the interpersonal distance measure were less pronounced when participants were rather good in perceiving their bodily signals. We therefore suggest that IS supports the effective down-regulation of negative affect and associated bodily changes occurring during social exclusion which might lead to a lower "cost" for the self. IS assessed in one physiological system (the cardiac system) relevantly mediates emotion regulation in situations evoking physiological activity such as social exclusion. As suggested by Füstös et al. (2013), being aware of one's bodily signals might therefore constitute a positive precondition for effective self-regulation of behavior. Supporting this interpretation, data obtained in a public speech paradigm showed that IS was associated with less self-reported state anxiety before and during such a task (Werner et al., 2009a). And also Lenggenhager et al. (2013) manipulated visceral feedback of their participants and reported that heightened feedback regarding one's own visceral processes increased a self-centered perspective and affected drive socioeconomic exchanges accordingly. In contrast to these studies, Van 't Wout et al. (2013) did not observe a reliably significant relationship between IS and the acceptance of unfair offers or habitual use of emotion regulation.

As expected, the need to affiliate as one coping mechanism after social exclusion was higher after social exclusion. This was only the case when no alternative explanation for the experienced social rejection was provided, as social exclusion with technical failure and social inclusion were associated with comparable distance measures. The behavioral measure—preferred interpersonal distance—was significantly smaller after social rejection and positively correlated with the experienced threat of needs. We interpret this result as an indicator for a stronger tendency to socially affiliate after rejection. Referring to the interaction between IS and interpersonal distance after social rejection, we assume that IS moderates affective processes and the coping with such negative emotions as well as the behavioral tendencies to deal with the outcome of such negative situations. In relation to our results, interoceptive processes might preserve the common resource used for self-regulatory processes. Self-regulation uses self-monitoring and affective self-reaction (Maes and Karoly, 2005) which might constitute abilities that are linked with bodily processes and the conscious feedback of these processes as operationalized by IS. Recent work supports this idea: Weiss et al. (2014) could demonstrate that IS was positively correlated with self-regulatory capacities as assessed by questionnaire. And also Koch and Pollatos (2014) showed that IS is positively correlated with greater adaptability as assessed by questionnaire in children. It can be followed that interoception helps to preserve limited resources involved in self-regulation, presumably by faster or more differentiated detection of bodily response changes occurring in significant situations such as social rejection and might help to constitute a feeling of higher control over one's negative experiences in everyday life.

Acknowledgments

We want to thank Dr. Stefan Pfattheicher for his support in data analyses and presentation.

References

Abler, B., and Kessler, H. (2009). Emotion regulation questionnaire—eine deutschsprachige fassung des ERQ von gross und john. *Diagnostica* 55, 144–152. doi: 10.1026/0012-1924.55.3.144

Aiken, L. S., and West, S. G. (1991). *Multiple Regression: Testing and Interpreting Interactions.* Newbury Park, CA: Sage.

Baumeister, R. F., and Leary, M. R. (1995). The need to belong: desire for interpersonal attachments as a fundamental human motivation. *Psychol. Bull.* 117, 497–529. doi: 10.1037/0033-2909.117.3.497

Bolling, D. Z., Pitskel, N. B., Deen, B., Crowley, M. J., cPartland, J. C., ayes, L. C., et al. (2011). Dissociable brain mechanisms for processing social exclusion and rule violation. *Neuroimage* 54, 2462–2471. doi: 10.1016/j.neuroimage.2010.10.049

Craig, A. D. (2004). Human feelings: why are some more aware than others? *Trends Cogn. Sci.* 8, 239–241. doi: 10.1016/j.tics.2004.04.004

Critchley, H. D., Wiens, S., Rotshtein, P., Ohman, A., and Dolan, R. J. (2004). Neural systems supporting interoceptive awareness. *Nat. Neurosci.* 7, 189–195. doi: 10.1038/nn1176

Dalbert, C. (1991). Subjektives Wohlbefinden junger erwachsener: theoretische und empirische analysen der struktur und stabilität. *Z. Differ. Diagn. Psychol.* 13, 207–220.

Damasio, A. R. (1994). *Descartes' Error: Emotion, Reason, and the Human Brain.* New York: Grosset/Putman.

Dunn, B. D., Dalgleish, T., Ogilvie, A. D., and Lawrence, A. D. (2007). Heartbeat perception in depression. *Behav. Res. Ther.* 45, 1921–1930. doi: 10.1016/j.brat.2006.09.008

Dunn, B. D., Galton, H. C., Morgan, R., Evans, D., Oliver, C., Meyer, M., et al. (2010). Listening to your heart: how interoception shapes emotion experience and intuitive decision making. *Psychol. Sci.* 21, 1835–1844. doi: 10.1177/0956797610389191

Eisenberger, N. I., Lieberman, M. D., and Williams, K. D. (2003). Does rejection hurt? An fMRI study of social exclusion. *Science* 302, 290–292. doi: 10.1126/science.1089134

Feldman Barrett, L., Gross, J., Christensen, T. C., and Benvenuto, M. (2001). Knowing what your're feeling and knowing what to do about it: mapping the relation between emotion differentiation and emotion regulation. *Cogn. Emot.* 15, 713–724. doi: 10.1080/02699930143000239

Ferri, F., Ardizzi, M., Ambrosecchia, M., and Gallese, V. (2013). Closing the ap between the inside and the outside: interoceptive sensitivity and social distances. *PLoS ONE* 8:e75758. doi: 10.1371/journal.pone.0075758

Füstös, J., Gramann, K., Herbert, B. M., and Pollatos, O. (2013). On the embodiment of emotion regulation: interoceptive awareness facilitates reappraisal. *Soc. Cogn. Affect. Neurosci.* 8, 911–917. doi: 10.1093/scan/nss089

Goldin, P. R., McRae, K., Ramel, W., and Gross, J. J. (2008). The neural bases of emotion regulation: reappraisal and suppression of negative emotion. *Biol. Psychiatry.* 63, 577–586. doi: 10.1016/j.biopsych.2007.05.031

Gross, J. J., and John, O. P. (2003). Individual differences in two emotion regulation processes: implications for affect, relationships, and well-being. *J. Pers. Soc. Psychol.* 85, 348–362. doi: 10.1037/0022-3514.85.2.348

Hawkley, L. C., Williams, K. D., and Cacioppo, J. T. (2011). Responses to ostracism across adulthood. *Soc. Cogn. Affect. Neurosci.* 6, 234–243. doi: 10.1093/scan/nsq045

James, W. (1884). What is an emotion? *Mind* 9, 188–205. doi: 10.1093/mind/os-IX.34.188

Jamieson, J. P., Harkins, S. G., and Williams, K. D. (2010). Need threat can motivate performance after ostracism. *Pers. Soc. Psychol. Bull.* 36, 690–702. doi: 10.1177/0146167209358882

Koch, A., and Pollatos, O. (2014). Cardiac sensitivity in children: sex differences and its relationship to parameters of emotional processing. *Psychophysiology* 51, 932–941. doi: 10.1111/psyp.12233

Lenggenhager, B., Azevedo, R. T., Mancini, A., and Aglioti, S. M. (2013). Listening to your heart and feeling yourself: effects of exposure to interoceptive signals during the ultimatum game. *Exp. Brain Res.* 230, 233–241. doi: 10.1007/s00221-013-3647-5

Maes, S., and Karoly, P. (2005). Self-Regulation assessment and intervention in physical health and illness: a review. *Appl. Psychol. Int. Rev.* 54, 267–299. doi: 10.1111/j.1464-0597.2005.00210.x

Ochsner, K. N., Ray, R. D., Cooper, J. C., Robertson, E. R., Chopra, S., Gabrieli, J. D. E., et al. (2004). For better or for worse: neural systems supporting the cognitive down- and up-regulation of negative emotion. *Neuroimage* 23, 483–499. doi: 10.1016/j.neuroimage.2004.06.030

Onoda, K., Okamoto, Y., Nakashima, K., Nittono, H., Yoshimura, S., Yamawaki, S., et al. (2010). Does low self-esteem enhance social pain? The relationship between trait self-esteem and anterior cingulate cortex activation induced by ostracism. *Soc. Cogn. Affect. Neurosci.* 5, 385–391. doi: 10.1093/scan/nsq002

Perry, A., Mankuta, D., and Shamay-Tsoory, S. G. (2015). OT promotes closer interpersonal distance among highly empathic individuals. *Soc. Cogn. Affect. Neurosci.* 10, 3–9. doi: 10.1093/scan/nsu017

Perry, A., Rubinsten, O., Peled, L., and Shamay-Tsoory, S. G. (2013). Don't stand so close to me: a behavioral and ERP study of preferred interpersonal distance. *Neuroimage* 83, 761–769. doi: 10.1016/j.neuroimage.2013.07.042

Pollatos, O., and Gramann, K. (2012). Attenuated modulation of brain activity accompanies emotion regulation deficits in alexithymia. *Psychophysiology* 49, 651–658. doi: 10.1111/j.1469-8986.2011.01348.x

Pollatos, O., Gramann, K., and Schandry, R. (2007a). Neural systems connecting interoceptive awareness and feelings. *Hum. Brain Mapp.* 28, 9–18. doi: 10.1002/hbm.20258

Pollatos, O., Herbert, B. M., Kaufmann, C., Auer, D. P., and Schandry, R. (2007b). Interoceptive awareness, anxiety and cardiovascular reactivity to isometric exercise. *Int. J. Psychophysiol.* 65, 167–173. doi: 10.1016/j.ijpsycho.2007.03.005

Pollatos, O., Kurz, A. L., Albrecht, J., Schreder, T., Kleemann, A. M., Schöpf, V., et al. (2008). Reduced perception of bodily signals in anorexia nervosa. *Eat. Behav.* 9, 381–388. doi: 10.1016/j.eatbeh.2008.02.001

Schandry, R. (1981). Heart beat perception and emotional experience. *Psychophysiology* 18, 483–488. doi: 10.1111/j.1469-8986.1981.tb02486.x

Sebastian, C., Viding, E., Williams, K. D., and Blakemore, S. J. (2010). Social brain development and the affective consequences of ostracism in adolescence. *Brain Cogn.* 72, 134–145. doi: 10.1016/j.bandc.2009.06.008

Van 't Wout, M., Faught, S., and Menino, D. (2013). Does interoceptive awareness affect the ability to regulate unfair treatment by others? *Front. Psychol.* 4:880. doi: 10.3389/fpsyg.2013.00880

Weiss, S., Sack, M., Henningsen, P., and Pollatos, O. (2014). On the interaction of self-regulation, interoception and pain perception. *Psychopathology* 47, 377–382. doi: 10.1159/000365107

Werner, N. S., Duschek, S., Mattern, M., and Schandry, R. (2009a). Interoceptive sensitivity modulates anxiety during public speaking. *J. Psychophysiol.* 23, 85–94. doi: 10.1027/0269-8803.23.2.85

Werner, N. S., Jung, K., Duschek, S., and Schandry, R. (2009b). Enhanced cardiac perception is associated with benefits in decision-making. *Psychophysiology* 46, 1123–1129. doi: 10.1111/j.1469-8986.2009.00855.x

Werner, N. S., Kerschreiter, R., Kindermann, N. K., and Duschek, S. (2013). Interoceptive awareness as a moderator of affective responses to social exclusion. *J. Psychophysiol.* 27, 39–50. doi: 10.1027/0269-8803/a000086

Whitehead, W. E., and Drescher, V. M. (1980). Perception of gastric contractions and self-control of gastric motility. *Psychophysiology* 17, 552–558. doi: 10.1111/j.1469-8986.1980.tb02296.x

Wiens, S. (2005). Interoception in emotional experience. *Curr. Opin. Neurol.* 18, 442–447. doi: 10.1097/01.wco.0000168079.92106.99

Williams, K. D. (2006). Ostracism. *Annu. Rev. Psychol.* 58, 425–452. doi: 10.1146/annurev.psych.58.110405.085641

Williams, L. M., Palmer, D., Liddell, B. J., Song, L., and Gordon, E. (2006). The 'when' and 'where' of perceiving signals of threat versus non-threat. *Neuroimage* 31, 458–467. doi: 10.1016/j.neuroimage.2005.12.009

Zöller, C., Maroof, P., Weik, U., and Deinzer, R. (2010). No effect of social exclusion on salivary cortisol secretion in women in a randomized controlled study. *Psychoneuroendocrinology* 35, 1294–1298. doi: 10.1016/j.psyneuen.2010.02.019

13

Emotional Experience and Awareness of Self: Functional MRI Studies of Depersonalization Disorder

Nick Medford[1,2], Mauricio Sierra[3], Argyris Stringaris[3], Vincent Giampietro[3], Michael J. Brammer[3] and Anthony S. David[3]*

[1] Department of Psychiatry, Brighton and Sussex Medical School, Brighton, UK, [2] Sackler Centre for Consciousness Science, University of Sussex, Brighton, UK, [3] Institute of Psychiatry, Psychology, and Neuroscience, London, UK

***Correspondence:**
Nick Medford
n.medford@bsms.ac.uk

This paper presents functional MRI work on emotional processing in depersonalization disorder (DPD). This relatively neglected disorder is hallmarked by a disturbing change in the quality of first-person experience, almost invariably encompassing a diminished sense of self and an alteration in emotional experience such that the sufferer feels less emotionally reactive, with emotions experienced as decreased or "damped down," so that emotional life seems to lack spontaneity and subjective validity. Here we explored responses to emotive visual stimuli to examine the functional neuroanatomy of emotional processing in DPD before and after pharmacological treatment. We also employed concurrent skin conductance measurement as an index of autonomic arousal. In common with previous studies we demonstrated that in DPD, there is attenuated psychophysiological response to emotional material, reflected in altered patterns of (i) regional brain response, (ii) autonomic responses. By scanning participants before and after treatment we were able to build on previous findings by examining the changes in functional MRI response in patients whose symptoms had improved at time 2. The attenuation of emotional experience was associated with reduced activity of the insula, whereas clinical improvement in DPD symptoms was associated with increased insula activity. The insula is known to be implicated in interoceptive awareness and the generation of feeling states. In addition an area of right ventrolateral prefrontal cortex emerged as particularly implicated in what may be "top-down" inhibition of emotional responses. The relevance of these findings to the wider study of emotion, self-related processes, and interoception is discussed.

Keywords: depersonalization, insula, interoception, self-awareness, fMRI

INTRODUCTION

Depersonalization disorder (DPD) is characterized by a persistent and distressing alteration in the quality of subjective experience, such that the individual experiences both themselves (depersonalization) and their surroundings (derealization) as oddly estranged and unreal. Although depersonalization and derealization are separable phenomena, in practice they usually co-occur (Sierra and David, 2011). In recent years there has been a growing appreciation that the phenomenology of DPD entails a sense of alienation and estrangement from experience in general, and in keeping with this, DPD patients often describe a reduced capacity for emotional response and a generalized reduction in the experience of bodily sensation. In phenomenological literature, the terms "de-affectualization" and "de-somatization" have been applied to these aspects of the condition. Two large factor analyses (Sierra et al., 2005; Simeon et al., 2008) of DPD symptomatology have confirmed emotional numbing (i.e., "de-affectualization") as a key symptom domain in DPD. There are sound theoretical reasons for believing that this alteration in emotional experience is central to the condition (Medford, 2012), and it is very rare to encounter a patient who fulfills criteria for a diagnosis of DPD but does not describe this deficit of subjective emotional response.

A number of previous fMRI and psychophysiological studies of DPD have focused specifically on this emotional aspect of the condition, showing that autonomic response to emotionally salient visual stimuli is significantly reduced in DPD (Sierra et al., 2002), and that the changes in regional brain activity seen in response to emotional stimuli in healthy controls are largely absent in DPD (Phillips et al., 2001; Medford et al., 2006). These findings can be seen as plausible biological correlates of the subjective deficit of emotional experience described by patients with DPD (Medford, 2012).

In this study, patients with primary DPD were studied using an emotional vs. neutral block-design fMRI paradigm employed in an earlier imaging study of DPD (Phillips et al., 2001), utilizing stimuli drawn from the International Affective Picture System (IAPS, Bradley and Lang, 1999). The IAPS is a library of images that have been given normative ratings for emotional salience and arousal and which are widely used in emotion research. Patients were scanned before and after receiving pharmacological treatment (lamotrigine) specifically tailored to DPD (Sierra et al., 2001, 2006). Thus patients were scanned twice at two different time points (14 patients scanned at time 1, of whom 10 were re-scanned at time 2). The idea of this approach is twofold: scanning at time 1 represents an attempt to replicate previous findings, in a group of as yet untreated (i.e., medication-free) patients with DPD. Given the paucity of literature on this condition, it is important to attempt this kind of replication in order to explore the robustness of previous findings. Secondly, re-scanning at time 2 allows an assessment of the effects of pharmacological treatment on the neural response to emotional stimuli in this patient group, and how this relates to changes in clinical state.

Specific hypotheses were as follows:

1. That functional MRI results in DPD patients at time 1 would essentially replicate previous findings using IAPS images in a DPD patient group (Phillips et al., 2001) in showing a reduced response to emotionally salient stimuli, and a relative lack of difference in the response to emotional material compared to neutral. More specifically, it was predicted that right ventrolateral prefrontal cortex (Brodmann area 47) would be significantly activated in the emotional phase in DPD patients at time 1, but not in the controls, as in the Phillips et al. study. It was also predicted that healthy controls would show a response to emotional images characterized by significant activation in the emotional phase in areas known to be involved in emotional processing, specifically anterior insula, and sensory (in this case visual) cortex activation related to modulation of sensory cortex by affective processing.
2. That at time 2, DPD patients reporting a significant reduction in their symptoms would show a pattern of neural responses to emotional stimuli more akin to that seen in healthy controls, whereas those reporting no such reduction would continue to show a response pattern similar to that seen at time 1. In particular, because of prior work implicating underactivity of the anterior insula in DPD (discussed in detail below), we hypothesized that DPD patients who had improved clinically would at time 2 show significantly increased activation, compared to time 1, of anterior insula in response to emotional material.
3. That DPD patients would show generally reduced skin conductance responses (SCRs) to IAPS stimuli compared to healthy controls, and that at time 2 this effect would be influenced by response to treatment in a similar manner to the fMRI findings i.e., that DPD patients who improved clinically would, at time 2, have an SCR profile resembling that seen in controls, with significantly more autonomic response to emotional stimuli than in the same patients at time 1 or non-improved patients at time 2.

The key hypothesis, in its simplest form, was that patients who show a significant response to treatment [defined as a 30% or greater reduction in Cambridge Depersonalization State Scale (CDSS) scores at time 2 relative to time 1 (Sierra et al., 2001, 2006), accompanied by subjective reported improvement at clinical follow-up] would also show a concomitant change in their neural and physiological response patterns.

PARTICIPANTS AND METHODS

Fourteen adult patients were recruited from the Depersonalization Clinic at the Maudsley Hospital. Every patient had a diagnostic assessment performed by a psychiatrist experienced in the evaluation and treatment of DPD (either NM or MS). Patients were recruited according to the following inclusion and exclusion criteria: (1) Meeting diagnostic criteria for primary DPD. Where other psychopathology (e.g., anxiety) was present, it was a requirement that this was secondary to chronic depersonalization, such that the latter was the predominant presenting symptom, and had been the dominant symptom for the duration of the condition (Medford et al., 2005). (2) On no psychotropic medication. Where patients had previously been on such medication, it was a study requirement that they had been medication-free for at least 6 months. However, toward the end of the study period, after four patients had dropped out of the study for various reasons, a decision was taken to include two further patients who were already on conventional antidepressant medication at time 1 (fluoxetine and venlafaxine, respectively). (3) Suitability for lamotrigine treatment, and having given informed consent for this treatment.

(4) No contra-indication to MRI scanning, and willingness to undergo scanning on two separate occasions. (5) No history of other psychiatric or neurological illness.

All patients gave informed consent, and were told they were free to withdraw from the study at any time, without having to give any reason. Patients were further informed that declining to participate in, or withdrawing from, the study, would have no bearing on the nature or duration of their clinical care. The study had ethical approval from the relevant university and UK National Health Service review boards.

Fourteen patients with DPD satisfying entry criteria gave informed consent to participate (11 male, 3 female, age range 23–59, mean 33.7, SD 8.9). Estimated verbal IQ, assessed using the NART (National Adult Reading Test, Nelson, 1982) ranged from 103 to 127, mean 115, SD 7.79. Duration of DPD symptoms ranged from 2 to over 40 years, mean 15 years, SD 11.42. In other words all cases were highly chronic. Psychiatric symptoms were quantified using the Cambridge Depersonalization Scale (CDS, Sierra and Berrios, 2000), the Beck Depression Inventory (BDI, Beck et al., 1988), and the Spielberger Anxiety Inventory (SAI; Spielberger, 1983). The CDS is a self-report scale in which respondents are asked to rate the frequency and duration of depersonalization-related experiences over the preceding 6 months (Sierra and Berrios, 2000). Because we were interested in changes between two time points, rather than symptoms over a 6 month period, we used an adapted "state" version of the CDS, i.e., CDSS, comprising 22 depersonalization-related statements (e.g., *Things around me are now looking 'flat' or 'lifeless', as if I were looking at a picture*). Each statement is rated on a simple visual analog scale between 0 and 100, such that the maximum possible score is (22 × 100) = 2200. The BDI is a widely used instrument for probing mood state and other mental and behavioral aspects of relevance to depression, and SAI is another widely used anxiety scale that has two sections, one probing anxiety-related experiences over the preceding 6 months (Trait), and one asking respondents to rate feelings or experiences in the moment (State) (Spielberger, 1983).

Participants underwent the first fMRI scanning session prior to commencing treatment with lamotrigine, which was initiated at the dose of 25 mg per day and then slowly increased in fortnightly increments (Medford et al., 2005). Provided there were no adverse events related to treatment, they remained on this medication for a sufficient period to reach a target dose in the range 200–400 mg per day. It was not felt realistic or desirable to specify that all participants should be on the same target dose, as previous clinical experience with using lamotrigine for DPD strongly suggests that, when lamotrigine is effective, the dose at which it exerts optimal therapeutic effect varies widely between individuals (as is true for many drugs used to treat psychiatric conditions; Sierra et al., 2006).

Four patients reported that the DPD symptoms had originally arisen in the immediate aftermath of illicit drug use [cannabis in three cases, MDMA ("Ecstasy") in the fourth, see **Table 1**]. It was not felt that this represented a significant confound, as those who described drug use preceding the onset of symptoms were otherwise inexperienced drug users who had avoided all illicit drugs since the first appearance of DPD symptoms. Moreover, in every case the episode of drug use was so far in the past as to make it very likely that, even if drug effects had played a role in the initial experience of depersonalization, the ongoing maintenance of the depersonalized state was due to psychological and biological processes underpinning DPD in general, rather than any distinct "drug-induced" variety of the condition. There is strong empirical evidence to support this assumption, from a study demonstrating that the course, nature and phenomenology of DPD is highly consistent irrespective of whether drug use was involved at onset (Medford et al., 2003).

Of the original clinical sample of 14, four patients did not return for scans at time 2 (see **Table 1**). Two of these four withdrew from the study soon after their time 1 scans, deciding that they did not wish to take psychotropic medication. One patient experienced persistent nausea after commencing lamotrigine and stopped taking it after a week. The fourth patient took lamotrigine as described and then contacted the clinic to say she felt her DPD symptoms had been completely abolished by it, and that she no longer needed to be seen, and did not wish to attend for the second scan as she lived a considerable distance from the clinic.

At time 2, the 10 remaining patients were classified either as "improved" or "not improved" on the basis of clinical assessment and their scores on the Cambridge Depersonalization Scale, state version (CDSS, see **Table 1**). In keeping with previous work (Sierra et al., 2001, 2006), clinical improvement was defined as a reduction of at least 30% in CDSS score between time 1 and time 2, with percentage change calculated by the formula [(Time 1 CDSS score–Time 2 CDSS score)/Time 1 CDSS score] × 100. In every case, the categorisation as "improved" or "not improved" was corroborated by descriptions of symptoms and general mental state given by patients at clinical interview. Using this method, five of the 10 patients were classified as "improved" and five as "non-improved." This 50% improvement rate is consistent with previous work examining lamotrigine as a potential treatment for DPD (Sierra et al., 2001, 2006). At this point it should be stressed that the current study does not represent a formal clinical trial of lamotrigine- aside from the small numbers, the study was not placebo-controlled or blinded, so the clinical improvement observed in some patients cannot be confidently ascribed to the effects of lamotrigine. Nevertheless, it is of note that placebo effects appear to be minimal in DPD (Sierra et al., 2001; Simeon et al., 2004), something particularly relevant here, given that patients had been experiencing symptoms for a prolonged period and in most cases had tried other psychotropic drug treatments (usually SSRIs) prior to referral to the DPD clinic, without any significant impact on their DPD symptoms.

In addition to the patient group, 25 healthy control subjects (14 male, 11 female) were also recruited (age range 23–46, mean 29.8, SD 5.68; estimated verbal IQ 107–127, mean 117.5, SD 5.47). Mean symptom scale scores for the control group were: CDSS 38.7 (SD 48.0), BDI 2.56 (2.73), SSAI 31.6 (SD 8.61).

Although the patient group contained relatively more males (11 M, 3 F vs. 14 M, 11 F), the difference in the gender composition of the two groups was not statistically significant (chi-square value 1.99, df = 1, $p = 0.16$). Unpaired t-tests were performed to check for other differences between the groups.

TABLE 1 | Patients at time 1 and time 2: rating scale measures, medication at time 2, and improvement status (1 = improved, 2 = not improved).

	Time 1				Medication (at time 2, daily dose)	Time 2			% change in CDSS score	Improved?
	BDI	SSAI	CDSS	Timelag (days)		BDI	SSAI	CDS (State)		
1	1	26	475	180	Lamotrigine 400 mg	0	26	140	−70.5	1
2	27	58	780	242	Lamotrigine 400 mg	30	66	1175	+50.6	2
3	1	39	985	180	Lamotrigine 300 mg	3	31	615	−37.6	1
4	6	38	295	154	Lamotrigine 200 mg	10	46	285	−3.4	2
5	15	47	700	–	–	–	–	–	–	–
6	14	42	520	–	–	–	–	–	–	–
7	3	32	589	223	Lamotrigine 400 mg	4	37	237	−59.8	1
8	4	44	685	221	Lamotrigine 250 mg	4	55	344	−49.8	1
9	21	54	1355	–	–	–	–	–	–	–
10	10	57	591	–	–	–	–	–	–	–
11	6	29	467	127	Lamotrigine 250 mg	11	27	564	+20.8	2
12	18	50	1007	119	Lamotrigine 400 mg Venlafaxine 75 mg	17	53	1608	+59.7	2
13	9	47	529	205	Lamotrigine 400 mg	9	40	588	+11.2	2
14	7	37	696	158	Lamotrigine 300 mg Fluoxetine 20 mg	5	22	260	−62.6	1
Mean	10.14	42.86	691.00	180.9	–	9.3	40.9	581.6	–	–
SD	7.86	10.03	272.42	41.74	–	8.77	14.58	467.70	–	–

Timelag, number of days between time 1 and time 2; BDI, Beck Depression Inventory; SSAI, Spielberger State Anxiety Inventory; DES, Dissociative Experiences Scale; CDS, Cambridge Depersonalization Scale (State Version). All 14 patients were scanned at time 1. Time 2 columns indicate which patients were subsequently re-scanned (10 of 14).

TABLE 2 | Group comparisons on SCR variables: controls vs. DPD patients (at time 1).

SC Measure	Group	Mean rank	Sum of ranks	U	p
WEflucs	NC	22.22	555.5	119.5	0.052
	DPDt1	16.04	224.5		
WEamp	NC	22.62	565.5	109.5	0.027*
	DPDt1	15.32	214.5		
WEmean	NC	22.24	556.0	119.0	0.052
	DPDt1	16.00	224.0		
E1flucs	NC	21.84	546.0	129.0	0.087
	DPDt1	16.71	234.0		
E1amp	NC	22.24	556.0	119.0	0.049*
	DPDt1	16.00	224.0		
E1mean	NC	22.28	557.0	118.0	0.049*
	DPDt1	15.93	223.0		
N1flucs	NC	23.70	592.5	82.5	0.002*
	DPDt1	13.39	187.5		
N1amp	NC	24.40	610.0	65.0	<0.001*
	DPDt1	12.14	170.0		
N1mean	NC	22.12	553.0	122.0	0.062
	DPDt1	16.21	227.0		

*P-values corrected for exactness within SPSS. P-values marked *represent findings significant at a threshold of p = 0.05.*

TABLE 3 | Areas of significantly greater activation in response to aversive images (upper part of table) and neutral images (lower part) in normal control subjects (*n* = 25).

No. of voxels	x	y	z	Region	Side
AREAS ACTIVATED BY EMOTIONAL > NEUTRAL					
48	43	−59	−24	Cerebellum	R
37	36	−78	−13	BA18 Secondary visual cortex	R
25	−43	−70	−7	BA19 Secondary visual cortex	L
23	−40	−67	−29	Cerebellum	L
18	40	7	20	BA44 dorsolateral prefrontal cortex	R
14	−36	7	26	BA44 dorsolateral prefrontal cortex	L
10	40	22	−2	Anterior insula	R
8	51	−44	37	BA40 supramarginal gyrus	R
6	−51	19	20	BA45 dorsolateral prefrontal cortex	L
AREAS ACTIVATED BY NEUTRAL > EMOTIONAL					
66	−14	−52	−29	Cerebellum	L
45	−58	−22	−7	BA21 middle temporal gyrus	L
23	47	−52	37	BA40 supramarginal gyrus	R

BA, Brodmann Area.

There were no significant differences between groups for age ($t = 1.66$, df = 37, $p = 0.11$) or estimated verbal IQ ($t = -1.18$, df = 37, $p = 0.24$). There were highly significant differences between groups on the three clinical rating scales: the BDI ($t = 4.41$, df = 37, $p < 0.001$), the SSAI ($t = 3.71$, df = 37, $p = 0.001$), and the CDSS ($t = 11.89$, df = 37, $p < 0.001$).

Experimental Procedure and Data Analysis

Subjects underwent fMRI scanning during which they viewed alternating blocks of neutral and aversive images selected

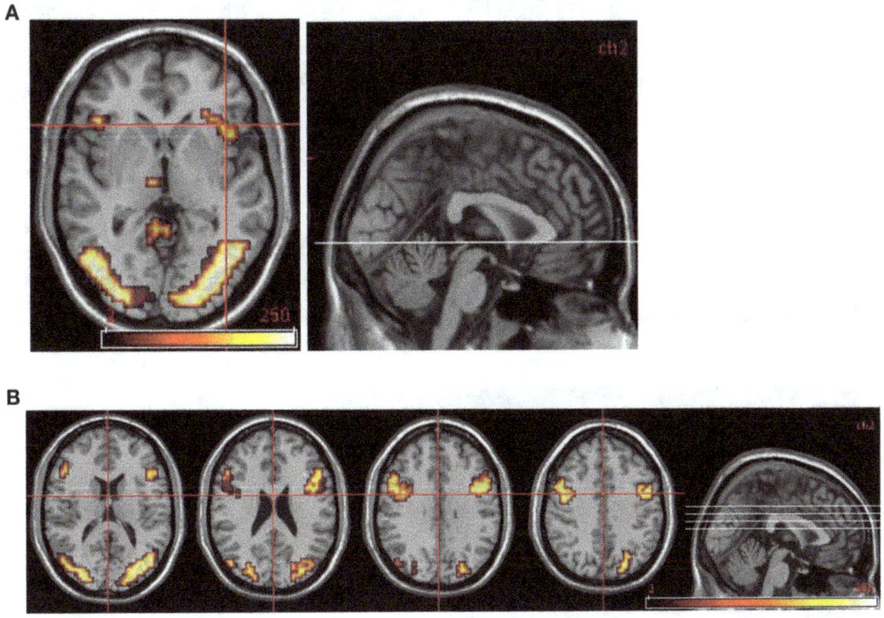

FIGURE 1 | Normal control group map. (A) Transverse slice at $z = -2$ showing activation in R anterior insula. **(B)** Transverse slices at $z = 16, 24, 32, 40$ showing bilateral activation in DLPFC and occipital cortex. All areas significantly more activated by viewing aversive images compared to neutral images (i.e., in emotional phase).

TABLE 4 | Areas of significantly greater activation in response to aversive images (upper part of table) and neutral images (lower part) in patients with DPD at time1 ($n = 14$).

Size	x	y	z	Region	Side
AREAS ACTIVATED By EMOTIONAL > NEUTRAL					
38	43	7	20	BA44 DLPFC	R
36	43	−70	7	BA37 primary visual cortex	R
32	0	22	26	BA 24/32 anterior cingulate cortex	R
29	−4	26	20	BA24 anterior cingulate cortex	L
28	40	4	26	BA44 DLPFC	R
27	43	19	15	BA 44/45 DLPFC	R
21	0	22	31	BA32 anterior cingulate cortex	R
17	−43	−70	−2	BA19/37 primary visual cortex	L
15	40	30	−7	BA47 VLPFC	R
13	−7	52	15	BA9 medial prefrontal cortex	L
11	40	30	9	BA 45/46 DLPFC	R
AREAS ACTIVATED BY NEUTRAL > EMOTIONAL					
224	58	−15	−13	BA21 middle temporal gyrus	R

BA, Brodmann Area; DLPFC, dorsolateral prefrontal cortex.

from the IAPS (Bradley and Lang, 1999). Each block of aversive or neutral images lasted 30 s and comprised five pictures, onscreen for 6 s each. Subjects were informed during the consent procedure that some images were potentially unpleasant or shocking, but were not given details as to the purpose of the experiment. For each image, subjects were asked to press a button to indicate whether the scene depicted was taking place outdoors or indoors. This simple task ensures cognitive processing of the stimulus image but does not involve explicit judgements of emotional arousal or salience: rather, the emotional processing is implicit. This is the same procedure used in our earlier fMRI study of depersonalization disorder utilizing IAPS images (Phillips et al., 2001).

During each scanning session, SCR was recorded continuously using standard silver-silver chloride electrodes 0.5 cm in diameter. Electrodes were attached to the first and second fingertips of the nondominant hand. All electrodes and cables were MR-compatible, allowing concurrent SCR recording during the fMRI scanning sessions. Using a skin conduction coupler, analog-to-digital conversion, and preamplification, the signal was processed by a PSYLAB (Cambridge, MA) SC5SA preamplifier, a SC5AL amplifier, and real-time PSYLAB online signal processing software. These devices employ a 24-bit A/D converter, filter the response signal at 10 Hz to prevent aliasing, and have an output range of 0–100 MicroSiemens. SCR was then sampled at 100 Hz without further compression and stored offline serially, including stimulus presentation points. Security measures were included to prevent electric shocks, radiofrequency (RF) burning, and to protect SCL signals from scanner-induced magnetic distortion using screened cables. Shielding of the wires received a 5.6-kOhm correction by resistors that produced a stable 1% shift of the signal. The cable was led from the magnetically shielded scanner environment through a Faraday cage, and the signal filtered at 1000 pF by capacitors to a copper cap before signal processing. Water soluble jelly (KY Jelly; Johnson and Johnson, Slough, England) was used as an electrolyte

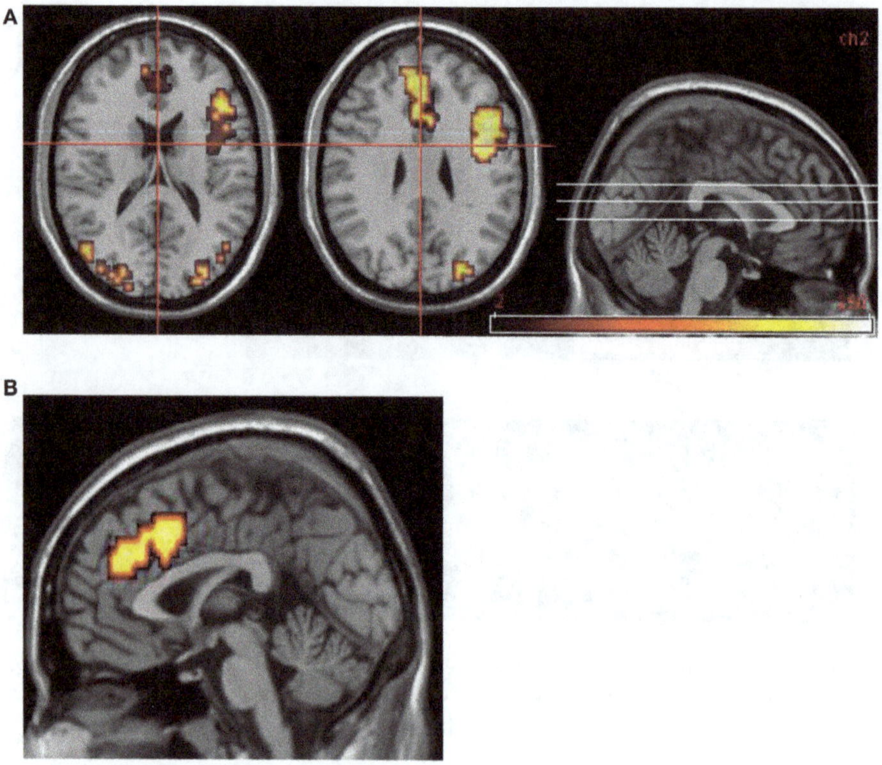

FIGURE 2 | Group data for all DPD patients at time 1. (A) Transverse slices at level $z = 18, 29$ showing activations in emotional phase in bilateral visual cortex, right VLPFC, and bilateral medial prefrontal cortex (including ACC bilaterally). **(B)** Midline sagittal slice (i.e., $x = 0$) showing location of medial prefrontal activation.

contact medium between electrodes and skin. Before the application of electrodes, all subjects washed their hands using a plain nonabrasive soap (as in Sierra et al., 2002) in order to standardize the dermo-gel-electrode interface across subjects as far as possible. During scanning, participants underwent approximately 12 min of structural scans prior to the commencement of the functional MRI experimental runs, allowing adequate time for habituation of SCR once in the scanning environment.

fMRI Hardware and Data Acquisition

Gradient echo echoplanar images were acquired on a GE Signa 1.5 T Neurovascular system (General Electric, Milwaukee, WI, USA) at the Maudsley Hospital, London. One hundred T2*-weighted images depicting BOLD (blood oxygenation level dependent) contrast (Ogawa et al., 1990) were acquired over 5 min (for each task) at each of 14 near-axial non-contiguous 5-mm thick planes parallel to the intercommissural (AC-PC) line: TE 40 ms, TR 3 s, in-plane resolution 5 mm, and interslice gap 0.5 mm. This EPI dataset provided complete coverage of the temporal lobes and almost complete coverage of the frontal, occipital, and parietal lobes (Simmons et al., 1999).

fMRI data were analyzed with the software XBAM, developed at the Institute of Psychiatry, using a non-parametric approach (see Fusar-Poli et al., 2010, for further discussion of this method and comparison with parametric mapping). Data were first

TABLE 5 | Areas of significantly greater activation in response to aversive images and neutral images in those DPD patients who at time 2 exhibited a significant decrease in CDSS score ($n = 5$).

Size	*x*	*y*	*z*	Region	Side
AREAS ACTIVATED BY EMOTIONAL > NEUTRAL					
45	22	−81	−7	BA18 visual cortex	R
38	29	−63	−29	Cerebellum	R
37	−22	−74	−29	Cerebellum	L
32	−7	−89	−13	BA17 lingual gyrus/visual cortex	L
29	−43	19	−7	Anterior insula/BA47	L
26	−51	19	9	BA45	L
19	−32	30	4	Anterior insula	L
11	−47	19	20	BA45 DLPFC	L
9	−40	0	31	BA44 DLPFC	L
AREAS ACTIVATED BY NEUTRAL > EMOTIONAL					
455	14	−52	−29	Cerebellum	R

processed (Bullmore et al., 1999a) to minimize motion related artifacts. A 3D volume consisting of the average intensity at each voxel over the whole experiment was calculated and used as a template. The 3D image volume at each timepoint was then realigned to this template by computing the combination of rotations (around the x, y, and z axes) and translations (in x, y, and z) that maximized the correlation between the image intensities of the volume in question and the template. Following

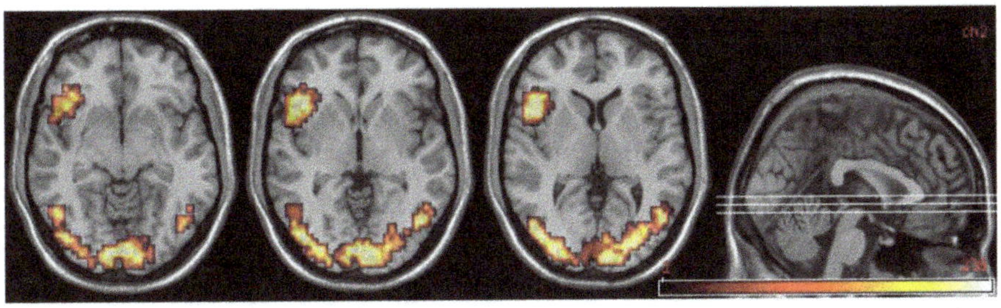

FIGURE 3 | DPD patients at time 2: group data for improvers (n = 5) only. Transverse slices at levels $z = -8, -2, 6$, showing activations in emotional phase in visual cortex bilaterally, and left anterior insula extending into DLPFC.

TABLE 6 | Areas of significantly greater activation in response to aversive images and neutral images in those DPD patients who at time 2 did not exhibit a significant decrease in CDSS score (*n* = 5).

Size	*x*	*y*	*z*	Region	Side
AREAS ACTIVATED BY EMOTIONAL > NEUTRAL					
74	36	−67	−18	Cerebellum	R
64	51	−63	−13	BA19/37 visual cortex	R
40	−40	−67	−24	Cerebellum	L
36	47	−67	−2	BA19 visual cortex	R
34	−43	−70	−7	BA19/37 visual cortex	L
26	43	30	15	BA45, DLPFC	R
14	43	30	−7	BA47, VLPFC	R
AREAS ACTIVATED BY NEUTRAL > EMOTIONAL					
87	−25	−33	−29	Cerebellum	L

realignment, data were then smoothed using a Gaussian filter (FWHM 7.2 mm) to improve the signal to noise characteristics of the images.

Responses to the experimental paradigms were then detected by first convolving each component of the experimental design with each of two gamma variate functions (peak responses at 4 and 8 s, respectively). The best fit between the weighted sum of these convolutions and the time series at each voxel was computed using a constrained BOLD effect model (Friman et al., 2003) This reduces the possibility of the model fitting procedure giving rise to mathematically plausible but physiologically implausible results. Following computation of the model fit, a goodness of fit statistic was computed. This consisted of the ratio of the sum of squares of deviations from the mean image intensity (over the whole time series) due to the model to the sum of squares of deviations due to the residuals (SSQratio). This statistic is used to overcome the problem inherent in the use of the F (variance ratio) statistic that the residual degrees of freedom are often unknown in fMRI time series due to the presence of colored noise in the signal. It has also been shown to behave equivalently to F under permutation testing (Edgington, 1995).

Following computation of the observed SSQratio at each voxel, the data are permuted by a wavelet-based method (Bullmore et al., 2001). Repeated application of this method at each voxel followed by recomputation of the SSQratio from the permuted data allows (by combination of results over all intracerebral voxels) the data-driven calculation of the null distribution of SSQratios under the assumption of no experimentally determined response. Using this distribution it is possible to calculate the critical value of SSQratio needed to threshold the maps at any desired type I error rate. In addition, detection of activated voxels is extended from voxel to cluster level (Bullmore et al., 1999b). In addition to the SSQratio, the size of the BOLD response to each experimental condition is computed for each individual at each voxel as a percentage of the mean resting image intensity level. In order to calculate the BOLD effect size, the difference between the maximum and minimum values of the fitted model for each condition is expressed as a percentage of the mean image intensity level over the whole time series. The observed and permuted SSQratio maps for each individual, as well as the BOLD effect size maps are transformed into the standard space (Talairach and Tournoux, 1988) using a two stage warping procedure (Brammer et al., 1997). This involves first computing the average image intensity map for each individual over the course of the experiment. The transformations required to map this image to the structural scan for each individual and then from "structural space" to the Talairach template are then computed by maximizing the correlation between the images at each stage. The SSQratio and BOLD effect size maps are then transformed into Talairach space using these transformations. Group activation maps are then computed by determining the median SSQratio at each voxel (over all individuals) in the observed and permuted data maps (medians are used to minimize outlier effects). The distribution of median SSQratios over all intracerebral voxels from the permuted data is then used to derive the null distribution of SSQratios and can be thresholded to produce group activation maps at any desired voxel or cluster-level type I error rate. Cluster level maps are thresholded at < 1 expected type I error cluster per brain. The computation of a standardized measure of effect SSQratio at the individual level, followed by analysis of the median SSQratio maps over all individuals treats intra and inter subject variations in effect separately, constituting a mixed effect approach to analysis.

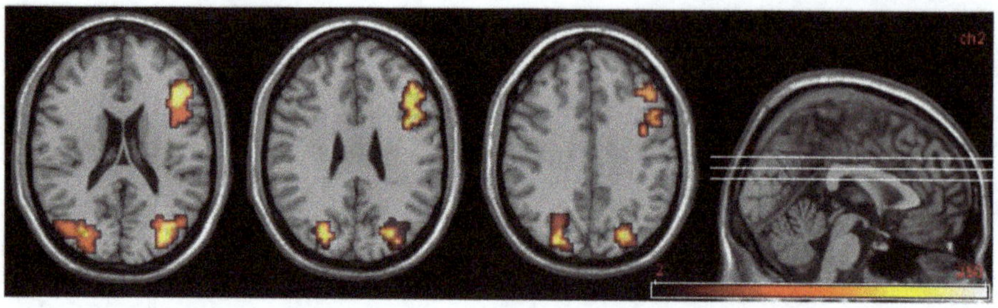

FIGURE 4 | DPD patients at time 2, group data for non-improvers ($n = 5$). Transverse slices at $z = 20, 28, 36$ showing activations in emotional phase in bilateral visual cortex and right DLPFC.

TABLE 7 | Comparison between controls and DPD patients at time 1.

Size	x	y	z	Region	Side	p
CONTROLS > DPD PATIENTS						
70	47	−63	−2	BA19 2viscx	R	0.005
55	−43	−67	−13	BA19 2vis cx	L	0.006
DPD PATIENTS > CONTROLS						
27	40	4	20	BA44 DLPFC	R	0.037
26	0	22	31	BA32 ACC	R/L	0.039

All activations shown were in the emotional phase. 2vis cx, secondary visual cortex; DLPFC, dorsolateral prefrontal cortex; ACC, anterior cingulate cortex.

Comparisons of responses between groups or experimental conditions were performed by fitting the data at each intraceberal voxel at which all subjects have non-zero data using a linear model of the type (Y = a + bX + e) where Y is the vector of BOLD effect sizes for each individual, X is the contrast matrix for the particular inter condition/group contrasts required, a is the mean effect across all individuals in the various conditions/groups, b is the computed group/condition difference and e is a vector of residual errors. The model is fitted by minimizing the sum of absolute deviations rather than the sums of squares to reduce outlier effects. The null distribution of b is computed by permuting data between conditions/groups (assuming the null hypothesis of no effect of experimental condition or group membership) and refitting the above model. Group difference maps are computed as described above at voxel or cluster level by appropriate thresholding of the null distribution of b. BOLD effect maps were used to compute significant group/condition differences rather than standardized measures such as SSQratio, F or t as these contain explicit noise components (error SSQ or error variance), raising the possibility that group differences resulting from F, SSQratio or t comparisons could reflect differences in noise rather than signal.

RESULTS

Behavioral Data

All subjects scored >95% correct responses when judging whether scenes depicted in IAPS images were taking place outdoors or indoors, with no difference between the groups.

Skin Conductance Response (SCR) Data

Group Comparisons: All Controls (n = 25) vs. DPD Patients at Time 1 (n = 14)

The following variables were examined: number of fluctuations (amplitude of over 0.02 microsiemens), amplitude of the largest fluctuation, and mean SCL (skin conductance level) across the epoch. These variables were computed for (i) the whole epoch (WE) (i.e., the entire scanning session), (ii) the first 30-s emotional block within each session (E1), and (iii) the first 30-s neutral block within each session (N1). As the experiment progresses, the likelihood of SCR responses in subsequent blocks being contaminated by responses to stimuli from earlier blocks increases. Therefore, SCR variables were not computed for individual blocks occurring after the first emotional and neutral blocks of each session. This resulted in a total of 9 SCR variables designated as WEflucs, WEamp, WEmean (whole epoch fluctuations, maximum amplitude, and mean SCR), E1flucs, E1amp, E1mean (the same variables computed for the first 30-s emotional block), and N1flucs, N1amp, N1mean (the same variables for the first 30-s neutral block).

Initial exploratory statistics performed on these nine variables, subdivided by group, showed that these data exhibited considerable variance that violated the assumptions of normality, even after log transformation. This was largely due to the data being positively skewed (e.g., four of the 14 DPD patients at time 1 showed no fluctuations greater than 0.02 microsiemens across the entire scanning epoch, meaning that for this group the distribution of values is skewed to the right of the normal distribution curve). For this reason, non-parametric tests were used to compare within and across groups.

Within the normal control group ($n = 25$), two-tailed Wilcoxon signed-rank tests were used to compare SCR responses in the first emotional and neutral blocks of each scanning epoch (E1 and N1, see above). Fluctuations in the initial emotional blocks (E1flucs) were significantly greater than those in initial neutral blocks (N1flucs; sum of ranks 34.5 for E1flucs, 118.5 for N1flucs, $p = 0.041$). There were no significant differences in amplitude of highest fluctuation or mean SCL. Similar results were obtained in the DPD patient group at time 1, $n = 14$, where E1flucs was significantly greater than N1flucs (sum of ranks 0 for E1flucs, 28.0 for N1flucs, $p = 0.017$), with no significant differences in mean SCL.

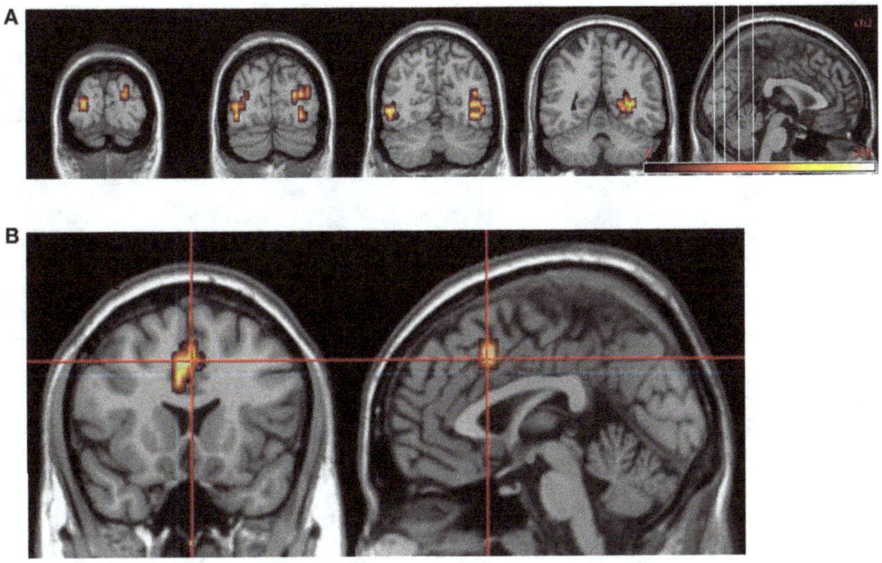

FIGURE 5 | Group comparisons between controls and DPD patients at time 1. (A) Areas significantly more active in emotional phase in controls: coronal slices showing extensive bilateral visual cortical activity. **(B)** Areas significantly more active in DPD patients: coronal and midline sagittal views showing location of anterior cingulate cortex activation.

TABLE 8 | DPD patients: comparison between improvers and non-improvers at time 2.

Size	x	y	z	Region	Side	p
AREAS ACTIVATED IN IMPROVERS > NON-IMPROVERS						
55	−43	19	−7	Anterior Insula	L	0.013
50	−7	−89	−18	Cerebellum	L	0.016
25	32	−74	4	BA19 visual cortex	R	0.029
AREAS ACTIVATED IN NON-IMPROVERS > IMPROVERS						
101	14	−52	−29	Cerebellum	R	0.003
15	18	−85	−12	BA18 visual cortex	R	0.042
12	−36	−74	4	BA19 visual cortex	L	0.026

All activations shown were in the emotional phase. 2vis cx, secondary visual cortex.

However, the key comparisons here are the group comparisons between controls and patients. Here, for each variable, group comparisons were performed using a Mann-Whitney U test, and in view of the relatively small sample sizes this statistic was corrected for exactness within SPSS. These data are summarized in **Table 2**. Tests of significance were one-tailed, as there was a clear *a priori* hypothesis that patients at time 1 would show diminished autonomic reactivity compared to controls. In the study by Sierra et al. (2002), the only measure on which DPD patients were found to be more reactive than controls was a shortened latency of response to non-specific stimuli (clap and sigh), while latency of response to aversive images of the type used in the current study was prolonged in the DPD patient group, and DPD patients also showed significantly reduced magnitude of SCR response to such pictures. Non-specific stimuli were not used in the current study, and latency of response was not computed as the methodology allows only for computation of responses across epochs or blocks, rather than response to individual events. For these reasons, the hypothesis for the current study was that at time 1, DPD patients would show consistently reduced autonomic activity (as indexed by SCR) compared to healthy controls, and this would be reflected across all variable types (number of fluctuations, amplitude, mean SCR level).

For all subsequent SCR data analyses, non-parametric tests, corrected for exactness, were used, according to the rationale above.

All group differences were either significant at the <0.05 level or approaching that threshold.

Group Comparisons: DPD Patients at Time 2, Improvers (n = 5) vs. Non-improvers (n = 5)

Here comparisons were made between improvers and non-improvers at time 2. There were no significant differences between the groups for any of the individual variables.

Functional MRI Data

Group Data: Controls (n = 25)

For healthy controls, the key within-group analysis was the group-level comparison of regional brain activation in response to aversive images compared to neutral images. Areas significantly more activated by viewing aversive images were cerebellum, bilateral visual cortical areas, bilateral dorsolateral prefrontal cortex (DLPFC, Brodmann areas 44 and 45), right supramarginal gyrus. Areas significantly more activated by viewing neutral images were cerebellum, left middle temporal gyrus, right supramarginal gyrus, and left insula. These results are summarized in **Table 3** (in all fMRI data tables that follow, for

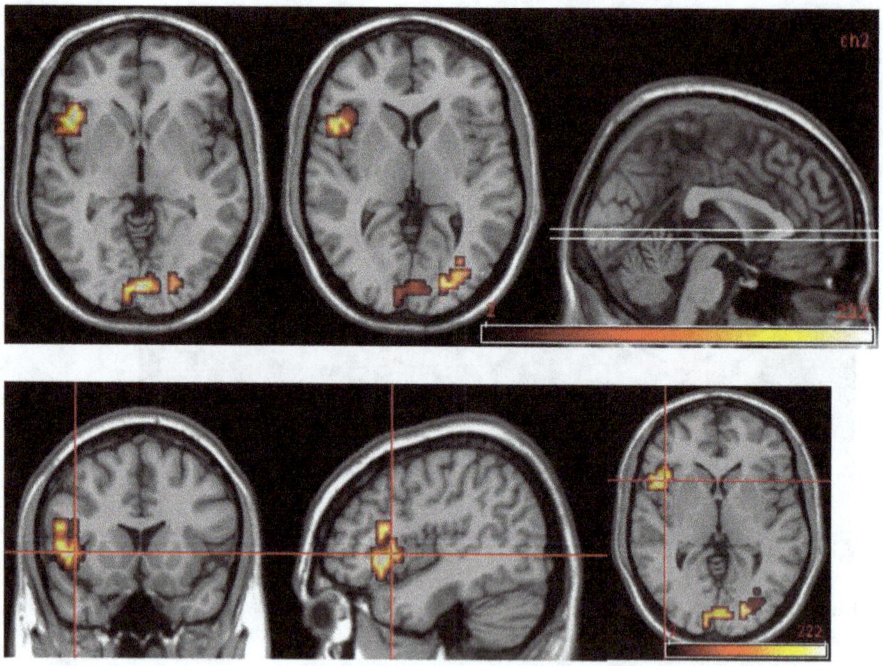

FIGURE 6 | Areas significantly more activated in emotional phase in improvers compared to non-improvers at time 2. Crosshair projection centered on left anterior insula.

TABLE 9 | DPD patients who showed a significant fall in CDSS score compared at time 1 and time 2.

Size	*x*	*y*	*z*	Region	Side	*p*
TIME 2 > TIME 1						
57	−26	−59	31	BA19 1vis cx	L	0.008
34	−7	−70	−29	Cerebellum	L	0.007
19	36	−63	31	BA19 1vis cx	R	0.019
14	−47	−63	−7	BA18/37 2vis cx	L	0.030
TIME 1 > TIME 2						
21	47	4	15	BA44 DLPFC	R	0.013
12	29	−26	−29	Cerebellum	R	0.039
12	−51	−7	26	BA4 precentral gyrus	L	0.036

Abbreviations as above.

each cluster the Talairach co-ordinates of the maximally activated voxel within that cluster are shown) and **Figure 1**.

Group Data: DPD Patients, Time 1 (n = 14)

Areas significantly more activated by viewing aversive images were areas of right lateral prefrontal cortex, bilateral primary visual cortex, bilateral anterior cingulate cortex (ACC), and left medial prefrontal cortex. Middle temporal gyrus was significantly activated by viewing neutral images in contrast to aversive images (see **Table 4** and **Figure 2**).

Group Data: DPD Patients at Time 2, Improvers (n = 5)

Areas significantly more activated by viewing aversive images were areas of bilateral visual cortex, left DLPFC, and left anterior insula (**Table 5**, **Figure 3**). One large cluster in cerebellum was significantly more active in the neutral phase.

Group Data: DPD Patients at Time 2, Non-improvers (n = 5)

Areas significantly more activated by viewing aversive images were areas of bilateral visual cortex, right DLPFC, right VLPFC (Brodmann area 47), and cerebellum. For the reverse comparison, one large cluster in cerebellum was observed (**Table 6**, **Figure 4**).

Group Comparisons: Controls (n = 25) vs. all DPD Patients at Time 1 (n = 14)

Findings from the controls vs. patients (at time 1) comparison are summarized in **Table 7** and **Figure 5**.

Group Comparisons: DPD Patients at Time 2, Improvers (n = 5) vs. Non-improvers (n = 5)

For this comparison, time 2 data for DPD patients who showed a clinically significant reduction in CDSS score ($n = 5$) was compared with time 2 data for those patients who did not show any such reduction. See **Table 8** and **Figure 6**.

Comparison Between Improvers at Time 1 and Time 2

These data are summarized in **Table 9** and **Figure 7**.

DISCUSSION

With regard to the rating scale data: it is unsurprising that patients and controls differed significantly on the anxiety

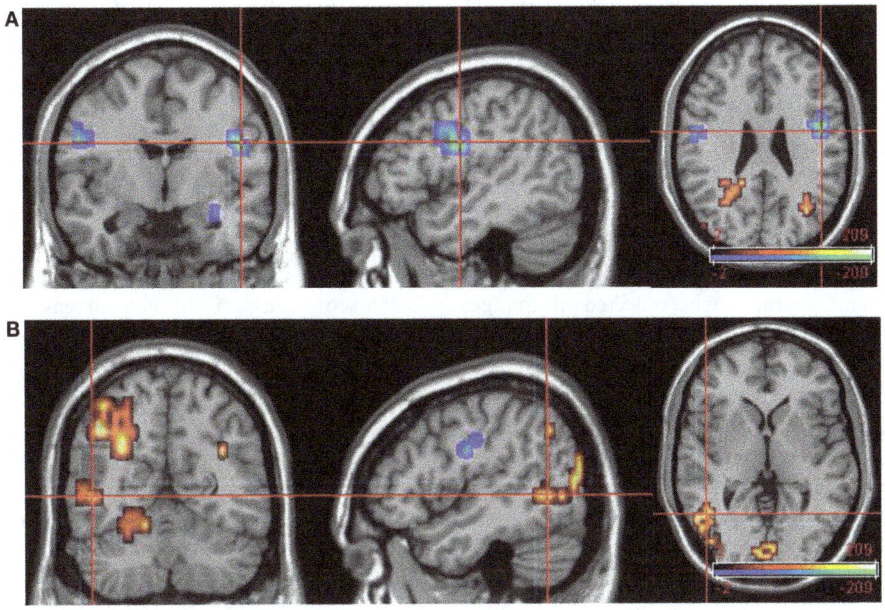

FIGURE 7 | Group data comparisons for improvers at time 1 and time 2. Areas significantly more active in emotional phase at time 1 shown in fire, areas significantly more active at time 2 in blue. **(A)** Transverse slice at $z = 26$ with crosshair projection centered on cluster in R DLPFC. Note activation in similar contralateral region. **(B)** Extensive posterior cortical activation at time 2 (> time 1). Transverse slice at $z = 1$, crosshair projection centered on cluster in L occipitotemporal cortex.

scale (SSAI) in view of the established relationship between depersonalization and anxiety (Medford et al., 2005, although DPD is not simply an anxiety disorder, see Sierra et al., 2012), and similar anxiety scores have been found in previous empirical studies of patients with primary DPD (e.g., Sierra et al., 2002). With regard to scores on the BDI, although DPD patients scored significantly higher than controls, it should nevertheless be noted that within the patient group the mean score is still well below the usual clinical cut-off, as are all the individual scores within the patient group, with the exception of one individual scoring 30 and another with a borderline score of 17.

SCR data was analyzed along three axes (number of fluctuations, amplitude of highest fluctuation, and mean SCR level) for the whole scanning session and, in addition, for the first 30-s aversive and neutral stimulus blocks, giving a total of nine dependent variables. Tests of the hypothesis that DPD patients (at time 1) would be generally less autonomically reactive than healthy controls produced significant results for five of the nine variables, and non-significant trends in the same direction for the other four variables. Taken as a whole, these findings provide strong evidence to support the above hypothesis. It should be noted that this effect was not specific to responses to emotional stimuli: the findings from neutral blocks are similar to those from emotional blocks. As discussed above, the study by Sierra et al. (2002), found decreased latency of response to physical (non-image) auditory stimuli (clap, sigh) in the DPD patient group compared to controls. The authors interpreted this as supporting a previous theoretical suggestion (Sierra and Berrios, 1998) that in DPD there is an inhibition of emotional response combined with a simultaneous state of heightened alertness or vigilance (this latter aspect was not specifically examined in the current study). However, in the same study, in addition to the finding that DPD patients had diminished responsiveness to aversive images, there was also weak evidence (not reaching statistical significance) of diminished responses to neutral images in the DPD patient group.

There is also older work supporting the idea that depersonalization is associated with diminished emotional arousal and autonomic reactivity. There are reports of individuals with episodic depersonalization in whom periods of depersonalization were associated with SCR response patterns that were strikingly different (in terms of showing greatly reduced amplitude, and lack of fluctuation) from SCR responses recorded from the same individuals at times when they were not experiencing depersonalization symptoms (Lader and Wing, 1966). Another study (Kelly and Walter, 1968) used forearm blood flow as an index of sympathetic activity and found that patients with depersonalization had the lowest baseline levels compared to various other psychiatric patient groups. Thus the finding of reduced autonomic reactivity in depersonalization is a consistent one, and in line with the results of the current study.

Improvements in clinical state at time 2, as indexed by CDSS score, were not, however, associated with any significant change in SCR variables. No significant differences were seen in individual SCR variables between improvers and non-improvers at time 2, or when comparing SCR results in improvers at time 1 and time 2. Caution must be exercised in interpretation of these findings, as these comparisons are between small groups (two groups of five in each case). Keeping this caveat in mind, it appears that while measures of SCR reactivity appear to reliably discriminate patients from controls, they do not appear to have predictive value in terms of response to pharmacotherapy or progress over time, and nor are improvements in clinical state necessarily mirrored by corresponding changes in autonomic responses. This may reflect the fact that at time 2, even those patients who exhibited a significant fall in CDSS score compared to time 1 nevertheless still had CDSS scores well above the clinical cut-off for this scale, with only one exception. In other words, four of these five patients still had definite symptoms of DPD, despite having improved clinically to a significant degree. The improvement appears to be reflected in changes in neural activation pattern in response to the IAPS images (discussed in detail below), but this is not mirrored by changes in autonomic response. This suggests that the inhibition of autonomic emotional response is a core feature of the DPD syndrome which is one of the last aspects to "normalize" when patients improve clinically, whether this improvement is spontaneous or in response to specific treatment interventions. Although patients who improved were designated as "improvers" for the purposes of this study, it cannot be definitively stated that their improvement was in fact due to the prescribed pharmacotherapy. The study was not designed as a formal treatment trial- rather it is an observational study of changes over time in a group of patients, and the cause of these changes cannot be definitely inferred. However, the response rate of 50% (among the 10 patients who were re-scanned at time 2) is consistent with a previous study of lamotrigine (either as monotherapy or in combination with antidepressant medication) (Sierra et al., 2006), and it should also be noted that all the patients had at time 1 reported chronic, intractable symptoms that had not responded to any previous interventions.

In contrast to the SCR data, the functional MRI results suggest a clear pattern of association between changes in mental state over time and corresponding changes in regional brain activity. At time 1, DPD patients showed significant activation in response to aversive images in right lateral prefrontal cortex (both ventrolateral and dorsolateral regions), bilateral primary visual cortex, bilateral ACC, and left medial prefrontal cortex. Visual cortical activations are frequently seen in functional neuroimaging studies when emotionally salient material is presented in the visual modality, as in the current study. This is thought to reflect the "modulation" of sensory cortex by back projections from areas involved in emotional processing (e.g., Morris et al., 1998; Tabbert et al., 2005). This process appears to still be in operation in the DPD patients, despite their otherwise reduced biological and experiential response to emotion. However, it is important to note that this modulatory effect appears to increase when symptoms of depersonalization are (at least partly) ameliorated. The evidence for this is that when group data for improvers at time 2 was contrasted with equivalent data from time 1 for the same subgroup of patients, there was significantly more activation in visual cortical areas (primary visual cortex bilaterally, and left secondary visual cortex) at time 2. This suggests that the modulatory effect seen in visual cortex at time 1 is less pronounced than would normally be expected, and this is confirmed by the fact that at time 1, controls showed significantly greater activation than DPD patients in visual cortex bilaterally. These activations were large (70 and 55 voxels, respectively) and highly statistically significant ($p < 0.01$). In contrast, no visual cortical areas were more active in DPD patients than in controls. However, in DPD patients at time 2, the pattern of results in visual cortex is more complex. Both improvers and non-improvers show bilateral visual cortical activations at time 2, with different areas being significantly more active in each of these two subgroups. However, it is of note that in the responder subgroup, visual cortex was significantly more active at time 2 than at time 1, suggesting that a reduction in CDSS score is associated with increased modulation of sensory cortex by emotional processing. It may be that as this modulatory effect increases, different areas of secondary sensory cortex are recruited, perhaps accounting for the different patterns of visual cortical activity seen in the two subgroups.

A network of right prefrontal regions was activated by viewing aversive images in the DPD patient group at time 1. This included a specific region (falling within Brodmann area 47) of right ventrolateral prefrontal cortex. In a previous fMRI study of DPD (Phillips et al., 2001), this same region was also found to be preferentially activated in DPD patients in response to aversive images compared to neutral images, and it was suggested that this region may have a key role in an involuntary "top-down" inhibition of brain regions involved in the generation of emotional response. The current study provides further evidence that this area plays a key role in the biological underpinnings of the depersonalization state. Firstly, it was activated in response to aversive images in patients at time 1, as detailed above. Secondly, the same area was activated by viewing aversive images in non-improvers at time 2, but not in improvers. This suggests that lessening in the severity of depersonalization symptoms is associated with reduced activity in this area during emotional processing. The implication of this is that right BA47 is a critical region for the inhibition of emotional responses in DPD. In this context, it is interesting to note that the same region has repeatedly been identified as an important inhibitory area in studies that have examined voluntary regulation of emotion in normal volunteers (Ochsner and Gross, 2005; Ohira et al., 2006). Taken together, these findings suggest that right BA47 is recruited when emotional responses are inhibited, regardless of whether this emotional inhibition is volitional (as in normal voluntary emotional self-regulation) or not (as in DPD).

Other lateral prefrontal regions activated in DPD patients by viewing aversive images were predominantly right-sided, with BA44, 45, and 46 all involved. An area of right DLPFC was significantly more active in DPD patients at time 1 compared to controls. This suggests that, in addition to the area of right VLPFC detailed above, more dorsal regions of the right lateral prefrontal cortex are also involved in the inhibition of emotional response formation in DPD.

Medial prefrontal areas, including bilateral ACC, were also activated by viewing aversive images in DPD patients at time 1. ACC is known to be a key area in the evaluation of emotional salience and regulation of emotional response (Medford and Critchley, 2010) and it might be seen as a candidate for an important role in the inhibition of emotional responses in DPD. The current study provides only partial support for this idea.

ACC was activated in patients at time 1, but ACC activation was not seen in either patient subgroup (improvers or non-improvers) at time 2, nor did ACC emerge as an area of significant difference between the two subgroups at time 2.

One possible objection to our study might be that the time 2 scan results could have been confounded by some effect of lamotrigine on regional brain perfusion and therefore on BOLD response, as there is one study suggesting that lamotrigine may reduce the BOLD response to somatosensory stimulation in rodents (Kida et al., 2006). More pertinent to the current work, however, is a more recent study showing no significant effect of lamotrigine on brain perfusion patterns in humans (Shcherbinin et al., 2015), suggesting that any such confound is unlikely. Furthermore, in a previous fMRI study of DPD in which some patients were on lamotrigine and some were unmedicated, subgroup analyses of unmedicated participants did not suggest a confounding effect of lamotrigine (Medford et al., 2006).

The left anterior insula emerged as a key region in this study. This region was activated in controls in response to aversive images, and this finding is consistent with a large body of literature implicating the anterior insula in the response to emotionally salient (particularly aversive) material (Phillips et al., 1997; Mataix-Cols et al., 2008; Meriau et al., 2009). Insula activation was not seen in response to emotional images in DPD patients at time 1. However, anterior insula was activated by viewing aversive images in improvers at time 2, but not in non-improvers at time 2, and the difference was statistically significant when the two groups were compared. This provides powerful evidence that a lack of anterior insula activity is related to the diminished emotional responsiveness seen in DPD, and that a "re-awakened" insula is seen when patients improve and de-affectualization symptoms (and DPD symptoms more generally) are ameliorated.

The insula is frequently reported as a key brain region in processes relating to emotion and bodily sensation (Craig, 2009). In particular it has received attention for its putative role in interoception: "our ability to sense ourselves" (Damasio et al., 2000; Craig, 2002, 2009; Damasio, 2003; Critchley et al., 2004; Medford and Critchley, 2010). Interoception encompasses the sensing of discrete bodily events such as heartbeats, in addition to a more general sense of bodily state thought to rely largely on sensation related to autonomic and visceromotor processes (Farb et al., 2015). It is widely argued that this general feeling state is closely related to emotional state (Craig, 2002, 2009; Medford and Critchley, 2010). This relationship of interoceptive processes to emotional state, in both healthy and clinical groups, has been the focus of considerable empirical and theoretical work in recent years (for recent reviews see Barrett and Simmons, 2015; Farb et al., 2015). One influential conceptualization suggests that ascending somatosensory information is collated and integrated ("represented") in posterior and mid insula (regarded as "primary interoceptive cortex") before then being "re-represented" by anterior insula to generate a consciously accessible feeling state ("how I feel") (Craig, 2009): a "sense of the whole" or "sense of the internal milieu." Theorists differ according to how much of a central role they attribute to the insula. For example it has been suggested (Craig, 2009) that the insula may be the seat of self-awareness itself, whereas a more recent formulation (Barrett and Simmons, 2015) argues that insular cortex is rather one of a number of interoceptive hubs, so that while it contributes to interoception, it is not crucial for it. Certainly the former, more dramatic, claim appears inconsistent with the fact that case studies of patients with extensive bilateral insula damage (Philippi et al., 2012; Damasio et al., 2013; Feinstein et al., 2015) show that fundamental aspects of self-awareness (such as the ability to recognize oneself, or to have apparently normal experiences of agency) and emotional response are preserved in such cases. However, the disturbance of self-awareness encountered in DPD is of a more subtle type: patients retain basic self-awareness, and do not become delusional, rather something about the quality of self-experience has changed. Patients with DPD will often use the phrase "..as if" when attempting to convey their experiences of the depersonalized state (e.g., ".. *as if* I were an automaton," ".. *as if* the world were artificial"), a state in which experience of both the self and the surrounding world feel oddly attenuated. Thus work in DPD may provide a clinical sounding-board for the notion that, while insular cortex may not be essential for interoceptive experience, its activity is important in some way that adds important higher-order experiential features (Damasio et al., 2013), aspects of subjective experience that can usually be taken for granted but which are compromised in DPD.

It is interesting that in our sample, DPD patients showing clinical improvement at time 2 showed no significant change in their sympathetic arousal as indexed by skin conductance. This suggests that their self-reported improved self-awareness, and concomitant changes in neural activity, may not be immediately related to changes in sympathetic arousal but rather to other interoceptive changes—here it is worth recalling that the bodily signals that contribute to interoceptive experience are manifold, including autonomic, visceromotor, hormonal, immunological, and humoral elements (Critchley and Harrison, 2013). With regard to interoception and DPD, a study of heartbeat awareness in DPD found no difference in performance between DPD patients and healthy controls (Michal et al., 2014), but more recent work suggests some anomalies in the cortical representation of bodily events in DPD (Schulz et al., 2015). In any event, in DPD the alteration in self-experience may not be best probed by studies measuring awareness of discrete events such as heartbeats, as patient self-reports strongly suggest that in DPD it is a more general sense of one's physical being that is somehow attenuated or compromised (Medford, 2012).

In this context, it is of particular interest that the insula should emerge as critical in DPD: in view of its role in the generation of feeling states, and the relationship between these and emotional experience, diminished (or inhibited) anterior insula activity is a plausible biological substrate for both desomatization symptoms and the phenomenon of de-affectualization which is the explicit target of this study. It may be, then, that the subjective alteration of experience that is the core of the depersonalized state is rooted in a lack of anterior insula activity. The fact that this lack of activity appears to be amenable to treatment- or, at least, that it is not fixed over time- is encouraging, as it suggests a potential neurobiological target for future treatment strategies in DPD.

AUTHOR CONTRIBUTIONS

NM recruited most of the participants, designed and ran the fMRI experiments, carried out the bulk of the data analysis and interpretation, prepared the figures, and wrote most of the manuscript. MS recruited the remaining participants and contributed to theoretical discussion of the findings. AS contributed to data analysis and theoretical discussion. VG and MB contributed to the analysis both theoretically (writing software routines) and practically (helping to write the Methods section). AD supervised the work and contributed to the conceptual background and theoretical discussion.

FUNDING

NM was supported by a Wellcome Trust Research Training Fellowship, ref 065799/Z/01/Z.

REFERENCES

Barrett, L. F., and Simmons, W. K. (2015). Interoceptive predictions in the brain. *Nat. Rev. Neurosci.* 16, 419–429. doi: 10.1038/nrn3950

Beck, A., Steer, R., and Garbin, M. (1988). Psychometric properties of the Beck Depression Inventory: twenty-five years of evaluation. *Clin. Psychol. Rev.* 8, 77–100. doi: 10.1016/0272-7358(88)90050-5

Bradley, M. M., and Lang, P. J. (1999). *Affective Norms for English Words (ANEW).* Gainesville, FL: NIMH Center for the Study of Emotion and Attention, University of Florida.

Brammer, M. J., Bullmore, E. T., Simmons, A., Williams, S. C. R., Grasby, P. M., Howard, R. J., et al. (1997). Generic brain activation mapping in functional magnetic resonance imaging: a nonparametric approach. *Magn. Reson. Imaging* 15, 763–770. doi: 10.1016/S0730-725X(97)00135-5

Bullmore, E. T., Brammer, M. J., Rabe-Hesketh, S., Curtis, V. A., Morris, R. G., Williams, S. C. R., et al. (1999a). Methods for diagnosis and treatment of stimulus-correlated motion in generic brain activation studies using fMRI. *Hum. Brain Mapp.* 7, 38–48.

Bullmore, E. T., Long, C., Suckling, J., Fadili, J., Calvert, G., Zelaya, F., et al. (2001). Colored noise and computational inference in fMRI time series analysis: resampling methods in time and wavelet domains. *Neuroimage* 13, S86. doi: 10.1016/S1053-8119(01)91429-6

Bullmore, E. T., Suckling, J., Overmeyer, S., Rabe-Hesketh, S., Taylor, E., and Brammer, M. J. (1999b). Global, voxel, and cluster tests, by theory and permutation, for a difference between two groups of structural MR images of the brain. *IEEE Trans. Med. Imaging* 18, 32–42. doi: 10.1109/42.750253

Craig, A. D. (2002). How do you feel? Interoception: the sense of the physiological condition of the body. *Nat. Rev. Neurosci.* 3, 655–666. doi: 10.1038/nrn894

Craig, A. D. (2009). How do you feel–now? The anterior insula and human awareness. *Nat. Rev. Neurosci.* 10, 59–70. doi: 10.1038/nrn2555

Critchley, H. D., and Harrison, N. A. (2013). Visceral influences on brain and behavior. *Neuron* 77, 624–638. doi: 10.1016/j.neuron.2013.02.008

Critchley, H. D., Wiens, S., Rotshtein, P., Ohman, A., and Dolan, R. J. (2004). Neural systems supporting interoceptive awareness. *Nat. Neurosci.* 7, 189–195. doi: 10.1038/nn1176

Damasio, A. (2003). Feelings of emotion and the self. *Ann. N.Y. Acad. Sci.* 1001, 253–261. doi: 10.1196/annals.1279.014

Damasio, A., Damasio, H., and Tranel, D. (2013). Persistence of feelings and sentience after bilateral damage of the insula. *Cereb. Cortex* 23, 833–846. doi: 10.1093/cercor/bhs077

Damasio, A. R., Grabowski, T. J., Bechara, A., Damasio, H., Ponto, L. L., Parvizi, J., et al. (2000). Subcortical and cortical brain activity during the feeling of self-generated emotions. *Nat. Neurosci.* 3, 1049–1056. doi: 10.1038/79871

Edgington, E. S. (1995). *Randomization Tests, 3rd Edn.* New York, NY: Dekker.

Farb, N., Daubenmier, J., Price, C. J., Gard, T., Kerr, C., Dunn, B. D., et al. (2015). Interoception, contemplative practice, and health. *Front. Psychol.* 6:763. doi: 10.3389/fpsyg.2015.00763

Feinstein, J. S., Khalsa, S. S., Salomons, T. V., Prkachin, K. M., Frey-Law, L. A., Lee, J. E., et al. (2015). Preserved emotional awareness of pain in a patient with extensive bilateral damage to the insula, anterior cingulate, and amygdala. *Brain Struct. Funct.* 221, 1499–1511. doi: 10.1007/s00429-014-0986-3

Friman, O., Borga, M., Lundberg, P., and Knutsson, H. (2003). Adaptive analysis of fMRI data. *Neuroimage* 19, 837–845. doi: 10.1016/S1053-8119(03)00077-6

Fusar-Poli, P., Bhattacharyya, S., Allen, P., Crippa, J. A., Borgwardt, S., Martin-Santos, R., et al. (2010). Effect of image analysis software on neurofunctional activation during processing of emotional human faces. *J. Clin. Neurosci.* 17, 311–314. doi: 10.1016/j.jocn.2009.06.027

Kelly, D. H., and Walter, C. J. (1968). The relationship between clinical diagnosis and anxiety, assessed by forearm blood flow and other measurements. *Br. J. Psychiatry* 114, 611–626. doi: 10.1192/bjp.114.510.611

Kida, I., Smith, A. J., Blumenfeld, H., Behar, K. L., and Hyder, F. (2006). Lamotrigine suppresses neurophysiological responses to somatosensory stimulation in the rodent. *Neuroimage* 29, 216–224. doi: 10.1016/j.neuroimage.2005.07.015

Lader, M. H., and Wing, L. (1966). Physiological Measures, Sedative Drugs, and Morbid Anxiety. Maudsley Monographs no. 14. (London, UK: Oxford University Press), 10:314. doi: 10.1016/0022-3999(66)90124-3

Mataix-Cols, D., An, S. K., Lawrence, N. S., Caseras, X., Speckens, A., and Giampietro, V., et al. (2008). Individual differences in disgust sensitivity modulate neural responses to aversive/disgusting stimuli. *Eur. J. Neurosci.* 27, 3050–3058. doi: 10.1111/j.1460-9568.2008.06311.x

Medford, N. (2012). Emotion and the unreal self: depersonalization disorder and de-affectualization. *Emot. Rev.* 4, 139–144. doi: 10.1177/1754073911430135

Medford, N., Baker, D., Hunter, E., Sierra, M., Lawrence, E., Phillips, M. L., et al. (2003). Chronic depersonalization following illicit drug use: a controlled analysis of 40 cases. *Addiction* 98, 1731–1736. doi: 10.1111/j.1360-0443.2003.00548.x

Medford, N., Brierley, B., Brammer, M., Bullmore, E. T., David, A. S., and Phillips, M. L. (2006). Emotional memory in depersonalization disorder: a functional MRI study. *Psychiatry Res.* 148, 93–102. doi: 10.1016/j.pscychresns.2006.05.007

Medford, N., and Critchley, H. D. (2010). Conjoint activity of anterior insular and anterior cingulate cortex: awareness and response. *Brain Struct. Funct.* 214, 535–549. doi: 10.1007/s00429-010-0265-x

Medford, N., Sierra, M., Baker, D., and David, A.S. (2005). Understanding and treating depersonalisation disorder. *Adv. Psychiatr. Treat.* 11, 92–100. doi: 10.1192/apt.11.2.92

Meriau, K., Wartenburger, I., Kazzer, P., Prehn, K., Villringer, A., and van der Meer, E., et al. (2009). Insular activity during passive viewing of aversive stimuli reflects individual differences in state negative affect. *Brain Cogn.* 69, 73–80. doi: 10.1016/j.bandc.2008.05.006

Michal, M., Reuchlein, B., Adler, J., Reiner, I., Beutel, M. E., Vogele, C., et al. (2014). Striking discrepancy of anomalous body experiences with normal interoceptive accuracy in depersonalization-derealization disorder. *PLoS ONE* 9:e89823. doi: 10.1371/journal.pone.0089823

Morris, J. S., Friston, K. J., and Dolan, R. J. (1998) Experience-dependent modulation of tonotopic neural responses in human auditory cortex. *Proc. Biol. Sci.* 265, 649–657. doi: 10.1098/rspb.1998.0343

Nelson, H. E. (1982). *National Adult Reading Test (NART): For the Assessment of Premorbid Intelligence in Patients with Dementia: Test Manual.* Windsor, UK: NFER-Nelson.

Ochsner, K. N., and Gross, J. J. (2005). The cognitive control of emotion. *Trends Cogn. Sci.* 9, 242–249. doi: 10.1016/j.tics.2005.03.010

Ogawa, S., Lee, T. M., Kay, A. R., and Tank, D. W. (1990). Brain magnetic-resonance-imaging with contrast dependent on blood oxygenation. *Proc. Natl. Acad. Sci. U.S.A.* 87, 9868–9872. doi: 10.1073/pnas.87.24.9868

Ohira, H., Nomura, M., Ichikawa, N., Isowa, T., Iidaka, T., Sato, A. et al. (2006). Association of neural and physiological responses during voluntary emotion suppression. *Neuroimage* 29, 721–733. doi: 10.1016/j.neuroimage.2005.08.047

Philippi, C. L., Feinstein, J. S., Khalsa, S. S., Damasio, A., Tranel, D., Landini, G., et al. (2012). Preserved self-awareness following extensive bilateral brain damage to the insula, anterior cingulate, and medial prefrontal cortices. *PLoS ONE* 7:e38413. doi: 10.1371/journal.pone.0038413

Phillips, M. L., Medford, N., Senior, C., Bullmore, E. T., Suckling, J., Brammer, M. J., et al. (2001). Depersonalization disorder: thinking without feeling. *Psychiatry Res.* 108, 145–160. doi: 10.1016/S0925-4927(01)00119-6

Phillips, M. L., Young, A. W., Senior, C., Brammer, M., Andrew, C., and Calder, A. J., et al. (1997). A specific neural substrate for perceiving facial expressions of disgust. *Nature* 389, 495–498. doi: 10.1038/39051

Schulz, A., Köster, S., Beutel, M. E., Schachinger, H., Vögele, C., Rost, S., et al. (2015). Altered patterns of heartbeat-evoked potentials in depersonalization/derealization disorder: neurophysiological evidence for impaired cortical representation of bodily signals. *Psychosom. Med.* 77, 506–516. doi: 10.1097/PSY.0000000000000195

Shcherbinin, S., Doyle, O., Zelaya, F. O., de Simoni, S., Mehta, M. A., and Schwarz, A. J. (2015). Modulatory effects of ketamine, risperidone and lamotrigine on resting brain perfusion in healthy human subjects. *Psychopharmacology (Berl).* 232, 4191–4204 doi: 10.1007/s00213-015-4021-z

Sierra, M., Baker, D., Medford, N., and David, A. S. (2005). Unpacking the depersonalization syndrome: an exploratory factor analysis on the Cambridge Depersonalization Scale. *Psychol. Med.* 35, 1523–1532. doi: 10.1017/S0033291705005325

Sierra, M., Baker, D., Medford, N., Lawrence, E., Patel, M., Phillips, M. L., et al. (2006). Lamotrigine as an add-on treatment for depersonalization disorder: a retrospective study of 32 cases. *Clin. Neuropharmacol.* 29, 253–258. doi: 10.1097/01.WNF.0000228368.17970.DA

Sierra, M. and Berrios, G. E. (1998) Depersonalization: neurobiological perspectives. *Biol. Psychiatry,* 44, 898–908.

Sierra, M. and Berrios, G. E. (2000). The Cambridge Depersonalization Scale: a new instrument for the measurement of depersonalization. *Psychiatry Res.* 93, 153–164.

Sierra, M., and David, A. S. (2011). Depersonalization: a selective impairment of self-awareness. *Conscious. Cogn.* 20, 99–108. doi: 10.1016/j.concog.2010.10.018

Sierra, M., Medford, N., Wyatt, G., and David, A. S. (2012). Depersonalization disorder and anxiety: a special relationship? *Psychiatry Res.* 197, 123–127. doi: 10.1016/j.psychres.2011.12.017

Sierra, M., Phillips, M. L., Lambert, M. V., Senior, C., David, A. S., and Krystal, J. H. (2001). Lamotrigine in the treatment of depersonalization disorder. *J. Clin. Psychiatry* 62, 826–827. doi: 10.4088/JCP.v62n1012b

Sierra, M., Senior, C., Dalton, J., McDonough, M., Bond, A., Phillips, M. L., et al. (2002). Autonomic response in depersonalization disorder. *Arch. Gen. Psychiatry* 59, 833–838. doi: 10.1001/archpsyc.59.9.833

Simeon, D., Guralnik, O., Schmeidler, J., and Knutelska, M. (2004). Fluoxetine therapy in depersonalisation disorder: randomised controlled trial. *Br. J. Psychiatry* 185, 31–36. doi: 10.1192/bjp.185.1.31

Simeon, D., Kozin, D. S., Segal, K., Lerch, B., DuJour, R., and Giesbrecht, T. (2008). De-constructing depersonalization: further evidence for symptom clusters. *Psychiatry Res.* 157, 303–306. doi: 10.1016/j.psychres.2007.07.007

Simmons, A., Moore, E., and Williams, S. C. R. (1999). Quality control for functional magnetic resonance imaging using automated data analysis and Shewhart charting. *Magn. Reson. Med.* 41, 1274–1278.

Spielberger, C. (1983). *Manual for the State-Trait Anxiety Inventory (STAI).* PaloAlto, CA: Consulting Psychologists Press.

Tabbert, K., Stark, R., Kirsch, P., and vaitl, D. (2005). Hemodynamic responses of the amygdala, the orbitofrontal cortex and the visual cortex during a fear conditioning paradigm. *Int. J. Psychophysiol.* 57, 15–23. doi: 10.1016/j.ijpsycho.2005.01.007

Talairach, J., and Tournoux, P. (1988). *Co-planar Stereotactic Atlas of the Human Brain.* Stuttgart: Thieme.

14

Interaction of Physical Activity and Interoception in Children

Eleana Georgiou [1], Ellen Matthias [1], Susanne Kobel [2], Sarah Kettner [2], Jens Dreyhaupt [3], Jürgen M. Steinacker [2] and Olga Pollatos [1]*

[1] Health Psychology, Institute of Psychology and Education, Ulm University, Ulm, Germany, [2] Research Group "Join the Healthy Boat – Primary School", Sports and Rehabilitation Medicine, Ulm University, Ulm, Germany, [3] Institute of Epidemiology and Medical Biometry, Ulm University, Ulm, Germany

***Correspondence:**
Eleana Georgiou,
Health Psychology, Institute of Psychology and Education, Ulm University, Albert-Einstein-Allee 41, 89069 Ulm, Germany
eleana.georgiou@uni-ulm.de

Background: Physical activity (PA) is associated with positive health outcomes, whereas physical inactivity is related to an increased risk for various health issues including obesity and cardiovascular diseases. Previous research indicates that interindividual differences in the perception of bodily processes (interoceptive sensitivity, IS) interact with the degree of PA in adults. Whether there is a similar relationship between PA and IS in children has not been investigated yet. Therefore, the aim of this study was to investigate the interaction between IS and PA during physical performance tasks and in everyday situations. **Methods:** IS was assessed using a heartbeat perception task in a sample of 49 children within the health promotion program "Join the Healthy Boat" which is implemented in several primary schools in the southwest of Germany. PA was examined using a physical performance task, assessing the distance covered during a standardized 6-min run. In a subsample of 21 children, everyday PA was measured by a multi-sensor device (Actiheart, CamNtech, Cambridge, UK) during five consecutive days with more than 10 h of daily data collection. **Results:** Children with higher IS performed better in the physical performance task. Additionally, based on energy expenditure defined as metabolic equivalents, IS was positively correlated with the extent of light PA levels in the morning and afternoon. **Conclusion:** Our findings reveal that IS interacts positively with the degree of PA in children supporting the idea that interoception is important for the self-regulation of health-related behavior.

Keywords: interoceptive sensitivity, physical performance, metabolic equivalent, childhood/youth, interoception, self-regulation

Introduction

The feedback of afferent signals arising from within the body to the brain and their perception is commonly known as *interoception* (Vaitl, 1996; Cameron, 2001), while the sensitivity in perceiving bodily signals is known as *interoceptive sensitivity* (IS). Referring back to the emotion theories of James (1884) as well as Schachter and Singer (1962), who postulated that there is no emotional experience without the perception of bodily changes, a substantial body of research demonstrates evidence for the suggested close link between interoception and emotional processes in recent decades (for reviews, see Wiens, 2005; Herbert and Pollatos, 2008). Furthermore, the somatic marker theory of Damasio (1994, 1999) emphasized the relevance of visceral and somatosensory feedback from the body to the brain, mapped in distinct higher brain areas, for the emergence of emotions and the regulation of one's behavior. Thus, Damasio inspired studies that demonstrated the benefit of a person's high degree of bodily sensitivity to regulate emotional decision making (e.g., Werner et al., 2009; Dunn et al., 2010, 2012) as well as to process emotions (Wiens et al., 2000; Barrett et al., 2004; Pollatos et al., 2005; Herbert et al., 2007a). Furthermore, the feedback of internal bodily signals is related to the self-regulation of physical activity (PA). The self-perception of bodily signals, like breathing or other internal bodily processes based on cardiovascular activity, might influence the amount of physical effort, when undertaking PA and is known as the self-control of physical load (Pennebaker and Lightner, 1980; Herbert et al., 2007b).

Moreover, one study investigated the regulation of PA in relation to interoceptive processes (Herbert et al., 2007b) and highlighted that these two processes might interact positively with each other. In this study, Herbert et al. (2007b) instructed their participants, aged 20–40, to cycle on a bicycle ergometer for a constant period of time (15 min), whilst being free to control their cycling pace. Results showed that good heartbeat perceivers set their physical endeavor to a lesser extent, demonstrating a more enhanced perception of their exhaustion and of their internal bodily states, than poor heartbeat perceivers. Furthermore, good heartbeat perceivers revealed lower changes in their heart rate, stroke volume and cardiac output. This suggested that participants with higher IS where those who at the same time regulated better their physical load in a performance task. Moreover, Herbert et al. (2007b) implemented this study in terms of regulating physical load in a performance task (cycling task). However, it remains a rather open question whether interoceptive processes do also interact with PA in terms of day to day PA (i.e., in terms of PA under free-living conditions). Lastly, in this study there was no evidence found showing that physical fitness does play a role when differentiating the good from the bad heartbeat perceivers.

On the other hand, but in line with this question, two older studies suggested that a higher state of fitness might be advantageous for better IS (Borg and Linderholm, 1967; Montgomery et al., 1984). Montgomery and Jones (1984) also found that the best predictor for IS in men is the body's fat content. In agreement with these results, Herbert and Pollatos (2014) also demonstrated an inverse relationship between the body mass index (BMI) and IS in adults, and that overweight and obese individuals are less accurate in perceiving bodily signals. These studies suggest that interoceptive processes and individual differences in IS are important for physical fitness or weight status in adults and are of high importance to health-related questions. Although being overweight is established in childhood and youth, there is scant research in examining these questions in children.

Most of the aforementioned studies assess and quantitate IS in terms of people's ability to perceive their own heartbeat. Assessment of heartbeat perception in adults has been shown to be sufficiently reliable (Schandry, 1981; Mussgay et al., 1999), to have substantial interindividual differences (Ehlers and Breuer, 1992; Domschke et al., 2010) and to correlate with the ability to detect changes in other autonomic innervated organs (Whitehead and Drescher, 1980; Harver et al., 1993; Herbert et al., 2012). One of the most frequently used methods to measure cardiac sensitivity is the *Mental Tracking Method* proposed by Schandry (1981; Dunn et al., 2007; Herbert et al., 2007b; Pollatos et al., 2009; Ainley et al., 2012). A recent study by Koch and Pollatos (2014a) adapted this method for children by shortening the intervals used. In spite of the great importance of interoceptive processes for health processes, only a few studies exist that assess IS in children. To our knowledge, there are only four studies that assess heartbeat perception in children, either using the original Schandry paradigm (three trials with the length of 35, 25, and 45 s; see Eley et al., 2004, 2007; Schmitz et al., 2012) or the adapted version (Koch and Pollatos, 2014a,b). The internal consistency of the shorter adapted version was found to be excellent (Cronbach's $\alpha = 0.91$) in a large sample of about 1350 children aged between 6 and 11 years (Koch and Pollatos, 2014a). Additionally, Koch and Pollatos (2014a) investigated the distribution of cardiac sensitivity in relation to general emotional processing and found correlations to interpersonal emotional intelligence and adaptability. Relevant findings were the associations between interoceptive processes and anxiety symptoms (panic and social anxiety) in small samples of eight to 12-year-old children (Eley et al., 2004, 2007; Schmitz et al., 2012). Furthermore, cardiac sensitivity in children can be seen as a dynamic developmental process, that has weaker but yet similar characteristics and relations to emotional parameters than found in adults (Koch and Pollatos, 2014a).

The scope of the present study was to investigate the interrelation between PA in a performance task and in day to day activities (under free-living conditions) with IS in a main sample of 49 primary school children and moreover in a subsample of 21 in the context of the "Baden Württemberg Study," which evaluated the effectiveness of the health promotion program "Join the Healthy Boat" in Southwestern Germany. Taking into account the majority of the studies regarding this matter in adults (Borg and Linderholm, 1967; Montgomery et al., 1984; Herbert and Pollatos, 2014) and the lack of studies in this field among children, we hypothesized that a healthy and fit physical state is associated with higher IS in children. This should be reflected in a better performance when participating in a physical task as well as in a higher level of daily PA in children. Furthermore, considering the results of the study from Herbert et al. (2007b), we further hypothesized that children with a higher IS will also demonstrate a finer ability to regulate their PA, in comparison to children with low IS.

Materials and Methods

Participants

Participants were children of the third wave of a larger on-going intervention study ("Join the Healthy Boat—Primary School", Schreiber et al., 2014). Our data derived from a subsample of the Baden-Württemberg Study, which evaluated the health promotion program "Join the Healthy Boat—Primary School" in the south-west of Germany. Protocol and study design of the Baden-Württemberg Study have been depicted elsewhere (Dreyhaupt et al., 2012). In total, 1047 children in the age of 9.59 (SD = 0.63) years from third to fourth grade were recruited from primary schools in the federal state of Baden-Württemberg, after legal guardians had provided written informed consent. Approval for the study was obtained from the Ministry of Education as well as from the University's Ethics Committee.

For logistical reasons (scope of measurements of the Baden-Württemberg study and distances between schools) objective PA assessments and IS measurements were only carried out in different schools in the region of Ulm and among fewer children. The sample used in the present analyses consisted of 49 children and the subsample of 21 respectively with complete data on all variables of interest. Children were tested individually with regard to IS in a separate room at school. Additionally, all participants filled out questionnaires regarding their health status, in order to fulfill all criteria for participating in this study. In total, there were 49 healthy participants, 20 girls (40.8%) and 29 boys (59.2%) with a mean age of $M = 9.72$ years (SD = 0.58) in the main sample. In the subsample of the study, there were 21 children in total, 10 boys (47.6%) and 11 girls (52.4%) with a mean age of $M = 9.97$ years (SD = 0.39).

Performance Task for Physical Fitness

For the measurement of children's physical fitness, the 6 min run-performance task was conducted. The Dordel-Koch-Test (Dordel and Koch, 2004; Graf et al., 2004) is conducted in order to measure the developmental state of the participant's basic motoric skills in accordance to intensity, coordination and endurance of the movement. The 6 min run is a subtask of the Dordel-Koch-Test, which assesses a person's aerobic endurance. It could be compared to a screening test, in which cardiopulmonary stamina is being measured. The participants of this task had to run around a volleyball field (round length 54 m) for 6 min. The last round was announced by the instructor of the task, who, afterward, marked the rounds already covered by the children. For the evaluation of this task, the aggregation of the number of completely covered rounds and the distance of the last round were calculated.

The 6-min-run task is a highly standardized assessment tool, which is frequently used not only because it is extremely time-saving, but also relatively simple to conduct and unambiguous. For this reason, it has a high level of objectivity. As far as the norm values are concerned, the reliability and objectivity are $r = 0.91$ for the total sample. Von Haaren et al. (2008) also reported reliability values that swayed between $r = 0.61$ and 0.92. Moreover, they compared the results regarding the VO_{2max} (among others an essential indicator for the aerobic stamina) between the 6-min-run performance and a treadmill test and they noticed a strong connection between those two performance tasks ($r = 0.69$).

Measurement of Daily Physical Activity

Within the context of the Baden-Württemberg study, daily PA was measured in a subsample of 21 primary school children, in order to examine if current PA recommendations of at least 60 min MVPA (moderate to vigorous PA) per day were met, not only during school time, but also during spare time. After the informed consent of the parents, PA was measured via a multi-sensor device (Actiheart®, CamNtech Ltd., Cambridge, UK).

The Actiheart® is a light multi-sensor device that combines recordings of the accelerometer (in counts per minute) and of heart rate (in beats per minute). This multi-sensor device was attached using two standard electrocardiograph electrode pads on the children's chest. The children were instructed to follow their daily routine while wearing the Actiheart® and not to remove the device even during sleep for five consecutive days (á 24 h). Energy expenditure based on metabolic equivalents (METs) was calculated by the Actiheart® software using a branched model approach, previously validated in children (Corder et al., 2007). For this specific recording, a 15-s epochs recording interval was selected. Besides heart rate and accelerometry data, gender, age, body weight, and height were also taken into account in order to assess MET levels. MET was divided into three categories: sedentary (<1.5 METs), light (1.5–3.0 METs), moderate to vigorous (>3.0–6.0 METs) (Pate et al., 1995).

For data assessment, some specific criteria had to be fulfilled in order for the data to be included in the analysis. Only recordings with at least 10 h of daily data collection were included in this sample. Moreover, recordings of 5 days in total were selected, in which at least 1 day was a weekend day and the rest were weekdays, so as to secure a high reliability among children (see also Trost et al., 2000). All recordings between 11 pm and 6 am were removed as this time was considered as normal sleeping time. Additionally, recordings that included more than 45 min of sedentary activity, between 9 pm and 11 pm, were also extracted. Furthermore, recording periods of 100 min or more with zero activity counts were also removed. Hesketh et al. (2014) suggested three possible day-segmentations that might correspond to the daily program of the preschool and school children in the UK. We chose a similar day-segmentation, because of the resemblance to the German primary school system. Therefore, each day was divided in three periods, as follows: morning (6 am–12 pm), afternoon (12–5 pm) and, lastly, evening (5–11 pm).

Heartbeat Perception Task

For the heartbeat perception task, cardiac activity was recorded using the mobile heart frequency monitor RS800CX (Polar Electro Oy, Kempele, Finland). Polar watches are mobile devices that enable easy, non-invasive recording of inter-beat-intervals and were used to assess IS in other studies with children (Koch and Pollatos, 2014a,b). Validity and reliability compared to alternative ECG measurement devices have been proven in children and adults (Radespiel-Tröger et al., 2003; Kingsley et al., 2005; Gamelin et al., 2008; Nunan et al., 2008). Due to the fact that the study took place in the school setting, the strap with the electrodes was attached to both hands and secured on a table. Signals were sampled at 1000 Hz and analyzed by the corresponding Polar ProTrainer five software (version 5.40.172).

The heartbeat perception task was performed following the Mental Tracking Method, proposed by Schandry (1981) and identical to the children's version implemented by Koch and Pollatos (2014a). After a short practice interval of about 10 s, there were three intervals of 15, 20, and 18 s, separated by two standard resting periods of 20 s. During each interval, participants were precisely instructed to silently count their own heartbeats by concentrating on their heart activity while not being allowed to check their pulse or to attempt any other physical manipulations (e.g., holding their breath) that could facilitate the detection of heartbeats. Participants were seated and were given no information as to the length of the intervals or their performance. The test supervisor signaled the beginning and the end of the counting phases by announcing "start" and "stop." Participants were asked to verbally report the number of counted heartbeats straight after the "stop" signal. IS was determined via the heartbeat perception score, which represents the mean score across the three intervals and which is calculated according to the following transformation:

$$1/3\Sigma[1 - (|\text{recorded heartbeats} - \text{counted heartbeats}|/\text{recorded heartbeats})]$$

Higher scores indicate higher IS, so that the maximum score of 1 indicates absolute accuracy of heartbeat perception.

Body Mass Index

Children's body weight and height were taken according to the International Society for the Advancement of Kinanthropy (ISAK; Stewart et al., 2011), with children wearing only underwear and no shoes. More specifically, body weight was measured to the nearest 0.05 kg, using calibrated electronic scales (Seca 862, weighing and measurement systems, Hamburg, Germany). Using a stadiometer, height was measured to the nearest 0.1 cm (Seca 213, weighing and measurement systems, Hamburg, Germany). BMI was calculated (kg/m^2) and converted to BMI percentiles (BMIPCT) based on national reference data for German children (Kromeyer-Hauschild et al., 2001).

Statistical Analyses

Continuous variables were summarized as mean and standard deviation; frequencies were used to analyze nominal and ordinal variables. Continuous variables were compared using the two-sample t-test and analysis of variance (ANOVA). Regression analyses as well as Pearson correlations and partial correlations were conducted to investigate the relationship between IS, age and gender, BMI and the activity measures.

As far as physical performance is concerned, one hierarchical regression analysis (linear mixed effects regression model, forward stepping) was carried out with IS, BMI and the interaction term IS × BMI as predictors and performed distance as criterion.

As far as everyday PA is concerned, we calculated partial correlations between mean activity at different activity levels and while controlling for age and BMI.

All statistical analyses were conducted using Statistical Package for Social Sciences (SPSS, version 21). Because of the explorative nature of this study, no adjustment for multiple testing was made (Altman, 1991; Victor et al., 2010). A p value less than 0.05 was considered significant. The results of all statistical tests are interpreted in an exploratory sense.

Results

Sample Descriptives

Sample characteristics on relevant variables are shown in **Table 1**.

TABLE 1 | Characteristics of the main and subsample.

	$N^1 = 49$	$N^2 = 21$
Female, n (%)	20 (40.8)	11 (52.4)
Age (years)	9.72 (0.56)	9.97 (0.39)
BMI (kg/m^2)	17.33 (2.57)	16.59 (2.26)
BMIPCT	48.04 (27.58)	44.19 (25.89)
Heartbeat perception score	0.59 (0.18)	0.56 (0.16)

All values are mean (SD) unless stated otherwise; N[1], main sample; N[2], sub-sample; BMI, body mass index; BMIPCT, BMI percentiles; SD, standard deviation.

Interoceptive Sensitivity as Assessed by Heartbeat Perception

The mean heartbeat perception score for the main sample was 0.59 (SD = 0.18). Moreover, there was a minimum score of 0.16 and a maximum of 0.91 regarding the heartbeat perception in the overall sample. Taking into account the possible interrelation between gender and heartbeat perception, a two-sample sample t-test was conducted, resulting in boys ($M = 0.62$, SD = 0.19) and girls ($M = 0.55$, SD = 0.17) not differing in their IS [$t(46) = 1.2, p = 0.24$]. The same finding occured after comparing gender and heartbeat perception in the subsample [$t(19) = 1.9, p = 0.07$]. No significant gender difference was found after considering BMIPCT as a covariate. Accordingly, a one way ANOVA was conducted in order to examine heartbeat perception and age, after dividing age in three categories. No differences were found between these variables [$F(2,45) = 0.46, p = 0.63$] in the main sample and in the subsample, respectively [$F(1,19) = 0.07, p = 0.79$].

Physical Fitness

In the physical performance task, boys covered a mean distance of 1003.6 meters (SD = 138.5), whereas girls' mean covered distance was 904.97 meters (SD = 131.77). Moreover, boys and girls showed significantly different physical performances in the 6-min-run [$t(45) = 2.45, p = 0.02, d = 0.73$]. The distance covered in 6 min were classified, according to Jouck (2008) in different grades from 1 to 6, taking into account not only the gender but also the age. Mark one refers to an excellent physical performance and mark six to a poor physical performance. As far as minimum and maximum marks of the participants go, there was a deviation ranging from 2 to 6. On the subject of differentiating the physically fit toward physically non-fit children, a dichotomisation of the median value was implemented, resulting in physically fit children having either 2 or 3 as a mark, in comparison to non-physically fit children, who gained the marks 4, 5, or 6. Finally, 31 participants in this study were classified as physically fit, and 18 as physically non-fit. The data showed no significant difference between the different age groups and physical fitness of the main sample [$F(2,44) = 1.32, p = 0.27$].

Relationship Between IS and Physical Fitness

In order to examine the relationship between IS and physical fitness, we conducted a linear mixed effects regression analysis (forward stepping) with the covered distance of the 6 min run-performance task as criterion, and BMI, IS and the interaction term of BMI × IS as predictors. The criterion was explained by BMI ($T = -2.07, \beta = -0.28, p = 0.04$), IS ($T = 2.02, \beta = 0.29, p = 0.04$) and the interaction between both [$T = 2.14, \beta = 0.32, p = 0.04; F(3,45) = 5.09, p = 0.00, R = 0.50, R^2 = 0.25$]. These effects are depicted in **Figure 1**.

More specifically, a higher IS is associated with a greater covered distance which implies a higher physical fitness. Moreover, a higher physical fitness is positively associated with the interaction between IS and BMI. Therefore, **Figure 1** depicts a positive interaction between high IS, covered distance and BMI, but a strong negative interrelation between low IS, covered distance and BMI. In other words, **Figure 1** reveals a positive correlation between physical fitness and BMI among good heartbeat perceivers and a negative correlation between physical fitness and BMI among bad heartbeat perceivers.

Daily Physical Activity

In total, 21 children, forming the subsample of the study, had valid PA data for five consecutive days with a mean daily wear time of 846 min (corresponding to about 14 h, $M = 846.2$ min, SD = 207.1; see **Table 2**).

Boys ($M = 237.5$, SD = 61.02) and girls ($M = 178.2$, SD = 63.10) differ as far as their light PA in the evening is concerned [$t(19) = 2.18$, $p = 0.04$, $d = 0.95$]. Moreover, the data showed a difference regarding sedentary activity [$t(19) = 1.27$, $p = 0.02$, $d = 1.1$], with girls ($M = 115.04$, SD = 88.15) being more sedentary in the evening in comparison to boys ($M = 35.8$, SD = 50.7). Concerning MVPA there was also a gender-specific difference [$t(19) = 3.9$, $p = 0.01$, $d = 1.7$]. MVPA in the morning was higher for boys ($M = 44.6$, SD = 30.2) in comparison to girls ($M = 5.5$, SD = 12.6). Concerning age, there was no significant difference between the three age groups and the different levels of PA.

Relationship Between IS and Daily Physical Activity

Examining the relationship between IS and daily PA, we conducted partial correlations after setting BMI and children's age as control variables. There was a statistically significant correlation between IS and light PA, more specifically in the morning ($r = 0.39$, $p = 0.04$) and in the afternoon ($r = 0.39$, $p = 0.04$), showing that higher IS was positively related to more light PA in the morning and afternoon hours. This interrelation is depicted in **Figure 2**.

Discussion

The present study investigated the interrelation between IS and PA among primary school children. Our results demonstrate, firstly, that IS is a determinable measure in children, indicating that primary school children differ considerably in their ability to perceive ongoing signals stemming from the heart. This is in line with recent data from a large representative sample of about 1350 children (Koch and Pollatos, 2014a). Secondly, further findings of this study highlight that physical fitness and IS are positively associated, showing that higher IS is related to a greater distance covered in the 6 min running performance task. These findings are

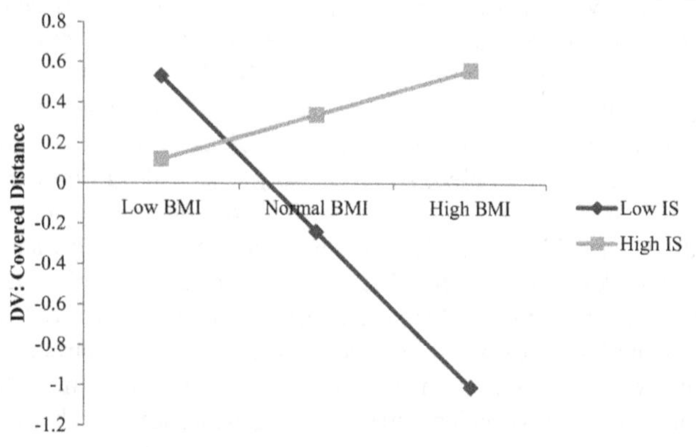

FIGURE 1 | Interaction between IS (low and high), covered distance and body mass index (BMI); Low BMI, normal BMI, high BMI: based on the national reference data for German children (Kromeyer-Hauschild et al., 2001); DV: dependent variable; classification of low and high IS according to the method of median split.

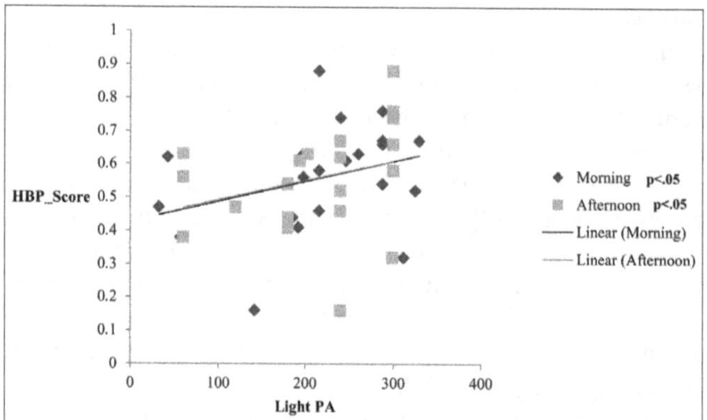

FIGURE 2 | Partial correlations between interoceptive Sensitivity (IS) and physical activity (PA) in the morning and in the afternoon, after controlling for body mass index (BMI) and age; HBP_Score: heartbeat perception score.

TABLE 2 | Children's average daily physical activity levels in minutes segmented across the day.

	Daily total (min/day) (SD)	Morning (min/day) (SD)	Afternoon (min/day) (SD)	Evening (min/day) (SD)
Recorded time	846.2 (207.1)	299.5 (75.8)	250.7 (82.8)	296.0 (85.0)
Sedentary PA	156.1 (70.4)	58.2 (77.6)	20.6 (29.9)	77.3 (81.8)
Active PA	690.1 (203.9)	241.3 (117.2)	230.1 (116.6)	218.7 (110.2)
Light PA	636.6 (187.9)	217.2 (87.9)	213 (81.2)	206.4 (67.7)
MVPA	53.5 (70.6)	24.1 (29.9)	17.1 (33.6)	12.3 (22.4)

All values depicted are in minutes/day; in parentheses: SD, standard deviation; PA, physical activity; Active PA, sum of light PA and MVPA; MVPA, moderate to vigorous physical activity; morning: 6 am–12 pm; afternoon: 12–5 pm; evening: 5–11 pm.

not in line with previous studies regarding IS and physical fitness among adults, for example the study from Herbert et al. (2007b), where good and poor heartbeat perceivers did not differentiate in their fitness level.

Furthermore, results concerning physical fitness and IS revealed also the role of the BMI in this motif. More specifically, these findings indicate that children with a normal BMI and a high physical fitness (according to the 6-min-run) showed a higher IS than those children with a normal BMI but a lower fitness state. Lastly, children with a high BMI and high physical fitness indicated also, surprisingly, a higher IS than those children with a high BMI and a lower physical fitness. Thus, good heartbeat perceivers were these children with a normal to high BMI and with a greater fitness state. These findings could be explained by the fact that BMI could play a role in the degree of IS but the most important factor is one's physical state. In other words, IS does not seem to decline when a child has a high BMI but also at the same time rather good physical fitness. This is in accordance with a few older studies in adults reporting that a higher state of fitness is advantageous for better IS (Borg and Linderholm, 1967; Montgomery et al., 1984). In contrast, children with a lower physical fitness but with higher BMI seemed, in this study, to be bad heartbeat perceivers. In respect to the possible limitations regarding this finding, the use of BMI instead of BMIPCT when referring to children and the absence of use of alternative anthropometric measurements, such as skin-fold thickness or body girth etc. (Chen et al., 2002), should not be neglected.

Moreover, our daily PA related outcomes reveal a positive association between IS and light daily PA in the morning as well as in the afternoon. No significant correlations were observed with the moderate to vigorous activity level, though. This result could be partly due to the fact that our sample was quite small and only 6 (28.6%) participants of the total sample met current PA recommendations of at least 60 min MVPA per day. Referring to former research, there are substantial methodological differences whether daily activity was assessed by self-report or by the use of objective measurements like the Actiheart device in the present study, and which cut-off points were used when determining MVPA (see Borracino et al., 2009; Aznar et al., 2011; Ekelund et al., 2011; Kettner et al., 2013). Being sensitive to one's bodily signals might constitute a positive precondition for effective self-regulation of behavior, as it was suggested for the field of emotion regulation (see Füstös et al., 2013). Taking both obtained results together, we can show that both in a performance task as well as in day to day life, IS interacts positively with PA suggesting that the ability to accurately perceive bodily signals is crucially associated with more fitness and daily activity in young children. Therefore, a link between IS and other health-related outcome variables is to be assumed, such as demonstrated in adults (Herbert and Pollatos, 2014) or in children (Koch and Pollatos, 2014a,b).

The fact that we observed positive correlations between IS, physical fitness and light daily PA generates further questions regarding the regulation of PA. A former study by Herbert et al. (2007b) demonstrated that when participants were instructed to cycle on a bicycle ergometer at a speed they felt comfortable with, good heartbeat perceivers covered a shorter distance as compared to poor heartbeat perceivers. The good heartbeat perceivers indicated a more self-controled physical workload, by perceiving better their internal bodily signals and regulating their fatigue. In contrast to the instruction given by Herbert and co-workers, the 6-min performance task used in this study focused on how fit children were with the clear instruction to run as fast and as far as possible in a certain time. The instruction when undertaking everyday PA was to keep on undertaking normal activity as usual, which implied that the participant was not in a situation where his/her performance was being evaluated. Therefore, we hypothesize that higher PA might favor the development of a better ability to identify internal body signals as assessed by IS. The exact developmental mechanisms of interoception remain yet unclear, due to the lack of prospective studies and in general studies concerning the distribution of cardiac sensitivity in children (Koch and Pollatos, 2014a). Future studies should focus on this research gap and assess PA and IS in a longitudinal fashion.

In accordance to former research, we found some gender differences in daily activity between boys and girls (Riddoch et al., 2004; Jago et al., 2005; Borracino et al., 2009). Moreover, in examining physical fitness, our results suggest that boys are more physically fit than girls. This can be explained by the physiology of the male body and more specifically, by the fact that males have a greater muscle mass than females, which implies a greater ability of achieving high levels of PA. This finding is in the same line with the study of Li et al. (2007), who concluded that males covered a greater 6 min walk distance than females in the performance task. Moving on to age and PA, our results reveal that 9–11-year-old participants undertook the same amount of PA. These results are in line with other studies, which suggest that children under the age of 13 indicate the same levels of PA (Strauss et al., 2001). We did not find any difference in IS according to gender, like other studies with adults (Katkin et al., 1981; Jones et al., 1984; Jones, 1995), while Koch and Pollatos (2014a,b) showed a small but significant difference with higher IS in boys. We assume that the sample size is small to show such an effect.

To sum up, our study demonstrated that IS is not only determinable but also diverse among primary school children, emphasizing the fact that IS is based on individual differences and age. Further studies could shed light on the developmentary processes of IS through the life span. Taking into account the lack of studies in this field, in the present study we tried to scrutinize the role of PA and IS in children and our findings demonstrate the first evidence regarding the interaction between physical fitness and IS in young children. Further research is necessary to examine the specific role of BMI as well as the direction of the observed interaction, e.g., by training physical fitness in children and assessing concomitant interoceptive processes, and by using alternative anthropometric measurements apart from BMI (such as skin-fold thickness etc.). In specific, we demonstrated that physical fitness could contribute to a higher IS and in its turn, IS might be trained using PA (Schandry and Weitkunat, 1990). We assume that improving children's perception of their body signals could contribute to a more effective way of regulating health-related behavior, e.g., by a finer ability to tune their physical load in everyday situations and to prevent exhaustion more effectively. Our results suggest that IS constitutes an important factor associated with PA during childhood. This issue deserves further exploration, by researching a larger sample of children, thanks to the great significance of IS and PA over the life span.

Acknowledgments

The school-based health promotion programme "Join the Healthy Boat" and its evaluation study were financed by the Baden-Württemberg Foundation, which had no influence on the content of this paper. Further, the authors would like to thank all members of the programme "Join the Healthy Boat" for their input, as well as: the Institute of Epidemiology and Medical Biometry, Ulm University, the Institute of Psychology and Pedagogy, Ulm University, all assistants who were involved in the performance of measurements, and especially all teachers and families for their participation.

References

Ainley, V., Tajadura-Jiménez, A., Fotopoulou, A., and Tsakiris, M. (2012). Looking into myself: changes in interoceptive sensitivity during mirror self-observation. *Psychophysiology* 49, 1504–1508. doi: 10.1111/j.1469-8986.2012.01468.x

Altman, D. G. (1991). *Practical Statistics for Medical Research*. London: Chapman & Hall/CRC Texts in Statistical Science.

Aznar, S., Naylor, P. J., Silva, P., Pérez, M., Angulo, T., Laguna, M., et al. (2011). Patterns of physical activity in Spanish children: a descriptive pilot study. *Child Health Care Dev.* 37, 322–328. doi: 10.1111/j.1365-2214.2010.01175.x

Barrett, L. F., Quigley, K. S., Bliss-Moreau, E., and Aronson, K. R. (2004). Interoceptive sensitivity and self-reports of emotional experience. *J. Pers. Soc. Psychol.* 87, 684–697. doi: 10.1037/0022-3514.87.5.684

Borg, G., and Linderholm, H. (1967). Perceived exertion and pulse rate during graded exercise in various age groups. *Acta Med. Scand.* 181, 194–206. doi: 10.1111/j.0954-6820.1967.tb12626.x

Borracino, A., Lemma, P., Ianotti, R., Zambon, A., Dalmasso, P., Lazzeri, G., et al. (2009). Socio-economic effects on meeting PA guidelines: comparisons among 32 countries. *Med. Sci. Sports Exerc.* 41, 749–756. doi: 10.1249/MSS.0b013e3181917722

Cameron, O. G. (2001). Interoception: the inside story—a model for psychosomatic processes. *Psychosomatic Med.* 63, 697–710. doi: 10.1097/00006842-200109000-00001

Chen, W., Lin., C. C., Peng, C. T., Li, C. I., Wu, H. C., Chiang, J., et al. (2002). Approaching health body mass index norms for children and adolescents from health-related physical fitness. *Obes. Rev.* 3, 225–232. doi: 10.1046/j.1467-789X.2002.00064.x

Corder, K., Brage, S., Mattocks, C., Ness, A., Riddoch, C., Wareham, N. J., et al. (2007). Comparison of two methods to assess PAEE during six activities in children. *Med. Sci. Sports Exerc.* 39, 2180–2188. doi: 10.1249/mss.0b013e318150dff8

Damasio, A. R. (1994). *Descartes's Error: Emotion, Reason and the Human Brain*. New York, NY: Grosset/Putnam.

Damasio, A. R. (1999). *The Feeling of What Happens: Body and Emotion in the Making of Consciousness*. New York, NY: Harcourt Brace.

Domschke, K., Stevens, S., Pfleiderer, B., and Gerlach, A. L. (2010). Interoceptive sensitivity in anxiety and anxiety disorders: an overview and integration of neurobiological findings. *Clin. Psychol. Rev.* 30, 1–11. doi: 10.1016/j.cpr.2009.08.008

Dordel, S., and Koch, B. (2004). *Dordel-Koch-Test: A Motor Test for the Assessment of Motor Performance in Childhood and Youth. Fitness Test According to Dordel and Koch*. Köln: Manual.

Dreyhaupt, J., Koch, B., Wirt, T., Schreiber, A., Brandstetter, S., Kesztyüs, D., et al. (2012). Evaluation of a health promotion program in children: study protocol and design of the cluster-randomized Baden Württemberg primary school study [DRKS-ID: DRKS00000494]. *BMC Public Health* 12:157. doi: 10.1186/1471-2458-12-157

Dunn, B. D., Dalgleish, T., Ogilvie, A. D., and Lawrence, A. D. (2007). Heartbeat perception in depression. *Behav. Res. Ther.* 45, 1921–1930. doi: 10.1016/j.brat.2006.09.008

Dunn, B. D., Evans, D., Makarova, D., White, J., and Clark, L. (2012). Gut feelings and the reaction to perceived inequity: the interplay between bodily responses, regulation, and perception shapes the rejection of unfair offers on the ultimatum game. *Cogn. Affect. Behav. Neurosci.* 12, 419–429. doi: 10.3758/s13415-012-0092-z

Dunn, B. D., Galton, H. C., Morgan, R., Evans, D., Oliver, C., Meyer, M., et al. (2010). Listening to your heart. How interoception shapes emotion experience and intuitive decision making. *Psychol. Sci.* 21, 1835–1844. doi: 10.1177/0956797610389191

Ehlers, A., and Breuer, B. (1992). Increased cardiac awareness in panic disorder. *J. Abnorm. Psychol.* 101, 371–382. doi: 10.1037/0021-843X.101.3.371

Ekelund, U., Tomkinson, G., and Armstrong, N. (2011). What proportion of youth are physically active? Measurement issues, levels and recent time trends. *Br. J. Sport Med.* 48, 859–865. doi: 10.1136/bjsports-2011-090190

Eley, T. C., Gregory, A. M., Clark, D. M., and Ehlers, A. (2007). Feeling anxious: a twin study of panic/somatic ratings, anxiety sensitivity and heartbeat perception in children. *J. Child Psychol. Psychiatry* 48, 1184–1191. doi: 10.1111/j.1469-7610.2007.01838.x

Eley, T. C., Stirling, L., Ehlers, A., Gregory, A. M., and Clark, D. M. (2004). Heartbeat perception, panic/somatic symptoms and anxiety sensitivity in children. *Behav. Res. Ther.* 42, 439–448. doi: 10.1016/S0005-7967(03)00152-9

Füstös, J., Gramann, K., Herbert, B. M., and Pollatos, O. (2013). On the embodiment of emotion regulation: interoceptive awareness facilitates reappraisal. *Soc. Cogn. Affect. Neurosci.* 8, 911–917. doi: 10.1093/scan/nss089

Gamelin, F. X., Baquet, G., Berthoin, S., and Bosquet, L. (2008). Validity of the Polar S810 to Measure R-R Intervals in Children. *Int. J. Sports Med.* 29, 134–138. doi: 10.1055/s-2007-964995

Graf, C., Koch, B., Dordel, S., Schindler-Marlow, S., Icks, A., Schüller, A., et al. (2004). Physical activity, leisure habits and obesity in first-grade children. *Eur. J. Cardiovasc. Prev. Rehabil.* 11, 284–290. doi: 10.1097/01.hjr.0000129740.30593.18

Harver, A., Katkin, E. S., and Bloch, E. (1993). Signal-detection outcomes on heartbeat and respiratory resistance detection tasks in male and female subjects. *Psychophysiology* 30, 223–230. doi: 10.1111/j.1469-8986.1993.tb03347.x

Herbert, B. M., Muth, E. R., Pollatos, O., and Herbert, C. (2012). Interoception across Modalities: on the relationship between cardiac awareness and the sensitivity for gastric functions. *PLoS ONE* 7:e36646. doi: 10.1371/journal.pone.0036646

Herbert, B. M., and Pollatos, O. (2008). Interozeptive sensitivität, gefühle und verhaltensregulation. *Z. Neuropsychol.* 19, 125–137. doi: 10.1024/1016-264X.19.3.125

Herbert, B. M., and Pollatos, O. (2014). Attenuated interoceptive sensitivity in overweight and obese individuals. *Eating Behav.* 15, 445–448. doi: 10.1016/j.eatbeh.2014.06.002

Herbert, B. M., Pollatos, O., and Schandry, O. (2007a). Interoceptive sensitivity and emotion processing: an EEG study. *Int. J. Psychophysiol.* 65, 214–227. doi: 10.1016/j.ijpsycho.2007.04.007

Herbert, B. M., Ulbrich, P., and Schandry, R. (2007b). Interoceptive sensitivity and physical effort: implications for the self-control of physical load in everyday life. *Psychophysiology* 44, 194–202. doi: 10.1111/j.1469-8986.2007.00493.x

Hesketh, K. R., McMinn, A. M., Ekelund, U., Sharp, S. J., Collings, P. J., Harvey, N. C., et al. (2014). Objectively measured physical activity in four-year-old British children: a cross sectional analysis of activity patterns segmented across the day. *Int. J. Behav. Nutr. Phys. Act.* 11:1. doi: 10.1186/1479-5868-11-1

Jago, R., Anderson, C., Baranowski, T., and Watson, K. (2005). Adolescent patterns of physical activity: differencies by gender, day and time of day. *Am. J. Prev. Med.* 28, 447–452. doi: 10.1016/j.amepre.2005.02.007

James, W. (1884). What is an emotion? *Mind* 9, 188–205. doi: 10.1093/mind/os-IX.34.188

Jones, G. E. (1995). "Constitutional and physiological factors in heartbeat perception," in *From the Heart to the Brain. The Psychophysiology of Circulation-Brain Interaction*, eds D. Vaitl and R. Schandry (Frankfurt: Peter Lang-Verlag). 173–192.

Jones, G. E., O'Leary, R. T., and Pipkin, B. L. (1984). A comparison of Brener-Jones and Whitehead procedures for assessing cardiac awareness. *Psychophysiology* 21, 143–148. doi: 10.1111/j.1469-8986.1984.tb00196.x

Jouck, S. (2008). *Dordel-Koch-Test (DKT)-Ein Test zur Erfassung der Motorischen Leistungsfähigkeit im Kindes- und Jugendalter.* Köln: Dissertation, Deutsche Sportschule.

Katkin, E. S., Blascovich, J., and Goldband, S. (1981). Empirical assessment of visceral self-perception: individual and sex differences in the acquisition of heartbeat discrimination. *J. Pers. Soc. Psychol.* 40, 1095–1101. doi: 10.1037/0022-3514.40.6.1095

Kettner, S., Kobel., S., Fischbach, N., Drenowatz, C., Dreyhaupt, J., Wirt, T., et al. (2013). Objectively determined physical activity levels of primary school children in south-west Germany. *BMC Public Health* 13:895. doi: 10.1186/1471-2458-13-895

Kingsley, M., Lewis, M. J., and Marson, R. E. (2005). Comparison of Polar 810s and an ambulatory ECG system for RR interval measurement during progressive exercise. *Int. J. Sports Med.* 26, 39–44. doi: 10.1055/s-2004-817878

Koch, A., and Pollatos, O. (2014a). Cardiac sensitivity in children: sex differences and its relationship to parameters of emotional processing. *Psychophysiology* 51, 932–941. doi: 10.1111/psyp.12233

Koch, A., and Pollatos, O. (2014b). Interoceptive sensitivity, body weight and eating behavior in children: a prospective study. *Front. Psychol.* 5:1003. doi: 10.3389/fpsyg.2014.01003

Kromeyer-Hauschild, K., Wabitsch, M., Kunze, D., Geller, F., Geiß, H. C., Hesse, V., et al. (2001). Percentiles of body mass index in children and adolescents evaluated from different regional German studies. *Monatsschr. Kinderheilkd.* 149, 807–818. doi: 10.1007/s001120170107

Li, M., Yin, J., Au, J. T., So, H. K., Tsang, T., Wong, E., et al. (2007). Standard reference for the six-minute-walk test in healthy children aged 7 to 16 years. *J. Respir. Crit. Care Med.* 176, 174–180. doi: 10.1164/rccm.200607-883OC

Montgomery, W. A., and Jones, G. E. (1984). Laterality, emotionality and heartbeat perception. *Psychophysiology* 21, 459–465. doi: 10.1111/j.1469-8986.1984.tb00227.x

Montgomery, W. A., Jones, G. E., and Hollandsworth, J. G. (1984). The effect of physical fitness and exercise on cardiac awareness. *Biol. Psychol.* 18, 11–22. doi: 10.1016/0301-0511(84)90022-X

Mussgay, L., Klinkenberg, N., and Rüddel, H. (1999). Heart beat perception in patients with depressive, somatoform, and personality disorders. *J. Psychophysiol.* 13, 27–36. doi: 10.1027//0269-8803.13.1.27

Nunan, D., Jakovljevic, D. G., Donovan, G., Hodges, L. D., Sandercock, G. R. H., and Brodie, D. A. (2008). Levels of agreement for RR intervals and short-term heart rate variability obtained from the Polar S810 and an alternative system. *Eur. J. Appl. Physiol.* 103, 529–537. doi: 10.1007/s00421-008-0742-6

Pate, R. R., Pratt, M., Blair, S. N., Haskell, W. L., Macera, C. A., Bouchard, C., et al. (1995). Physical activity and public health. A recommendation from the Centers for Disease Control and Prevention and the American College of Sports Medicine. *J. Am. Med. Assoc.* 273, 402–407. doi: 10.1001/jama.1995.03520290054029

Pennebaker, J. W., and Lightner, J. M. (1980). Competition of internal and external information in an exercise setting. *J. Pers. Soc. Psychol.* 39, 165–174. doi: 10.1037/0022-3514.39.1.165

Pollatos, O., Kirsch, W., and Schandry, R. (2005). On the relationship between interoceptive awareness, emotional experience and brain processes. *Cogn. Brain Res.* 25, 948–962. doi: 10.1016/j.cogbrainres.2005.09.019

Pollatos, O., Traut-Mattausch, E., and Schandry, R. (2009). Differential effects of anxiety and depression on interoceptive accuracy. *Depress. Anxiety* 26, 167–173. doi: 10.1002/da.20504

Radespiel-Tröger, M., Rauh, R., Mahlke, C., Gottschalk, T., and Mück-Weymann, M. (2003). Agreement of two different methods for measurement of heart rate variability. *Clin. Auton. Res.* 13, 99–102. doi: 10.1007/s10286-003-0085-7

Riddoch, C. J., Andersen, L. B., Wedderkopp, N., Harro, M., Klasson-Heggebo, L., Sardinha, L. B., et al. (2004). Physical activity levels and patterns of 9-and 15-yr-old european children. *Med. Sci. Sports Exerc.* 36, 86–92. doi: 10.1249/01.MSS.0000106174.43932.92

Schachter, S., and Singer, J. E. (1962). Cognitive, social and physiological determinants of emotional state. *Psychol. Rev.* 69, 379–399. doi: 10.1037/h0046234

Schandry, R. (1981). Heart beat perception and emotional experience. *Psychophysiology* 18, 483–488. doi: 10.1111/j.1469-8986.1981.tb02486.x

Schandry, R., and Weitkunat, R. (1990). Enhancement of heartbeat-related brain potentials through cardiac awareness training. *Int. J. Neurosci.* 53, 243–253. doi: 10.3109/00207459008986611

Schmitz, J., Blechert, J., Krämer, M., Asbrand, J., and Tuschen-Caffier, B. (2012). Biased perception and interpretation of bodily anxiety symptoms in childhood social anxiety. *J. Clin. Child Adolesc. Psychol.* 41, 92–102. doi: 10.1080/15374416.2012.632349

Schreiber, A., Kesztyus, D., Wirt, T., Erkelenz, N., Kobel, S., Steinacker, J., et al. (2014). Why do mothers encourage their children to control their weight? A cross-sectional study of possible contributing factors. *BMC Public Health* 14:450. doi: 10.1186/1471-2458-14-450

Stewart, A., Marfell-Jones, M., Olds, T., and Mike Marfell-Jon. (2011). *International Standards for Anthropometric Assessment.* Potchefstroom: International Society for the Advancement of Kinanthropometry.

Strauss, R. S., Rodzilsky, D., Burack, G., and Colin, M. (2001). Psychosocial correlates of physical activity in healthy children. *Arch. Pediatr. Adolesc. Med.* 155, 897–902. doi: 10.1001/archpedi.155.8.897

Trost, S. G., Pate, R. R., Freedson, P. S., Sallis, J. F., and Taylor, W. C. (2000). Using objective physical activity measures with youth: how many days of monitoring are needed? *Med. Sci. Sports Exerc.* 32, 426–431. doi: 10.1097/00005768-200002000-00025

Vaitl, D. (1996). Interoception. *Biol. Psychol.* 42, 1–27. doi: 10.1016/0301-0511(95)05144-9

Victor, A., Elsäßer, A., Hommel, G., and Blettner, M. (2010). Judging a plethora of p-values how to contend with the problem of multiple testing—part 10 of a series on evaluation of scientific publications. *Dtsch. Arztebl. Int.* 107, 50–56.

Von Haaren, B., Härtel, S., Seidel, I., Schlenker, L., and Bös, K. (2008). Die validität des 6-Minuten-Laufs und 20m shutthle runs bei 9- bis 11-jährigen Kindern. *Dtsch. Z. Sportmed.* 62, 351–355.

Werner, N. S., Jung, K., Duschek, S., and Schandry, R. (2009). Enhanced cardiac perception is associated with benefits in decision-making. *Psychophysiology* 46, 1123–1129. doi: 10.1111/j.1469-8986.2009.00855.x

Whitehead, W. E., and Drescher, V. M. (1980). Perception of gastric contractions and self-control of gastric motility. *Psychophysiology* 17, 552–558. doi: 10.1111/j.1469-8986.1980.tb02296.x

Wiens, S. (2005). Interoception in emotional experience. *Curr. Opin. Neurol.* 18, 442–447. doi: 10.1097/01.wco.0000168079.92106.99

Wiens, S., Mezzacappa, E. S., and Katkin, E. S. (2000). Heartbeat detection and the experience of emotion. *Cogn. Emot.* 14, 417–427. doi: 10.1080/026999300378905

15

Interoception and Stress

André Schulz* and Claus Vögele

Institute for Health and Behaviour, Integrative Research Unit on Social and Individual Development, University of Luxembourg, Walferdange, Luxembourg

***Correspondence:**
André Schulz,
Institute for Health and Behaviour,
Integrative Research Unit on Social
and Individual Development,
University of Luxembourg,
11, Porte des Sciences, L-4366
Esch-sur-Alzette, Luxembourg
andre.schulz@uni.lu

Afferent neural signals are continuously transmitted from visceral organs to the brain. Interoception refers to the processing of visceral-afferent neural signals by the central nervous system, which can finally result in the conscious perception of bodily processes. Interoception can, therefore, be described as a prominent example of information processing on the ascending branch of the brain–body axis. Stress responses involve a complex neuro-behavioral cascade, which is elicited when the organism is confronted with a potentially harmful stimulus. As this stress cascade comprises a range of neural and endocrine pathways, stress can be conceptualized as a communication process on the descending branch of the brain–body axis. Interoception and stress are, therefore, associated via the bi-directional transmission of information on the brain–body axis. It could be argued that excessive and/or enduring activation (e.g., by acute or chronic stress) of neural circuits, which are responsible for successful communication on the brain–body axis, induces malfunction and dysregulation of these information processes. As a consequence, interoceptive signal processing may be altered, resulting in physical symptoms contributing to the development and/or maintenance of body-related mental disorders, which are associated with stress. In the current paper, we summarize findings on psychobiological processes underlying acute and chronic stress and their interaction with interoception. While focusing on the role of the physiological stress axes (hypothalamic-pituitary-adrenocortical axis and autonomic nervous system), psychological factors in acute and chronic stress are also discussed. We propose a positive feedback model involving stress (in particular early life or chronic stress, as well as major adverse events), the dysregulation of physiological stress axes, altered perception of bodily sensations, and the generation of physical symptoms, which may in turn facilitate stress.

Keywords: chronic stress, HPA axis, interoception, SAM axis, somatization, stress disorders, sympathetic nervous system, symptom perception

Introduction

Interoception, i.e., the perception of bodily processes, plays an important role for symptom generation in body-related mental disorders, such as panic disorder (PD; Ehlers and Breuer, 1996), somatoform disorders (SDs; Bogaerts et al., 2010; Pollatos et al., 2011b), dissociative disorders (Michal et al., 2014; Sedeno et al., 2014; Schulz et al., 2015b), or eating disorders (Pollatos et al., 2008; Herbert and Pollatos, 2014). Stress responses involve a complex neuro-behavioral cascade, which is elicited when the organism is confronted with a potentially harmful stimulus, and includes cognitive (e.g., facilitation of selective attention; Chajut and Algom, 2003), affective (e.g., anxiety or fear; McEwen, 2000), and physiological changes [e.g., activation of the autonomic nervous system (ANS) and the hypothalamic-pituitary-adrenocortical (HPA) axis; Chrousos and Gold, 1992]. Exposure to an acute stressor of limited duration does not normally affect an organism's health. Nevertheless, the experience of early life stress, chronic stress or major adverse events may represent one of the most important risk factors for mental disorders, cardiovascular diseases, and auto-immune disorders. Previously, it has been proposed that stress alters the central representation of bodily processes (Craig, 2002). Although interoception may be affected differentially across disorders, it is likely that the role of stress in the etiology of these disorders is mediated by interoception. As the knowledge on the interaction of interoception, stress and mental disorders associated with physical symptoms is only fragmentary (e.g., limited to a single interoceptive indicator, to one stress test or one physiological stress axis, or to one mental disorder), the objective of the current review is to summarize and synthesize existing knowledge from both 'normal' and 'dysregulated' interoception in mental disorders. This review of the existing literature may help to identify yet unanswered questions in this field and our synthesized framework model as developed in this review may stipulate further research in the area.

Structure of this Review

In the current paper we describe the psychobiology of interoception and stress, as well as a framework to demonstrate why both concepts are interconnected and of immediate relevance for health and disease (see The Ascending and the Descending Branch of the Brain-Body Axis). First, we briefly describe both physiological stress axes with specific attention on processes, which may be relevant for the interaction of interoception and stress (see Physiological Stress Axes). We then review the literature on the relationship between interoception and stress, as well as on the role of interoception in mental disorders, which are associated with both dysregulations of physiological stress axes and physical symptoms (see Method and Results). Before we present the results of our literature search, we summarize methodological approaches to assess facets of interoception, which have been used in the relevant literature (see Methodology to Assess Interoception). The summary of findings on interoception and stress (see Synthesis of Findings on Interoception and Stress) is separated into effects of acute and chronic stress on interoception, and into effects of both physiological stress axes on interoception. The synthesis of findings results in a postulation of a framework model incorporating a positive feedback loop that describes the relationship of (chronic) stress, dysregulation of stress axes, altered perception of bodily sensations and physical symptom. Further, we integrate findings on interoception and stress into existing knowledge on the neurobiology of interoception and speculate on the possibly underlying neuro-endocrine signal circuitry (see Neuroendocrine Pathways). Finally, we identify a number of psychological mechanisms (e.g., attention, learning, intuitive decision making), which are known to be affected by interoception and stress and which may play an additional role in the proposed framework model (Psychobiological Mechanisms Involved in Interoception and Stress).

Physiological Stress Axes

Physiological systems to provide resources to cope with the confrontation with a potentially harmful stimulus (e.g., elicitation of a fight-or-flight response) can be sub-divided into two partially independent stress axes: (1) The sympatho-adreno-medullary (SAM) axis, including components of the ANS and (2) the HPA axis (for a detailed review: see Chrousos and Gold, 1992; Chrousos, 2009).

The SAM axis represents a heterogeneous network of neural and endocrine functions, which are interconnected to activate sympathetic processes. The release of corticotropin releasing factor (CRF) as neurotransmitter in the locus coeruleus (LC) leads to the activation of medullary centers, which control the sympathetic nervous system. When activated, sympathetic processes may stimulate two mechanisms with different pathways: (a) a neural pathway via ganglia, which innervate effector organs over mainly noradrenergic synapses and (b) an endocrine pathway that elicits the release of catecholamines (e.g., epinephrine and norepinephrine) into the circulation by the adrenal glands. Circulating catecholamines stimulate effector organs via specific adrenergic receptors (e.g., β1-adrenergic receptors at the myocardium). The LC is part of the central noradrenergic system and connected with structures in the limbic system (e.g., amygdalae) and the frontal cortex. Central α2-adrenergic receptors in the noradrenergic system, mainly located in the LC and the nucleus tractus solitarius (NTS), may down-regulate sympathetic activation and thus represent a negative feedback mechanism in the SAM axis (Isaac, 1980).

The HPA axis involves three consecutive stages: when encountering a stressor, (a) CRF is released from the hypothalamus into the blood circuit, which elicits (b) a release of adrenocorticotropic hormone (ACTH) from the pituitary. Circulating ACTH is registered by (c) the adrenal cortex that releases cortisol in humans, or corticosterone in rodents. Cortisol (or corticosterone) may inhibit the production of both CRF and ACTH and thus constitute a negative feedback loop. Receptors for glucocorticoids, such as cortisol, are distributed over all cells in the entire body. Nuclear mineralocorticoid and glucocorticoid receptors may slowly induce changes in gene transcription over a time frame of hours (De Kloet et al., 1998). Furthermore, glucocorticoids may also elicit rapid, non-genomic effects on cells (mediated via membrane receptors) within several minutes (de Kloet et al., 2008; Groeneweg et al., 2011). As every cell is potentially affected by circulating cortisol, it is a challenge for stress research to understand signaling pathways. Cortisol, for example, has been shown to affect metabolism, immune function and various CNS processes, such as sleep and activity, learning and memory, and attention. Nevertheless, CRF and ACTH may also affect CNS processes, as is evident from their role in anxiety and depression (Arborelius et al., 1999; Strohle et al., 2000). Given the interdependence between CRF, ACTH and cortisol via negative feedback mechanisms, all *in vivo* relationships between cortisol and psychological processes cannot be solely attributed to cortisol. However, since the majority of findings on HPA axis activity focuses on cortisol effects, the current review will focus on the latter.

It should be noted that the two axes are not fully independent from each other. For instance, circulating CRF may inhibit central noradrenergic processes, while noradrenergic activity inhibits CRF production in the hypothalamus (Chrousos and Gold, 1992). There is evidence for between-individual specificity in stress response patterns (Fahrenberg and Foerster, 1982), which is assumed to play a role for individuals' vulnerability for certain disorders (Cohen and Hamrick, 2003). This partial independence of the two stress axes may also imply that a response in one stress axis may not necessarily be associated with the other, depending on individual and environmental factors.

The Ascending and the Descending Branch of the Brain–Body Axis

The strong relationship between stress and interoception is illustrated by the fact that both processes reflect the communication between the CNS (i.e., the brain) and the periphery (i.e., the body). On the one hand, interoception describes the processing and perception of internal bodily states, which are transmitted from peripheral organs, presumably via afferent nerve fibers, to the brain. Interoception can, therefore, be seen as an example for ascending information on the brain–body axis. On the other hand, stress, can be seen as a prominent example of communication on the brain–body axis in descending direction (Schulz, 2015), as physiological stress responses affect the activity of peripheral organs, and metabolic and immune functions via neural and endocrine pathways.

It is important to note that ascending and descending information between the brain and the body is continuously exchanged, and not only during interoception and stress. Homeostatic regulation of peripheral physiological processes via neural and endocrine feedback loops typically work without the involvement of higher CNS functions, unless they are working beyond a 'normal' range of functioning. If they exceed a certain threshold, they may be pushed into awareness of the organism. A physiological stress response elicits the activation of the SAM and HPA axes, which is far above the 'normal' range of homeostatic regulation, a process that has been named 'allostasis' (McEwen, 2004). Stress can, therefore, be conceptualized as descending information on the brain–body axis of increased amplitude as compared to a state of homeostasis, which may shift into focus of awareness. Interoception implicates that awareness is focused on bodily functions, even if they work in a normal, homeostatic range.

In summary, the relationship between interoception and stress involves a bi-directional communication on the brain–body axes. The aims of this review are to elucidate (1) the bi-directional communication between the brain and the body and (2) its role in the etiology of mental disorders. Using a systematic literature review we investigated the relationship between descending (stress) and ascending (interoception) transmission on the brain–body axis. We further address the question whether the dysregulation of bi-directional communication on the brain–body axis facilitates the generation of physical sensations. Furthermore, we defined psycholological processes (anxiety, attention, learning, decision making) that have previously been shown to be associated with stress.

Method

A systematic literature review was performed on the relationship between interoception and stress, and between interoception and mental disorders, which are associated with stress and physical symptoms (i.e., depression, PD, SDs, dissociative disorders) using Pubmed/Medline, PsycInfo and PSYNDEX. Primary keywords used were "interoception," "heartbeat perception/detection," "visceral perception" and "symptom perception." To address research objective (1), primary keywords were combined with secondary keywords "stress," "anxiety," "autonomic/sympathetic nervous system," "nor/epinephrine," "HPA axis" and "cortisol." To address objective (2), primary keywords were combined with keywords "depression/depressive," "panic/agoraphobia," "somatoform," "fucntional disorder/syndrome/complaint," "medically unexplained symptoms," "dissociation/dissociative" and "depersonalization/derealization." On abstract-based search, studies were identified to contain at least one interoceptive indicator and either an experimental task that is considered inducing stress in a laboratory environment, and/or a player in physiological stress axes, and/or self-reported stress, and/or the inclusion of at least one of the respective mental disorders. Secondary literature as provided by the identified papers was also screened according to these criteria. For supplementary sections addressing the relationship between interoception psychological processes, which are affected by stress (anxiety, attention, learning, decision making), non-exhaustive literature searches were performed. Therefore, primary keywords were combined with secondary keywords dependent on the topic of the respective section. Based on the identified literature, interoceptive research methods are summarized and discussed in Section "Terminology and Definition," which precedes the integration of the main results. Extraction of literature was performed independently by both authors.

Results

The search on the relationship between interoception and stress resulted in 24 studies, which are exhaustively reported below (see **Table 1**). Furthermore, we identified nine studies on interoception in depression (one review), 14 studies on interoception in PD (three reviews), 12 studies on interoception and somatoform/functional disorders, and five studies on interoception in depersonalization disorder (DPD; one hypothesis paper). Due to space limitations the search was restricted to studies investigating the relationship between interoception and stress or stress-related processes.

Terminology and Definition

Sensory information from the body can originate from different sources, i.e., (1) from exteroceptors, e.g., located in the

TABLE 1 | Summary on empirical research papers addressing acute or chronic stress and interoception.

Reference	Stress intervention	Interoceptive indicator	Study type	Sample size	Main findings on stress and interoception
Durlik et al. (2014)	Anticipation of public speakting (10 min)	Schandry-based heartbeat perception task	Control-/stress group (between design)	62 (42 f)	Increase of IA during anticipation
Eichler and Katkin (1994)	Mental arithmetic task (1 min; 3 min rest)	Whitehead-/Katkin-based heartbeat perception tasks	Baseline-/stress period (within factor); fixed order; good vs. poor heartbeat perceivers (quasi-experimental factor)	48 m; 23 good vs. 25 poor perceivers	Good perceivers show higher PEP and HI, and marginally higher CO stress response
Elsenbruch et al. (2010a)	Public speaking paradigm (5 min prep., 5 min speaking), fMRI compatible	BOLD response to painful and non-painful rectal stimulation	Stress-/relaxation period (within design)	15 (f) IBS patients, 12 (f) healthy controls	During stress increased activation of insula, midcingulate cortex, and ventrolateral prefrontal cortex in IBS
Fairclough and Goodwin (2007)	4 × mental arithmetic task (3 min each)	Whitehead-based heartbeat perception task	Stress-/relaxation session (within design); counterbalanced order	40 (20 f)	Decrease of IA after stress in females
Gupta et al. (2014)	No intervention; early life stress assessed via Early Trauma Inventory	Functional connectivity in six resting state networks (BOLD)	IBS/healthy control group (between design); correlative design (early life stress and brain network activity)	58 (28 f) IBS patients, 110 (72 f) healthy controls	Correlation between early life stress and activation of salience/executive control network in IBS patients
Gray et al. (2007)	Mental arithmetic task	Heartbeat-evoked potentials (arithmetic/control task)	Baseline-/stress period (within design)	10 m with cardiac dysfunction	Change of cardiac output correlated with HEP changes during stress; no effect of stress on HEPs
Herbert et al. (2010)	Mental arithmetic task (first 5 min)	Schandry-based heartbeat perception task	Baseline-/stress period (within factor); fixed order; good vs. poor heartbeat perceivers (quasi-experimental factor)	38 (19 f); 19 good vs. 19 poor perceivers	Stress-induced increase of HR, PEP and HI correlated to IA; good perceivers show higher HR and PEP stress response
Jones and Hollandsworth (1981)	Physical exercise (bicycle; to achieve 75% increase in heart rate)	Identification of correct or false heart rate feedback	Baseline-/exercise period (within factor); exercise: tennis players, distance runners, control (quasi-experimental)	36 (18 f)	Distance runners had highest IA during baseline; IA in tennis and control group after exercise
Kindermann and Werner (2014)	3 × 3 min mental arithmetic test (PASAT)	Schandry-based heartbeat perception task	Good vs. poor heartbeat perceivers (quasi-experimental factor)	20 good vs. 20 poor perceivers	Good perceivers show higher negative affect during stress; no difference in heart rate response
Moor et al. (2005)	Epinephrine; esmolol; norepinephrine; sodium-nitroprusside (dose-response)	Whitehead-based heartbeat perception task	Placebo-controlled study (within design); fixed order	24 m	Nitroprussid and epinephrine increased, esmolol decreased IA
Pohl et al. (1998)	Preparation for public speech	Report of hypoglycemia symptoms after insulin bolus	Placebo-controlled study (2 × 2 between design: insulin vs. placebo; stress vs. control intervention)	40 m	Less accurate detection of insulin adminstration and recudes hypoglycemia symptoms after stress
Pollatos et al. (2007b)	Isometric handgrip exercise	Schandry-based heartbeat perception task	Baseline-/exercise period (within factor); high vs. low cardiovascular responders (quasi-experimental factor)	18 m; 9 high vs. 9 low cardiovascular responders	Higher IA in high than low responders; IA correlated with response in HR, SBP, CO and PEP
Richards et al. (1996)	Physical exercise (stepping machine; 1 min)	Correlation between actual and perceived changes in heart rate	Relaxation-/exercise condition (within); panic vs. control individuals (group); intra-correlation design	26 panic patients (14 f); 14 healthy controls (9 f)	Higher IA (intra-correlation) after exercise, no interaction with group factor
Rosenberger et al. (2009)	Public speaking paradigm, fMRI compatible	BOLD response to painful and non-painful rectal stimulation	baseline-/stress period (within design); randomized order	14 f	Stress induces differences in activity of right posterior cingulate and S1, and left thamalus during painful stimulations
Schandry and Specht (1981)	Public speaking test	Schandry-based heartbeat perception task	Baseline-/stress period (within design)	41	Increase of IA after stress

(Continued)

TABLE 1 | Continued

Reference	Stress intervention	Interoceptive indicator	Study type	Sample size	Main findings on stress and interoception
Schandry et al. (1993)	0–90° tilt; ergometric bicycle exercise (0, 25, 50, 75 W)	Schandry-based heartbeat perception task	0, 25, 50, 75 W; fixed order conditions (within design)	25 (14 f)	IA correlated with HR, SV, HI and momentum over all conditions
Schmitz et al. (2012)	Public speaking test (3 min)	Schandry-based heartbeat perception task	Low vs. high social anxiety (quasi-experimental factor)	40 (21 f) children; 20 high vs. 20 low socially anxious	After stress high socially anxious show higher IA than low socially anxious children
Schulz et al. (2011)	3-min socially-evaluated cold pressor test (0–3°C)	Cardiac modulation of startle	Control-/stress group (between design)	38 (24 f)	Earlier CMS effect after stress
Schulz et al. (2013a)	3-min socially-evaluated cold pressor test (0–3°C)	Schandry-/Whitehead-based heartbeat perception task	Control-/stress group (between design)	42 (29 f)	Higher Schandry-, lower Whitehead-based IA after stress
Schulz et al. (2013b)	4 mg of intraveneous cortisol	Heartbeat-evoked potentials (rest)	Placebo-controlled study (within design), counterbalanced order	16 m	Higher HEPs after cortisol in open than in closed eyes
Shao et al. (2011)	10-min cold pressor test (10°C)	Heartbeat-evoked potentials (rest/control task/CP)	Baseline-/stress period (within design); randomized order	21 (9 f)	Decrease of HEPs during CP
Steptoe and Vogele (1992)	5-min mental arithmetic, 3-min cold pressor test	Correlation between actual and perceived physiology	Baseline-/arithmetic/cold pressor condition (intra-correlation design)	30 f	No effect of stress on IA reported; IA correlated with information-seeking coping style
Stevens et al. (2011)	Anticipation of public speaking	Schandry-based heartbeat perception task	Resting-/anticipation period (within factor); low vs. high social anxiety (quasi-experimental factor)	48 (25 f); 24 high vs. 24 low socially anxious	Marginal increase of IA after stress; no interaction between stress and anxiety groups
Sturges and Goetsch (1996)	7-min mental arithmetic task	Schandry-based heartbeat perception task	Baseline-/stress period (within factor); low vs. high anxiety sensitivity (between factor)	59 f; 29 high vs. 30 low anxiety sensitive	Higher IA after stress in high than in low anxiety sensitivity

skin, which transmit somatosensory information, (2) from proprioceptors, e.g., receptors in the spindles of skeletal muscles, and (3) from interoceptors, e.g., mechano-, chemo-, thermo-, or metabo-receptors within visceral organs (Schulz, 2015). In fact, signals from interoceptive and somatosensory sources are likely to be integrated in CNS structures to construct a representation of bodily processes (Craig, 2002, 2009; Wiens, 2005). This could explain, for instance, why patients with damaged or degenerated afferent autonomic nerves show impaired perception of visceral sensations (Pauli et al., 1991; Leopold and Schandry, 2001; Schulz et al., 2009a), while largely maintaining their affective experience (Cobos et al., 2004; Heims et al., 2004), which is assumed to be associated with interoception. However, the current review makes the pre-assumption that visceral-afferent signal transmission plays an important, yet not an exclusive role in the central representation of bodily signals.

There has been a debate in the literature on the taxonomy of facets of interoception. Garfinkel and Critchley (2013) have suggested separating the subjective tendency to be focused on interoceptive sensations from the actual accuracy in interoceptive tasks. They call the former 'interoceptive sensibility' and the latter 'interoceptive sensitivity.' In between, a third level was proposed, which was hypothesized to represent the degree of predictive value between interoceptive sensibility and sensitivity, which was named 'interoceptive awareness.' This taxonomy is in partial disagreement with earlier studies, which consider accuracy in heartbeat perception tasks to be an index of 'interoceptive awareness.' It is undisputed that awareness is required to perform in heartbeat perception tasks, but empirical evidence clearly shows that 'awareness' to interoceptive sensations and performance in heartbeat perception are partially unrelated (Khalsa et al., 2008; Ceunen et al., 2013). Although the differentiation between 'sensibility' and 'sensitivity' is plausible, in our opinion the measure indexed by interoceptive tasks should be called 'interoceptive accuracy' (IA) instead. 'Interoceptive sensitivity' originates from signal detection theory and implies the minimum threshold to detect an interoceptive signal from background noise (Farb et al., 2015). In contrast, 'IA' refers to the objective performance in counting interoceptive sensations or in discriminating and interoceptive from an exteroceptive sensation (see below; Ceunen et al., 2013). 'IA' does not suggest that a specific score reflects 'sensitivity' or 'awareness,' as some individuals may be hyper-sensitive or show extreme awareness to interoceptive sensations, which leads to misinterpretation of signals and thus to lower accuracy, while others may rarely focus awareness toward interoceptive stimuli, but may be highly accurate in performing heartbeat perception tasks (Schulz, 2015). In a revised version of the theory, performance in heartbeat perception tasks was now called 'IA' (Garfinkel et al., 2015), which is also reflected in this review.

The perception of bodily processes involves at least three consecutive stages: (1) afferent neural signals from the body, such as visceral organs, (2) the direction of attention toward bodily sensations, and (3) the evaluation of these signals and their integration into psychological processes. As suggested by Vaitl (1996) these three stages can be further differentiated. For instance, visceral-afferent signals involve the stimulation of interoceptors at the peripheral organ, the neural transmission of these signals from interoceptors to CNS structures, and finally the CNS representation of these signals. The evaluation of bodily sensations and their integration into psychological processes are associated with subjective reports of physical sensations, and an individual learning history concerning these sensations.

Methodology to Assess Interoception

Interoceptive Sensibility

The meta-cognitive tendency (i.e., interoceptive sensibility) to focus on interoceptive sensations is commonly assessed via questionnaires. The majority of existing questionnaires is designed to assess physical symptoms and clinically altered interoception in mental disorders, such as eating disorders (Eating Disorder Inventory-2: Garner, 1991) or SDs (Somatosensory Amplification Scale: Barsky et al., 1990). Instruments for the assessment of 'normal' interoception are scarce. Several existing studies have focused on the Body Perception Questionnaire by Porges (1993). However, there are no sufficient estimations for psychometric properties available for this questionnaire. The recently developed Multidimensional Assessment of Interoceptive Awareness (MAIA; Mehling et al., 2012) incorporates a number of interoceptive aspects (noticing, non-distracting, not-worrying, attention regulation, emotional awareness, self-regulation, body-listening, trusting) and is thus designed to study 'normal' and clinically dysregulated interoception.

Interoceptive Accuracy

The majority of empirical data on interoception is based on 'IA' as assessed by experimental paradigms that require subjective reports of physical sensations. Regarding the cardiovascular system, there are two main categories of tasks: (1) heartbeat counting tasks and (2) heartbeat discrimination tasks, as summarized elsewhere (Schulz, 2015). Heartbeat counting tasks were developed by Schandry (1981), who named the task 'mental tracking test.' The task consists of time intervals of different duration (original version: 35, 45, and 55 s), during which participants are instructed to silently count the number of their heartbeats. This number is later compared to the actual number of heartbeats in this interval. The absolute value of the difference between both is divided by the actual number of heartbeats and this 'inaccuracy' index is then subtracted from 1, which results in the accuracy score of the heartbeat counting task. It has been repeatedly demonstrated that the accuracy score depends on the wording of the instruction given. The comparison between a standard instruction ("count all heartbeats you feel in the body") and a strict instruction ("count only those heartbeats about which you are sure of") suggests that an individual's knowledge on their heart rate and capacity to accurately estimate the duration of time intervals may be important factors for the accuracy in heartbeat counting tasks (Ehlers et al., 1995; Ehlers and Breuer, 1996). Heartbeat discrimination tasks were originally developed by Brener and Jones (1974) and further elaborated by others (Whitehead et al., 1977; Katkin et al., 1981; Brener and Kluvitse, 1988). In these tasks participants are asked to judge whether a set of consecutive exteroceptive stimuli (e.g., lights, tones, tactile stimuli) appear simultaneously with their own heartbeats (S+ trials) or not (S− trials). The available variants of this task mainly differ in the setup of S− trials, since some present exteroceptive stimuli with a fixed delay to heartbeats, while others simulate a set of artificial stimuli without any relation to the actual heartbeats. Previous research has yielded mixed findings regarding the question whether IA assessed by both tasks families are correlated: some report a moderate positive correlation (Knoll and Hodapp, 1992; Schaefer et al., 2012; Hart et al., 2013), while others did not find any association (Phillips et al., 1999; Schulz et al., 2013a; Michal et al., 2014).

Beyond heartbeat perception tasks, subjective reports are also required for paradigms incorporating the perception of respiratory resistances (Pappens et al., 2013; Tsai et al., 2013; Petersen et al., 2014), of gastric distensions (Rosenberger et al., 2009; Elsenbruch et al., 2010b), and of non-specific skin conductance fluctuations (Andor et al., 2008). In our view, the outcomes of these tasks reflect all stages of interoception at any given point in time (i.e., learning, report, awareness, CNS representation, visceral-afferent signal transmission, and stimulation of interoceptors). We would argue, therefore, that in addition to these measures, methods are required that enable the differentiation of interoception stages, e.g., between those at lower and higher levels of awareness.

Psychophysiological Methods

To date, at least three psychophysiological indicators are available to assess interoceptive signal transmission below the threshold of consciousness: (1) Baroreflex-sensitivity (BRS) quantifies the changes in heart period in responses to changes in arterial blood pressure (Robbe et al., 1987). As this brainstem reflex requires intact baro-afferent signals transmission, BRS is considered an indicator of the integrity of afferent autonomic nerves (Frattola et al., 1997). The disadvantage of this indicator is the fact that it reflects both the afferent and efferent branch of the baroreflex at a time. (2) Using the 'cardiac modulation of startle' (CMS) paradigm we could show that visceral-afferent neural signals from the cardiovascular system may affect acoustic startle responses (Schulz et al., 2009a,b,c, 2011). This modulation is largely diminished in individuals with degeneration of afferent autonomic nerves (Schulz et al., 2009a) and may also involve afferent signals from other organ systems (Schachinger et al., 2009; Schulz et al., 2013b, 2015c). The CMS effect can be interpreted as an indicator of visceral-afferent neural signal processing at brainstem level, as it probably only involves brainstem mechanisms and occurs below the threshold of consciousness, (3) viscerally evoked brain potentials are considered indicators for the cortical representation of afferent signals from visceral organs. They can be further differentiated into heartbeat-evoked potentials (HEPs), which are related to the processing of heartbeats (Schandry et al., 1986; Leopold and Schandry, 2001), and respiratory-related potentials that are observed when an inspiratory or expiratory occlusion is presented (Davenport et al., 1986; Chan and Davenport, 2010; von Leupoldt et al., 2010).

Another psychophysiological indicator of interoception is the single trial covariance between event-related brain potentials (N300) and changes in heart period (~3–4 s after stimulus) in response to feedback in decision making tasks (Mueller et al., 2010a, 2012, 2013). This indicator is limited, however, to the use with decision making paradigms involving feedback. The theoretical background for this effect is provided by Damasio's (1994) "somatic marker hypothesis" positing that afferent information from visceral organs is integrated in the evaluation of alternatives in decision making. It is plausible that the covariation between cortical and cardiac responses is specifically relevant for decision making paradigms tasks, although it is for future research to show whether this effect also extends to other embodied cognitions.

Synthesis of Findings on Interoception and Stress

Based on the findings from the current review, we propose a conceptual framework that allows for the explanation of the role of the dysregulated association between interoception and stress for the generation of physical symptoms (see **Figure 1**). In this model we postulate that stress is the initial point of a positive feedback cascade. In case of an acute stressor, the cascade will follow the pathway b–d–e. In particular, acute stress will activate physiological stress axes, such as the release of cortisol and activation of peripheral organs by sympathetic mechanisms. The altered stimulation of peripheral interoceptors and central effects of cortisol and noradrenergic structures will affect the perception of bodily sensations (e.g., perception of tachycardia or positive cardiac inotropy). Alterations in the perception of bodily sensations may temporarily feed into the experience of physical symptom during acute stress, for instance, palpitations, nausea or breathlessness. The perception of these symptoms could then be perceived as stressful and thus contribute to the maintenance of a stress response. However, due to the limited duration of an acute stress response, the cascade will be disrupted after the stress-eliciting stimulus has disappeared. In case of the confrontation with a chronic, early or major life stressor, the dysregulation of physiological stress axes (e.g., chronic hyper-activation of the HPA axis or the SAM axis) is implied and the model will follow the pathway a-c-d-e. Once the manifestation of a stress axis disorder has occurred, the cascade cannot be easily disrupted anymore, even if the initial stressor has disappeared. In detail, if the organism experiences chronic dysregulation of a stress axis, this state may permanently induce altered perception of bodily sensations and contribute to the manifestation of physical symptoms, whose perception consolidates the experience of stress. This model has an intended similarity to the model of somatosensory amplification by Barsky et al. (1988) and Barsky (1992), which emphasizes the importance of positive feedback mechanisms in activation, stress and physical symptoms. In contrast to the model by Barsky et al. (1988) however, the current model makes no assumptions about the role of attention to bodily sensations, although the latter could be conceived of as one of several psychobiological mechanisms. The core assumption in

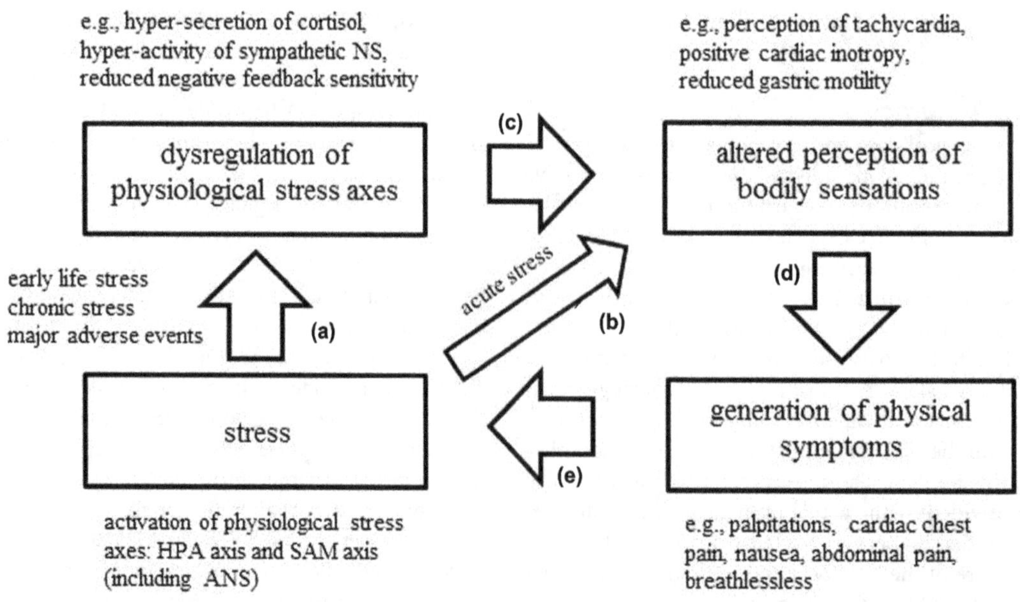

FIGURE 1 | Synthesis of findings on acute and chronic stress, dysregulation of physiological stress axes, altered interoception and the generation of physical symptoms into model comprising a positive feedback loop.

the current model concerns the interrelation of ascending and descending signals on the brain body axis in a positive feedback fashion. The following paragraphs will summarize empirical findings supporting this model.

Chronic Stress and Dysregulation of Physiological Stress Axes (Path *a*)

It is well documented in the literature that prolonged exposure to psychosocial stress, early life adversity and/or major adverse events represent major risk factors for the dysregulation of the physiological stress axes. The regulation of the HPA axis can be differentiated into three different mechanisms (Li-Tempel et al., submited): (1) baseline activity and circadian rhythmicity, such as morning or daily profiles in cortisol level, (2) reactivity of CRF, ACTH, and cortisol in response to acute stress, and (3) (negative) feedback sensitivity, such as the suppression of cortisol release by dexamethasone. In general, similar mechanisms also exist for the SAM axis. In particular, baseline concentrations of catecholamines can be assessed from blood or urine, indirectly from salivary samples via alpha-amylase, or psychophysiological indicators, such as pre-ejection period. The investigation of circadian rhythmicity is limited due to the very short plasma half-life of catecholamines (~1–3 min.). Furthermore, the same indicators can be investigated in response to an acute stressor. Negative feedback mechanisms in the SAM axis are, for instance, central α2-adrenergic receptors. These receptors are critically involved in brainstem-relayed reflex circuits that regulate the homeostasis of sympathetic and parasympathetic output, e.g., the arterial baroreflex (Sved et al., 1992). As the players involved in SAM axis regulation involve complex neural and endocrine pathways and are, therefore, more heterogeneous, the same is true for their indicators. For example, some parameters, such as low frequency heart rate and blood pressure variability are more sensitive to central sympathetic activation, while others, such as pre-ejection period are indicative of peripheral sympathetic activation (e.g., circulating catecholamines; Schachinger et al., 2001).

Early life and chronic stress, as well as major adverse life events, are consequently associated with dysregulation in all three types of mechanisms and in both stress axes (McEwen, 2000). Regarding HPA axis baseline and circadian rhythmicity, for instance, chronic stress, resulting from caregiving to family members, is reflected in altered cortisol awakening response (CAR; de Vugt et al., 2005), reduced cortisol release during daytime (Bella et al., 2011), and elevated hair cortisol levels (Stalder et al., 2014). In chronic stress associated with peer bullying and victimization reduced CAR was observed (Knack et al., 2011). Aberrant cortisol responses to acute stressors have been also found in family caregivers (De Andres-Garcia et al., 2012) and those exposed to peer victimization (Knack et al., 2011). Feedback sensitivity by dexamethasone suppression is altered in individuals exposed to high work stress, as indicated by self-reported burnout symptomatology (Pruessner et al., 1999), and after major traumatic experiences (Yehuda et al., 1993). Regarding the SAM axis, chronic stress by caregiving is associated with baseline indices of peripheral sympathetic activation (Cacioppo et al., 2000). Work stress as indicated by over-commitment to work is related to lower baseline norepinephrine levels (Wirtz et al., 2008), but increased activity in psychophysiological indicators of peripheral epinephrine circulation (Vrijkotte et al., 2004) and lower central parasympathetic output (Chandola et al., 2010). Furthermore, lower norepinephrine responses to acute stress have been found in individuals high in over-commitment, suggesting high work-related stress (Wirtz et al., 2008). Individuals exposed to early life stress show blunted reactivity to acute stress in indicators reflecting peripheral sympathetic activation (i.e., pre-ejection period; McLaughlin et al., 2014). Finally, there is evidence from animal models of reductions in the expression of α2-adrenergic receptors (which are involved in SNS-down-regulation) in response to chronic psychosocial stress (Flugge et al., 2003). The arterial baroreflex circuit, which is partially mediated by α2-adrenoceptors (Sved et al., 1992), shows reduced functioning under conditions of chronic work stress (Thomas et al., 2004).

Chronic stress and the associated dysregulation of physiological stress responses represent important factors for the development and maintenance of mental disorders, in particular those, which are associated with symptoms of altered perception of bodily sensations (e.g., depression, PD, SDs, dissociative disorders). Nevertheless, the current literature lacks a discussion of the mechanisms underlying the relationship between stress, interoception and physical symptoms. For example, it has been repeatedly suggested that somatic syndromes, e.g., chronic fatigue syndrome (CFS) or fibromyalgia (FMS), are associated with lower cortisol responsiveness or increased feedback sensitivity (Cleare, 2004; Tanriverdi et al., 2007; Wingenfeld et al., 2008; Tak et al., 2011), but only few hypotheses addressing the mechanisms of how cortisol may contribute to the generation of physical symptoms. One such hypothesis addresses immunological changes that are associated with dysregulation of the HPA axis (Fries et al., 2005). It is possible that the modulation of pro- and anti-inflammatory cytokines may contribute to somatic syndromes, such as FMS. Another explanation focuses on the importance of hypocortisolism in CFS and argues that cortisol mobilizes energy resources to overcome daily demands, which are insufficiently available in CFS (Tak and Rosmalen, 2010). However, both explanations are focused on peripheral bodily processes and do not incorporate possible effects of stress on cortical representation and perception of bodily sensations. The possibility that cortisol could also modulate the perception of bodily sensations was first raised by Rief et al. (1998b).

Acute Stress and Interoception (Path *b*)

Laboratory Stress Tests

A broad variety of methods has been used to induce acute stress in laboratory settings (Steptoe and Vogele, 1991). Mental arithmetic tasks elicit 'psychological' stress generally by inducing mental load and specifically by involving central control mechanisms over autonomic processes (Moriguchi et al., 1992; Sloan et al., 1995). Psychosocial stressors that emphasize public speaking challenges, such as the Trier Social Stress Test (TSST), are known to provoke an intense response of the HPA axis (indexed e.g., by an increase of salivary cortisol; (Kirschbaum et al., 1993; Kudielka et al., 2004) and activation of the cardiovascular system by sympathetic mechanisms. Given the TSST's comparatively long duration of ~15 min on the one hand, and the very fast acting release and effects of catecholamines (i.e., seconds) on the other hand, specific sympathetic processes involved in the TSST are difficult to disentangle. As the cold pressor test induces stress associated with the experience of ischemic pain, these results in specific cardiovascular response patterns, which can partially be attributed to the experience of pain, but also to the vaso-constricting effect of cold water (Streff et al., 2010). The socially-evaluated cold pressor task (SECPT) attempts to integrate psychosocial and physical stressors, and has been demonstrated to elicit significant increases in saliva cortisol (Schwabe et al., 2008; Lass-Hennemann et al., 2010, 2011; Larra et al., 2014), as well as sympathetic activation, as indicated by an increase of heart rate, and systolic and diastolic blood pressure (Schwabe et al., 2008; Schulz et al., 2011, 2013a) and a decrease of pulse-transit time and BRS (Schulz et al., 2011). To understand possibly diverging results on the impact of acute stressors on indicators of interoceptive signal processing and IA, differences in the methods used to induce stress and to assess interoception have to be taken into account (Steptoe and Vogele, 1992).

Interoceptive Accuracy Assessed with Heartbeat Perception Tasks

Interoceptive accuracy as estimated by Schandry-based heartbeat tracking tasks typically increase under acute stress. While this effect has been observed in anticipation of (Durlik et al., 2014) and after a public speaking test (Schandry and Specht, 1981), one study only found a marginal increase in IA during anticipation (Stevens et al., 2011). The importance of the amplitude of cardiovascular responsiveness during stress for the enhancement of IA is underlined by the positive correlation between IA and responsiveness in heart rate, pre-ejection period and the Heather index in a mental arithmetic task (Herbert et al., 2010). A similar association between cardiovascular responsiveness and IA could also be shown for the heartbeat discrimination task (Eichler and Katkin, 1994). Acute responses to stress include sympathetic effects on the cardiovascular system, and a delayed increase in cortisol secretion, if a psychosocial challenge (e.g., public speaking) is involved. There are also attentional and affective changes in response to stress (see Psychobiological Mechanisms Involved in Interoception and Stress), which may interact with interoception. For instance, individuals with high IA in heartbeat perception show higher negative affect during stress than individuals with low IA (Kindermann and Werner, 2014). Furthermore, it has been shown that IA increases after physical exercise (Jones and Hollandsworth, 1981; Richards et al., 1996), and that the same correlation between cardiovascular reactivity measures (cardiac output, heart rate, systolic blood pressure, Heather index, momentum, stroke volume) to exercise and IA exists (Schandry et al., 1993; Pollatos et al., 2007b). It could be argued, therefore, that peripheral sympathetic activation during stress represents the core factor of this relationship. In contrast to these results using Schandry-based heartbeat tracking tasks, IA as assessed with heartbeat discrimination tasks, is reduced in women performing a mental arithmetic task (Fairclough and Goodwin, 2007). Using the SECPT we could replicate these seemingly contradictory findings, i.e., an increase in IA after stress as assessed by the Schandry-heart beat tracking task and a decrease in IA when using a visual heartbeat discrimination task (Schulz et al., 2013a). As the results obtained in these studies seems to be IA-paradigm dependent the observed differences are most likely associated with task specific characteristics: while heartbeat tracking tasks require participants to only focus their attention on visceral sensations, the heartbeat discrimination task involves the concurrent monitoring of visceral sensations and external signals. According to the competition-of-cues model by Pennebaker and Lightner (1980), interoceptive and exteroceptive signals compete for a limited resource, i.e., attention. As acute stress may narrow attentional resources and favor attention to task-relevant stimuli (Chajut and Algom, 2003; Plessow et al., 2011), the multisensory integration of information as required by the heartbeat discrimination task may be impaired by stress exposure. This assumption is further supported by the observation that after the preparation for public speech, individuals exhibited reduced accuracy in detecting an insulin-induced hypoglycemic state (Pohl et al., 1998), which also requires attention focused on multiple symptoms. In summary, there is evidence for two mechanisms relevant for IA in response to acute stress: (1) peripheral sympathetic activation, which induces an increased stimulation of cardiac interoceptors (e.g., arterial baroreceptors) and intensifies afferent signal transmission from the cardiovascular system, and (2) attention focus on visceral sensations, which may improve the detection of visceral signals, but diminish the integration of signals from other sensory modalities. When both effects are opposed, as e.g., in the heartbeat discrimination task, it appears that attention represents the more important determinant of IA. This dissociation emphasizes the need for the separate assessment of visceral-afferent signal transmission and representation on the one hand, and attention to visceral signals on the other hand.

Psychophysiological Indicators of Interoceptive Signal Processing

While HEPs and CMS may serve as indicators of visceral-afferent signal transmission independent of attention to these signals, empirical data on stress effects on these indicators is scarce. In one study we investigated the impact of the SECPT on CMS (Schulz et al., 2011). After stress exposure we observed CMS to occur earlier within the cardiac cycle (0–200 ms instead of 200–300 ms after the R-wave); yet, contrary to our expectations, there was no *amplification* of the CMS effect. Nevertheless, as the CMS reflects the intact transmission of afferent neural signals from the cardiovascular system at brainstem level (Schulz et al., 2009a), these findings suggest that the *amplitude* of representation of afferent signals may be unchanged by stress. Similar findings have also been reported for HEPs, with reduced amplitudes during a long-term, but mild cold pressor test (10-min, 10°C) and a return to baseline levels after the cold pressor test (Shao et al., 2011). This response pattern may be explained by participants focusing their attention during the cold pressor primarily on the pain experience, and thus away from cardiovascular sensations. Nevertheless, after termination of stress exposure, the expected sympathetic response did not affect HEP amplitudes. Gray et al. (2007) investigated the impact of a mental arithmetic task on HEPs in 12 men with cardiac dysfunction, and found a positive relationship between changes in cardiac output and HEPs, but no effect of the stress task on HEP amplitudes. To the best of our knowledge, there is only one study reporting an effect of acute stress on a psychophysiological indicator of interoception: after an MRT-compatible public speaking test, the BOLD response to painful rectal distensions was different in the right posterior cingulate and right somatosensory area, as well as the left thalamus (Rosenberger et al., 2009). Despite the large methodological differences across these studies in terms of laboratory stress tasks employed, organ systems investigated and derivation of interoceptive indicators, it may be speculated that acute stress specifically changes the processing of visceral-afferent signals from the gastrointestinal system. As the majority of studies demonstrate no *quantitative* change of raw representations of visceral-afferent signals by stress, it is still plausible that stress induces a *qualitative* change in interoception, as suggested by our findings (Schulz et al., 2011). It can be summarized that, contrary to expectations, current results implicate that there is no main effect of acute laboratory stress on the *amplitude* of psychophysiological indicators of interoception after the stressor. Possible reasons for the null findings in these studies could involve the limited effectiveness of mental arithmetic and 10°C cold pressor tasks, as well as the limited time frame after stress during which responses were monitored. Given the heterogeneity of mechanisms induced by acute laboratory stress, it seems necessary to determine the differential role of the physiological stress axes on interoception. **Table 1** provides an overview of studies addressing the relationship of stress and indicators of interoception.

The Sympatho-Adreno-Medullary Axis and Interoception

As described earlier, reactivity in pre-ejection period or Heather index to laboratory stress, (which may reflect peripheral sympathetic activation) is positively related to heartbeat perception (Eichler and Katkin, 1994; Herbert et al., 2010). However, as neither study derived heartbeat perception scores from a post-stress period, it can only be concluded that individuals with a general tendency for responding with peripheral sympathetic activation show higher IA in heartbeat perception tasks. To elucidate the contribution of both the central and peripheral sympathetic branch to heartbeat perception, the impact of different adrenergic agents on accuracy in a heartbeat discrimination task has been investigated by (Moor et al., 2005). In particular, sodium nitroprusside as α1-adrenergic antagonist, and norepinephrine as α1-agonist, as well as the β1-agonist epinephrine and the β1-antagonist esmolol were employed. Since β1-adrenoreceptors are located in the myocardium and are sensitive to circulating catecholamines, especially epinephrine, selective stimulation or de-stimulation represents a pharmacological model for peripheral sympathetic activity (Schachinger et al., 2001). Epinephrine increased and esmolol decreased IA as assessed by a heartbeat discrimination task as compared to placebo (Moor et al., 2005). This finding suggests that peripheral sympathetic activation enhances IA. Since epinephrine cannot cross the blood–brain barrier, two alternative ways of signal transmission are possible: first, β-adrenergic receptors localized at vagal nerve endings may directly be stimulated by circulating epinephrine (Mravec, 2011). Second, increased cardiac contractility may cause increased stimulation of cardiac interoceptors (e.g., baroreceptors), whose neural signals are transmitted over the nervus glossopharyngeus (Jänig, 2006). α1-adrenoreceptors are primarily located in the vascular musculature. Their stimulation induces an increase in vascular resistance and, therefore, in blood pressure. This information is relayed via the arterial baroreflex circuit in order to decrease heart rate accordingly for the maintenance of blood pressure level. The dis-stimulation of α1-adrenoreceptors causes the opposite effect. This pharmacological design is thus suitable to investigate the selective loading and unloading of arterial baroreceptors and the subsequent central sympathetic activation induced by this baroreceptor stimulation (Schachinger et al., 2001). Baroreceptor unloading by sodium nitroprusside resulted in an increase in IA as compared to placebo, while norepinephrine had no effect on heartbeat perception (Moor et al., 2005), which could be explained by increased central sympathetic output due to baroreceptor unloading. Central α2-adrenoreceptors, which are mainly located in the LC and NTS act as negative feedback mechanism in the central noradrenergic system (Rockhold and Caldwell, 1980). Hence, administration of a α2-adrenergic antagonist causes a concurrent activation of central and peripheral sympathetic mechanisms (Isaac, 1980; Philippsen et al., 2007). In a pilot study we found the α2-antagonist yohimbine to suppress the CMS effect (Schulz et al., 2007). This finding implies that concurrent central and peripheral sympathetic activation combined with increased arousal and vigilance as caused by noradrenergic activation may diminish the central processing of visceral-afferent signals at a low, presumably brainstem-associated level. Our own observations on α2-antagonism and IA suggest reduced IA after the administration of yohimbine. This observation may implicate that central noradrenergic activation, including increased alertness and vigilance, may also impair the cortical processing of visceral-afferent signals. Taken together, it can be concluded that there is a strong positive relationship between peripheral sympathetic activation and cardiac IA, but only a limited association between central sympathetic tone and IA. A strong activation of central noradrenergic mechanisms may even inhibit interoception, although concurrent peripheral activation would suggest a more intense stimulation of interoceptors.

The HPA Axis and Interoception

Cortisol has been found to modulate HEP amplitudes within a timeframe of 1–17 min after infusion (Schulz et al., 2013c). More specifically, 4 mg of cortisol resulted in higher HEPs under open- compared with closed-eyes conditions, i.e., states of high vs. low alertness. This effect could eventually feed into a vicious circle of increased attention focus on physical symptoms, increased anxiety and higher levels of cortisol, and represent a psychobiological mechanism underlying positive feedback models of somatosensory amplification (Barsky et al.,

1988). In this study HEPs were assessed during rest, without the conscious perception of heartbeats. The effect of cortisol administration on resting HEPs suggests that cortisol may affect the raw representation of visceral-afferent signals, independent of the conscious perception of heartbeats. Moreover, we did not observe any effect of cortisol on cardiovascular activation. This finding suggests that cortisol may selectively modulate the central, presumably cortical, representation of visceral-afferent signals, while the peripheral origination of these signals remain unaffected. Interestingly, in a complementary study no effect of 1.5 mg cortisol on the CMS was observed within the same time frame (Schulz et al., 2010). Since the CMS is assumed to reflect visceral-afferent signal transmission at brainstem level, one could speculate that the effects of cortisol on visceral-afferent signal relaying are restricted to the cortex. However, it needs to be acknowledged that the dosages of administered cortisol were not identical. Furthermore, the first study covered a time frame of up to 37 min and therefore includes possible genomic effects, while the second study focused on non-genomic effects only. The importance of cortisol for the cortical representation of visceral-afferent signals is further emphasized by the negative relationship between basal cortisol level and HEP amplitudes of $r = -0.29$ (Schulz et al., 2015b). Despite the fact that this result was based on a mixed sample of healthy individuals and patients with DPD, the correlation was unaffected by the diagnosis, or depression and anxiety scores. It can be summarized that acute cortisol administration tends to increase HEP amplitudes (when eyes are open), presumably via a non-genomic mechanism, while basal cortisol level shows a negative relationship with HEPs. One may speculate that the effect of cortisol on the cortical representation of visceral-afferent signals reverses into long-term cortisol elevations. Interestingly, these opposite effects contrast with findings in pain research: in experimental short-term manipulation of cortisol, an oral administration of 40 mg of cortisol reduces pain sensitivity (Michaux et al., 2012), while cortisol blockade by metyrapone intensifies pain perception (Kuehl et al., 2010). Meanwhile, in chronic pain syndromes, such as FMS, also reduced cortisol baseline levels and hyper-suppression by dexamethasone were observed (Wingenfeld et al., 2007, 2008). This possible dissociation in cortisol relationships between 'normal' visceral-afferent neural transmission and pain has to be acknowledged, since interoception and pain processing only partially share neural structures (Craig, 2002). Future research should clarify whether acute cortisol release may selectively favor the representation of 'normal' visceral-afferent transmission, which may reverse in long-term increases of cortisol levels. The majority of the existing literature on processes of body, symptom and pain perception concentrates on the role of cortisol, the final product of HPA axis activation. However, it needs to be taken into account that the role of CRF and ACTH on interoceptive processes remains unclear and possible differences in acute and chronic cortisol levels may also be attributed to feedback-induced changes in CRF or ACTH.

Taken together, there is considerable evidence to show that acute stress and players of physiological stress axes may alter the perception of bodily sensations. Based on our model, we propose that acute stress may induce an acute alteration of physical sensations and transient symptoms specific to the stress, which constitute the subjective experience of the stress response (e.g., tachycardia, palpitations, nausea, etc.). However, these symptoms disappear when the stress is of limited duration.

Dysregulation of Physiological Stress Axes and Interoception (Path *c*)

So far there is only partial and indirect evidence for a direct effect of dysregulated stress axes on altered perception of bodily sensations. Earlier findings suggest that chronic stress in healthy individuals is not related to heartbeat perception accuracy (Schulz et al., 2013a) or sensitivity for gastric stimulations (Rosenberger et al., 2009). The direct pathway between chronic stress and altered perception of bodily sensations was, therefore, omitted in our model. It should be noted, however, that moderately elevated levels of self-reported chronic stress in healthy individuals, as reported in the former studies, are unlikely to be accompanied by a dysregulation of the physiological stress axes.

Our assumption that the dysregulation of the physiological stress axes may induce altered perception of bodily sensations is mainly based on the following observations:

(1) In numerous mental disorders that are accompanied by physical symptoms, altered interoception and dysregulation of physiological stress axes are reported. In detail, (a) individuals with major depression (MD) and depressive symptoms exhibit reduced IA and HEP amplitudes (Dunn et al., 2007; Terhaar et al., 2012). Concerning the activity of the HPA axis, previous research has shown differences between the melancholic and the atypical sub-type (Gold and Chrousos, 2002; O'Keane et al., 2012; Gold, 2015): on the one hand, the atypical sub-type exhibited normal basal cortisol levels and increased dexamethasone-induced feedback sensitivity (Levitan et al., 2002; O'Keane et al., 2005; O'Keane et al., 2012). On the other hand, the more frequent melancholic sub-type is characterized by hypersecretion of CRF (Nemeroff et al., 1984; Wong et al., 2000), and reduced concentration and sensitivity of CRF neurons, resulting in a blunted ACTH response to CRF administration (Gold et al., 1988; Lesch et al., 1988; Ehlert et al., 2001), elevated basal cortisol levels (Gold et al., 1988; Ehlert et al., 2001; Gold, 2015), blunted cortisol responsiveness to acute stressors and reduced feedback sensitivity as provoked by dexamethasone (Gold et al., 1995; Pariante and Lightman, 2008). (b) Patients with PD show increased IA compared to healthy individuals (Ehlers and Breuer, 1992, 1996; Ehlers et al., 1995), whereas a meta-analysis has pointed out that this difference may be due to a minority within the PD group (Willem Van der Does et al., 2000). However, PD patients exhibit an increased coupling of feedback-evoked EEG amplitude and heart rate changes than healthy individuals (Mueller et al., 2014), which has previously been shown to be an indicator of neuro-visceral connectivity (Mueller et al., 2010a). Regarding the physiological stress axes, PD patients do not differ from healthy individuals in baseline morning or diurnal cortisol release or negative feedback sensitivity (Ising et al., 2012), but they show blunted

cortisol responses to psychosocial stress (Petrowski et al., 2010, 2013). In terms of SAM axis activity existing studies have yielded mixed findings. While some failed to find differences in indicators of central (e.g., low frequency heart rate variability/HRV) or peripheral sympathetic activation (α-amylase) between PD and healthy controls (Tanaka et al., 2012), others observed reduced central (McCraty et al., 2001), but increased peripheral activation (Marshall et al., 2002). However, it has been proposed that dysfunction in α2-adrenergic regulation of the ANS represents an important neurophysiological correlate of PD (Bremner et al., 1996). (c) In SDs previous studies have yielded mixed findings, which may be partially explained by the heterogeneity of symptomatology collapsed in this diagnostic category. Some studies have failed to find differences in IA between patients with SD and healthy individuals (Barsky et al., 1995; Mussgay et al., 1999; Schaefer et al., 2012), while others observed exaggerated report of bodily sensations (Bogaerts et al., 2010), resulting in an overall decrease of IA (Pollatos et al., 2011a; Weiss et al., 2014). SD patients show both alterations in baseline cortisol levels, such as the CAR (Rief et al., 1998b) or daily profile (Tak et al., 2009; Tak and Rosmalen, 2010), and aberrant cortisol responses to psychosocial stress (Janssens et al., 2012). The question, therefore, is not *whether* dysregulation of physiological stress axes, altered interoception and physical symptoms are associated, but *how* they are related and *which direction* these relationships have.

(2) In pharmacological studies, the acute administration of catecholamines (Moor et al., 2005) or cortisol (Schulz et al., 2013c) affects interoception and interoceptive signal transmission. Despite the fact that chronically altered levels of adrenergic stress hormones or cortisol are not fully comparable to an acute administration, it is unlikely that chronic alterations of these hormones do not affect interoception at all. We interpret these findings as support for the assumed direction of physiological stress axes affecting interoception in our model.

Previous studies have addressed mental disorders with physical symptoms, interoception, and indicators of only *one* physiological stress axis in the same sample. Pollatos et al. (2011a) found lower IA in patients with SD to be accompanied with decreased autonomic balance, as indicated by low/high frequency HRV ratio. In a study by Ehlers et al. (1995) higher heart rates were observed in panic patients than in healthy individuals, suggesting increased sympathetic tone in patients with PD. As both stress axes may differentially affect interoceptive processes, however, future studies are needed that include indicators of *both* stress axes at a time. Currently, there is only one study using multiple indicators of *both* physiological stress axes in the investigation of physical symptoms associated with a mental disorder (Schulz et al., 2015b). Results show an insensitivity of HEPs for attention focused on heartbeats in patients with DPD, but not in healthy individuals. This difference was associated with higher basal level of salivary α-amylase, an indicator for peripheral sympathetic activation (Chatterton et al., 1996; Nater et al., 2005). Furthermore, across DPD patients and healthy individuals there was a negative correlation between basal cortisol level and HEP amplitude (Schulz et al., 2015b). In the same sample IA did not differ between DPD and healthy control individuals (Michal et al., 2014), although differences between DPD and healthy individuals have been reported in another study (Sedeno et al., 2014).

To elucidate the direction of the relationship between dysregulated stress responses and interoception, prospective long-term observations may be required. There is a notable lack of reports on interoception in chronically stressed individuals, who already show a dysregulation of both physiological stress axes, but do not yet fulfill the diagnostic criteria for a mental disorder associated with physical symptoms. To date, there is only one study investigating the relationship of early life stress and brain activity in healthy individuals and patients with irritable bowel syndrome (IBS): Gupta et al. (2014) report altered activation of brain networks associated with pain processing and early life stress in IBS patients. The importance of chronic stress for altered interoception in IBS is further underlined by the fact that IBS patients also show altered activation of brain regions associated with interoception during stress (Elsenbruch et al., 2010a).

The question remains whether there is a direct relationship between the dysregulation of physiological stress axes and the generation of physical symptoms without the mediating effect of interoception. Based on the currently available evidence this question cannot be unequivocally answered. We do not assume a direct pathway for the following reasons: (1) There is currently no model to explain the psychobiological processes connecting the dysregulation of physiological stress axes and the generation of physical symptoms, despite a broad empirical basis showing stress axes dysregulation in mental disorders to be associated with bodily symptoms, e.g., in depression (Holsboer, 2000; Ehlert et al., 2001; Gold and Chrousos, 2002; Pariante and Lightman, 2008), FMS (Tanriverdi et al., 2007; Wingenfeld et al., 2007, 2008); IBS (Chang et al., 2009; Suarez-Hitz et al., 2012), or CFS (Cleare, 2004; Nater et al., 2008). (2) Some direct relationships between stress axes and physical symptoms are even more heterogeneous than those observed between interoception and physical symptoms, e.g., in SD. Previous studies addressing interoception in SD have either produced null findings or shown reduced IA. Meanwhile, studies investigating indicators of stress axes in SD find directly opposing findings, ranging from higher (Rief et al., 1998a; Rief and Barsky, 2005), identical (Rief and Auer, 2000) to reduced cortisol output in SD patients compared to healthy controls (Heim et al., 2000; Tak and Rosmalen, 2010; Janssens et al., 2012). A popular way of explaining these potentially conflicting findings is to emphasize the heterogeneity of symptoms collapsed into SD classification. However, an alternative explanation would be a mediating factor between stress axes and physical symptoms, i.e., altered interoceptive processes.

Interoception and the Generation of Physical Symptoms (Path *d*)

Many mental disorders that are associated with physical symptoms are characterized by altered interoception, as discussed earlier in this paper. Among those, PD is a prominent example, where the experience of (frightening) physical symptoms is a defining diagnostic criterion. It is, therefore, not surprising that PD has had perhaps the longest standing history in the investigation of interoceptive processes, however, with very mixed results. There are reports of increased IA in panic patients (Ehlers and Breuer, 1992; Ehlers et al., 1995), but also those failing to find any differences in IA between panic patients and healthy individuals (Antony et al., 1995; Van der Does et al., 1997; Wolk et al., 2014). This mixed picture of results gave rise to the assumption that methodological differences (e.g., the precise wording of instructions for the most often employed heart-beat tracking task) between studies may account for these inconsistent findings (Ehlers et al., 1995; Ehlers and Breuer, 1996). Alternatively, the possibility of differences in IA between PD patients has been discussed (i.e., a PD subgroup high in IA; (Ehlers et al., 1995; Ehlers and Breuer, 1996; Van der Does et al., 1997; Willem Van der Does et al., 2000). Furthermore, the question arose whether IA may represent a risk factor for the development of PD symptoms. Supporting this assumption, IA was found to be higher in patients with maintained or relapsed PD and infrequent panic attacks, than in remitted patients (Ehlers, 1995). Moreover, exposure treatment did not change IA (Ehlers et al., 1995).

In contrast to PD, IA has been shown to be lower in SD (Bogaerts et al., 2010; Pollatos et al., 2011a; Weiss et al., 2014), and there also seems to be a positive association between symptom severity and IA impairment (Schaefer et al., 2012). Accordingly, reduced IA could be a risk factor for the development of SD. This hypothesis is supported in a recent study by Schaefer et al. (2014), showing that state symptom perception is reduced after cardiac IA training. Although the course of symptom severity in dependence of IA may suggest the direction of the assumed pathway, the quasi-experimental design of these studies prevents from causal inferences to be drawn. In fact, all of these studies included individuals who had already been diagnosed with a mental disorder at the time of the investigation and, therefore, a pre-morbid indicator of IA was not available. Again, to address this potential shortcoming, prospective studies in the general population or in specific high-risk groups (e.g., exposed to chronic stress) are required.

In addition to SD, reduced IA has also been observed in patients with anorexia nervosa (Pollatos et al., 2008) and obesity (Herbert and Pollatos, 2014), whereas intuitive eating is positively associated with IA (Herbert et al., 2013). The direction of the pathway between interoception and the generation of physical symptoms may depend on the type of mental disorder. As almost all studies in this field are cross-sectional and quasi-experimental in design, it is difficult to come to any conclusions on the direction of this relationship. Beyond clinically relevant eating behavior, the experimental manipulation of eating behavior using short-term food deprivation induces an increase in IA as assessed by heartbeat perception (Herbert et al., 2012) and HEP amplitudes (Schulz et al., 2015a). As some neuroendocrinological parameters, such as sympathetic tone or peptide YY output reverse from hyper- to hypo-activation in long-term fasting, one may argue that eating behavior could serve as a coping mechanism to regulate the perception of bodily sensations. The direction of the relationship between interoception and physical symptoms in eating disorders may, therefore, be reversed in disordered eating (path *d*) or even bi-directional. It needs to be acknowledged that this interpretation is based on experimentally manipulated eating behavior and its translation to pathological eating behavior (i.e., eating disorders) should be done with caution. In the interest of greater simplicity of our suggested model, we have refrained, however, from including this reversed path *d*, and acknowledge, therefore, that the current model is limited to physical symptoms and disorders that imply the passive experience of altered bodily sensations (e.g., MD, PD, SD, DPD), and does not apply to physical symptoms that are partially induced by disordered eating behavior.

Physical Symptoms and Stress (Path *e*)

It is widely accepted that the repeated and enduring experience of physical symptoms is (probably causally) related to the experience of stress, as denoted by the term 'symptom distress.' The majority of reported physical 'somatization' symptoms involve the experience of pain (Rief et al., 2001). Chronic pain is described in the literature as severe 'inescapable stress.' In animal models chronic pain has been shown to induce depressive-like symptoms, which may provide a model for the co-occurrence of chronic stress, pain and depression in humans (Blackburn-Munro and Blackburn-Munro, 2001). In support of this notion, individuals with chronic pain exhibit higher perceived chronic stress and higher cortisol as assessed from hair (Van Uum et al., 2008), an indicator sensitive to chronic stress (Russell et al., 2012). Symptom distress may not necessarily have sufficient quality and intensity to be comparable to severe chronic psychosocial stress, such as caregiving to family members or harassment at working place. Nevertheless, the current model posits that the repeated exposure to physical symptoms may induce stress, thereby increasing and adding to the experience of already existing stress. In SD, symptom distress is largely affected by SD-typical forms of behavior, such as inadequate reassurance or negative interactions between patient and doctor (Rief and Broadbent, 2007). As symptom distress is associated with the automatic negative evaluation of afferent somatosensory signals (Witthoft et al., 2012), it could be hypothesized that specific cognitive styles may contribute to the positive feedback mechanism relating stress and physical symptoms. The correlational design of this study, however, does not allow for any such causal interpretations, i.e., whether automatic negative evaluation is a cause or consequence of symptom distress. In the current model, negative evaluation style may be seen as a moderator of the pathway between physical symptoms and stress. In summary, we propose that via the impact of physical symptoms on stress, the vicious circle between stress and symptoms are completed, which is characteristic of a self-maintaining positive feedback mechanism.

Neuroendocrine Pathways

Comprehensive reviews on neural structures supporting interoceptive signal processing, and original research papers including neuroimaging data are available elsewhere (Cameron, 2001; Craig, 2002; Critchley et al., 2004; Pollatos et al., 2007a,c). In the following we provide a summary of neural pathways that are important for transmitting and processing afferent signals from the body to the brain, and which may also be involved in the experience of stress. The majority of the available literature focuses on interoception of cardiac sensations. As previously described (Schulz et al., 2013c), visceral-afferent neural signals from the cardiovascular system are relayed over the NTS, the major sensory center for visceral-afferent neural signals in the brainstem (Jänig, 2006). The NTS projects onto the parabrachial nucleus and the LC, from where hypothalamic and thalamic nuclei are reached (Cameron, 2001). Cortical structures that process visceral-afferent neural signals include the anterior cingulate cortex (ACC), the frontal cortex, the somatosensory cortex and the right insula (Cameron, 2001; Critchley et al., 2004; Pollatos et al., 2007a,c). In a dipole localization study it was demonstrated that HEPs originate from exactly these four brain areas (Pollatos et al., 2005). When focusing on brain regions that are involved in the elicitation of a stress response, specific attention is paid to the ACC and the right insula. Blood perfusion in both areas is increased after a mental arithmetic task in individuals with high self-reported stress (Wang et al., 2005). The replication of these results suggests that these areas are specifically sensitive to psychological stress in women (Wang et al., 2007). The importance of the ACC in mediating stress responses has been demonstrated in numerous studies: functional connectivity in the ACC is related to cortisol release in response to a combined dexamethasone/CRF administration in healthy individuals (Kiem et al., 2013), while ACC connectivity may be reduced in traumatic stress experience (Kennis et al., 2014). Altered insular activity can be observed in early life (Mueller et al., 2010b) and traumatic stress (Bruce et al., 2012). However, these studies may not disentangle whether the ACC and the insula are involved in the up-regulation of physiological stress axes or if their altered activity and connectivity is a result of altered afferent input from visceral organs due to the dysregulation of stress responses.

Brain regions that are sensitive to the effects of stress hormones, and play a role in interoception, involve the thalamus, and other limbic structures, such as the amygdala and hippocampus, which are important for learning and memory (see below). In particular, blood perfusion in the thalamus is reduced within a time frame of 17 min after cortisol infusion, implying a rapid, non-genomic effect of cortisol (Strelzyk et al., 2012). In contrast, when investigating a later time period, in which genomic and non-genomic mechanisms overlap, cortisol affects activity in the amygdala and hippocampus, whereas no effect on the thalamus could be observed (Lovallo et al., 2010). The thalamus is a major relay center for sensory information that processes and integrates intero- and exteroceptive signals. Cortisol may rapidly affect thalamic activity and could eventually favor the processing of interoceptive signals at the cost of others, while this effect disappears over time. Future studies should clarify if thalamic activity may be chronically altered in HPA axis dysregulation, which may play a role in the perception of physical symptoms.

At receptor level, special attention has been paid to the role of α2-adrenoceptors. As repeatedly demonstrated, early life stress (Caldji et al., 1998; Liu et al., 2000) and chronic stress (Flugge et al., 2003) may impair the expression of α2-receptors in the NTS, which plays an important role in the down-regulation of the sympathetic nervous system (Flugge, 1999) and the sensory relaying of visceral-afferent neural signals (Jänig, 2006). α2-adrenoceptors are involved in the adequate processing of visceral-afferent signals from the cardiovascular (Sved et al., 1992) and the gastrointestinal system (Myers et al., 2005). It could be argued, therefore, that the reduced density and functionality of α2-adrenoceptors under chronic stress may reduce the individual's capacity for the adequate processing of visceral-afferent signals, which may eventually result in the generation of physical symptoms.

In SD patients there is evidence for altered activity in medullary control mechanisms for the sympathetic nervous system (Laederach-Hofmann et al., 2008), the ACC (Klug et al., 2011) and insula (Gundel et al., 2008). PD may be accompanied by reduced sensitivity of α2-adrenoceptors (Bremner et al., 1996), while reduced gray matter in the insula (Uchida et al., 2008) and altered ACC activity (Asami et al., 2008; Shin et al., 2013) could also be observed. Anorexia nervosa is characterized by changes in insular activity (Oberndorfer et al., 2013; Strigo et al., 2013). We propose that alterations in receptor sensitivity, volume, blood-flow or connectivity of the respective brain regions could be a result of stress system dysregulation and represent an important factor in altered interoception in these disorders.

Taken together, acute stress, the release of stress hormones and chronic stress may affect the processing of visceral-afferent neural signals at different brain levels, which are important for interoception, such as the NTS, the thalamus, the ACC and the insula. It remains for future research to elucidate whether certain mental disorders, which are associated with physical symptoms, can be differentiated by specific neurobiological patterns of dysregulated receptor- and brain area functioning.

Psychobiological Mechanisms Involved in Interoception and Stress

While the present review and the proposed model focus on the role of physiological stress responses and their dysregulation for altered interoception and the generation of physical symptoms, stress responses include psychological processes that are constituent factors of the experience of stress. In the following we, therefore, briefly discuss psychological processes that may be important for interoception and stress, without making specific reference to these in the proposed model.

Anxiety and Interoception

Individual stress responses are accompanied by mood changes that include anxiety or fear. There is indeed such an overlap of concepts, methodologies (including experimental paradigms) and physiological correlates of acute stress and anxiety that it seems all but impossible to differentiate between the two constructs (Stevens et al., 2011; Schmitz et al., 2012; Durlik et al.,

2014). At least part of the literature discussed in the Section on "Acute Stress and Interoception (Path *b*)" could, therefore, also be subsumed under the title "Anxiety and Interoception." To understand similarities and differences between both concepts, their theoretical definitions have to be evaluated. On the one hand, anxiety is a distinct emotion, which is characterized by a limited duration, an antecedent appraisal of a stimulus, and multiple components including specific affective, behavioral, cognitive, and physiological response patterns (Russell, 2003; Ekkekakis, 2013). Anxiety has a state component, i.e., a transitory condition of anxiety, and a trait component, which describes a relative stable disposition to respond to stimuli with state anxiety (Spielberger and Reheiser, 2004). On the other hand, psychological (dis-)stress is characterized as core affect (component of negative affect), which has a long duration, a variable intensity, it does not require either an antecedent appraisal or an eliciting stimulus, and is restricted to the affective component (Russell, 2003; Ekkekakis, 2013). It has to be acknowledged, however, that this definition may be limited to affective response patterns as to be expected in enduring or chronic stress, but it does not include specific physiological response patterns that are implied by chronic exposure to stress. In stress research, the definition of acute stress is commonly based on a neuro-psycho-endocrine perspective. This perspective emphasizes physiological response patterns, as earlier described, but also includes behavioral, cognitive and affective components of stress, such as increased anxiety (McEwen, 2000). In contrast to the core affect of psychological distress, in research on acute stress as induced by established laboratory stressors, an antecedent appraisal of a specific stimulus exists, which may be an additional overlap with the emotional state of anxiety. Important differences between state anxiety and acute stress would be the broader range of distinct emotions that could be elicited during acute stress (e.g., anxiety, fear, anger, etc.) that imply a more diffuse pattern of behavioral and physiological responses, and the longer duration of an acute stress response. Trait anxiety, anxiety sensitivity, and anxiety disorders are all strongly associated with interoception. For example, healthy individuals with high social anxiety show higher IA than those with low social anxiety (Stevens et al., 2011). Furthermore, anxiety sensitivity may moderate the effects of acute stress on interoception (Sturges and Goetsch, 1996). As argued by Domschke et al. (2010), anxiety sensitivity may increase the vulnerability to anxiety disorders by increasing the perceptual basis for catastrophic interpretations of physical symptoms. Although we do not explicate the role of trait anxiety or anxiety sensitivity for interoception in the current model, it may be speculated that either aspect of anxiety could potentially moderate almost all proposed variables (e.g., the probability and intensity of stress experience, physical symptoms) and pathways, e.g., between physiological stress axes and interoception (path *c*), between interoception and perception of physical symptoms (*d*) and between symptoms and stress experience (*e*).

Attention and Interoception

In models on altered interoception in PD (Ehlers and Breuer, 1996; Willem Van der Does et al., 2000) or SD (Rief and Barsky, 2005; Rief and Broadbent, 2007) attention focused on physical symptoms plays an essential role. While there is largely agreement in the literature on attentional bias to body- and health-related sensations in patients with PD and SD, evidence on the factors or processes contributing to attentional bias (e.g., acute and chronic stress) and, therefore, physical symptom generation, is lacking. Regarding the SAM axis, the central noradrenergic system, including the LC, is of particular importance. Early life or chronic stress may decrease the expression of α2-adrenoceptors in the LC (Caldji et al., 1998; Liu et al., 2000), which are involved in the regulation of the 'alertness' component of attention. As described above, α2-adrenoceptors are involved in a negative feedback mechanism that down-regulates sympathetic activation and associated processes, such as alertness and vigilance. Consequently, the acute activation of α2-adrenergic receptors increases vigilance and alertness (Berridge and Foote, 1991), while their deactivation results in chronic hypervigilance and alertness (Aston-Jones et al., 1999; Usher et al., 1999). Regarding the HPA axis, the relationship between stress hormones and attentional processes may be more complex, since cortisol may affect all neurons in the entire brain. As summarized by Erickson et al. (2003), glucocorticoids facilitate focused attention at the cost of irrelevant stimuli and alter the perceptual threshold of external stimuli (Fehm-Wolfsdorf and Nagel, 1996). These effects are probably due to increased availability of norepinephrine, as caused by glucocorticoids (Irwin et al., 1986; Kvetnansky et al., 1993). Future research should clarify whether this change in perceptual threshold may also apply to interoceptive signals. Mental stress tests that have been shown to affect both stress axes, also induce increased selective attention on primary tasks at the expense of cognitive flexibility (Chajut and Algom, 2003; Plessow et al., 2011). The differential effects of stress on IA as assessed by the Schandry- and Whitehead-based heartbeat perception tasks (Schulz et al., 2013a) indicate that different facets of attention are of relevance for the impact of stress on interoception. It is plausible, for example, that focused attention on interoceptive signals is facilitated (Schandry task), while the integration of intero- and exteroceptive signals is attenuated (Whitehead task) during stress. With respect to the proposed model, attention may moderate the relationship between physiological stress axes and interoception (path *c*), between interoception and the perception of physical symptoms (*d*), and between symptoms and the experience of stress (*e*).

The Role of Interoceptive Signal Processing in Stress Effects on Memory and Learning

Stress affects memory at acquisition, consolidation and recall. Interoceptive signal processing may contribute to some of these mechanisms (Schulz, 2015). McGaugh (2000) and Roozendaal et al. (2006) postulated the requirement of two physiological processes for the enhancement of memory consolidation after stress: (1) glucocorticoids that bind on glucocorticoid receptors within the NTS, the LC and the basolateral amygdala (BLA) and (2) peripherally circulating epinephrine, which cannot cross the blood-brain barrier, but activates visceral organs, whose afferent signals are transmitted via the vagus nerve to the NTS and LC. From these structures, via noradrenergic pathways, β-adrenergic synapses in the BLA are activated, which induces the release of cyclic adenosine monophosphate (cAMP) and cAMP-dependent protein kinase. Both substances can enhance memory consolidation. Postsynaptic efficacy of β-adrenergic synapses is increased by glucocorticoids. In animal studies it could be shown that both processes glucocorticoid secretion (Quirarte et al., 1997) and afferent signal transmission from visceral organs (Roozendaal et al., 1999) are necessary for improvement of memory consolidation. In humans, the increase of baro-afferent neural feedback induced by norepinephrine-infusion can improve long-term memory (Moor et al., 2005). Interoceptive neural signals, therefore, play an important role for the stress-induced improvement of memory consolidation. It is plausible that increased visceral-afferent input could also facilitate dysfunctional learning processes that contribute to the generation of physical symptoms, such as the formation of pain memory.

Intuitive Decision Making

The influential somatic marker hypothesis by Damasio (1994) postulates that interoceptive signals are regularly integrated into processes of intuitive decision making. It is assumed that in every context, in which intuitive decisions are made, the outcomes of possible action alternatives are anticipated. This anticipation produces a specific visceral response to be integrated in affective responses to the expected outcome. These responses, either transmitted over somatosensory or visceral-afferent circuits, are called 'somatic markers.' Dunn et al. (2006) have provided a comprehensive review of the somatic marker hypothesis. In support of this theory IA positively correlates with performance in decision-making tasks (Werner et al., 2009; Wolk et al., 2014). Stress affects executive functions and thus decision making: cortisol may reduce performance in intuitive decision making tasks both in humans and rodents (van den Bos et al., 2009; Koot et al., 2013). This effect was explained by rapid, presumably non-genomic glucocorticoid effects in the orbitofrontal and insular cortex (Koot et al., 2013), which are also involved in interoceptive signal processing. Furthermore, a quadratic relationship exists between peripheral SAM axis activity and performance in decision making tasks, implying a performance increase in moderate sympathetic activation (Pabst et al., 2013). Again, since catecholamines do not cross the blood–brain barrier, afferent neural signals from visceral organs (e.g., transmitted over the vagus nerve) due to increased stimulation of interoceptors, are likely to be involved in this effect. All effects of peripheral sympathetic activation on brain functions, therefore, require neural signal transmission from visceral organs, which can be considered 'interoceptive' signals.

Integration and Outlook

In the earlier introduced model of interoception by Vaitl (1996) different levels of interoception were proposed. The model implies that interoception is a bottom–up process, whose effects at a low hierarchical level consecutively affect higher levels. The current review summarizes the state-of-the-art knowledge

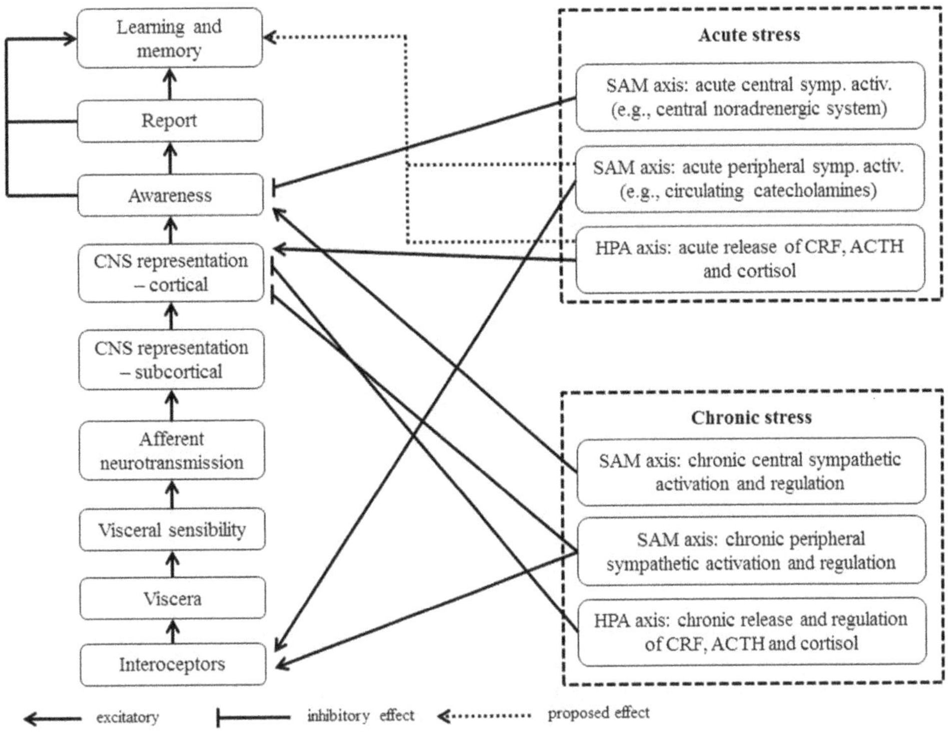

FIGURE 2 | Overview of existing findings on interoception and stress based on the model of interoception by Vaitl (1996).

on the effects of stress (research objective 1), mainly based on research investigating afferent signals from the cardiovascular system. In **Figure 2** we integrate the reported findings on interoception and stress into a model, specifically providing links to different levels of interoception, as suggested by Vaitl (1996). In contrast to the existing model, and based on evidence of dissociations between interoceptive indicators for cortical (HEP; Schulz et al., 2013c) and brainstem processes (CMS; Schulz et al., 2010), in the current model we separated 'CNS representation' into cortical and sub-cortical processes. Furthermore, we focus on visceral-afferent signals and do not address the possible contribution of somatosensory information to interoception. As some conditions are associated with selective degeneration of afferent autonomic nerves (e.g., diabetes mellitus), the level 'afferent neurotransmission' was newly introduced into the model. Although the same mechanisms (e.g., stress axes) are involved in mediating acute stress responses and chronic stress conditions, the complex interplay between stress hormones, binding sites and feedback regulation requires their independent consideration. Acute peripheral sympathetic activation clearly increases stimulation of cardiac interoceptors (Moor et al., 2005). Given the clear physiological relationship between circulating catecholamines and stimulation of cardiac interoceptors (e.g., arterial baroreceptors), we assume that this relationship persists also in chronic elevations of catecholamine output. To our knowledge, selective effects of stress on afferent neurotransmission or sub-cortical CNS representation concerning interoception are unknown, as of to date. A first study by our group suggests that there is no effect of cortisol on the CMS (Schulz et al., 2010), possibly reflecting brainstem processing of visceral-afferent neural signals (Schulz et al., 2009a). On the one hand, acute administration of cortisol affects the cortical representation of visceral-afferent signals (Schulz et al., 2013c) as indicated by larger HEP amplitudes (when eyes are open). On the other hand, basal cortisol levels are negatively associated with HEPs (Schulz et al., 2015b). The additional finding that HEP sensitivity for attention focused on heartbeats is negatively associated with α-amylase may implicate that repeated stimulation of interoceptors could reduce sensitivity of CNS representation at a cortical level. Central sympathetic activation has particularly strong effects on attention and awareness. The overall activation of the central noradrenergic system (via α2-adrengergic antagonists) facilitates arousal and alertness, which may impair the adequate processing of visceral-afferent signals. This relationship may eventually be translated to chronic states of central sympathetic activation. It has repeatedly been demonstrated that chronic stress (Flugge, 1999) and mental disorders associated with physical symptoms are related to the dysregulation of α2-adrenoceptors (Nutt, 1989; Bremner et al., 1996). We thus assume that the chronically inadequate processing of visceral-afferent signals may reverse in increased awareness of physical sensations, which may facilitate the generation of physical symptoms. The collective effect of sympathetic activation and cortisol on learning and memory is conceptualized as 'proposed' effect in this model, since it has so far only been demonstrated for declarative memory (Roozendaal et al., 2004; McGaugh, 2006), although it may also play a role in the formation of memory for physical symptoms. With this model, we hope to stimulate future research on the yet unknown relationships between physiological stress systems and levels of interoceptive signal processing.

Author Contributions

AS conceived the idea to conduct this review; AS and CV performed literature search; AS and CV interpreted the data and integrated them into the proposed framework models; and AS and CV authored the manuscript.

Acknowledgments

This work is in parts supported by means of the project "Interoception in medically unexplained symptoms" (INMEDEX) (grant to Dr. André Schulz), funded by the University of Luxembourg.

References

Andor, T., Gerlach, A. L., and Rist, F. (2008). Superior perception of phasic physiological arousal and the detrimental consequences of the conviction to be aroused on worrying and metacognitions in GAD. *J. Abnorm. Psychol.* 117, 193–205. doi: 10.1037/0021-843X.117.1.193

Antony, M. M., Brown, T. A., Craske, M. G., Barlow, D. H., Mitchell, W. B., and Meadows, E. A. (1995). Accuracy of heartbeat perception in panic disorder, social phobia and nonanxious subjects. *J. Anxiety Disord.* 5, 355–371. doi: 10.1016/0887-6185(95)00017-I

Arborelius, L., Owens, M. J., Plotsky, P. M., and Nemeroff, C. B. (1999). The role of corticotropin-releasing factor in depression and anxiety disorders. *J. Endocrinol.* 160, 1–12. doi: 10.1677/joe.0.1600001

Asami, T., Hayano, F., Nakamura, M., Yamasue, H., Uehara, K., Otsuka, T., et al. (2008). Anterior cingulate cortex volume reduction in patients with panic disorder. *Psychiatry Clin. Neurosci.* 62, 322–330. doi: 10.1111/j.1440-1819.2008.01800.x

Aston-Jones, G., Rajkowski, J., and Cohen, J. (1999). Role of locus coeruleus in attention and behavioral flexibility. *Biol. Psychiatry* 46, 1309–1320. doi: 10.1016/S0006-3223(99)00140-7

Barsky, A. J. (1992). Amplification, somatization, and the somatoform disorders. *Psychosomatics* 33, 28–34. doi: 10.1016/S0033-3182(92)72018-0

Barsky, A. J., Brener, J., Coeytaux, R. R., and Cleary, P. D. (1995). Accurate awareness of heartbeat in hypochondriacal and non-hypochondriacal patients. *J. Psychosom. Res.* 39, 489–497. doi: 10.1016/0022-3999(94)00166-3

Barsky, A. J., Goodson, J. D., Lane, R. S., and Cleary, P. D. (1988). The amplification of somatic symptoms. *Psychosom. Med.* 50, 510–519. doi: 10.1097/00006842-198809000-00007

Barsky, A. J., Wyshak, G., and Klerman, G. L. (1990). The somatosensory amplification scale and its relationship to hypochondriasis. *J. Psychiatr. Res.* 24, 323–334. doi: 10.1016/0022-3956(90)90004-A

Bella, G. P., Garcia, M. C., and Spadari-Bratfisch, R. C. (2011). Salivary cortisol, stress, and health in primary caregivers (mothers) of children with cerebral palsy. *Psychoneuroendocrinology* 36, 834–842. doi: 10.1016/j.psyneuen.2010.11.005

Berridge, C. W., and Foote, S. L. (1991). Effects of locus coeruleus activation on electroencephalographic activity in neocortex and hippocampus. *J. Neurosci.* 11, 3135–3145.

Blackburn-Munro, G., and Blackburn-Munro, R. E. (2001). Chronic pain, chronic stress and depression: coincidence or consequence? *J. Neuroendocrinol.* 13, 1009–1023. doi: 10.1046/j.0007-1331.2001.00727.x

Bogaerts, K., Van Eylen, L., Li, W., Bresseleers, J., Van Diest, I., De Peuter, S., et al. (2010). Distorted symptom perception in patients with medically unexplained symptoms. *J. Abnorm. Psychol.* 119, 226–234. doi: 10.1037/a0017780

Bremner, J. D., Krystal, J. H., Southwick, S. M., and Charney, D. S. (1996). Noradrenergic mechanisms in stress and anxiety: II. *Clin. Stud. Synapse* 23, 39–51. doi: 10.1002/(SICI)1098-2396(199605)23:1<39::AID-SYN5>3.0.CO;2-I

Brener, J., and Jones, J. M. (1974). Interoceptive discrimination in intact humans: detection of cardiac activity. *Physiol. Behav.* 13, 763–767. doi: 10.1016/0031-9384(74)90259-5

Brener, J., and Kluvitse, C. (1988). Heartbeat detection: judgments of the simultaneity of external stimuli and heartbeats. *Psychophysiology* 25, 554–561. doi: 10.1111/j.1469-8986.1988.tb01891.x

Bruce, S. E., Buchholz, K. R., Brown, W. J., Yan, L., Durbin, A., and Sheline, Y. I. (2012). Altered emotional interference processing in the amygdala and insula in women with Post-Traumatic Stress Disorder. *Neuroimage Clin.* 2, 43–49. doi: 10.1016/j.nicl.2012.11.003

Cacioppo, J. T., Burleson, M. H., Poehlmann, K. M., Malarkey, W. B., Kiecolt-Glaser, J. K., Berntson, G. G., et al. (2000). Autonomic and neuroendocrine responses to mild psychological stressors: effects of chronic stress on older women. *Ann. Behav. Med.* 22, 140–148. doi: 10.1007/BF02895778

Caldji, C., Tannenbaum, B., Sharma, S., Francis, D., Plotsky, P. M., and Meaney, M. J. (1998). Maternal care during infancy regulates the development of neural systems mediating the expression of fearfulness in the rat. *Proc. Natl. Acad. Sci. U.S.A.* 95, 5335–5340. doi: 10.1073/pnas.95.9.5335

Cameron, O. G. (2001). Interoception: the inside story–a model for psychosomatic processes. *Psychosom. Med.* 63, 697–710. doi: 10.1097/00006842-200109000-00001

Ceunen, E., Van Diest, I., and Vlaeyen, J. W. (2013). Accuracy and awareness of perception: related, yet distinct (commentary on Herbert et al., 2012). *Biol. Psychol.* 92, 426–427. doi: 10.1016/j.biopsycho.2012.09.012

Chajut, E., and Algom, D. (2003). Selective attention improves under stress: implications for theories of social cognition. *J. Pers. Soc. Psychol.* 85, 231–248. doi: 10.1037/0022-3514.85.2.231

Chan, P. Y., and Davenport, P. W. (2010). Respiratory related evoked potential measures of cerebral cortical respiratory information processing. *Biol. Psychol.* 84, 4–12. doi: 10.1016/j.biopsycho.2010.02.009

Chandola, T., Heraclides, A., and Kumari, M. (2010). Psychophysiological biomarkers of workplace stressors. *Neurosci. Biobehav. Rev.* 35, 51–57. doi: 10.1016/j.neubiorev.2009.11.005

Chang, L., Sundaresh, S., Elliott, J., Anton, P. A., Baldi, P., Licudine, A., et al. (2009). Dysregulation of the hypothalamic-pituitary-adrenal (HPA) axis in irritable bowel syndrome. *Neurogastroenterol. Motil.* 21, 149–159. doi: 10.1111/j.1365-2982.2008.01171.x

Chatterton, R. T. Jr., Vogelsong, K. M., Lu, Y. C., Ellman, A. B., and Hudgens, G. A. (1996). Salivary alpha-amylase as a measure of endogenous adrenergic activity. *Clin. Physiol.* 16, 433–448. doi: 10.1111/j.1475-097X.1996.tb00731.x

Chrousos, G. P. (2009). Stress and disorders of the stress system. *Nat. Rev. Endocrinol.* 5, 374–381. doi: 10.1038/nrendo.2009.106

Chrousos, G. P., and Gold, P. W. (1992). The concepts of stress and stress system disorders. Overview of physical and behavioral homeostasis. *Jama* 267, 1244–1252. doi: 10.1001/jama.1992.03480090092034

Cleare, A. J. (2004). The HPA axis and the genesis of chronic fatigue syndrome. *Trends Endocrinol. Metab.* 15, 55–59. doi: 10.1016/j.tem.2003.12.002

Cobos, P., Sánchez, M., Pérez, N., and Vila, J. (2004). Effects of spinal cord injuries on the subjective component of emotions. *Cogn. Emot.* 18, 281–287. doi: 10.1080/02699930244000471

Cohen, S., and Hamrick, N. (2003). Stable individual differences in physiological response to stressors: implications for stress-elicited changes in immune related health. *Brain Behav. Immun.* 17, 407–414. doi: 10.1016/S0889-1591(03)00110-7

Craig, A. D. (2002). How do you feel? Interoception: the sense of the physiological condition of the body. *Nat. Rev. Neurosci.* 3, 655–666. doi: 10.1038/nrn894

Craig, A. D. (2009). How do you feel–now? The anterior insula and human awareness. *Nat. Rev. Neurosci.* 10, 59–70. doi: 10.1038/nrn2555

Critchley, H. D., Wiens, S., Rotshtein, P., Ohman, A., and Dolan, R. J. (2004). Neural systems supporting interoceptive awareness. *Nat. Neurosci.* 7, 189–195. doi: 10.1038/nn1176

Damasio, A. (1994). *Descartes' Error: Emotion, Reason and the Human Brain.* New York: Grosset/Putnam.

Davenport, P. W., Friedman, W. A., Thompson, F. J., and Franzen, O. (1986). Respiratory-related cortical potentials evoked by inspiratory occlusion in humans. *J. Appl. Physiol.* 60, 1843–1848.

De Andres-Garcia, S., Moya-Albiol, L., and Gonzalez-Bono, E. (2012). Salivary cortisol and immunoglobulin A: responses to stress as predictors of health complaints reported by caregivers of offspring with autistic spectrum disorder. *Horm. Behav.* 62, 464–474. doi: 10.1016/j.yhbeh.2012.08.003

de Kloet, E. R., Karst, H., and Joels, M. (2008). Corticosteroid hormones in the central stress response: quick-and-slow. *Front. Neuroendocrinol.* 29:268. doi: 10.1016/j.yfrne.2007.10.002

De Kloet, E. R., Vreugdenhil, E., Oitzl, M. S., and Joels, M. (1998). Brain corticosteroid receptor balance in health and disease. *Endocr. Rev.* 19, 269–301. doi: 10.1210/er.19.3.269

de Vugt, M. E., Nicolson, N. A., Aalten, P., Lousberg, R., Jolle, J., and Verhey, F. R. (2005). Behavioral problems in dementia patients and salivary cortisol patterns in caregivers. *J. Neuropsychiatry Clin. Neurosci.* 17, 201–207. doi: 10.1176/jnp.17.2.201

Domschke, K., Stevens, S., Pfleiderer, B., and Gerlach, A. L. (2010). Interoceptive sensitivity in anxiety and anxiety disorders: an overview and integration of neurobiological findings. *Clin. Psychol. Rev.* 30, 1–11. doi: 10.1016/j.cpr.2009.08.008

Dunn, B. D., Dalgleish, T., and Lawrence, A. D. (2006). The somatic marker hypothesis: a critical evaluation. *Neurosci. Biobehav. Rev.* 30, 239–271. doi: 10.1016/j.neubiorev.2005.07.001

Dunn, B. D., Dalgleish, T., Ogilvie, A. D., and Lawrence, A. D. (2007). Heartbeat perception in depression. *Behav. Res. Ther.* 45, 1921–1930. doi: 10.1016/j.brat.2006.09.008

Durlik, C., Brown, G., and Tsakiris, M. (2014). Enhanced interoceptive awareness during anticipation of public speaking is associated with fear of negative evaluation. *Cogn. Emot.* 28, 530–540. doi: 10.1080/02699931.2013.832654

Ehlers, A. (1995). A 1-year prospective study of panic attacks: clinical course and factors associated with maintenance. *J. Abnorm. Psychol.* 104, 164–172. doi: 10.1037/0021-843X.104.1.164

Ehlers, A., and Breuer, P. (1992). Increased cardiac awareness in panic disorder. *J. Abnorm. Psychol.* 101, 371–382. doi: 10.1037/0021-843X.101.3.371

Ehlers, A., and Breuer, P. (1996). How good are patients with panic disorder at perceiving their heartbeats? *Biol. Psychol.* 42, 165–182. doi: 10.1016/0301-0511(95)05153-8

Ehlers, A., Breuer, P., Dohn, D., and Fiegenbaum, W. (1995). Heartbeat perception and panic disorder: possible explanations for discrepant findings. *Behav. Res. Ther.* 33, 69–76. doi: 10.1016/0005-7967(94)E0002-Z

Ehlert, U., Gaab, J., and Heinrichs, M. (2001). Psychoneuroendocrinological contributions to the etiology of depression, posttraumatic stress disorder, and stress-related bodily disorders: the role of the hypothalamus-pituitary-adrenal axis. *Biol. Psychol.* 57, 141–152. doi: 10.1016/S0301-0511(01)00092-8

Eichler, S., and Katkin, E. S. (1994). The relationship between cardiovascular reactivity and heartbeat detection. *Psychophysiology* 31, 229–234. doi: 10.1111/j.1469-8986.1994.tb02211.x

Ekkekakis, P. (2013). *The Measurement of Affect, Mood, and Emotion.* Cambridge: Cambridge University Press. doi: 10.1017/CBO9780511820724

Elsenbruch, S., Rosenberger, C., Bingel, U., Forsting, M., Schedlowski, M., and Gizewski, E. R. (2010a). Patients with irritable bowel syndrome have altered emotional modulation of neural responses to visceral stimuli. *Gastroenterology* 139, 1310–1319. doi: 10.1053/j.gastro.2010.06.054

Elsenbruch, S., Rosenberger, C., Enck, P., Forsting, M., Schedlowski, M., and Gizewski, E. R. (2010b). Affective disturbances modulate the neural processing of visceral pain stimuli in irritable bowel syndrome: an fMRI study. *Gut* 59, 489–495. doi: 10.1136/gut.2008.175000

Erickson, K., Drevets, W., and Schulkin, J. (2003). Glucocorticoid regulation of diverse cognitive functions in normal and pathological emotional states. *Neurosci. Biobehav. Rev.* 27, 233–246. doi: 10.1016/S0149-7634(03)00033-2

Fahrenberg, J., and Foerster, F. (1982). Covariation and consistency of activation parameters. *Biol. Psychol.* 15, 151–169. doi: 10.1016/0301-0511(82)90039-4

Fairclough, S. H., and Goodwin, L. (2007). The effect of psychological stress and relaxation on interoceptive accuracy: implications for symptom perception. *J. Psychosom. Res.* 62, 289–295. doi: 10.1016/j.jpsychores.2006.10.017

Farb, N. A., Daubenmier, J. J., Price, C. J., Gard, T., Kerr, C. E., Dunn, B. D., et al. (2015).Interoception, contemplative practice, and health. *Front. Psychol.* 6:763 doi: 10.3389/fpsyg.2015.00763

Fehm-Wolfsdorf, G., and Nagel, D. (1996). Differential effects of glucocorticoids on human auditory perception. *Biol. Psychol.* 42, 117–130. doi: 10.1016/0301-0511(95)05150-3

Flugge, G. (1999). Effects of cortisol on brain alpha2-adrenoceptors: potential role in stress. *Neurosci. Biobehav. Rev.* 23, 949–956. doi: 10.1016/S0149-7634(99)00028-7

Flugge, G., Van Kampen, M., Meyer, H., and Fuchs, E. (2003). Alpha2A and alpha2C-adrenoceptor regulation in the brain: alpha2A changes persist after chronic stress. *Eur. J. Neurosci.* 17, 917–928. doi: 10.1046/j.1460-9568.2003.02510.x

Frattola, A., Parati, G., Gamba, P., Paleari, F., Mauri, G., Di Rienzo, M., et al. (1997). Time and frequency domain estimates of spontaneous baroreflex sensitivity provide early detection of autonomic dysfunction in diabetes mellitus. *Diabetologia* 40, 1470–1475. doi: 10.1007/s001250050851

Fries, E., Hesse, J., Hellhammer, J., and Hellhammer, D. H. (2005). A new view on hypocortisolism. *Psychoneuroendocrinology* 30, 1010–1016. doi: 10.1016/j.psyneuen.2005.04.006

Garfinkel, S. N., and Critchley, H. D. (2013). Interoception, emotion and brain: new insights link internal physiology to social behaviour. Commentary on:: "Anterior insular cortex mediates bodily sensibility and social anxiety" by Terasawa et al. (2012). *Soc. Cogn. Affect. Neurosci.* 8, 231–234. doi: 10.1093/scan/nss140

Garfinkel, S. N., Seth, A. K., Barrett, A. B., Suzuki, K., and Critchley, H. D. (2015). Knowing your own heart: distinguishing interoceptive accuracy frominteroceptive awareness. *Biol. Psychol.* 104, 65–74. doi: 10.1016/j.biopsycho.2014.11.004

Garner, D. M. (1991). *Eating Disorder Inventory-2 Professional Manual.* Odessa: Psychological Resources.

Gold, P. W. (2015). The organization of the stress system and its dysregulation in depressive illness. *Mol. Psychiatry* 20, 32–47. doi: 10.1038/mp.2014.163

Gold, P. W., and Chrousos, G. P. (2002). Organization of the stress system and its dysregulation in melancholic and atypical depression: high vs low CRH/NE states. *Mol. Psychiatry* 7, 254–275. doi: 10.1038/sj.mp.4001032

Gold, P. W., Goodwin, F. K., and Chrousos, G. P. (1988). Clinical and biochemical manifestations of depression. Relation to the neurobiology of stress (2). *New Engl. J. Med.* 319, 413–420. doi: 10.1056/NEJM198808183190706

Gold, P. W., Licinio, J., Wong, M. L., and Chrousos, G. P. (1995). Corticotropin releasing hormone in the pathophysiology of melancholic and atypical depression and in the mechanism of action of antidepressant drugs. *Ann. N. Y. Acad. Sci.* 771, 716–729. doi: 10.1111/j.1749-6632.1995.tb44723.x

Gray, M. A., Taggart, P., Sutton, P. M., Groves, D., Holdright, D. R., Bradbury, D., et al. (2007). A cortical potential reflecting cardiac function. *Proc. Natl. Acad. Sci. U.S.A.* 104, 6818–6823. doi: 10.1073/pnas.0609509104

Groeneweg, F. L., Karst, H., De Kloet, E. R., and Joels, M. (2011). Rapid non-genomic effects of corticosteroids and their role in the central stress response. *J. Endocrinol.* 209, 153–167. doi: 10.1530/JOE-10-0472

Gundel, H., Valet, M., Sorg, C., Huber, D., Zimmer, C., Sprenger, T., et al. (2008). Altered cerebral response to noxious heat stimulation in patients with somatoform pain disorder. *Pain* 137, 413–421. doi: 10.1016/j.pain.2007.10.003

Gupta, A., Kilpatrick, L., Labus, J., Tillisch, K., Braun, A., Hong, J. Y., et al. (2014). Early adverse life events and resting state neural networks in patients with chronic abdominal pain: evidence for sex differences. *Psychosom. Med.* 76, 404–412. doi: 10.1097/PSY.0000000000000089

Hart, N., Mcgowan, J., Minati, L., and Critchley, H. D. (2013). Emotional regulation and bodily sensation: interoceptive awareness is intact in borderline personality disorder. *J. Personal. Disord.* 27, 506–518. doi: 10.1521/pedi_2012_26_049

Heim, C., Ehlert, U., and Hellhammer, D. H. (2000). The potential role of hypocortisolism in the pathophysiology of stress-related bodily disorders. *Psychoneuroendocrinology* 25, 1–35. doi: 10.1016/S0306-4530(99)00035-9

Heims, H. C., Critchley, H. D., Dolan, R., Mathias, C. J., and Cipolotti, L. (2004). Social and motivational functioning is not critically dependent on feedback of autonomic responses: neuropsychological evidence from patients with pure autonomic failure. *Neuropsychologia* 42, 1979–1988. doi: 10.1016/j.neuropsychologia.2004.06.001

Herbert, B. M., Blechert, J., Hautzinger, M., Matthias, E., and Herbert, C. (2013). Intuitive eating is associated with interoceptive sensitivity. Effects on body mass index. *Appetite* 70C, 22–30. doi: 10.1016/j.appet.2013.06.082

Herbert, B. M., Herbert, C., Pollatos, O., Weimer, K., Enck, P., Sauer, H., et al. (2012). Effects of short-term food deprivation on interoceptive awareness, feelings and autonomic cardiac activity. *Biol. Psychol.* 89, 71–79. doi: 10.1016/j.biopsycho.2011.09.004

Herbert, B. M., and Pollatos, O. (2014). Attenuated interoceptive sensitivity in overweight and obese individuals. *Eat. Behav.* 15, 445–448. doi: 10.1016/j.eatbeh.2014.06.002

Herbert, B. M., Pollatos, O., Flor, H., Enck, P., and Schandry, R. (2010). Cardiac awareness and autonomic cardiac reactivity during emotional picture viewing and mental stress. *Psychophysiology* 47, 342–354. doi: 10.1111/j.1469-8986.2009.00931.x

Holsboer, F. (2000). The corticosteroid receptor hypothesis of depression. *Neuropsychopharmacology* 23, 477–501. doi: 10.1016/S0893-133X(00)00159-7

Irwin, J., Ahluwalia, P., Zacharko, R. M., and Anisman, H. (1986). Central norepinephrine and plasma corticosterone following acute and chronic stressors: influence of social isolation and handling. *Pharmacol. Biochem. Behav.* 24, 1151–1154. doi: 10.1016/0091-3057(86)90471-5

Isaac, L. (1980). Clonidine in the central nervous system: site and mechanism of hypotensive action. *J. Cardiovasc. Pharmacol.* 2(Suppl. 1), S5–S19. doi: 10.1097/00005344-198000021-00002

Ising, M., Hohne, N., Siebertz, A., Parchmann, A. M., Erhardt, A., and Keck, M. (2012). Stress response regulation in panic disorder. *Curr. Pharm. Des.* 18, 5675–5684. doi: 10.2174/138161212803530880

Jänig, W. (2006). *The Integrative Action of the Autonomic Nervous System.* New York: Cambridge University Press. doi: 10.1017/CBO97805115 41667

Janssens, K. A., Oldehinkel, A. J., Verhulst, F. C., Hunfeld, J. A., Ormel, J., and Rosmalen, J. G. (2012). Symptom-specific associations between low cortisol responses and functional somatic symptoms: the TRAILS study. *Psychoneuroendocrinology* 37, 332–340. doi: 10.1016/j.psyneuen.2011.06.016

Jones, G. E., and Hollandsworth, J. G. (1981). Heart rate discrimination before and after exercise-induced augmented cardiac activity. *Psychophysiology* 18, 252–257. doi: 10.1111/j.1469-8986.1981.tb03029.x

Katkin, E. S., Blascovich, J., and Goldband, S. (1981). Empirical assessment of visceral self-perception: individual and sex differences in the acquisition of heartbeat discrimination. *J. Pers. Soc. Psychol.* 40, 1095–1101. doi: 10.1037/0022-3514.40.6.1095

Kennis, M., Rademaker, A. R., Van Rooij, S. J., Kahn, R. S., and Geuze, E. (2014). Resting state functional connectivity of the anterior cingulate cortex in veterans with and without post-traumatic stress disorder. *Hum. Brain Mapp.* 36, 99–109. doi: 10.1002/hbm.22615

Khalsa, S. S., Rudrauf, D., Damasio, A. R., Davidson, R. J., Lutz, A., and Tranel, D. (2008). Interoceptive awareness in experienced meditators. *Psychophysiology* 45, 671–677. doi: 10.1111/j.1469-8986.2008.00666.x

Kiem, S. A., Andrade, K. C., Spoormaker, V. I., Holsboer, F., Czisch, M., and Samann, P. G. (2013). Resting state functional MRI connectivity predicts hypothalamus-pituitary-axis status in healthy males. *Psychoneuroendocrinology* 38, 1338–1348. doi: 10.1016/j.psyneuen.2012.11.021

Kindermann, N. K., and Werner, N. S. (2014). Cardiac perception enhances stress experience. *J. Psychophysiol.* 28, 225–232. doi: 10.1027/0269-8803/a000114

Kirschbaum, C., Pirke, K. M., and Hellhammer, D. H. (1993). The 'trier social stress test'–a tool for investigating psychobiological stress responses in a laboratory setting. *Neuropsychobiology* 28, 76–81. doi: 10.1159/000119004

Klug, S., Anderer, P., Saletu-Zyhlarz, G., Freidl, M., Saletu, B., Prause, W., et al. (2011). Dysfunctional pain modulation in somatoform pain disorder patients. *Eur. Arch. Psychiatry Clin. Neurosci.* 261, 267–275. doi: 10.1007/s00406-010-0148-4

Knack, J. M., Jensen-Campbell, L. A., and Baum, A. (2011). Worse than sticks and stones? Bullying is associated with altered HPA axis functioning and poorer health. *Brain Cogn.* 77, 183–190. doi: 10.1016/j.bandc.2011.06.011

Knoll, J. F., and Hodapp, V. (1992). A comparison between two methods for assessing heartbeat perception. *Psychophysiology* 29, 218–222. doi: 10.1111/j.1469-8986.1992.tb01689.x

Koot, S., Baars, A., Hesseling, P., Van Den Bos, R., and Joels, M. (2013). Time-dependent effects of corticosterone on reward-based decision-making in a rodent model of the Iowa Gambling Task. *Neuropharmacology* 70, 306–315. doi: 10.1016/j.neuropharm.2013.02.008

Kudielka, B. M., Schommer, N. C., Hellhammer, D. H., and Kirschbaum, C. (2004). Acute HPA axis responses, heart rate, and mood changes to psychosocial stress (TSST) in humans at different times of day. *Psychoneuroendocrinology* 29, 983–992. doi: 10.1016/j.psyneuen.2003.08.009

Kuehl, L. K., Michaux, G. P., Richter, S., Schachinger, H., and Anton, F. (2010). Increased basal mechanical pain sensitivity but decreased perceptual wind-up in a human model of relative hypocortisolism. *Pain* 149, 539–546. doi: 10.1016/j.pain.2010.03.026

Kvetnansky, R., Fukuhara, K., Pacak, K., Cizza, G., Goldstein, D. S., and Kopin, I. J. (1993). Endogenous glucocorticoids restrain catecholamine synthesis and release at rest and during immobilization stress in rats. *Endocrinology* 133, 1411–1419. doi: 10.1210/en.133.3.1411

Laederach-Hofmann, K., Ruddel, H., and Mussgay, L. (2008). Pathological baroreceptor sensitivity in patients suffering from somatization disorders: do they correlate with symptoms? *Biol. Psychol.* 79, 243–249. doi: 10.1016/j.biopsycho.2008.06.003

Larra, M. F., Schulz, A., Schilling, T. M., Ferreira De Sa, D. S., Best, D., Kozik, B., et al. (2014). Heart rate response to post-learning stress predicts memory consolidation. *Neurobiol. Learn. Mem.* 109C, 74–81. doi: 10.1016/j.nlm.2013.12.004

Lass-Hennemann, J., Deuter, C. E., Kuehl, L. K., Schulz, A., Blumenthal, T. D., and Schachinger, H. (2010). Effects of stress on human mating preferences: stressed individuals prefer dissimilar mates. *Proc. Biol. Sci.* 277, 2175–2183. doi: 10.1098/rspb.2010.0258

Lass-Hennemann, J., Kuehl, L. K., Schulz, A., Oitzl, M. S., and Schachinger, H. (2011). Stress strengthens memory of first impressions of others' positive personality traits. *PLoS ONE* 6:e16389. doi: 10.1371/journal.pone.0016389

Leopold, C., and Schandry, R. (2001). The heartbeat-evoked brain potential in patients suffering from diabetic neuropathy and in healthy control persons. *Clin. Neurophysiol.* 112, 674–682. doi: 10.1016/S1388-2457(01)00480-1

Lesch, K. P., Widerlov, E., Ekman, R., Laux, G., Schulte, H. M., Pfuller, H., et al. (1988). Delta sleep-inducing peptide response to human corticotropin-releasing hormone (CRH) in major depressive disorder. Comparison with CRH-induced corticotropin and cortisol secretion. *Biol. Psychiatry* 24, 162–172. doi: 10.1016/0006-3223(88)90271-5

Levitan, R. D., Vaccarino, F. J., Brown, G. M., and Kennedy, S. H. (2002). Low-dose dexamethasone challenge in women with atypical major depression: pilot study. *J. Psychiatry Neurosci.* 27, 47–51.

Liu, D., Caldji, C., Sharma, S., Plotsky, P. M., and Meaney, M. J. (2000). Influence of neonatal rearing conditions on stress-induced adrenocorticotropin responses and norepinepherine release in the hypothalamic paraventricular nucleus. *J. Neuroendocrinol.* 12, 5–12. doi: 10.1046/j.1365-2826.2000.00422.x

Lovallo, W. R., Robinson, J. L., Glahn, D. C., and Fox, P. T. (2010). Acute effects of hydrocortisone on the human brain: an fMRI study. *Psychoneuroendocrinology* 35, 15–20. doi: 10.1016/j.psyneuen.2009.09.010

Marshall, R. D., Blanco, C., Printz, D., Liebowitz, M. R., Klein, D. F., and Coplan, J. (2002). A pilot study of noradrenergic and HPA axis functioning in PTSD vs. panic disorder. *Psychiatry Res.* 110, 219–230. doi: 10.1016/S0165-1781(02)00126-9

McCraty, R., Atkinson, M., Tomasino, D., and Stuppy, W. P. (2001). Analysis of twenty-four hour heart rate variability in patients with panic disorder. *Biol. Psychol.* 56, 131–150. doi: 10.1016/S0301-0511(01)00074-6

McEwen, B. S. (2000). The neurobiology of stress: from serendipity to clinical relevance. *Brain Res.* 886, 172–189. doi: 10.1016/S0006-8993(00)0 2950-4

McEwen, B. S. (2004). Protection and damage from acute and chronic stress: allostasis and allostatic overload and relevance to the pathophysiology of psychiatric disorders. *Ann. N. Y. Acad. Sci.* 1032, 1–7. doi: 10.1196/annals.1314.001

McGaugh, J. L. (2000). Memory–a century of consolidation. *Science* 287, 248–251. doi: 10.1126/science.287.5451.248

McGaugh, J. L. (2006). Make mild moments memorable: add a little arousal. *Trends Cogn. Sci.* 10, 345–347. doi: 10.1016/j.tics.2006.06.001

McLaughlin, K. A., Sheridan, M. A., Alves, S., and Mendes, W. B. (2014). Child maltreatment and autonomic nervous system reactivity: identifying dysregulated stress reactivity patterns by using the biopsychosocial model of challenge and threat. *Psychosom. Med.* 76, 538–546. doi: 10.1097/PSY.0000000000000098

Mehling, W. E., Price, C., Daubenmier, J. J., Acree, M., Bartmess, E., and Stewart, A. (2012). The Multidimensional assessment of interoceptive awareness (MAIA). *PLoS ONE* 7:e48230. doi: 10.1371/journal.pone.0048230

Michal, M., Reuchlein, B., Adler, J., Reiner, I., Beutel, M. E., Vogele, C., et al. (2014). Striking discrepancy of anomalous body experiences with normal interoceptive accuracy in depersonalization-derealization disorder. *PLoS ONE* 9:e89823. doi: 10.1371/journal.pone.0089823

Michaux, G. P., Magerl, W., Anton, F., and Treede, R. D. (2012). Experimental characterization of the effects of acute stresslike doses of hydrocortisone in human neurogenic hyperalgesia models. *Pain* 153, 420–428. doi: 10.1016/j.pain.2011.10.043

Moor, T., Mundorff, L., Bohringer, A., Philippsen, C., Langewitz, W., Reino, S. T., et al. (2005). Evidence that baroreflex feedback influences long-term incidental visual memory in men. *Neurobiol. Learn. Mem.* 84, 168–174. doi: 10.1016/j.nlm.2005.07.003

Moriguchi, A., Otsuka, A., Kohara, K., Mikami, H., Katahira, K., Tsunetoshi, T., et al. (1992). Spectral change in heart rate variability in response to mental arithmetic before and after the beta-adrenoceptor blocker, carteolol. *Clin. Auton. Res.* 2, 267–270. doi: 10.1007/BF01819547

Mravec, B. (2011). Role of catecholamine-induced activation of vagal afferent pathways in regulation of sympathoadrenal system activity: negative feedback loop of stress response. *Endocr. Regul.* 45, 37–41.

Mueller, E. M., Evers, E. A., Wacker, J., and Van Der Veen, F. (2012). Acute tryptophan depletion attenuates brain-heart coupling following external feedback. *Front. Hum. Neurosci.* 6:77. doi: 10.3389/fnhum.2012.00077

Mueller, E. M., Panitz, C., Nestoriuc, Y., Stemmler, G., and Wacker, J. (2014). Panic disorder and serotonin reuptake inhibitors predict coupling of cortical and cardiac activity. *Neuropsychopharmacology* 39, 507–514. doi: 10.1038/npp.2013.224

Mueller, E. M., Stemmler, G., Hennig, J., and Wacker, J. (2013). 5-HTTLPR and anxiety modulate brain-heart covariation. *Psychophysiology* 50, 441–453. doi: 10.1111/psyp.12016

Mueller, E. M., Stemmler, G., and Wacker, J. (2010a). Single-trial electroencephalogram predicts cardiac acceleration: a time-lagged P-correlation approach for studying neurovisceral connectivity. *Neuroscience* 166, 491–500. doi: 10.1016/j.neuroscience.2009.12.051

Mueller, S. C., Maheu, F. S., Dozier, M., Peloso, E., Mandell, D., Leibenluft, E., et al. (2010b). Early-life stress is associated with impairment in cognitive control in adolescence: an fMRI study. *Neuropsychologia* 48, 3037–3044. doi: 10.1016/j.neuropsychologia.2010.06.013

Mussgay, L., Klinkenberg, N., and Ruddel, H. (1999). Heart beat perception in patients with depressive, somatoform, and personality disorders. *J. Psychophysiol.* 13, 27–36. doi: 10.1027//0269-8803.13.1.27

Myers, E. A., Banihashemi, L., and Rinaman, L. (2005). The anxiogenic drug yohimbine activates central viscerosensory circuits in rats. *J. Comp. Neurol.* 492, 426–441. doi: 10.1002/cne.20727

Nater, U. M., Maloney, E., Boneva, R. S., Gurbaxani, B. M., Lin, J. M., Jones, J. F., et al. (2008). Attenuated morning salivary cortisol concentrations in a population-based study of persons with chronic fatigue syndrome and well controls. *J. Clin. Endocrinol. Metab.* 93, 703–709. doi: 10.1210/jc.2007-1747

Nater, U. M., Rohleder, N., Gaab, J., Berger, S., Jud, A., Kirschbaum, C., et al. (2005). Human salivary alpha-amylase reactivity in a psychosocial stress paradigm. *Int. J. Psychophysiol.* 55, 333–342. doi: 10.1016/j.ijpsycho.2004.09.009

Nemeroff, C. B., Widerlov, E., Bissette, G., Walleus, H., Karlsson, I., Eklund, K., et al. (1984). Elevated concentrations of CSF corticotropin-releasing factor-like immunoreactivity in depressed patients. *Science* 226, 1342–1344. doi: 10.1126/science.6334362

Nutt, D. J. (1989). Altered central alpha 2-adrenoceptor sensitivity in panic disorder. *Arch. Gen. Psychiatry* 46, 165–169. doi: 10.1001/archpsyc.1989.01810020067011

O'Keane, V., Dinan, T. G., Scott, L., and Corcoran, C. (2005). Changes in hypothalamic-pituitary-adrenal axis measures after vagus nerve stimulation therapy in chronic depression. *Biol. Psychiatry* 58, 963–968. doi: 10.1016/j.biopsych.2005.04.049

O'Keane, V., Frodl, T., and Dinan, T. G. (2012). A review of atypical depression in relation to the course of depression and changes in HPA axis organization. *Psychoneuroendocrinology* 37, 1589–1599. doi: 10.1016/j.psyneuen.2012.03.009

Oberndorfer, T., Simmons, A., Mccurdy, D., Strigo, I., Matthews, S., Yang, T., et al. (2013). Greater anterior insula activation during anticipation of food images in women recovered from anorexia nervosa versus controls. *Psychiatry Res.* 214, 132–141. doi: 10.1016/j.pscychresns.2013.06.010

Pabst, S., Brand, M., and Wolf, O. T. (2013). Stress and decision making: a few minutes make all the difference. *Behav. Brain Res.* 250, 39–45. doi: 10.1016/j.bbr.2013.04.046

Pappens, M., Van Den Bergh, O., Vansteenwegen, D., Ceunen, E., De Peuter, S., and Van Diest, I. (2013). Learning to fear obstructed breathing: comparing interoceptive and exteroceptive cues. *Biol. Psychol.* 92, 36–42. doi: 10.1016/j.biopsycho.2011.05.004

Pariante, C. M., and Lightman, S. L. (2008). The HPA axis in major depression: classical theories and new developments. *Trends Neurosci.* 31, 464–468. doi: 10.1016/j.tins.2008.06.006

Pauli, P., Hartl, L., Marquardt, C., Stalmann, H., and Strian, F. (1991). Heartbeat and arrhythmia perception in diabetic autonomic neuropathy. *Psychol. Med.* 21, 413–421. doi: 10.1017/S0033291700020523

Pennebaker, J. W., and Lightner, J. M. (1980). Competition of internal and external information in an exercise setting. *J. Pers. Soc. Psychol.* 39, 165–174. doi: 10.1037/0022-3514.39.1.165

Petersen, S., Schroijen, M., Molders, C., Zenker, S., and Van Den Bergh, O. (2014). Categorical Interoception: perceptual organization of sensations from inside. *Psychol. Sci.* 25, 1059–1066. doi: 10.1177/0956797613519110

Petrowski, K., Herold, U., Joraschky, P., Wittchen, H. U., and Kirschbaum, C. (2010). A striking pattern of cortisol non-responsiveness to psychosocial stress in patients with panic disorder with concurrent normal cortisol awakening responses. *Psychoneuroendocrinology* 35, 414–421. doi: 10.1016/j.psyneuen.2009.08.003

Petrowski, K., Wintermann, G. B., Schaarschmidt, M., Bornstein, S. R., and Kirschbaum, C. (2013). Blunted salivary and plasma cortisol response in patients with panic disorder under psychosocial stress. *Int. J. Psychophysiol.* 88, 35–39. doi: 10.1016/j.ijpsycho.2013.01.002

Philippsen, C., Hahn, M., Schwabe, L., Richter, S., Drewe, J., and Schachinger, H. (2007). Cardiovascular reactivity to mental stress is not affected by alpha2-adrenoreceptor activation or inhibition. *Psychopharmacology* 190, 181–188. doi: 10.1007/s00213-006-0597-7

Phillips, G. C., Jones, G. E., Rieger, E. J., and Snell, J. B. (1999). Effects of the presentation of false heart-rate feedback on the performance of two common heartbeat-detection tasks. *Psychophysiology* 36, 504–510. doi: 10.1017/S0048577299980071

Plessow, F., Fischer, R., Kirschbaum, C., and Goschke, T. (2011). Inflexibly focused under stress: acute psychosocial stress increases shielding of action goals at the expense of reduced cognitive flexibility with increasing time lag to the stressor. *J. Cogn. Neurosci.* 23, 3218–3227. doi: 10.1162/jocn_a_00024

Pohl, J., Frenzel, G., Kerner, W., and Fehm-Wolfsdorf, G. (1998). Acute stress modulates symptom awareness and hormonal counterregulation during insulin-induced hypoglycemia in healthy individuals. *Int. J. Behav. Med.* 5, 89–105. doi: 10.1207/s15327558ijbm0502_1

Pollatos, O., Gramann, K., and Schandry, R. (2007a). Neural systems connecting interoceptive awareness and feelings. *Hum. Brain Mapp.* 28, 9–18. doi: 10.1002/hbm.20258

Pollatos, O., Herbert, B. M., Kaufmann, C., Auer, D. P., and Schandry, R. (2007b). Interoceptive awareness, anxiety and cardiovascular reactivity to isometric exercise. *Int. J. Psychophysiol.* 65, 167–173. doi: 10.1016/j.ijpsycho.2007.03.005

Pollatos, O., Schandry, R., Auer, D. P., and Kaufmann, C. (2007c). Brain structures mediating cardiovascular arousal and interoceptive awareness. *Brain Res.* 1141, 178–187. doi: 10.1016/j.brainres.2007.01.026

Pollatos, O., Herbert, B. M., Wankner, S., Dietel, A., Wachsmuth, C., Henningsen, P., et al. (2011a). Autonomic imbalance is associated with reduced facial recognition in somatoform disorders. *J. Psychosom. Res.* 71, 232–239. doi: 10.1016/j.jpsychores.2011.03.012

Pollatos, O., Werner, N. S., Duschek, S., Schandry, R., Matthias, E., Traut-Mattausch, E., et al. (2011b). Differential effects of alexithymia subscales on autonomic reactivity and anxiety during social stress. *J. Psychosom. Res.* 70, 525–533. doi: 10.1016/j.jpsychores.2010.12.003

Pollatos, O., Kirsch, W., and Schandry, R. (2005). Brain structures involved in interoceptive awareness and cardioafferent signal processing: a dipole source localization study. *Hum. Brain Mapp.* 26, 54–64. doi: 10.1002/hbm.20121

Pollatos, O., Kurz, A. L., Albrecht, J., Schreder, T., Kleemann, A. M., Schopf, V., et al. (2008). Reduced perception of bodily signals in anorexia nervosa. *Eat. Behav.* 9, 381–388. doi: 10.1016/j.eatbeh.2008.02.001

Porges, S. W. (1993). *Body Perception Questionnaire. Laboratory of Developmental Assessment*. Baltimore, MD: University of Maryland.

Pruessner, J. C., Hellhammer, D. H., and Kirschbaum, C. (1999). Burnout, perceived stress, and cortisol responses to awakening. *Psychosom. Med.* 61, 197–204. doi: 10.1097/00006842-199903000-00012

Quirarte, G. L., Roozendaal, B., and Mcgaugh, J. L. (1997). Glucocorticoid enhancement of memory storage involves noradrenergic activation in the basolateral amygdala. *Proc. Natl. Acad. Sci. U.S.A.* 94, 14048–14053. doi: 10.1073/pnas.94.25.14048

Richards, J., Edgar, L., and Gibbon, P. (1996). Cardiac acuity in panic disorder. *Cogn. Ther. Res.* 20, 261–376. doi: 10.1007/BF02228039

Rief, W., and Auer, C. (2000). Cortisol and somatization. *Biol. Psychol.* 53, 13–23. doi: 10.1016/S0301-0511(00)00042-9

Rief, W., and Barsky, A. J. (2005). Psychobiological perspectives on somatoform disorders. *Psychoneuroendocrinology* 30, 996–1002. doi: 10.1016/j.psyneuen.2005.03.018

Rief, W., and Broadbent, E. (2007). Explaining medically unexplained symptoms-models and mechanisms. *Clin. Psychol. Rev.* 27, 821–841. doi: 10.1016/j.cpr.2007.07.005

Rief, W., Hessel, A., and Braehler, E. (2001). Somatization symptoms and hypochondriacal features in the general population. *Psychosom. Med.* 63, 595–602. doi: 10.1097/00006842-200107000-00012

Rief, W., Hiller, W., and Margraf, J. (1998a). Cognitive aspects of hypochondriasis and the somatization syndrome. *J. Abnorm. Psychol.* 107, 587–595. doi: 10.1037/0021-843X.107.4.587

Rief, W., Shaw, R., and Fichter, M. M. (1998b). Elevated levels of psychophysiological arousal and cortisol in patients with somatization syndrome. *Psychosom. Med.* 60, 198–203. doi: 10.1097/00006842-199803000-00016

Robbe, H. W., Mulder, L. J., Ruddel, H., Langewitz, W. A., Veldman, J. B., and Mulder, G. (1987). Assessment of baroreceptor reflex sensitivity by means of spectral analysis. *Hypertension* 10, 538–543. doi: 10.1161/01.HYP.10.5.538

Rockhold, R. W., and Caldwell, R. W. (1980). Cardiovascular effects following clonidine microinjection into the nucleus tractus solitarii of the rat. *Neuropharmacology* 19, 919–922. doi: 10.1016/0028-3908(80)90094-5

Roozendaal, B., Hahn, L. H., Nathan, S. V., De Quervain, D. J.-F., and Mcgaugh, J. L. (2004). Glucocorticoid effect on memory retrieval require concurrent noradrenergic activity in the hippocampus and basolateral amygdala. *J. Neurosci.* 34, 8161–8169. doi: 10.1523/JNEUROSCI.2574-04.2004

Roozendaal, B., Okuda, S., De Quervain, D. J., and Mcgaugh, J. L. (2006). Glucocorticoids interact with emotion-induced noradrenergic activation in influencing different memory functions. *Neuroscience* 138, 901–910. doi: 10.1016/j.neuroscience.2005.07.049

Roozendaal, B., Williams, C. L., and Mcgaugh, J. L. (1999). Glucocorticoid receptor activation in the rat nucleus of the solitary tract facilitates memory consolidation: involvement of the basolateral amygdala. *Eur. J. Neurosci.* 11, 1317–1323. doi: 10.1046/j.1460-9568.1999.00537.x

Rosenberger, C., Elsenbruch, S., Scholle, A., De Greiff, A., Schedlowski, M., Forsting, M., et al. (2009). Effects of psychological stress on the cerebral processing of visceral stimuli in healthy women. *Neurogastroenterol. Motil.* 21, e740–e745. doi: 10.1111/j.1365-2982.2009.01295.x

Russell, E., Koren, G., Rieder, M., and Van Uum, S. (2012). Hair cortisol as a biological marker of chronic stress: current status, future directions and unanswered questions. *Psychoneuroendocrinology* 37, 589–601. doi: 10.1016/j.psyneuen.2011.09.009

Russell, J. A. (2003). Core affect and the psychological construction of emotion. *Psychol. Rev.* 110, 145–172. doi: 10.1037/0033-295X.110.1.145

Schachinger, H., Degen, L., and Beglinger, C. (2009). Subliminal gastric distension enhances startle [Abstract]. *Psychophysiology* 46, S17.

Schachinger, H., Weinbacher, M., Kiss, A., Ritz, R., and Langewitz, W. (2001). Cardiovascular indices of peripheral and central sympathetic activation. *Psychosom. Med.* 63, 788–796. doi: 10.1097/00006842-200109000-00012

Schaefer, M., Egloff, B., Gerlach, A. L., and Witthoft, M. (2014). Improving heartbeat perception in patients with medically unexplained symptoms reduces symptom distress. *Biol. Psychol.* 101, 69–76. doi: 10.1016/j.biopsycho.2014.05.012

Schaefer, M., Egloff, B., and Witthoft, M. (2012). Is interoceptive awareness really altered in somatoform disorders? Testing competing theories with two paradigms of heartbeat perception. *J. Abnorm. Psychol.* 121, 719–724. doi: 10.1037/a0028509

Schandry, R. (1981). Heart beat perception and emotional experience. *Psychophysiology* 18, 483–488. doi: 10.1111/j.1469-8986.1981.tb02486.x

Schandry, R., Bestler, M., and Montoya, P. (1993). On the relation between cardiodynamics and heartbeat perception. *Psychophysiology* 30, 467–474. doi: 10.1111/j.1469-8986.1993.tb02070.x

Schandry, R., Sparrer, B., and Weitkunat, R. (1986). From the heart to the brain: a study of heartbeat contingent scalp potentials. *Int. J. Neurosci.* 30, 261–275. doi: 10.3109/00207458608985677

Schandry, R., and Specht, G. (1981). The influence of psychological and physical stress on the perception of heartbeats. [Abstract]. *Psychophysiology* 18, 154.

Schmitz, J., Blechert, J., Kramer, M., Asbrand, J., and Tuschen-Caffier, B. (2012). Biased perception and interpretation of bodily anxiety symptoms in childhood social anxiety. *J. Clin. Child Adolesc. Psychol.* 41, 92–102. doi: 10.1080/15374416.2012.632349

Schulz, A. (2015). "Interoception," in *International Encyclopedia of the Social and Behavioral Sciences*, 2nd Rev. Edn, ed. J. D. Wright (Amsterdam: Elsevier), 614–620. doi: 10.1016/B978-0-08-097086-8.14145-5

Schulz, A., Ferreira De Sá, D. S., Dierolf, A. M., Lutz, A., Van Dyck, Z., Vögele, C., et al. (2015a). Short-term food deprivation increases amplitudes of heartbeat-evoked potentials. *Psychophysiology* 52, 695–703. doi: 10.1111/psyp.12388

Schulz, A., Köster, S., Beutel, M. E., Schächinger, H., Vögele, C., Rost, S., et al. (2015b). Altered patterns of heartbeat-evoked potentials in depersonalization/derealization disorder: neurophysiological evidence for impaired cortical representation of bodily signals. *Psychosom. Med.* 77, 506–516. doi: 10.1097/PSY.0000000000000195

Schulz, A., Van Dyck, Z., Lutz, A., and Vögele, C. (2015c). "Gastrische modulation der Schreckreaktion: eine prä-attentive methode zur abbildung afferenter signale aus dem gastrointestinalen system. [Gastric modulation of startle: a pre-attentive method to assess afferent signals originating from the gastrointestinal system]," in: *The Proceeding of 41. Annual Meeting Psychologie und Gehirn [Psychology and Brain]*, Frankfurt.

Schulz, A., Lass-Hennemann, J., Nees, F., Blumenthal, T. D., Berger, W., and Schachinger, H. (2009a). Cardiac modulation of startle eye blink. *Psychophysiology* 46, 234–240. doi: 10.1111/j.1469-8986.2008.00768.x

Schulz, A., Lass-Hennemann, J., Richter, S., Romer, S., Blumenthal, T. D., and Schachinger, H. (2009b). Lateralization effects on the cardiac modulation of acoustic startle eye blink. *Biol. Psychol.* 80, 287–291. doi: 10.1016/j.biopsycho.2008.10.006

Schulz, A., Reichert, C. F., Richter, S., Lass-Hennemann, J., Blumenthal, T. D., and Schachinger, H. (2009c). Cardiac modulation of startle: effects on eye blink and higher cognitive processing. *Brain Cogn.* 71, 265–271. doi: 10.1016/j.bandc.2009.08.002

Schulz, A., Lass-Hennemann, J., Sutterlin, S., Schächinger, H., and Vögele, C. (2013a). Cold pressor stress induces opposite effects on cardioceptive accuracy dependent on assessment paradigm. *Biol. Psychol.* 93, 167–174. doi: 10.1016/j.biopsycho.2013.01.007

Schulz, A., Schilling, T. M., and Schachinger, H. (2013b). "Modulation der schreckreaktion durch viszeral-afferente signale aus dem kardialen und respiratorischen system. [Modulation of startle by visceral-afferent signals originating from the cardiovascular and respiratory system]," in *The Proceeding of 39. Annual Meeting Psychologie und Gehirn [Psychology and Brain]*, Würzburg.

Schulz, A., Strelzyk, F., Ferreira De Sa, D. S., Naumann, E., Vögele, C., and Schächinger, H. (2013c). Cortisol rapidly affects amplitudes of heartbeat-evoked brain potentials-Implications for the contribution of stress to an altered perception of physical sensations? *Psychoneuroendocrinology* 38, 2686–2693. doi: 10.1016/j.psyneuen.2013.06.027

Schulz, A., Philippsen, C., and Schächinger, H. (2007). The influence of central Alpha2-adrenoreceptor agonism and antagonism on the cardiac modulation of acoustic startle. *Psychophysiology* 44, S29.

Schulz, A., Plein, D. E., Richter, S., Blumenthal, T. D., and Schachinger, H. (2011). Cold pressor stress affects cardiac attenuation of startle. *Int. J. Psychophysiol.* 79, 385–391. doi: 10.1016/j.ijpsycho.2010.12.008

Schulz, A., Richter, S., and Schächinger, H. (2010). "Evidence that cortisol-induced increase in baroreflex sensitivity of heart rate control is not due to afferent visceral neurotransmission," in *The Proceeding of 7th World Congress on Stress*. Leiden.

Schwabe, L., Haddad, L., and Schachinger, H. (2008). HPA axis activation by a socially evaluated cold-pressor test. *Psychoneuroendocrinology* 33, 890–895. doi: 10.1016/j.psyneuen.2008.03.001

Sedeno, L., Couto, B., Melloni, M., Canales-Johnson, A., Yoris, A., Baez, S., et al. (2014). How do you feel when you can't feel your body? Interoception, functional connectivity and emotional processing in depersonalization-derealization disorder. *PLoS ONE* 9:e98769. doi: 10.1371/journal.pone.0098769

Shao, S., Shen, K., Wilder-Smith, E. P., and Li, X. (2011). Effect of pain perception on the heartbeat evoked potential. *Clin. Neurophysiol.* 122, 1838–1845. doi: 10.1016/j.clinph.2011.02.014

Shin, Y. W., Dzemidzic, M., Jo, H. J., Long, Z., Medlock, C., Dydak, U., et al. (2013). Increased resting-state functional connectivity between the anterior cingulate cortex and the precuneus in panic disorder: resting-state connectivity in panic disorder. *J. Affect. Disord.* 150, 1091–1095. doi: 10.1016/j.jad.2013.04.026

Sloan, R. P., Shapiro, P. A., Bagiella, E., Gorman, J. M., and Bigger, J. T. Jr. (1995). Temporal stability of heart period variability during a resting baseline and in response to psychological challenge. *Psychophysiology* 32, 191–196. doi: 10.1111/j.1469-8986.1995.tb03311.x

Spielberger, C. D., and Reheiser, E. C. (2004). "Measuring anxiety, anger, depression, and curiosity as emotional states and personality traits with the STAI, STAXI, and STPI," in *Comprehensive Handbook of Psychological Assessment*, eds M. J. Hilsenroth and D. L. Segal (Hoboken: Wiley), 70–86.

Stalder, T., Tietze, A., Steudte, S., Alexander, N., Dettenborn, L., and Kirschbaum, C. (2014). Elevated hair cortisol levels in chronically stressed dementia caregivers. *Psychoneuroendocrinology* 47, 26–30. doi: 10.1016/j.psyneuen.2014.04.021

Steptoe, A., and Vogele, C. (1991). Methodology of mental stress testing in cardiovascular research. *Circulation* 83, II14–II24.

Steptoe, A., and Vogele, C. (1992). Individual differences in the perception of bodily sensations: the role of trait anxiety and coping style. *Behav. Res. Ther.* 30, 597–607. doi: 10.1016/0005-7967(92)90005-2

Stevens, S., Gerlach, A. L., Cludius, B., Silkens, A., Craske, M. G., and Hermann, C. (2011). Heartbeat perception in social anxiety before and during speech anticipation. *Behav. Res. Ther.* 49, 138–143. doi: 10.1016/j.brat.2010.11.009

Streff, A., Kuehl, L. K., Michaux, G., and Anton, F. (2010). Differential physiological effects during tonic painful hand immersion tests using hot and ice water. *Eur. J. Pain* 14, 266–272. doi: 10.1016/j.ejpain.2009.05.011

Strelzyk, F., Hermes, M., Naumann, E., Oitzl, M., Walter, C., Busch, H. P., et al. (2012). Tune it down to live it up? Rapid, nongenomic effects of cortisol on the human brain. *J. Neurosci.* 32, 616–625. doi: 10.1523/JNEUROSCI.2384-11.2012

Strigo, I. A., Matthews, S. C., Simmons, A. N., Oberndorfer, T., Klabunde, M., Reinhardt, L. E., et al. (2013). Altered insula activation during pain anticipation in individuals recovered from anorexia nervosa: evidence of interoceptive dysregulation. *Int. J. Eat. Disord.* 46, 23–33. doi: 10.1002/eat.22045

Strohle, A., Holsboer, F., and Rupprecht, R. (2000). Increased ACTH concentrations associated with cholecystokinin tetrapeptide-induced panic attacks in patients with panic disorder. *Neuropsychopharmacology* 22, 251–256. doi: 10.1016/S0893-133X(99)00115-3

Sturges, L. V., and Goetsch, V. L. (1996). Psychophysiological reactivity and heartbeat awareness in anxiety sensitivity. *J. Anxiety Disord.* 10, 283–294. doi: 10.1016/0887-6185(96)00012-6

Suarez-Hitz, K. A., Otto, B., Bidlingmaier, M., Schwizer, W., Fried, M., and Ehlert, U. (2012). Altered psychobiological responsiveness in women with irritable bowel syndrome. *Psychosom. Med.* 74, 221–231. doi: 10.1097/PSY.0b013e318244fb82

Sved, A. F., Tsukamoto, K., and Schreihofer, A. M. (1992). Stimulation of alpha 2-adrenergic receptors in nucleus tractus solitarius is required for the baroreceptor reflex. *Brain Res.* 576, 297–303. doi: 10.1016/0006-8993(92)90693-4

Tak, L. M., Bakker, S. J., and Rosmalen, J. G. (2009). Dysfunction of the hypothalamic-pituitary-adrenal axis and functional somatic symptoms: a longitudinal cohort study in the general population. *Psychoneuroendocrinology* 34, 869–877. doi: 10.1016/j.psyneuen.2008.12.017

Tak, L. M., Cleare, A. J., Ormel, J., Manoharan, A., Kok, I. C., Wessely, S., et al. (2011). Meta-analysis and meta-regression of hypothalamic-pituitary-adrenal axis activity in functional somatic disorders. *Biol. Psychol.* 87, 183–194. doi: 10.1016/j.biopsycho.2011.02.002

Tak, L. M., and Rosmalen, J. G. (2010). Dysfunction of stress responsive systems as a risk factor for functional somatic syndromes. *J. Psychosom. Res.* 68, 461–468. doi: 10.1016/j.jpsychores.2009.12.004

Tanaka, Y., Ishitobi, Y., Maruyama, Y., Kawano, A., Ando, T., Imanaga, J., et al. (2012). Salivary alpha-amylase and cortisol responsiveness following electrical stimulation stress in panic disorder patients. *Neurosci. Res.* 73, 80–84. doi: 10.1016/j.neures.2012.01.006

Tanriverdi, F., Karaca, Z., Unluhizarci, K., and Kelestimur, F. (2007). The hypothalamo-pituitary-adrenal axis in chronic fatigue syndrome and fibromyalgia syndrome. *Stress* 10, 13–25. doi: 10.1080/10253890601130823

Terhaar, J., Viola, F. C., Bar, K. J., and Debener, S. (2012). Heartbeat evoked potentials mirror altered body perception in depressed patients. *Clin. Neurophysiol.* 123, 1950–1957. doi: 10.1016/j.clinph.2012.02.086

Thomas, K. S., Nelesen, R. A., Ziegler, M. G., Bardwell, W. A., and Dimsdale, J. E. (2004). Job strain, ethnicity, and sympathetic nervous system activity. *Hypertension* 44, 891–896. doi: 10.1161/01.HYP.0000148499.54730.0d

Tsai, H. W., Chan, P. Y., Von Leupoldt, A., and Davenport, P. W. (2013). The impact of emotion on the perception of graded magnitudes of respiratory resistive loads. *Biol. Psychol.* 93, 220–224. doi: 10.1016/j.biopsycho.2013.02.008

Uchida, R. R., Del-Ben, C. M., Busatto, G. F., Duran, F. L., Guimaraes, F. S., Crippa, J. A., et al. (2008). Regional gray matter abnormalities in panic disorder: a voxel-based morphometry study. *Psychiatry Res.* 163, 21–29. doi: 10.1016/j.pscychresns.2007.04.015

Usher, M., Cohen, J. D., Servan-Schreiber, D., Rajkowski, J., and Aston-Jones, G. (1999). The role of locus coeruleus in the regulation of cognitive performance. *Science* 283, 549–554. doi: 10.1126/science.283.5401.549

Vaitl, D. (1996). Interoception. *Biol. Psychol.* 42, 1–27. doi: 10.1016/0301-0511(95)05144-9

van den Bos, R., Harteveld, M., and Stoop, H. (2009). Stress and decision-making in humans: performance is related to cortisol reactivity, albeit differently in men and women. *Psychoneuroendocrinology* 34, 1449–1458. doi: 10.1016/j.psyneuen.2009.04.016

Van der Does, A. J., Van Dyck, R., and Spinhoven, P. (1997). Accurate heartbeat perception in panic disorder: fact and artefact. *J. Affect. Disord.* 43, 121–130. doi: 10.1016/S0165-0327(96)01414-0

Van Uum, S. H., Sauve, B., Fraser, L. A., Morley-Forster, P., Paul, T. L., and Koren, G. (2008). Elevated content of cortisol in hair of patients with severe chronic pain: a novel biomarker for stress. *Stress* 11, 483–488. doi: 10.1080/10253890801887388

von Leupoldt, A., Keil, A., Chan, P. Y., Bradley, M. M., Lang, P. J., and Davenport, P. W. (2010). Cortical sources of the respiratory-related evoked potential. *Res. Physiol. Neurobiol.* 170, 198–201. doi: 10.1016/j.resp.2009.12.006

Vrijkotte, T. G., Van Doornen, L. J., and De Geus, E. J. (2004). Overcommitment to work is associated with changes in cardiac sympathetic regulation. *Psychosom. Med.* 66, 656–663. doi: 10.1097/01.psy.0000138283.65547.78

Wang, J., Korczykowski, M., Rao, H., Fan, Y., Pluta, J., Gur, R. C., et al. (2007). Gender difference in neural response to psychological stress. *Soc. Cogn. Affect. Neurosci.* 2, 227–239. doi: 10.1093/scan/nsm018

Wang, J., Rao, H., Wetmore, G. S., Furlan, P. M., Korczykowski, M., Dinges, D. F., et al. (2005). Perfusion functional MRI reveals cerebral blood flow pattern under psychological stress. *Proc. Natl. Acad. Sci. U.S.A.* 102, 17804–17809. doi: 10.1073/pnas.0503082102

Weiss, S., Sack, M., Henningsen, P., and Pollatos, O. (2014). On the interaction of self-regulation, interoception and pain perception. *Psychopathology* 47, 377–382. doi: 10.1159/000365107

Werner, N. S., Jung, K., Duschek, S., and Schandry, R. (2009). Enhanced cardiac perception is associated with benefits in decision-making. *Psychophysiology* 46, 1123–1129. doi: 10.1111/j.1469-8986.2009.00855.x

Whitehead, W. E., Drescher, V. M., Heiman, P., and Blackwell, B. (1977). Relation of heart rate control to heartbeat perception. *Biofeedback Self Regul.* 2, 371–392. doi: 10.1007/BF00998623

Wiens, S. (2005). Interoception in emotional experience. *Curr. Opin. Neurol.* 18, 442–447. doi: 10.1097/01.wco.0000168079.92106.99

Willem Van der Does, A. J., Antony, M. M., Ehlers, A., and Barsky, A. J. (2000). Heartbeat perception in panic disorder: a reanalysis. *Behav. Res. Ther.* 38, 47–62. doi: 10.1016/S0005-7967(98)00184-3

Wingenfeld, K., Heim, C., Schmidt, I., Wagner, D., Meinlschmidt, G., and Hellhammer, D. H. (2008). HPA axis reactivity and lymphocyte glucocorticoid sensitivity in fibromyalgia syndrome and chronic pelvic pain. *Psychosom. Med.* 70, 65–72. doi: 10.1097/PSY.0b013e31815ff3ce

Wingenfeld, K., Wagner, D., Schmidt, I., Meinlschmidt, G., Hellhammer, D. H., and Heim, C. (2007). The low-dose dexamethasone suppression test in fibromyalgia. *J. Psychosom. Res.* 62, 85–91. doi: 10.1016/j.jpsychores.2006.06.006

Wirtz, P. H., Siegrist, J., Rimmele, U., and Ehlert, U. (2008). Higher overcommitment to work is associated with lower norepinephrine secretion before and after acute psychosocial stress in men. *Psychoneuroendocrinology* 33, 92–99. doi: 10.1016/j.psyneuen.2007.10.003

Witthoft, M., Basfeld, C., Steinhoff, M., and Gerlach, A. L. (2012). Can't suppress this feeling: automatic negative evaluations of somatosensory stimuli are related to the experience of somatic symptom distress. *Emotion* 12, 640–649. doi: 10.1037/a0024924

Wolk, J., Sutterlin, S., Koch, S., Vogele, C., and Schulz, S. M. (2014). Enhanced cardiac perception predicts impaired performance in the Iowa Gambling Task in patients with panic disorder. *Brain Behav.* 4, 238–246. doi: 10.1002/brb3.206

Wong, M. L., Kling, M. A., Munson, P. J., Listwak, S., Licinio, J., Prolo, P., et al. (2000). Pronounced and sustained central hypernoradrenergic function in major depression with melancholic features: relation to hypercortisolism and corticotropin-releasing hormone. *Proc. Natl. Acad. Sci. U.S.A.* 97, 325–330. doi: 10.1073/pnas.97.1.325

Yehuda, R., Southwick, S. M., Krystal, J. H., Bremner, D., Charney, D. S., and Mason, J. W. (1993). Enhanced suppression of cortisol following dexamethasone administration in posttraumatic stress disorder. *Am. J. Psychiatry* 150, 83–86. doi: 10.1176/ajp.150.1.83

16

Brain Mechanisms of Short-Term Habituation and Sensitization Toward Dyspnea

*M. Cornelia Stoeckel[1], Roland W. Esser[1], Matthias Gamer[1], Christian Büchel[1] and Andreas von Leupoldt[1,2]**

[1] *Department of Systems Neuroscience, University Medical Center Hamburg-Eppendorf, Hamburg, Germany,*

[2] *Research Group Health Psychology, Faculty of Psychology and Educational Sciences, University of Leuven, Leuven, Belgium*

Correspondence:
Andreas von Leupoldt,
Research Group Health Psychology,
Faculty of Psychology
and Educational Sciences, University
of Leuven, Tiensestraat 102,
3000 Leuven, Belgium
andreas.vonleupoldt@ppw.kuleuven.be

Dyspnea is a prevalent and threatening cardinal symptom in many diseases including asthma. Whether patients suffering from dyspnea show habituation or sensitization toward repeated experiences of dyspnea is relevant for both quality of life and treatment success. Understanding the mechanisms, including the underlying brain activation patterns, that determine the dynamics of dyspnea perception seems crucial for the improvement of treatment and rehabilitation. Toward this aim, we investigated the interplay between short-term changes of dyspnea perception and changes of related brain activation. Healthy individuals underwent repeated blocks of resistive load induced dyspnea with parallel acquisition of functional magnetic resonance imaging data. Late vs. early ratings on dyspnea intensity and unpleasantness were correlated with late vs. early brain activation for both, dyspnea anticipation and dyspnea perception. Individual trait and state anxiety were determined using questionnaire data. Our results indicate an involvement of the orbitofrontal cortex (OFC), midbrain/periaqueductal gray (PAG) and anterior insular cortex in habituation/sensitization toward dyspnea. Changes in the anterior insular cortex were particularly linked to changes in dyspnea unpleasantness. Changes of both dyspnea intensity and unpleasantness were positively correlated with state and trait anxiety. Our findings are in line with the suggested relationship between the anterior insular cortex and dyspnea unpleasantness. They further support the notion that habituation/sensitization toward dyspnea is influenced by anxiety. Our study extends the known role of the midbrain/PAG in anti-nociception to an additional involvement in habituation/sensitization toward dyspnea and suggests an interplay with the OFC.

Keywords: dyspnea, breathlessness, habituation, sensitization, fMRI, unpleasantness, insular cortex

Introduction

The experience of dyspnea (breathlessness) is the cardinal symptom in prevalent diseases such as asthma and chronic obstructive pulmonary disease (COPD). It is also common in other pathologies including cardiovascular and neuromuscular diseases, and panic and anxiety disorders (Parshall et al., 2012; Laviolette et al., 2014). Notably, the perception of dyspnea is not linearly related to objective lung function or sensory input. Instead, dyspnea perception has been shown to be modulated by several psychological factors, including attention, expectation, learning, categorization and comparison processes, emotional predispositions, and current mood (e.g., Janssens et al., 2009; Lansing et al., 2009; Herigstad et al., 2011; Petersen et al., 2011, 2014). Via their influence on dyspnea perception these factors also have a strong impact on coping strategies, disease management and disease progression (Hayen et al., 2013).

Furthermore repeated exposure to dyspnea can result in either increasing (sensitization) or decreasing (habituation) dyspnea perception (Bloch-Salisbury et al., 1996; Carrieri-Kohlman et al., 2001; Wan et al., 2008, 2009; von Leupoldt et al., 2011a; Hayen et al., 2015). Habituation toward dyspnea seems favorable in conditions such as COPD and panic disorder. In both conditions the experience of dyspnea might result in unfavorable avoidance behavior and habituation toward dyspnea might reduce this behavior. In COPD, in particular, avoidance of physical activity is one potential mechanism that accelerates disease progression as reduced physical fitness leads to unfavorable systemic consequences and has a negative effect on dyspnea severity ending in a spiral of decline (e.g., Troosters et al., 2013). In asthma patients, however, habituation to dyspnea might result in delayed treatment and critical under-medication (von Leupoldt et al., 2009, 2011a,b). Here, sensitization toward dyspnea seems more favorable as far as it might improve self-management by supporting the early initiation of actions during the onset of asthma exacerbations and by heightening the compliance with prescriptions.

Both habituation and sensitization toward dyspnea have been shown to interact with psychological factors such as negative affect. When healthy individuals repeatedly underwent hypercapnic rebreathing, more anxious individuals showed less habituation (Li et al., 2006). This reduction in habituation was more pronounced for the affective dimension of dyspnea (=unpleasantness) as compared to the sensory dimension of dyspnea (=intensity) (Wan et al., 2012).

Understanding the interplay of affective traits and -states with habituation vs. sensitization toward dyspnea seems crucial for the improvement of disease management and rehabilitation in those suffering from dyspnea. In this regard, a better knowledge of the underlying neural mechanisms appears important. However, the neural mechanisms involved in habituation vs. sensitization toward dyspnea have rarely been investigated (von Leupoldt et al., 2011a). In a previous study we observed reduced dyspnea unpleasantness ratings in patients with asthma to be related to reduced activations in the insular cortex and increased activations as well as gray matter volume in the anti-nociceptive periaqueductal gray (PAG) when compared with healthy controls (von Leupoldt et al., 2009, 2011b). These neural patterns were partly correlated with disease duration. Although these findings were interpreted as habituation toward dyspnea over time, the study did not directly examine habituation to repeated experiences of dyspnea.

In the present study, we repeatedly induced dyspnea in a set of healthy volunteers with parallel acquisition of functional magnetic resonance imaging (fMRI) data. We hypothesized that habituation/sensitization toward dyspnea unpleasantness as compared to dyspnea intensity would be correlated with activation in brain structures thought to be involved in the processing of dyspnea unpleasantness, in particular the insular cortex and the amygdala (von Leupoldt et al., 2008, 2009; Paulus et al., 2012). The amygdala has also been demonstrated to be involved in sensitization toward repetitive pain exposure (Stankewitz et al., 2013). Additional candidate areas were derived from the anti-nociceptive network involving the anterior cingulate cortex (ACC), and the midbrain/PAG (e.g., Petrovic et al., 2002; Bingel et al., 2007). A study investigating habituation toward aversive visceral stimulation demonstrated changes in the connectivity between rACC, and amygdala over repeated painful stimulation (Labus et al., 2009). A more recent study using repeated presentations of similar stimuli showed either increases or decreases of brain activation in the insular and cingulate cortex and the amygdala (Lowén et al., 2015).

We furthermore expected to see habituation/sensitization-related brain activation already during the anticipation of dyspnea. Although the anticipation of dyspnea has rarely been studied, one study on the anticipation of hyperventilation (Holtz et al., 2012) suggests an involvement of the orbitofrontal cortex (OFC) and the dorsomedial prefrontal cortex (dmPFC) along with insula and ACC. These areas also showed expectancy effects in placebo-studies on the anticipation/perception of pain (e.g., Hsieh et al., 1999) and were, therefore, of particular interest. Finally, we expected an interplay between anxiety-related personality traits and states with the dynamics of dyspnea ratings over late vs. early trials.

Materials and Methods

Participants

We re-analyzed data from 46 healthy individuals from a previous study (to be published elsewhere) with a specific focus on habituation/sensitization toward dyspnea. Normal lung function of participants (mean age 28.5 years, 18 females) was confirmed by standard spirometry (Miller et al., 2005). All participants negated any history of neurological, psychiatric, or respiratory disease. None of the subjects showed any anatomical anomaly of the throat. The average body-mass index was 23.4 (range 19.4–28.7). Trait- and state-anxiety were assessed using the State-Trait-Anxiety Inventory (STAI-T, STAI-S), Version X (Spielberger et al., 1983). Questionnaire data were analyzed after completion of the study. Written informed consent was obtained prior to the study. The study protocol was approved by the local medical ethics committee.

Induction of Dyspnea and Measurement of Respiratory Parameters

Volunteers breathed through a tightly fitted face mask that was connected to a breathing circuit. Dyspnea was induced by the introduction of MRI-compatible resistive loads to the inspiratory end of the breathing circuit. In a pre-test before entering the scanner, subjects were placed in a supine position and presented with loads of increasing magnitude. We explained dyspnea to our participants as a sensation of difficult and uncomfortable breathing. Then, each load was presented for 24 s and dyspnea intensity subsequently rated on a Borg-scale (0 = "not noticeable" to 10 = "maximally imaginable"). Load magnitude was increased until subjects reliably reported a sensation of "severe" dyspnea (Borg Score > 5). The respective load was then used to induce severe dyspnea during scanning (mean/SD = 2.23/1.18 kPa/l/s). For the baseline condition of mild dyspnea the smallest resistive load that was reliably rated higher than unloaded breathing was used (mean/SD = 0.25/0.18 kPa/l/s).

The breathing circuit provided ports for continuous recordings of end-tidal CO_2 pressure (PET_{CO2}) and inspiratory mouth pressure (P_I). Ports were connected with an MRI compatible pneumotachograph (ZAN 600 unit, ZAN Messgeräte GmbH, Oberhulba, Germany). A Y-valve with open expiratory port prevented re-breathing of CO_2 while a 2.6 m tube attached to the inspiratory port allowed for the easy introduction and removal of resistive loads in the scanner environment. PET_{CO2}, P_I, tidal volume (V_T), breathing frequency (f), minute ventilation (V_E), and inspiratory time (T_I) were continuously measured with the ZAN unit.

fMRI Data Acquisition

Imaging was performed on a 3-Tesla TRIO-Magnetom Scanner (Siemens, Medical Solutions, Erlangen, Germany) using a standard 32-channel head-coil. For each data volume we acquired 48 continuous axial-slices in descending order with 2 mm × 2 mm in-plane resolution, 2 mm slice thickness and a 1 mm gap using T2*-weighted echoplanar imaging (TR = 2870 ms, TE = 25 ms, flip angle = 80°, field of view = 208 mm × 208 mm). The first five volumes were discarded to allow for T1-saturation. Following the experiment we also acquired a high-resolution T1-weighted structural brain scan using a standard MP-RAGE sequence (1 mm × 1 mm × 1 mm spatial resolution, 240 slices). It took subjects 13–18 min to complete the paradigm, depending on the time subjects required to complete the ratings. Consequently, the number of volumes acquired varied between 275 and 374.

Experimental Protocol

Before starting the experiment in the scanner, subjects learned the association of visual cues and experimental conditions (see below) using standardized computer based instructions and practiced navigation through the Borg-scales on dyspnea intensity and unpleasantness. The subjects then entered the scanner with the face mask tightly fitted. A mirror attached to the head coil allowed the subjects to see the cues and scales that were projected into the bore. Before image acquisition started, subjects were allowed to familiarize with the scanner environment and the MRI-compatible button-box response-system. A test-run ensured the full visibility of all cues and scales and the tight fitting of the mask.

During the experiment, 10 blocks of mildly loaded breathing ("baseline") alternated with 10 blocks of severely loaded breathing ("dyspnea") using the individually pre-selected loads. Each block was visually cued for 6 s by a thin cross (red indicating baseline, green indicating dyspnea). After 6 s the thin cross changed into a solid cross and the load was introduced for 24 s. Each block of loaded breathing was followed by ratings on two Borg-scales, presented in random order: one for the unpleasantness of dyspnea and one for the intensity of dyspnea as perceived during the preceding block.

All experimental events were presented and logged using Presentation software (Neurobehavioral Systems, Inc., Albany, CA, USA). The ZAN-system, collecting the respiratory data, received triggers for the beginning of each experimental event.

Data Analysis

Ratings for dyspnea intensity and dyspnea unpleasantness were averaged across the five early and across the five late blocks of dyspnea, respectively. The development of the ratings over time was expressed as delta (Δ) by subtracting the average across early ratings from the average across late ratings. Thus, a positive Δ indicated increasing ratings over time, interpreted as "sensitization" while a negative Δ indicated decreased ratings, interpreted as "habituation." Δ intensity and Δ unpleasantness were tested for correlations with questionnaire data. These statistical analyzes were calculated using SPSS 20.0 software (SPSS Inc., Chicago, IL, USA).

Preprocessing and statistical analysis of fMRI data were carried out using SPM8 software[1]. Data were unwarped and realigned to the first image using six affine spatial transformation parameters, then normalized to the SPM standard template and finally smoothed using a 12 mm × 12 mm × 12 mm full-width at half-maximum Gaussian filter. Data were further filtered with a temporal highpass cut-off of 128 s. Statistical analysis on the first level was carried out within the framework of a general linear model using separate regressors for cue baseline, baseline, cue dyspnea, dyspnea, and ratings. The mean BOLD signal intensity of each volume and PET_{CO2} time logged to the beginning of each scan were included as covariates-of-no-interest. On the first level, we contrasted severe late dyspnea (blocks 6–10) with severe early dyspnea (blocks 1–5) and late cue dyspnea conditions with the early cue dyspnea conditions using the respective mild conditions as baseline. The second level analysis correlated the beta-values of the two contrast images obtained on the first level [(Δ cue dyspnea vs. cue baseline) and (Δ dyspnea vs. baseline)] with Δ intensity and Δ unpleasantness using separate models. All models included individual Δ breathing frequency and Δ inspiratory mouth pressure (averaged across cue dyspnea vs. cue baseline and dyspnea vs. baseline conditions, respectively) as covariates-of-no-interest as these breathing parameters showed slight, but statistically significant changes over time (see Results).

Correlations throughout the brain were accepted as significant if exceeding a family-wise-error (FWE) corrected threshold of $p < 0.05$. Given our *a priori* hypotheses for habituation/sensitization-related correlations during dyspnea anticipation und perception we conducted further region-of interest (ROI) analyses. Masks for the insula, the amygdala, the ACC, OFC, and dmPFC were generated from the automated anatomical labeling (AAL) template described by Tzourio-Mazoyer et al. (2002). A midbrain ROI centered on PAG was defined using a 10 mm sphere around the average coordinates for PAG activation reported by Linnman et al. (2012). Activation within these ROIs was considered significant, if exceeding as

[1] www.fil.ion.ucl.ac.uk/spm

threshold of $p < 0.05$, corrected for multiple comparisons within each ROI.

An additional ROI covering the visual cortex was included as control area. Changes in brain activation within this area were not expected to show any significant correlation with changes in dyspnea intensity and unpleasantness ratings over time. For this ROI we employed a more liberal threshold of $p < 0.001$ uncorrected.

Results

Ratings

In 30.5% of all subjects we observed decreasing ratings over time indicating habituation for both dyspnea intensity and dyspnea unpleasantness. In 54.5% we observed increasing ratings with regard to dyspnea intensity, while 63% showed increasing dyspnea unpleasantness ratings indicating sensitization. In the remainder of subjects (seven and three subjects, respectively) Δ intensity and Δ unpleasantness equaled zero (**Table 1**). The Δ for dyspnea intensity ranged from −1.3 to +3 while Δ unpleasantness ranged from −1.7 to +3 (**Figure 1**, see Supplementary Figure S1 for absolute ratings).

Personality Traits

Both, state and trait anxiety as measured using the STAI-T and STAI-S questionnaires, showed a significant positive correlation with Δ unpleasantness (both $r = 0.46$, $p = 0.001$) and Δ intensity ($r = 0.36$ and 0.32, respectively, $p = 0.007$ and 0.017, respectively). As higher scores indicate higher anxiety levels, a positive correlation indicates that individuals with higher anxiety were more likely to show sensitization while subjects with low anxiety were more likely to show habituation (**Figure 2**). When early and late ratings instead of difference scores were correlated with anxiety scores, only late dyspnea intensity and unpleasantness ratings showed a significant relation with anxiety (Supplementary Figure S2).

Respiratory Parameters

The changes of respiratory parameters between dyspnea and baseline blocks were comparable for late vs. early blocks with the exception of ΔPET_{CO2}, Δf, and ΔP_I (**Table 2**). The relation between PET_{CO2} during dyspnea as compared to baseline showed a slight but significant increase over time, while P_I decreased slightly over time during dyspnea as compared to baseline. The difference of f between cue dyspnea and cue baseline decreased from early to late blocks. While fluctuations in PET_{CO2} were accounted for on the first level analysis, fluctuations of P_I and f were included into the second level analysis as covariates-of-no-interest (see Materials and Methods).

fMRI

There were no significant differences of either dyspnea anticipation or dyspnea perception between the first (blocks 1–5) and second (blocks 6–10) half of the experiment. The

TABLE 1 | Portion of subjects (%) showing sensitization, habituation, and no changes for Δ intensity and Δ unpleasantness.

	Sensitization	No change	Habituation
Δ intensity	25 (54.5%)	7 (15%)	14 (30.5%)
Δ unpleasantness	29 (63%)	3 (6.5%)	14 (30.5%)

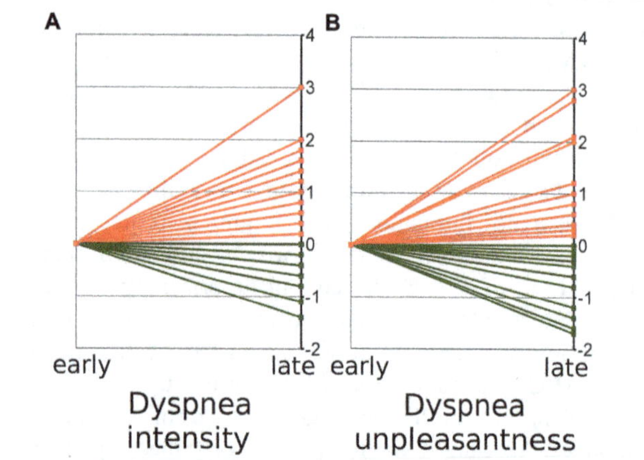

FIGURE 1 | Early (mean of block 1–5) and late (mean of block 6–10) ratings on dyspnea intensity (A) and unpleasantness (B). The figure shows the change from early ratings (mean set to zero) to late ratings. Red lines thus indicate increases of ratings over time ("sensitization") while green lines indicate decreases ("habituation").

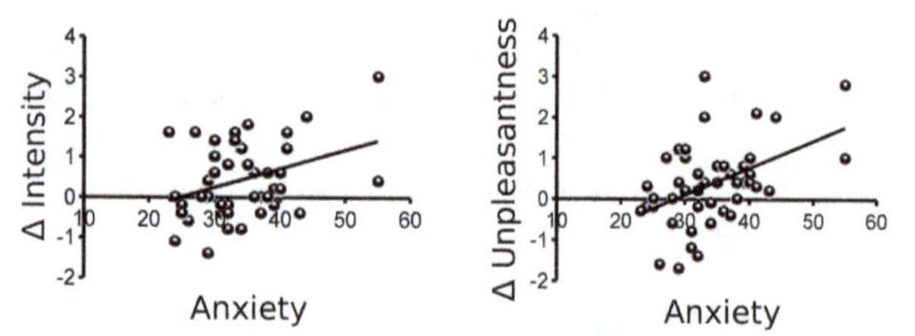

FIGURE 2 | Trait anxiety as measured by the State-Trait-Anxiety Inventory (STAI-T) correlates significantly with both Δ intensity ($r = 0.36$, $p = 0.007$) and Δ unpleasantness ($r = 0.46$, $p = 0.001$) of dyspnea ratings (mean late−mean early).

TABLE 2 | Mean (SD) Δ of breathing parameters between early and late experimental blocks (Δ cue dyspnea vs. cue baseline and Δ dyspnea vs. baseline averaged across subjects).

	Mean (SD) cue dyspnea vs. cue baseline	Mean (SD) dyspnea vs. baseline
Δ PET_{CO2} (mmHG)	0.08 (1.34)	0.27 (0.56)*
Δ V_T (L)	0.004 (0.18)	0.03 (0.16)
Δ V_E (L/min)	−0.03 (1.57)	0.09 (1.32)
Δ T_I (s)	0.08 (0.46)	0.1 (0.29)
Δ f (breaths/min)	−0.88 (1.71)*	−0.31 (0.9)
Δ P_I (mbar)	−0.07 (0.38)	−0.81 (1.76)*

*$p < 0.05$ for one-sample t-tests, corrected for multiple comparisons.
PET_{CO2}, end-tidal CO_2 pressure; V_T, tidal volume; V_E, minute ventilation; T_I, inspiratory time; f, breathing frequency; P_I, peak inspiratory mouth pressure.

whole-brain analysis based on a FWE-corrected $p < 0.05$ showed no significant correlations of Δ intensity or Δ unpleasantness during dyspnea anticipation (Δ cue dyspnea vs. cue baseline) or dyspnea perception (Δ dyspnea vs. baseline). The visual cortex as a control area showed no significant correlations with either Δ intensity or Δ unpleasantness during either dyspnea anticipation or dyspnea perception at a liberal threshold of p uncorr < 0.001.

For dyspnea anticipation, the ROI-based analysis showed a significant negative correlation of Δ intensity with the Δ of brain activation (Δ cue dyspnea vs. cue baseline) within the right OFC, extending into the ACC (**Figure 3A**, Supplementary Figure S3A). This indicated that sensitization toward dyspnea intensity was associated with decreased OFC activation, while habituation was associated with increasing OFC activation. For Δ unpleasantness a significant negative correlation was found within the midbrain/PAG (**Figure 3B**, Supplementary Figure S3B) while positive correlations were found for the anterior insular cortex bilaterally (**Figure 3C**, Supplementary Figures S3C–E). Sensitization toward dyspnea unpleasantness was, thus, associated with decreasing activation in the midbrain/PAG and increasing activation in the anterior insula. Conversely, habituation was associated with increasing midbrain/PAG activation and decreasing anterior insula activation.

For dyspnea perception (Δ dyspnea vs. baseline), we found a significant negative correlation within the right OFC for both Δ intensity and Δ unpleasantness, which extended into the ACC (**Figures 4A,B**, Supplementary Figures S3F,G). Thus, as during dyspnea anticipation, increasing OFC activation was associated with habituation, while decreasing OFC activation was associated with sensitization toward dyspnea, for both, intensity and unpleasantness. The Δ of brain activation (Δ dyspnea vs. baseline) within the right anterior insular cortex showed a significant positive correlation with Δ unpleasantness (**Figure 4C**, Supplementary Figure S3H), indicating that increasing activation of the anterior insular cortex during dyspnea perception was associated with sensitization and decreasing activation with habituation. Coordinates, Z-, r-, and p-values are summarized in **Table 3**.

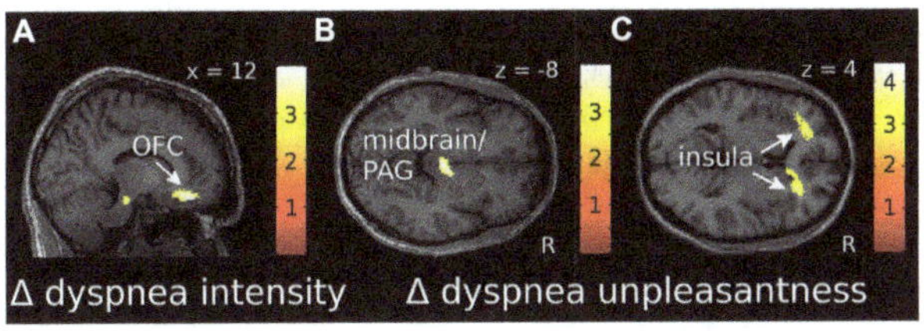

FIGURE 3 | Localization of correlations between Δ intensity (A) and Δ unpleasantness (B,C) ratings and changes in fMRI activation over time for dyspnea anticipation [(cue dyspnea late–cue baseline late)–(cue dyspnea early–cue baseline early)]. (A) and **(B)** show negative correlations, while **(C)** shows positive correlations. Correlations displayed at p uncorr < 0.005 are superimposed on a representative single subjects T1-weighted MR image. Color-bars indicate T-values. OFC, orbitofrontal cortex; PAG, periaqueductal gray; R, right.

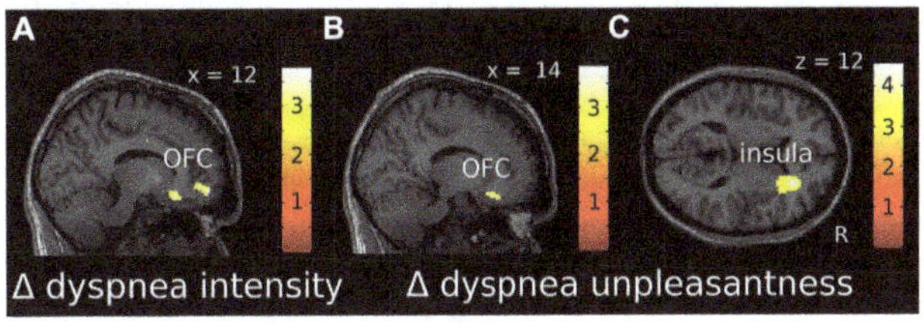

FIGURE 4 | Localization of correlations between Δ intensity (A) and Δ unpleasantness (B,C) ratings and changes in fMRI activation over time for dyspnea perception [(dyspnea late–baseline late)–(dyspnea early–baseline early)]. (A) and **(B)** show negative correlations, while **(C)** shows positive correlations. Correlations displayed at p uncorr < 0.005 are superimposed on a representative single subjects T1-weighted MR image. Color-bars indicate T-values. OFC, orbitofrontal cortex; R, right.

To investigate whether brain activation changes were rather related to general anxiety than changes in dyspnea intensity or unpleasantness perception, we included individual STAI-T scores as additional covariate-of -no-interest in a *post hoc* analysis, correlating extracted beta-values from the brain areas reported in **Table 3** with Δ intensity and Δ unpleasantness, respectively, (Supplementary Table S1). Correlations within all areas maintained significant in this analysis. When STAI-T scores were directly correlated with the extracted beta-values, mainly cue-related beta-values showed significant correlations with general anxiety. These correlations were attenuated or disappeared when changes in breathing parameters and dyspnea ratings were added as control variables (Supplementary Table S2).

Discussion

In this study we investigated the development of perceived dyspnea intensity and unpleasantness over repeated blocks of dyspnea together with parallel changes of brain activation. We observed significant correlations of late vs. early dyspnea ratings (Δ intensity and Δ unpleasantness) with late vs. early brain activity during dyspnea anticipation and dyspnea perception. These correlations were restricted to a subset of our ROIs, namely the OFC, midbrain/PAG, and the anterior insular cortex. A control area covering the visual cortex failed to show any significant rating-related changes indicating the specificity of our findings. While brain activity changes in the OFC and the midbrain/PAG were negatively correlated with Δ intensity and/or Δ unpleasantness, respectively, the anterior insular cortex showed positive correlations. Correlations were furthermore specific to the dimension of perceived dyspnea: Only Δ unpleasantness ratings showed significant positive correlations with the anterior insular cortex, and a negative correlation with the midbrain/PAG. While negative correlations of brain activation changes in the midbrain/PAG with Δ unpleasantness were limited to the anticipation period, the significant correlations of the OFC and the anterior insula were observed for both, dyspnea anticipation and dyspnea perception.

TABLE 3 | MNI-space coordinates, *Z*-, *r*-, and small-volume corrected *p*-values for peak voxels within areas showing a significant correlation with Δ intensity and Δ unpleasantness (partial correlations controlled for changes in breathing parameters P_I and *f*).

		x	*y*	*z*	*Z*	*r*	*p**
Dyspnea anticipation (Δ cue dyspnea vs. cue baseline)							
with Δ intensity	OFC R	12	42	−20	3.47	−0.5	0.04
with Δ unpleasantness	Midbrain/PAG	2	−20	−8	3.10	−0.45	0.04
	Insula R	26	32	4	3.40	0.49	0.05
	Insula L	−34	22	14	3.87	0.55	0.01
		−36	34	8	3.47	0.5	0.04
Dyspnea Perception (Δ dyspnea vs. baseline)							
with Δ intensity	OFC R	14	56	−14	3.34	−0.47	0.05
with Δ unpleasantness	OFC R	16	30	−20	3.32	−0.46	0.05
	Insula R	26	30	12	4.01	0.57	0.01

**Corrected for multiple comparisons in respective ROI.*
MNI, Montreal Neurological Institute; OFC, orbitofrontal cortex; PAG, periaqueductal gray; R, right, L, left.

Both, Δ intensity and Δ unpleasantness, were significantly correlated with state and trait anxiety levels. This is in line with the notion that high anxious as compared to low anxious individuals are less likely to show short-term habituation toward dyspnea, which has previously been shown for hypercapnic rebreathing (Li et al., 2006). In addition, data from Wan et al. (2008, 2012) suggest that the effect of anxiety on short-term habituation is more pronounced for dyspnea unpleasantness as compared to dyspnea intensity. This is partly supported by our data, as correlations for state and trait anxiety with Δ unpleasantness were more pronounced ($r = 0.46$) compared to correlations with Δ intensity (0.32 and 0.36, respectively). Moreover, while the portion of individuals showing habituation toward dyspnea was identical for the two dimensions of dyspnea perception, the portion of subjects showing sensitization was higher for dyspnea unpleasantness (63%) as compared to dyspnea intensity (54.5%).

The close relationship of Δ intensity and Δ unpleasantness with anxiety raises the question, whether any correlation of changes in brain activity might be better explained by individual anxiety scores rather than rating dynamics. However, *post hoc* partial correlation analyses yielded no support for this notion.

The functional neuroimaging data presented here mirror the higher impact of trial repetition on dyspnea unpleasantness as compared to dyspnea intensity. While significant correlations of Δ intensity with changes in brain activation were limited to the OFC, Δ unpleasantness was significantly correlated with changes in the midbrain/PAG and the bilateral anterior insular cortex as well. This observation of short-term habituation/sensitization in healthy individuals parallels results from a previous study in patients with asthma, which linked long-term habituation toward dyspnea unpleasantness with similar brain areas (von Leupoldt et al., 2009). More specifically, that study observed reduced insula activation and increased PAG activation to resistive load induced dyspnea in asthma patients compared to healthy controls, which were paralleled by reduced dyspnea unpleasantness ratings in the patient group and correlated with asthma duration (von Leupoldt et al., 2009). Interestingly, longer asthma duration and reduced dyspnea unpleasantness in the same patients were also correlated with structural brain changes in terms of increased gray matter volume in the PAG, which was interpreted as another potential mechanism of long-term habituation to dyspnea (von Leupoldt et al., 2011b).

The PAG, in particular, is thought to be a key area of the so called anti-nociceptive network as repeatedly shown by studies on pain modulation (e.g., Fairhurst et al., 2007; for review see Tracey and Mantyh, 2007). Interestingly, midbrain/PAG activation during pain anticipation has been demonstrated to have a significant effect on subsequent pain perception (Brodersen et al., 2012). This is in line with our results, showing a significant correlation of Δ unpleasantness with midbrain/PAG activation changes for dyspnea anticipation only. For the OFC, there was a significant effect on pain perception for both, brain activation during pain anticipation and during painful simulation (Brodersen et al., 2012). This is also in line with the data presented here, as a significant correlation of changes in dyspnea ratings with changes in brain activation was found during anticipation and perception of dyspnea. Increasing activation in both areas, midbrain/PAG and OFC, were related to decreasing ratings of dyspnea unpleasantness indicating that increased activation in these areas support habituation. Further studies are required to investigate whether habituation is effectuated, e.g., by top–down (OFC) and/or bottom–up (midbrain/PAG) inhibition of other brain areas relevant for dyspnea perception (von Leupoldt and Dahme, 2005; Davenport and Vovk, 2009; Evans, 2010). However, one candidate target area for inhibition is the insular cortex, as this area showed the reversed correlation pattern: increasing activation in the insular cortex was related to increasing ratings of dyspnea unpleasantness. A similar pattern of insular and midbrain/PAG activation was described for the anticipation of pain by Ploner et al. (2010).

Changes of anterior insular activation over time were significantly linked to changes in dyspnea unpleasantness only. Therefore, the present findings support the notion, that activation of the anterior insular cortex is particularly relevant for the perception of the affective dimension of aversive events in general (Seeley et al., 2007) and dyspnea unpleasantness in particular (von Leupoldt et al., 2008; Paulus et al., 2012).

The positive correlations of changes in dyspnea ratings with brain activation changes during the preceding anticipation periods suggest an effect of expectancy and/or learning. As our results are limited to short-term habituation/sensitization in healthy individuals, future studies on long-term habituation and sensitization in patients with chronic dyspnea are required to investigate whether this association between anticipatory brain activations and subsequent dyspnea experience provides a potential target to correct for unfavorable learning of either avoidance behavior (maladaptive sensitization) or under-medication (maladaptive habituation).

In this study we present a novel approach for the investigation of habituation and sensitization toward dyspnea by using the correlation between late vs. early ratings on perceived dyspnea and late vs. early brain activation patterns. This approach allowed the inclusion of all participants irrespective of individual rating patterns as the variance across subjects allowed to differentiate between habituation and sensitization processes. However, correlation studies are limited in their ability to derive causal or directional conclusions. The investigation of connectivity patterns between the brain areas identified as critical in this study and by others might be a useful next step to clarify the way in which relevant areas interact and modulate each other. The connection between state and trait anxiety and habituation/sensitization toward dyspnea suggested by others (Li et al., 2006; Wan et al., 2008, 2012) and confirmed by this study provides another promising target for future studies on treatment and rehabilitation optimization.

Author Contributions

MCS: Substantial contributions to the design of the work, data acquisition, data analysis, drafting the manuscript and revising it critically for important intellectual content, final approval of the version to be published, agreement to be accountable for all aspects of the work in ensuring that questions related to the accuracy or integrity of any part of the work are appropriately investigated and resolved.

RE: Substantial contributions to the design of the work, data acquisition, revising the manuscript critically for important intellectual content, final approval of the version to be published, agreement to be accountable for all aspects of the work in ensuring that questions related to the accuracy or integrity of any part of the work are appropriately investigated and resolved.

MG: Substantial contributions to the design of the work, interpretation of data for the work, data analysis, revising the manuscript critically for important intellectual content, final approval of the version to be published, agreement to be accountable for all aspects of the work in ensuring that questions related to the accuracy or integrity of any part of the work are appropriately investigated and resolved.

CB: Substantial contributions to the design of the work, interpretation of data for the work, revising the manuscript critically for important intellectual content, final approval of the version to be published, agreement to be accountable for all aspects of the work in ensuring that questions related to the accuracy or integrity of any part of the work are appropriately investigated and resolved.

AvL: Substantial contributions to the conception and design of the work, interpretation of data for the work, revising the manuscript critically for important intellectual content, final approval of the version to be published, agreement to be accountable for all aspects of the work in ensuring that questions related to the accuracy or integrity of any part of the work are appropriately investigated and resolved.

Acknowledgments

We wish to thank all our participants and Timo Krämer, Kathrin Wendt, and Katrin Müller for technical assistance. This study was supported by grants from the German research foundation (DFG LE 1843/9-2, 10-1, and 10-3) to AvL.

References

Bingel, U., Schoell, E., Herken, W., Büchel, C., and May, A. (2007). Habituation to painful stimulation involves the antinociceptive system. *Pain* 131, 21–30. doi: 10.1016/j.pain.2006.12.005

Bloch-Salisbury, E., Shea, S. A., Brown, R., Evans, K., and Banzett, R. B. (1996). Air hunger induced by acute increase in PCO2 adapts to chronic elevation of PCO2 in ventilated humans. *J. Appl. Physiol.* 81, 949–956.

Brodersen, K. H., Wiech, K., Lomakina, E. I., Lin, C.-S., Buhmann, J. M., Bingel, U., et al. (2012). Decoding the perception of pain from fMRI using multivariate pattern analysis. *Neuroimage* 63, 1162–1170. doi: 10.1016/j.neuroimage.2012.08.035

Carrieri-Kohlman, V., Gormley, J. M., Eiser, S., Demir-Deviren, S., Nguyen, H., Paul, S. M., et al. (2001). Dyspnea and the affective response during exercise training in obstructive pulmonary disease. *Nurs. Res.* 50, 136–146. doi: 10.1097/00006199-200105000-00002

Davenport, P. W., and Vovk, A. (2009). Cortical and subcortical central neural pathways in respiratory sensations. *Respir. Physiol. Neurobiol.* 167, 72–86. doi: 10.1016/j.resp.2008.10.001

Evans, K. C. (2010). Cortico-limbic circuitry and the airways: insights from functional neuroimaging of respiratory afferents and efferents. *Biol. Psychol.* 84, 13–25. doi: 10.1016/j.biopsycho.2010.02.005

Fairhurst, M., Wiech, K., Dunckley, P., and Tracey, I. (2007). Anticipatory brainstem activity predicts neural processing of pain in humans. *Pain* 128, 101–110. doi: 10.1016/j.pain.2006.09.001

Hayen, A., Herigstad, M., and Pattinson, K. T. S. (2013). Understanding dyspnea as a complex individual experience. *Maturitas* 76, 45–50. doi: 10.1016/j.maturitas.2013.06.005

Hayen, A., Herigstad, M., Wiech, K., and Pattinson, K. T. S. (2015). Subjective evaluation of experimental dyspnoea – Effects of isocapnia and repeated exposure. *Respir. Physiol. Neurobiol.* 208, 21–28. doi: 10.1016/j.resp.2014.12.019

Herigstad, M., Hayen, A., Wiech, K., and Pattinson, K. T. S. (2011). Dyspnoea and the brain. *Respir. Med.* 105, 809–817. doi: 10.1016/j.rmed.2010.12.022

Holtz, K., Pané-Farré, C. A., Wendt, J., Lotze, M., and Hamm, A. O. (2012). Brain activation during anticipation of interoceptive threat. *Neuroimage* 61, 857–865. doi: 10.1016/j.neuroimage.2012.03.019

Hsieh, J. C., Stone-Elander, S., and Ingvar, M. (1999). Anticipatory coping of pain expressed in the human anterior cingulate cortex: a positron emission tomography study. *Neurosci. Lett.* 262, 61–64. doi: 10.1016/S0304-3940(99)00060-9

Janssens, T., Verleden, G., De Peuter, S., Van Diest, I., and Van den Bergh, O. (2009). Inaccurate perception of asthma symptoms: a cognitive-affective framework and implications for asthma treatment. *Clin. Psychol. Rev.* 29, 317–327. doi: 10.1016/j.cpr.2009.02.006

Labus, J. S., Naliboff, B. D., Berman, S. M., Suyenobu, B., Vianna, E. P., Tillisch, K., et al. (2009). Brain networks underlying perceptual habituation to repeated aversive visceralstimuli in patients with iritable bowel syndrome. *Neuroimage* 47, 952–960. doi: 10.1016/j.neuroimage.2009.05.078

Lansing, R. W., Gracely, R. H., and Banzett, R. B. (2009). The multiple dimensions of dyspnea: review and hypotheses. *Respir. Physiol. Neurobiol.* 167, 53–60. doi: 10.1016/j.resp.2008.07.012

Laviolette, L., Laveneziana, P., and ERS Research Seminar Faculty. (2014). Dyspnoea: a multidimensional and multidisciplinary approach. *Eur. Respir. J.* 43, 1750–1762. doi: 10.1183/09031936.00092613

Li, W., Daems, E., Van de Woestijne, K. P., Van Diest, I., Gallego, J., De Peuter, S., et al. (2006). Air hunger and ventilation in response to hypercapnia: effects of repetition and anxiety. *Physiol. Behav.* 88, 47–54. doi: 10.1016/j.physbeh.2006.03.001

Linnman, C., Moulton, E. A., Barmettler, G., Becerra, L., and Borsook, D. (2012). Neuroimaging of the periaqueductal gray: state of the field. *Neuroimage* 60, 505–522. doi: 10.1016/j.neuroimage.2011.11.095

Lowén, M. B. O., Mayer, E., Tillisch, K., Labus, J., Naliboff, B., Lundberg, P., et al. (2015). Deficient habituation to repeated rectal distensions in irritable bowel syndrome patients with visceral hypersensitivity. *Neurogastroenterol. Motil.* 27, 646–655. doi: 10.1111/nmo.12537

Miller, M. R., Hankinson, J., Brusasco, V., Burgos, F., Casaburi, R., Coates, A., et al. (2005). Standardisation of spirometry. *Eur. Respir. J.* 26, 319–338. doi: 10.1183/09031936.05.00034805

Parshall, M. B., Schwartzstein, R. M., Adams, L., Banzett, R. B., Manning, H. L., Bourbeau, J., et al. (2012). An official American Thoracic Society statement: update on the mechanisms, assessment, and management of dyspnea. *Am. J. Respir. Crit. Care Med.* 185, 435–452. doi: 10.1164/rccm.201111-2042ST

Paulus, M. P., Flagan, T., Simmons, A. N., Gillis, K., Kotturi, S., Thom, N., et al. (2012). Subjecting elite athletes to inspiratory breathing load reveals behavioral and neural signatures of optimal performers in extreme environments. *PLoS ONE* 7:e29394. doi: 10.1371/journal.pone.0029394

Petersen, S., Schroijen, M., Mölders, C., Zenker, S., and Van den Bergh, O. (2014). Categorical interoception: perceptual organization of sensations from inside. *Psychol. Sci.* 25, 1059–1066. doi: 10.1177/0956797613519110

Petersen, S., van den Berg, R. A., Janssens, T., and Van den Bergh, O. (2011). Illness and symptom perception: a theoretical approach towards an integrative measurement model. *Clin. Psychol. Rev.* 31, 428–439. doi: 10.1016/j.cpr.2010.11.002

Petrovic, P., Kalso, E., Petersson, K. M., and Ingvar, M. (2002). Placebo and opioid analgesia- imaging a shared neuronal network. *Science* 295, 1737–1740. doi: 10.1126/science.1067176

Ploner, M., Lee, M. C., Wiech, K., Bingel, U., and Tracey, I. (2010). Prestimulus functional connectivity determines pain perception in humans. *Proc. Natl. Acad. Sci. U.S.A.* 1071, 355–360. doi: 10.1073/pnas.0906186106

Seeley, W. W., Menon, V., Schatzberg, A. F., Keller, J., Glover, G. H., Kenna, H., et al. (2007). Dissociable intrinsic connectivity networks for salience processing and executive control. *J. Neurosci.* 279, 2349–2356. doi: 10.1523/JNEUROSCI.5587-06.2007

Spielberger, C. D., Gorsuch, R. L., Lushene, R., Vagg, P. R., and Jacobs, G. A. (1983). *Manual for the State-Trait Anxiety Inventory.* Palo Alto, CA: Consulting Psychologists Press.

Stankewitz, A., Valet, M., Schulz, E., Wöller, A., Sprenger, T., Vogel, D., et al. (2013). Pain sensitisers exhibit grey matter changes after repetitive pain exposure: a longitudinal voxel-based morphometry study. *Pain* 154, 1732–1737. doi: 10.1016/j.pain.2013.05.019

Tracey, I., and Mantyh, P. W. (2007). The cerebral signature for pain perception and its modulation. *Neuron* 55, 377–391. doi: 10.1016/j.neuron.2007.07.012

Troosters, T., van der Molen, T., Polkey, M., Rabinovich, R. A., Vogiatzis, I., Weisman, I., et al. (2013). Improving physical activity in COPD: towards a new paradigm. *Respir. Res.* 14, 115. doi: 10.1186/1465-9921-14-115

Tzourio-Mazoyer, N., Landeau, B., Papathanassiou, D., Crivello, F., Etard, O., Delcroix, N., et al. (2002). Automated anatomical labeling of activations in SPM using a macroscopic anatomical parcellation of the MNI MRI single-subject brain. *Neuroimage* 15, 273–289. doi: 10.1006/nimg.2001.0978

von Leupoldt, A., and Dahme, B. (2005). Cortical substrates for the perception of dyspnea. *Chest* 128, 345–354. doi: 10.1378/chest.128.1.345

von Leupoldt, A., Sommer, T., Kegat, S., Baumann, H. J., Klose, H., Dahme, B., et al. (2008). The unpleasantness of perceived dyspnea is processed in the anterior insula and amygdala. *Am. J. Respir. Crit. Care Med.* 177, 1026–1032. doi: 10.1164/rccm.200712-1821OC

von Leupoldt, A., Sommer, T., Kegat, S., Eippert, F., Baumann, H. J., Klose, H., et al. (2009). Down-regulation of insular cortex responses to dyspnea and pain in asthma. *Am. J. Respir. Crit. Care Med.* 180, 232–238. doi: 10.1164/rccm.200902-0300OC

von Leupoldt, A., Vovk, A., Bradley, M. M., Lang, P. J., and Davenport, P. W. (2011a). Habituation in neural processing and subjective perception of respiratory sensations. *Psychophysiology* 48, 808–812. doi: 10.1111/j.1469-8986.2010.01141.x

von Leupoldt, A., Brassen, S., Baumann, H. J., Klose, H., and Büchel, C. (2011b). Structural brain changes related to disease duration in patients with asthma. *PLoS ONE* 6:e23739. doi: 10.1371/journal.pone.0023739

Wan, L., Stans, L., Bogaerts, K., Decramer, M., and Van den Bergh, O. (2012). Sensitization in medically unexplained dyspnea: differential effects on intensity and unpleasantness. *Chest* 141, 989–995. doi: 10.1378/chest.11-1423

Wan, L., Van Diest, I., De Peuter, S., Bogaerts, K., Oyen, N., Hombroux, N., et al. (2008). Repeated experiences of air hunger and ventilatory behavior in response to hypercapnia in the standardized rebreathing test: effects of anxiety. *Biol. Psychol.* 77, 223–232. doi: 10.1016/j.biopsycho.2007.10.013

Wan, L., Van Diest, I., De Peuter, S., Bogaerts, K., and Van den Bergh, O. (2009). Repeated breathlessness experiences induced by hypercapnia: differential effects on intensity and unpleasantness. *Chest* 135, 455–461. doi: 10.1378/chest.08-1226

17

Embodied Cognitive Flexibility and Neuroplasticity Following Quadrato Motor Training

Tal D. Ben-Soussan[1,2], Aviva Berkovich-Ohana[3,4], Claudia Piervincenzi[4,5], Joseph Glicksohn[1,6] and Filippo Carducci[4]*

[1] The Leslie and Susan Gonda (Goldschmied) Multidisciplinary Brain Research Center, Bar-Ilan University, Ramat Gan, Israel, [2] Research Institute for Neuroscience, Education and Didactics, Patrizio Paoletti Foundation, Assisi, Italy, [3] Department of Neurobiology, Weizmann Institute of Science, Rehovot, Israel, [4] Department of Physiology and Pharmacology, Sapienza University of Rome, Rome, Italy, [5] Institute for Advanced Biomedical Technologies, Università degli Studi Gabriele D'Annunzio, Chieti, Italy, [6] Department of Criminology, Bar-Ilan University, Ramat Gan, Israel

***Correspondence:**
Tal D. Ben-Soussan,
Research Institute for Neuroscience, Education and Didactics, Patrizio Paoletti Foundation, via Cristoforo Cecci 2, 06081-Santa Maria degli Angeli, Assisi, Perugia, Italy
research@fondazionepatriziopaoletti.org

Quadrato Motor Training (QMT) is a whole-body movement contemplative practice aimed at increasing health and well-being. Previous research studying the effect of one QMT session suggested that one of its means for promoting health is by enhancing cognitive flexibility, an important dimension of creativity. Yet, little is known about the effect of a longer QMT practice on creativity, or the relative contribution of the cognitive and motor aspects of the training. Here, we continue this line of research in two inter-related studies, examining the effects of prolonged QMT. In the first, we investigated the effect of 4-weeks of daily QMT on creativity using the Alternate Uses (AUs) Task. In order to determine whether changes in creativity were driven by the cognitive or the motor aspects of the training, we used two control groups: *Verbal Training* (VT, identical cognitive training with verbal response) and *Simple Motor Training* (SMT, similar motor training with reduced choice requirements). Twenty-seven participants were randomly assigned to one of the groups. Following training, cognitive flexibility significantly increased in the QMT group, which was not the case for either the SMT or VT groups. In contrast to one QMT session, ideational fluency was also significantly increased. In the second study, we conducted a pilot longitudinal structural magnetic resonance imaging and diffusion tensor imaging (4-weeks QMT). We report gray matter volume and fractional anisotropy changes, in several regions, including the cerebellum, previously related to interoceptive accuracy. The anatomical changes were positively correlated with cognitive flexibility scores. Albeit the small sample size and preliminary nature of the findings, these results provide support for the hypothesized creativity-motor connection. The results are compared to other contemplative studies, and discussed in light of theoretical models integrating cognitive flexibility, embodiment and the motor system.

Keywords: Quadrato Motor Training, creativity, cerebellum, MRI, embodiment

Introduction

Creativity, Training, and Health

The lexeme in the English word *creativity* comes from the Latin term creō, meaning "to create, make." Thus, creativity means bringing into being, as it involves generation of novelty and transformation of existent information (Chávez-Eakle et al., 2007). Creativity requires, as well as generates, new information that transcends informational boundaries, yet is integrated with existing information in a manner exhibiting value (Horan, 2007). Behaviorally, the first component, namely the generation of new information, can be measured by divergent thinking tests; whereas the second component, concerning existing information constraints, can be measured by convergent thinking tests. Here we focus on divergent thinking, studying *ideational fluency* and *cognitive flexibility,* two important measures of creativity. Specifically, cognitive flexibility is diminished in several neurodegenerative conditions, such as Parkinson's (Tomer et al., 2002). Thus, sustaining and improving cognitive flexibility may serve a significant role in maintaining cognitive and emotional well-being and health (Schmid, 2005). In the current study, we aimed at investigating the link between flexibility of behavior and cognitive flexibility. To this end, we employ *Quadrato Motor Training* (QMT by Patrizio Paoletti – see below), which requires constant flexibility in the movement and behavior, and assess its impact on cognitive flexibility and ideational fluency using the AUs Task (Chermahini and Hommel, 2010).

The QMT is a whole-body movement meditation, aimed at improving well-being, by enhancing attention, coordination and cognitive flexibility (Ben-Soussan et al., 2013, 2015). The QMT requires a state of enhanced attention, as it combines dividing attention to the motor response and cognitive processing for producing the correct direction of movement to the next point in the Quadrato space (Ben-Soussan et al., 2013; see **Figure 1**). QMT can further be viewed as *'Mindful movement,'* due to the increased awareness it requires to the body and its location in space. *Mindful movement* is a general term for practices that involve bringing awareness to the detailed experience of movement (Kabat-Zinn, 2009), such as walking meditation, Yoga and Tai Chi. Similar to other Mindful movement practices, QMT requires balance control, which is known to integrate inputs from the motor cortex, cerebellum, and basal ganglia, as well as feedback from vestibular and proprioceptive systems required to maintain an upright posture (Chang et al., 2010, 2014; Wayne et al., 2014). In line with that, a previous study demonstrated that a month of daily QMT practice significantly enhanced cerebellar oscillatory function (Ben-Soussan et al., 2014a). Yet, in comparison to other Mindful movement practices, QMT has the advantage of being a relatively short training (possibly several minutes), and can be practiced in limited spaces. These unique aspects render the QMT a technique warranting scientific exploration, with the future aim of implementing this technique in various health-promoting and educational setups.

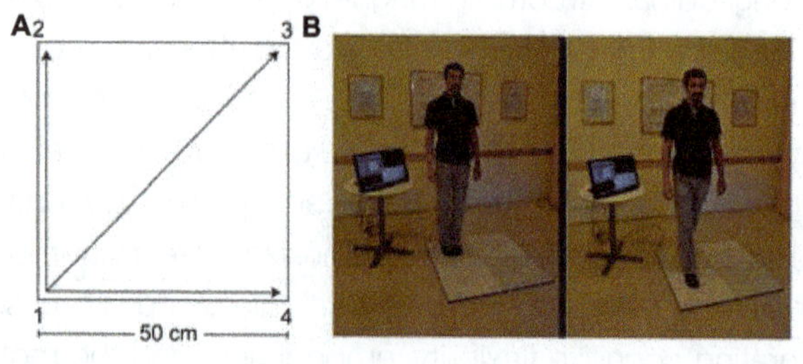

FIGURE 1 | The Quadrato Motor Training (QMT). (A) A graphical illustration of the QMT. **(B)** A participant during the QMT while waiting for the next instruction (left) and following the instruction (right).

We have previously reported that a single session of QMT increased cognitive flexibility, but not fluency, as well as improved spatial cognition and reflectivity (Ben-Soussan et al., 2013, 2014b) in contrast to two different control groups controlling for cognitive and motor load.

Long-term effects of a month of daily QMT on divergent thinking were previously studied in comparison to simple walking training (WT; Venditti et al., 2014). Both fluency and flexibility increased only in the QMT group (Venditti et al., 2014). These significant differences might have stemmed from the cognitive aspect of the contemplative movement (QMT), relatively missing from the non-contemplative simple walking (WT).

This has led us in the present study (Study A) to compare 4-weeks of daily QMT to two control groups, tapping on the cognitive and motor aspects of the QMT: Verbal Training (VT, identical cognitive training with verbal response) and Simple Motor Training (SMT, similar motor training with reduced choice requirements). At the same time, the present study replicates the 1-month QMT in a different language. In this study, we predicted that: (1) cognitive flexibility would increase following QMT compared to the other control groups, while (2) fluency is predicted to increase in both the QMT and the VT groups.

A direct continuation of Study A is to start uncovering the possible underlying mechanism mediating QMT-induced enhancement in divergent thinking. Indeed, a fascinating and under-explored aspect of creativity is its possible connection to the motor system, suggested by several authors (Cotterill, 2001; Dietrich, 2004; Chávez-Eakle et al., 2007; Vandervert et al., 2007; Carruthers, 2011; Koziol et al., 2012). Building on the previously hypothesized creativity-motor connection, we set out in the pilot Study B to examine a possible correlation between change in divergent thinking measures and structural changes in brain regions related to motor activation. To this end, we employed structural Magnetic Resonance Imaging (sMRI), and investigated structural changes and AU scores following 4-weeks of daily QMT. Here, we predicted that: (1) QMT would induce anatomical changes in motor regions; and (2) anatomical changes in motor regions would be correlated with increased cognitive flexibility.

Study A

Methods

Participants and Design

Twenty-seven female students (mean ± SD age: 24 ± 3 years) participated in the study, none of whom practiced QMT before. All were right-handed with no medical history that might affect their performance. Since gender-dependent differences have been frequently observed in both the motor and the cognitive realms (Baron-Cohen and Hammer, 1997) we chose to focus here on females. The study was conducted in the EEG/MEG unit at the Gonda Brain Research Center, and was approved by the ethics committee of Bar-Ilan University.

Upon entering the lab, the participant signed a written informed consent. Subsequently, participants were seated in a quiet room, in front of a computer screen and completed the AU Task. All data were collected both before and after 4 weeks of daily practice. Participants were randomly allocated to one of three groups: (1) Quadrato Motor Training (QMT- three choices and whole-body response); (2) SMT (one choice and whole-body response); and (3) VT (three choices and verbal response). Although the initial group sizes were identical ($n = 9$ each), the final number of participant finishing the 4-weeks training varied between the groups (QMT, $n = 6$; SMT, $n = 7$; and VT, $n = 5$).

Training Groups

Quadrato Motor Training

The QMT, by Patrizio Paoletti, requires standing at one corner of a 0.5 m×0.5 m square and making movements in response to verbal instructions given by an audio tape recording. There were three optional directions of movement. The instructions direct participants to keep the eyes focused straight ahead, hands loose at the side of the body. They are also told to immediately continue with the next instruction and not to stop due to mistakes. At each corner, there are three possible directions to move (for example, from corner 1 the participant can move to corner 2, to corner 3 or to corner 4). The training thus consists of 12 possible movements (3 directions × 4 corners): 2 forward, 2 backward, 2 left, 2 right and 4 diagonals. The participant is required to move from one corner to another according to the number on the recording. For example, if the sequence required is 1, 2, 1, 2, 1, 2, 3, 2, 4, 3, 1.... this means moving to the first corner, then to the second, then back to the first, and so on (see **Figure 1**). The daily training consisted of a sequence of 69 commands, lasting 7 min. For additional data regarding the training groups, see Ben-Soussan et al. (2013).

Simple Motor Training

The SMT group moved from corner to corner on the square in exactly the same manner as the QMT group (pace, duration, auditory cue), but their movement was consistently 1-2-3-4-1 etc. This group also practiced with the same recordings as the QMT group. However, while the QMT group was told that each number represented a different corner of the square, the SMT group was told to simply begin at a certain corner and to continue to the next corner clockwise in response to the instructions. That is, regardless of the number specified on the tape, they always moved in the same sequence. This reduced the uncertainty regarding the direction of the movement, compared to the QMT group. The SMT group thus provided a control of similar motor performance but with reduced cognitive demands.

Verbal Training

The VT group was designed to control for the motor load while keeping the same cognitive load and uncertainty. The participants, who were instructed to only make verbal responses, stood 1 m in front of the square, but did not move on the corners at all. Instead, they responded to the taped commands verbally by stating what direction of movement would be required in order to reach the corner specified by the command. For example, for a movement from corner 1 to corner 2, they were required to say "straight." All other training parameters were kept identical to the QMT (pace, duration, auditory cue).

The Alternate Uses (AUs) Task

The AU Task is an established measure assessing creativity (Guilford et al., 1978). Two main features of creativity are *fluency*, defined as the total number of ideas generated, and *flexibility*, defined as the tendency to generate a heterogeneous pool of responses, or to use a variety of categories and themes when producing ideas (Guilford, 1950; Guilford et al., 1978; Runco, 1986). Flexibility conveys information that is not conveyed by fluency (Guilford, 1968; Runco, 1986). This task was chosen here as it was previously used to assess change in cognitive flexibility following motor training (Netz et al., 2007; Venditti et al., 2014).

In a previous study with the aim of examining changes in creativity following training (reported in Ben-Soussan et al., 2013), we clustered a 908-word database developed by Levy-Drori and Henik (2006), using hierarchical cluster analysis, and grouped those words having similar ratings of concreteness, availability of context and familiarity assigned by Levy-Drori and Henik. Concreteness was rated by them on a scale from one to seven, where "1" indicated very low concreteness (very abstract word) to "7" which indicated a very concrete word. Availability of context was defined as ranging from 1 to 7, with "1" indicating that it is very difficult to think of a context and "7" indicating that it is very easy. Familiarity was defined as ranging from "1" indicating that it is not very familiar and "7" indicating that it is very familiar. We then chose 18 words from the three largest clusters for which the level of concreteness, similarity on familiarity and availability features were highest (see Supplementary Table S1). These 18 words were divided into two groups (nine words in each group). A total of 60 participants received one of the two lists, and were asked to produce as many AUs as possible, 1 min being allocated for each item. Each word was shown on a single page on which the participant had to write down the various uses. The scores for each word were analyzed by counting the number of AUs produced. The words were then divided into six sets of three words so that each set had a similar rating for familiarity, concreteness and availability. One of the sets (set d) resulted in higher fluency scores (see Supplementary data, where corrections maintain all reported results).

The presentation order of each set of words was counterbalanced using a Latin square (see Supplementary Table S2). Three names of objects were shown on a computer screen before the training and after the training (at the beginning and at the end of the month). Six pairs of sets (e.g., a–f) were used with internal order counterbalanced across six participants. The internal order of the three items per set (Table S1 in supplementary material) was presented in a random manner. Importantly, sets a, b, c, d, e, and f each appear once in each ordinal position.

In this task, the participant is required to name as many different ways in which a given item might be used, within a certain time frame (1 min). The fluency score was defined as the mean number of uses given by the participant for the three items. On the basis of all the uses made by the participants, 10 independent categories were defined across all the items. These included broad categories of usage such as "a weapon" or "a costume." The flexibility score was defined as the mean number of different categories employed by the participant across all three words presented (Russ, 1998). Hence, in order to calculate the flexibility score, all responses for a given item were first divided into the different independent categories. Two independent judges who were naïve to the identity of the participants and their training groups scored the test independently for flexibility, and consistency between judges was tested. We examined the correlation between the scores of the two judges using a two-tailed Pearson correlation coefficient test. A high correlation was found between scores given by the two judges both in the scores before the training and after the training: $r = 0.88$ and 0.86, $n = 18$, respectively.

Statistical Analysis

We ran a Group (QMT, SMT, VT) × Training (pre, post) analysis of variance (ANOVA) for creativity (separately on fluency and flexibility scores). Then, *post hoc* paired *t*-tests were conducted.

Results

The first ANOVA, conducted for fluency, revealed a main effect for Training [$F(1,15) = 8.80$, MSE = 17.25, $p = 0.01$], with post-training being generally higher compared to pre-training (**Figure 2A**). Albeit the Group × Training interaction was not significant [$F(2,15) < 1$], we tested directly each group's effect on fluency, to better evaluate the results compared to our previous studies. *Post hoc t*-tests showed that fluency significantly increased only in the QMT group [$t(5) = -4.21$, $p < 0.01$], and not in the SMT or VT groups [$t(6) = -0.31$, $t(4) = -1.33$, *ns*, respectively]. The second ANOVA conducted for flexibility similarly yielded a main effect for Training [$F(1,15) = 8.01$, MSE = 5.81, $p < 0.05$]. In addition, a significant Group × Training interaction was found [$F(2,15) = 5.20$, MSE = 0.727, $p < 0.05$]. For the QMT group, flexibility significantly increased [$t(5) = -3.20$, $p < 0.05$] in contrast to the SMT and VT groups who showed no change following training (see **Figure 2B**).

Importantly, correcting for AU set d did not change these results (see supplementary materials).

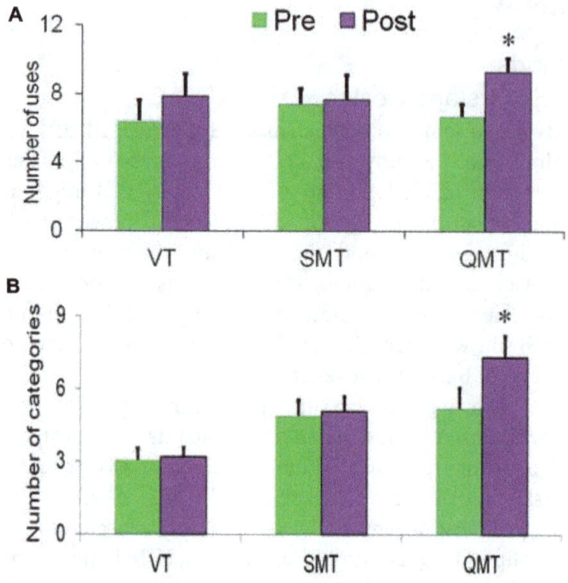

FIGURE 2 | Change in creativity as a function of Group and Training. **(A)** Fluency and **(B)** Flexibility. Data are expressed as mean ± SEM, *$p < 0.05$.

Study B

Methods

Participants and Design

Three healthy women participated in this pilot study (mean ± SD age: 41.5 ± 11.4 years), none of whom practiced QMT before, with no previous head injury which might have affected their brain structure. In this study, the QMT sequence consisted of 258 commands, and lasted 12 min. The reason for the change in procedure was due to the fact that this study is a part of a larger study conducted with the longer QMT practice paradigm aimed at examining longer sequences for neurodegenerative disease. The study took place at the St. Andrea Hospital, Rome. Upon entering the lab, the participant signed a written informed consent. sMRI was acquired immediately after the AU Task. The study was approved by the ethical committee of Università Campus Bio-Medico di Roma. The AUs task was employed similarly to the one reported in Study A, with different sets given to the participants before and following a month of QMT training. The AU task was completed prior to entering the magnet.

MRI Data

MRI Scans

For each subject, high-resolution 3D T1-weighted sMRI and diffusion tensor imaging (DTI) data were acquired at the beginning (pre-QMT) and after 4-weeks of daily QMT practice (post-QMT) using a Siemens MAGNETOM Sonata (Erlangen, Germany) 1.5 T scanner (sMRI: 3D T1-weighted MP-RAGE sequence, TR = 3000 ms, TE = 4.38 ms, flip angle = 15°, matrix = 192 × 192, FOV = 240 mm^2, 160 sagittal slices, voxel size = 1.25 mm × 1.25 mm × 1.20 mm;

DTI: 12 non-collinear direction sequence, TR = 7100 ms, TE = 94 ms, flip angle = 90°, matrix = 256 × 192, FOV = 240 mm × 320 mm, b = 0 and 900 s/mm^2, 48 axial slices, voxel size = 1.25 mm × 1.25 mm × 3.0 mm). We computed two anatomical measures: gray matter (GM) volume, and fractional anisotropy (FA). GM volume is a measure of the amount of cortical GM corresponding to each region and is computed using both surface area and cortical thickness (Frye et al., 2010). FA is a marker of white matter integrity, which is thought to reflect anatomical features of white matter, such as axon caliber, fiber density and myelination (Scholz et al., 2009).

Image Analysis

The sMRI data were analyzed using the voxel-based morphometry technique (VBM, Good et al., 2001). DTI data were analyzed using the Tract-Based Spatial Statistics (TBSS, Smith et al., 2006); SMRI were longitudinally processed with VBM2 toolbox[1] of SPM2[2]. In short, pre-QMT images were aligned to the T1 template and, then, post-QMT images were co-registered to the pre-QMT T1-aligned scans. All scans were bias corrected and segmented into gray and white, and CSF compartments. GM maps were normalized to MNI atlas space (1 mm × 1 mm × 1 mm voxels) and smoothed using an 8 mm FWHM Gaussian kernel. Preprocessing of DTI data was conducted with FSL[3]. DTIs were corrected for motion and eddy currents distortions and then skull-stripped using the Brain Extraction Tool (BET; Smith, 2002). Maps of FA were computed by fitting a tensor model to the raw diffusion data using the FMRIB's Diffusion Toolbox (FDT). These maps were projected onto a mean FA tract skeleton, before applying voxelwise within-subject statistics to compare them between time points, as described elsewhere (Scholz et al., 2009).

Statistical Analysis

Voxel-based morphometry technique data were analyzed using a general linear model (GLM). Anatomical localization was carried out with the MSU-MNI Space Utility toolbox of SPM using the AtlasQuery FSL tool. Statistical analysis of TBSS data was performed by using a permutation-based inference tool for non-parametric statistical thresholding. Pre- and post-QMT FA and VBM data were compared by using a paired t-test, with subject's age as a covariate, controlling for the potential effect of this variable (Luders et al., 2009). Significance level for the t-tests was set with a minimum cluster size of 700 voxels for the GM volume (Brubaker et al., 2010) and one of 100 voxels or more for the FA as the cluster-size threshold (Ocklenburg et al., 2013).

Results

Alternate Uses Task

Similar to the results of Study A, flexibility significantly improved following 4-weeks of daily QMT [$t(2) = -6.05$, $p < 0.05$; **Figure 3**]. Although fluency also increased, it did not reach significance [$t(2) = -0.39$, *ns*].

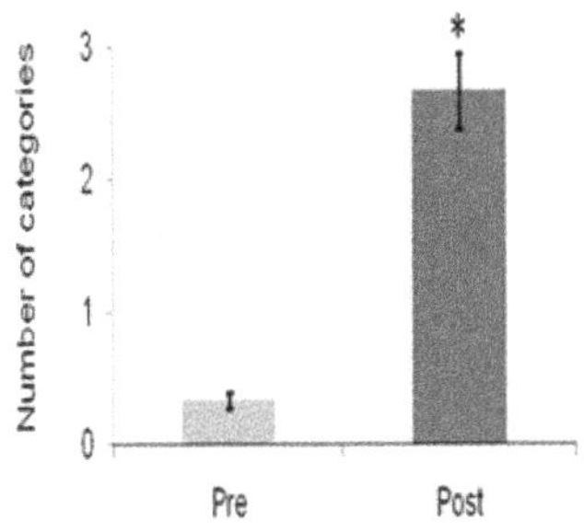

FIGURE 3 | Change in creativity as a function of Training for Flexibility. Data are expressed as mean ± SEM, *$p < 0.05$.

Brain Anatomy Results

Voxel-based morphometry technique analysis results showed significant ($p < 0.001$, uncorrected) GM volume increases, localized in left and right cerebellum, and frontal areas, mainly in the inferior frontal and middle frontal gyri (**Table 1**; **Figure 4A**). FA analysis results showed significant ($p < 0.01$, uncorrected) increases localized mainly in the middle cerebellar peduncles (**Table 2**; **Figure 4B**).

We then examined the correlation between the anatomical changes and changes in flexibility, both of which significantly increased post-training. Multiple regression analyses were performed to investigate whether GM and FA map changes were correlated with the change in flexibility, as previously reported for a similar group size (Ben-Soussan et al., 2015). We used the results of VBM and TBSS analyses as explicit masks to ensure that only the appropriate GM and fiber ROI which changed following the training were included in the analysis (Denier et al., 2013). These masks were subsequently inserted as explicit masks in two correlation analyses with flexibility, one for the GM and one for the FA.

As seen in **Figure 5A**, a positive correlation ($p < 0.005$, $n = 3$) was found between change in flexibility and the GM increment in the right cerebellum and the superior frontal gyrus (**Table 3**). In addition, a positive correlation ($p < 0.05$) was found between increased flexibility and FA changes, mostly located in the left corticospinal tract and the middle cerebellar peduncles (**Table 4**; **Figure 5B**).

Discussion

QMT Improves Divergent Thinking Creativity

The results of Study A show that 4-weeks of daily QMT practice induce increased cognitive flexibility and ideational fluency, which was not the case in either the SMT group or in the VT group, representing the motor and cognitive aspects of the training, respectively (**Figure 2**). This is in line with our first hypothesis, and with two previous studies. In the first relevant study, we reported findings regarding the effects of a single session of QMT (Ben-Soussan et al., 2013), where a similar increase in flexibility was observed in contrast to SMT or VT.

[1] http://dbm.neuro.uni-jena.de/vbm/
[2] http://www.fil.ion.ucl.ac.uk/spm/
[3] http://www.fmrib.ox.ac.uk/fsl

TABLE 1 | Significant increases in GM volume after 4-weeks of daily QMT.

			Coordinates						
k	*p*-value	*Z*-value	*x*	*y*	*z*	Hemisphere	Lobe	Region	BA
1584	0.000	3.63	24	−52	−22	Right cerebellum	Anterior lobe	Culmen	
							Posterior lobe	Cerebellar tonsil	
								Declive	
								Tuber	
								Uvula	
1128	0.000	3.79	−6	13	65	Left cerebrum	Frontal lobe	Medial frontal gyrus	6
								Middle frontal gyrus	6
								Superior frontal gyrus	8
851	0.000	4.07	−15	21	−23	Left cerebrum	Frontal lobe	Inferior frontal gyrus	47
								Middle frontal gyrus	11
								Orbital gyrus	47
								Subcallosal gyrus	47
								Superior frontal gyrus	11
706	0.000	4.34	−17	−37	−29	Left cerebellum	Anterior lobe	Culmen	
580	0.000	4.32	−32	−62	−33	Left cerebellum	Posterior lobe	Cerebellar tonsil	
								Culmen	
								Declive	
								Pyramis	
								Tuber	
								Uvula	
							Anterior lobe	Culmen	
471	0.000	3.88	−37	−5	−7	Left cerebrum	Sub-lobar	Claustrum	
								Extra-nuclear	13
							Temporal lobe	Superior temporal gyrus	38
297	0.000	4.10	−54	−37	−3	Left cerebrum	Temporal lobe	Middle temporal gyrus	21
								Superior temporal gyrus	22
262	0.000	4.18	11	28	−33	Right cerebrum	Frontal lobe	Orbital gyrus	11
								Rectal gyrus	11
102	0.001	3.19	21	−6	−38	Right cerebrum	Limbic lobe	Parahippocampal gyrus	35
								Uncus	36

Cluster extension (k), which represents the number of contiguous voxels passing the threshold of ≥100. All clusters meet the significance level set at $p < 0.001$ uncorrected. Neuroanatomical labels are referred to the Talairach coordinates x, y, z, by means of the MSU SPM tool.

TABLE 2 | Significant increases in FA values after 4-weeks of daily QMT.

		Coordinates			
k	*p*-value	*x*	*y*	*z*	*WM structures*
283	0.010	102	102	41	Middle cerebellar peduncle
59	0.004	94	107	38	Corticospinal tract L
31	0.008	100	62	40	Anterior thalamic radiation L
27	0.008	54	71	74	Posterior thalamic radiation R

The number of voxels (k) represents the number of voxels belonging to certain WM structures. Significance level set at $p < 0.01$ uncorrected. Neuroanatomical localization was obtained using AtlasQuery FSL tool with JHU White-Matter Tractography and the JHU ICBM-DTI-81 White-Matter Labels as reference atlases.

However, no significant change was reported for fluency. Our results show that the increase in flexibility was not significantly different between one session and 4-weeks of training, possibly suggesting that participants might have reached a ceiling effect in the AU task after one session, underscoring the need to utilize in the future different tasks to assess QMT-induced cognitive flexibility. Yet, as opposed to one session of training (Ben-Soussan et al., 2013), long-term QMT also increased ideational fluency, which is probably mediated by an accumulated effect of daily QMT sessions. Yet, the fluency results should be treated with caution as the Group × Training interaction was not significant. We further acknowledge the reduction in statistical power given participant attrition, in part because participants who successfully complete all follow-up measurements may have differed from those respondents lost to attrition. Nevertheless, no baseline differences were found in the AU scores between those who continued and those who dropped out (see Supplementary data and Table S4).

In another recent study, we examined the effects of a month of daily QMT as opposed to simple WT, tapping the motor aspect of QMT (Venditti et al., 2014). This study reported increased flexibility and fluency following QMT but not WT. Our results replicate the previous results in the QMT group, as expected. This demonstrates that there is no language bias

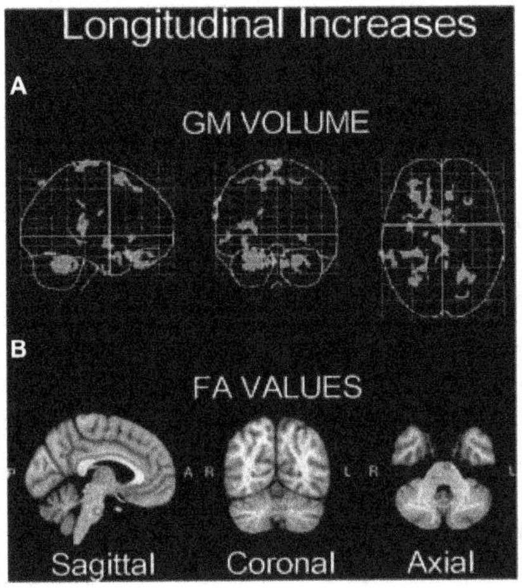

FIGURE 4 | Neuroplasticity following longitudinal (4-weeks) daily QMT. (A) Significant increases in GM volume ($p < 0.001$, uncorrected) projected onto the glass brain; **(B)** Significant increases in FA values ($p < 0.01$, uncorrected) onto brain template. Red-yellow color scale is related to the *p*-values (in which yellow is highest).

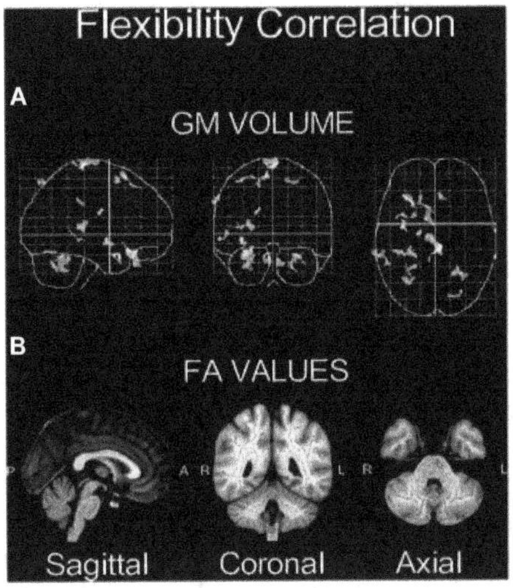

FIGURE 5 | Significant correlations between change and GM volume (A,B) FA values. Red-yellow color bar range is based on min-max statistical threshold. Refer to corresponding supplementary tables for more neuroanatomical and statistical details ($n = 3$).

affecting the results, as the current study was conducted with Hebrew speakers, whereas the previous study was conducted with Italian speakers. Importantly, while we hypothesized an increase in fluency following VT, tapping the cognitive aspect of the QMT, no change in fluency occurred in both the VT and SMT groups. This suggests that the combination of the cognitive and motor aspects of the QMT, and not the cognitive aspect *per se*, renders the fluency enhancement. Put differently, we suggest that this stems from the complexity of the movement, indicating that it

TABLE 3 | Positive correlation between flexibility values and GM maps showing volume increases after 4-weeks of daily QMT.

			Coordinates						
k	*p*-value	*Z*-value	*x*	*y*	*z*	Hemisphere	Lobe	Region	BA
612	0.000	3.73	21	−56	−37	Right cerebellum	Anterior lobe	Culmen	
							Posterior lobe	Cerebellar tonsil	
								Declive	
301	0.001	4.41	−6	18	62	Left cerebrum	Frontal lobe	Superior frontal gyrus	6, 8
292	0.001	3.47	−29	−64	−32	Left cerebellum	Posterior lobe	Cerebellar tonsil	
								Pyramis	
								Tuber	
								Uvula	
								Culmen	
								Declive	
							Anterior lobe	Culmen	
248	0.002	3.83	−37	−2	−7	Left cerebrum	Sub-lobar	Extra-nuclear	13
								Claustrum	
235	0.000	3.76	−25	22	−18	Left cerebrum	Frontal lobe	Inferior frontal gyrus	47
								Middle frontal gyrus	11
180	0.005	3.55	−37	9	58	Left cerebrum	Frontal lobe	Middle frontal gyrus	6
								Superior frontal gyrus	6
133	0.002	3.83	−37	−4	−5	Left cerebrum	Temporal lobe	Middle temporal gyrus	21
								Superior temporal gyrus	22
65	0.001	3.35	5	31	−28	Right cerebrum	Frontal lobe	Orbital gyrus	11
								Rectal gyrus	11

$k \geq 100$; $p < 0.005$ uncorrected. Refer to ***Table 1*** *for a detailed explanation of the table layout.*

TABLE 4 | Positive correlation between flexibility values and FA maps showing increases after 4-weeks of daily QMT.

		Coordinates			
k	*p*-value	*x*	*y*	*z*	WM structures
21	0.018	96	101	37	Corticospinal tract L
5	0.047	102	101	41	Middle cerebellar peduncle

Significance level set at $p < 0.05$ uncorrected. Refer to ***Table 2*** *for a detailed explanation of the table layout.*

is the combination of the motor and cognitive aspects embedded in QMT, which is important for increasing fluency.

In Study B, flexibility again significantly increased, as expected (**Figure 3**). Yet, although a trend was observed for fluency, it did not reach significance, possibly due to the small sample size. However, the possibility that the discrepancy between the results of Study A and Study B in terms of fluency is due to the altered QMT sequence in the second study (12 min in Study B vs. 7 min in Study A) cannot be ruled out. Therefore, the effect of QMT sequence length on different aspects of creativity should be further examined, underlying the importance of examining additional sequences of the QMT.

QMT as a Mindful Practice

Our results are in line with the proposal that contemplative practices increase creativity (Horan, 2007), albeit early evidence is inconsistent with that, possibly due to the wide variety of meditation techniques and creativity measures employed (Horan, 2007; Lippelt et al., 2014). Although there are many meditative techniques, we adopt here the conceptualization of Lutz et al. (2008), grossly dividing meditation into two forms: focused attention - learned control over the focus of one's attention by using a stable object with the goal of quieting the mind, and open monitoring (OM) - maximizing the breadth and clarity of maintained attention in order to bring higher momentary awareness to internal processes (the latter includes for example Mindfulness and Zen). Importantly, when focusing only on studies investigating the effect of OM techniques on creativity, the overall picture is of OM practices increasing divergent thinking (Cowger and Torrance, 1982; Zabelina et al., 2011; Colzato et al., 2012; Greenberg et al., 2012; Ding et al., 2014).

Here, we wish to draw attention to the mindful nature of QMT, supported by the similarity of the training effects in enhancing divergent thinking. Mindful practices have been noted in the literature using varied terminology depending on the discipline called upon, including "the act of becoming more aware," "a reflective act," or "mindfulness" (Depraz et al., 2000). It has been claimed that the mindful act has three interdependent phases: a first phase of suspension from the habitual act of allowing the mind and body to go where they want; a second phase of redirection of attention inwardly; and a third phase of receptivity toward the experience (Depraz et al., 2000). Mindfulness meditation, as conceptualized in the Western tradition, incorporates all three phases (Kabat-Zinn, 2009). Similarly, QMT can be thought of as training in all the above three phases. First, one suspends the automatic movement. Then, one redirects attention toward the external cue and the internally generated movement. Finally, one quietly stands in a receptive manner in between instructions, without correcting motor or decision errors. QMT is also dependent on the moment by-moment re-investment of attention. In support of that, we reported increased ability to respond in a non-habitual fashion following 5–12 min QMT (Moore and Malinowski, 2009; Ben-Soussan et al., 2014b). To sum, QMT can be considered to be a Mindful Movement practice, and similarly to other OM practices, leads to increased divergent thinking.

Structural Changes Following QMT

Structural changes in GM volume were found mainly in the left and right cerebellum, as well as in the left frontal lobe, whereas the main FA increase was found in the middle cerebellar peduncle (**Figure 4**). The changes in the cerebellum and the frontal cortex could be explained by their role in the acquisition of motor sequences (Exner et al., 2002). The cerebellum is important for the integration of somatosensory and motor information relevant to coordinate context-dependent planning and execution of coordinate motion (Ivry, 2000; Tesche and Karhu, 2000). Although interoception was not directly measured here, it is important to note in the context of this special issue, that the cerebellum is closely linked to interoceptive awareness reflecting explicit awareness of bodily processes (Critchley et al., 2004; Mercader et al., 2010). The left frontal regions, mainly in the inferior frontal and middle frontal gyri, are related to motor learning, action observation and intention understanding (Exner et al., 2002; Dapretto et al., 2005). In addition, both the inferior frontal and the middle frontal gyri (see **Table 1**) have been consistently related to working memory and response inhibition (Leung et al., 2002; Morin and Michaud, 2007; Swick et al., 2008) and selection among competing alternatives (Paulesu et al., 1993) which may increase following QMT. Notably, the spatial proximity of the GM and FA increases in the cerebellum following the training suggests that the QMT-induced increase in GM volume is related to an altered organization of underlying white matter pathways. Importantly, we have previously reported that following 3-months QMT cerebellar GM volume and FA increased, in positive correlation with increased brain-derived neurotrophic factor (BDNF) level (Ben-Soussan et al., 2015), a neurotrophin closely linked to interoceptive awareness (Mercader et al., 2010). In addition, QMT was previously reported to increase cerebellar low-rhythm activity (Ben-Soussan et al., 2014a). This has led to a theoretical account emphasizing the cerebellum as an important candidate which possibly mediates the QMT effects on well-being (Ben-Soussan et al., 2015). The current anatomical results provide further support to such a proposition, and show that 1-month QMT is sufficient to induce measurable cerebellar anatomical alterations.

The current results are in line with previous meditation studies which have found activation in areas closely linked to motor learning, such as the frontal gyrus and the cerebellum.

For example, Pagnoni and Cekic (2007) reported increased GM volume in Zen meditation practitioners compared to matched controls, especially in the putamen which is closely involved in the control of voluntary movement (Nambu et al., 2002). This result was interpreted as being related to the cognitive processes engaged by meditation, and especially to the conscious regulation of attention and posture (Pagnoni and Cekic (2007). Vestergaard-Poulsen et al. (2009) examined the effects of Tibetan Buddhist meditation, and found that in addition to increased GM density in the medulla oblongata, left superior and inferior frontal gyri, and left fusiform gyrus, an increment was found in the anterior lobe of the cerebellum in the meditator group compared to the control group. In addition, Hölzel et al. (2011) studied the effect of mindfulness-based stress reduction (MBSR) in meditation-naïve participants, before and after 8-weeks of training. In addition to other regions, they found increased GM density in the cerebellum in the MBSR group compared to controls. Hölzel et al. (2011) address their cerebellar-related results in relation to its role in the regulation of emotion and cognition. They further mention the claim made by Schmahmann (2004), that in the same way that the cerebellum regulates the rate, force, rhythm, and accuracy of movements, it also regulates the speed, capacity, consistency, and appropriateness of cognitive and emotional processes.

Compared to sitting meditation, the anatomical examination of whole-body movement-based contemplative practices is surprisingly rare. One exception is a recent study examining the effects of Tai Chi Chuan (TCC), a whole-body contemplative practice involving movement, such as weight-shifting between the right and left legs and asymmetrical diagonal leg movements (Wei et al., 2013). Wei et al. (2013) demonstrated that TCC practitioners showed greater cortical thickness in prefrontal cortex and temporal cortex relative to the matched control group. Interestingly, neuroimaging studies have consistently shown similar findings following aerobic exercise, as this was shown to lead to increased gray and white matter volume in the prefrontal cortex of older adults (Colcombe et al., 2006). In addition, greater amounts of physical activity, such as walking, are associated with sparing of prefrontal and temporal brain regions of late adulthood (Erickson et al., 2010). Taken together, the above studies show that both sitting and movement contemplative practices induce neuroplasticity in motor regions. This strengthens the claim that the results shown here might not be due to the movement *per se*, but involve higher cognitive modulation.

Creativity and the Motor System

Albeit the small sample size and the preliminary nature of the findings, a positive correlation was found between increased flexibility and cerebellar changes, both in the GM volume and the FA values (**Figure 5**). Due to the low power of the correlation analysis, this finding should be treated as being suggestive, but can still guide the hypotheses of future larger studies. This correlation supports the previously suggested link between creativity and the motor system (Dietrich, 2004; Carruthers, 2011; Koziol et al., 2012). For example, Cotterill (2001) suggested that: "If cognition is linked to overt or covert movement, intelligence becomes the ability to consolidate individual motor elements into more complex patterns, and creativity is the outcome of a race-to-threshold process which centers on the motor areas." (p. 1). Similarly, Carruthers (2011) wrote: "creativity lies in the assembly and activation of action-schemata, with creative thoughts arising subsequently from the mental rehearsal of those actions" (p. 437). The creativity-motor connection has been advanced to the point that it has been provocatively claimed that: "we were not born to think. We were born to move. Human creative ideas are nothing in the absence of the manual dexterity that allows tools to be made, complex architecture to be constructed, art to be created, and instruments to be played" (Koziol et al., 2012, p. 515).

The motor-creativity hypothesized connection was previously supported by several reports. For example, a positive correlation was found between figural and verbal creativity and cerebral blood flow in the right cerebellum (Chávez-Eakle et al., 2007). Takeuchi et al. (2010) measured GM volume using voxel-based morphometry and found a positive correlation between divergent thinking and the right dorsolateral prefrontal cortex and the bilateral striata. In accord with our anatomical results, other reports have also emphasized the involvement of the cerebellum and the precentral gyrus in creative processes (Cotterill, 2001; Chávez-Eakle et al., 2007; Vandervert et al., 2007). The current findings also provide preliminary insight regarding the possible relationship between anatomical changes in motor-related areas and creativity.

References

Baron-Cohen, S., and Hammer, J. (1997). Is autism an extreme form of the "male brain"? *Adv. Infancy Res.* 11, 193–218.

Ben-Soussan, T. D., Avirame, K., Glicksohn, J., Goldstein, A., Harpaz, Y., and Ben-Shachar, M. (2014a). Changes in cerebellar activity and inter-hemispheric coherence accompany improved reading performance following Quadrato Motor Training. *Front. Syst. Neurosci.* 8:81. doi: 10.3389/fnsys.2014.00081

Ben-Soussan, T. D., Berkovich-Ohana, A., Glicksohn, J., and Goldstein, A. (2014b). A suspended act: increased reflectivity and gender-dependent electrophysiological change following Quadrato Motor Training. *Front. Psychol.* 5:55. doi: 10.3389/fpsyg.2014.00055

Ben-Soussan, T. D., Berkovich-Ohana, A., and Glicksohn, J., (2015). From cerebellar activation and connectivity to cognition: a review of the Quadrato Motor Training. *Biomed. Res. Int.* 1–11.

Ben-Soussan, T. D., Glicksohn, J., Goldstein, A., Berkovich-Ohana, A., and Donchin, O. (2013). Into the square and out of the box: the effects of Quadrato Motor Training on creativity and alpha coherence. *PLoS ONE* 8:e55023. doi: 10.1371/journal.pone.0055023

Ben-Soussan, T. D., Piervincenzi, C., Venditti, S., Verdone, L., Caserta, M., and Carducci, F. (2015). Increased cerebellar volume and BDNF level following Quadrato Motor Training. *Synapse* 69, 1–6. doi: 10.1002/syn.21787

Brubaker, C. J., Dietrich, K. N., Lanphear, B. P., and Cecil, K. M. (2010). The influence of age of lead exposure on adult gray matter volume. *Neurotoxicology* 31, 259–266. doi: 10.1016/j.neuro.2010.03.004

Carruthers, P. (2011). Creative action in mind. *Philos. Psychol.* 24, 437–461. doi: 10.1080/09515089.2011.556609

Chang, Y. K., Nien, Y. H., Chen, A. G., and Yan, J. (2014). Tai Ji Quan, the brain, and cognition in older adults. *J. Sport Health Sci.* 3, 36–42. doi: 10.1016/j.jshs.2013.09.003

Chang, Y. K., Nien, Y. H., Tsai, C. L., and Etnier, J. L. (2010). Physical activity and cognition in older adults: the potential of Tai Chi Chuan. *J. Aging Phys. Act.* 18, 451–472.

Chávez-Eakle, R. A., Graff-Guerrero, A., García-Reyna, J. C., Vaugier, V., and Cruz-Fuentes, C. (2007). Cerebral blood flow associated with creative performance: a comparative study. *Neuroimage* 38, 519–528. doi: 10.1016/j.neuroimage.2007.07.059

Chermahini, S. A., and Hommel, B. (2010). The (b)link between creativity and dopamine: spontaneous eye blink rates predict and dissociate divergent and convergent thinking. *Cognition* 115, 458–465. doi: 10.1016/j.cognition.2010.03.007

Colcombe, S. J., Erickson, K. I., Scalf, P. E., Kim, J. S., Prakash, R., McAuley, E., et al. (2006). Aerobic exercise training increases brain volume in aging humans. *J. Gerontol. A Biol. Sci. Med. Sci.* 61, 1166–1170. doi: 10.1093/gerona/61.11.1166

Colzato, L. S., Ozturk, A., and Hommel, B. (2012). Meditate to create: the impact of focused-attention and open-monitoring training on convergent and divergent thinking. *Front. Psychol.* 3:116. doi: 10.3389/fpsyg.2012.00116

Cotterill, R. M. (2001). Cooperation of the basal ganglia, cerebellum, sensory cerebrum and hippocampus: possible implications for cognition, consciousness, intelligence and creativity. *Prog. Neurobiol.* 64, 1–33. doi: 10.1016/S0301-0082(00)00058-7

Cowger, E. L., and Torrance, E. P. (1982). Further examination of the quality of changes in creative functioning resulting from meditation (Zazen) trraining. *Creat. Child Adult Q.* 7, 211–217.

Critchley, H. D., Wiens, S., Rotshtein, P., Öhman, A., and Dolan, R. J. (2004). Neural systems supporting interoceptive awareness. *Nat. Neurosci.* 7, 189–195. doi: 10.1038/nn1176

Dapretto, M., Davies, M. S., Pfeifer, J. H., Scott, A. A., Sigman, M., Bookheimer, S. Y., et al. (2005). Understanding emotions in others: mirror neuron dysfunction in children with autism spectrum disorders. *Nat. Neurosci.* 9, 28–30. doi: 10.1038/nn1611

Denier, N., Schmidt, A., Gerber, H., Schmid, O., Riecher-Rössler, A., Wiesbeck G. A., et al. (2013). Association of frontal gray matter volume and cerebral perfusion in heroin addiction: a multimodal neuroimaging study. *Front. Psychiatry* 4:135. doi: 10.3389/fpsyt.2013.00135

Depraz, N., Varela, F. J., and Vermersch, P. (2000). "The gesture of awareness: an account of its structural dynamics," in *Investigating Phenomenal Consciousness*, ed. M. Velmans (Philadelphia, PA: John Benjamins), 121–136. doi: 10.1075/aicr.13.10dep

Dietrich, A. (2004). The cognitive neuroscience of creativity. *Psychon. Bull. Rev.* 11, 1011–1026. doi: 10.3758/BF03196731

Ding, X., Tang, Y.-Y., Tang, R., and Posner, M. I. (2014). Improving creativity performance by short-term meditation. *Behav. Brain Funct.* 10, 1–8. doi: 10.1186/1744-9081-10-9

Erickson, K. I., Raji, C. A., Lopez, O. L., Becker, J. T., Rosano, C., Newman, A. B., et al. (2010). Physical activity predicts gray matter volume in late adulthood: the Cardiovascular Health Study. *Neurology* 75, 1415–1422. doi: 10.1212/WNL.0b013e3181f88359

Exner, C., Koschack, J., and Irle, E. (2002). The differential role of premotor frontal cortex and basal ganglia in motor sequence learning: evidence from focal basal ganglia lesions. *Learn. Mem.* 9, 376–386. doi: 10.1101/lm.48402

Frye, R. E., Liederman, J., Malmberg, B., McLean, J., Strickland, D., and Beauchamp, M. S. (2010). Surface area accounts for the relation of gray matter volume to reading-related skills and history of dyslexia. *Cereb. Cortex* 20, 2625–2635. doi: 10.1093/cercor/bhq010

Good, C. D., Johnsrude, I., Ashburner, J., Henson, R. N., Friston, K. J., and Frackowiak, R. S. (2001). Cerebral asymmetry and the effects of sex and handedness on brain structure: a voxel-based morphometric analysis of 465 normal adult human brains. *Neuroimage* 14, 685–700. doi: 10.1006/nimg.2001.0857

Greenberg, J., Reiner, K., and Meiran, N. (2012). "Mind the trap": mindfulness practice reduces cognitive rigidity. *PLoS ONE* 7:e36206. doi: 10.1371/journal.pone.0036206

Guilford, J. P. (1950). Creativity. *Am. Psychol.* 5, 444–454. doi: 10.1037/h0063487

Guilford, J. P. (1968). *Creativity, Intelligence, and their Educational Implications.* San Diego, CA: Robert R. Knapp.

Guilford, J. P., Christensen, P. R., Merrifield, P. R., and Wilson, R. C. (1978). *Alternate Uses: Manual of Instructions and Interpretation.* Orange, CA: Sheridan Psychological Services.

Hölzel, B. K., Carmody, J., Vangel, M., Congleton, C., Yerramsetti, S. M., Gard, T., et al. (2011). Mindfulness practice leads to increases in regional brain gray matter density. *Psychiatry Res.* 191, 36–43. doi: 10.1016/j.pscychresns.2010.08.006

Horan, R. (2007). The relationship between creativity and intelligence: a combined yogic-scientific approach. *Creat. Res. J.* 19, 179–202. doi: 10.1080/10400410701397230

Ivry, R. (2000). Exploring the role of the cerebellum in sensory anticipation and timing: commentary on Tesche and Karhu. *Hum. Brain Mapp.* 9, 115–118. doi: 10.1002/(SICI)1097-0193(200003)9:3<115::AID-HBM1>3.0.CO;2-5

Kabat-Zinn, J. (2009). *Full Catastrophe Living: Using the Wisdom of Your Body and Mind to Face Stress, Pain, and Illness.* New York, NY: Delacour.

Koziol, L. F., Budding, D. E., and Chidekel, D. (2012). From movement to thought: executive function, embodied cognition, and the cerebellum. *Cerebellum* 11, 505–525. doi: 10.1007/s12311-011-0321-y

Leung, H. C., Gore, J. C., and Goldman-Rakic, P. S. (2002). Sustained mnemonic response in the human middle frontal gyrus during on-line storage of spatial memoranda. *J. Cogn. Neurosci.* 14, 659–671. doi: 10.1162/08989290260045882

Levy-Drori, S., and Henik, A. (2006). Concreteness and context availability in lexical decision tasks. *Am. J. Psychol.* 119, 45–65. doi: 10.2307/20445318

Lippelt, D. P., Hommel, B., and Colzato, L. S. (2014). Focused attention, open monitoring and loving kindness meditation: effects on attention, conflict monitoring, and creativity – a review. *Front. Psychol.* 5:1083. doi: 10.3389/fpsyg.2014.01083

Luders, E., Gaser, C., Narr, K. L., and Toga, A. W. (2009). Why sex matters: brain size independent differences in gray matter distributions between men and women. *J. Neurosci.* 29, 14265–14270. doi: 10.1523/JNEUROSCI.2261-09.2009

Lutz, A., Slagter, H. A., Dunne, J. D., and Davidson, R. J. (2008). Attention regulation and monitoring in meditation. *Trends Cogn. Sci.* 12, 163–169. doi: 10.1016/j.tics.2008.01.005

Mercader, J. M., Fernandez-Aranda, F., Gratacòs, M., Aguera, Z., Forcano, L., Ribasés, M., et al. (2010). Correlation of BDNF blood levels with interoceptive awareness and maturity fears in anorexia and bulimia nervosa patients. *J. Neural Transm.* 117, 505–512. doi: 10.1007/s00702-010-0377-8

Moore, A., and Malinowski, P. (2009). Meditation, mindfulness and cognitive flexibility. *Conscious. Cogn.* 18, 176–186. doi: 10.1016/j.concog.2008.12.008

Morin, A., and Michaud, J. (2007). Self-awareness and the left inferior frontal gyrus: inner speech use during self-related processing. *Brain Res. Bull.* 74, 387–396. doi: 10.1016/j.brainresbull.2007.06.013

Nambu, A., Kaneda, K., Tokuno, H., and Takada, M. (2002). Organization of corticostriatal motor inputs in monkey putamen. *J. Neurophysiol.* 88, 1830–1842.

Netz, Y., Tomer, R., Axelrad, S., Argov, E., and Inbar, O. (2007). The effect of a single aerobic training session on cognitive flexibility in late middle-aged adults. *Int. J. Sports Med.* 28, 82–87. doi: 10.1055/s-2006-924027

Ocklenburg, S., Hugdahl, K., and Westerhausen, R. (2013). Structural white matter asymmetries in relation to functional asymmetries during speech perception and production. *Neuroimage* 83, 1088–1097. doi: 10.1016/j.neuroimage.2013.07.076

Pagnoni, G., and Cekic, M. (2007). Age effects on gray matter volume and attentional performance in Zen meditation. *Neurobiol. Aging* 28, 1623–1627. doi: 10.1016/j.neurobiolaging.2007.06.008

Paulesu, E., Frith, C. D., and Frackowiak, R. S. (1993). The neural correlates of the verbal component of working memory. *Nature* 362, 342–345. doi: 10.1038/362342a0

Runco, M. A. (1986). Flexibility and originality in children's divergent thinking. *J. Psychol.* 120, 345–352. doi: 10.1080/00223980.1986.9712632

Russ, S. W. (1998). Play, creativity, and adaptive functioning: implications for play interventions. *J. Clin. Child Psychol.* 27, 469–480. doi: 10.1207/s15374424jccp2704_11

Schmahmann, J. D. (2004). Disorders of the cerebellum: ataxia, dysmetria of thought and the cerebellar cognitive affective syndrome. *J. Neuropsychiatry Clin. Neurosci.* 16, 367–378. doi: 10.1176/appi.neuropsych.16.3.36

Schmid, T. (ed.). (2005). "Promoting health through creativity: an introduction," in *Promoting Health through Creativity: for Professionals in Health, Arts and Education*, ed. T. Schmid (London: Whurr Publishers).

Scholz, J., Klein, M. C., Behrens, T. E., and Johansen-Berg, H. (2009). Training induces changes in white-matter architecture. *Nat. Neurosci.* 12, 1370–1371. doi: 10.1038/nn.2412

Smith, S. M. (2002). Fast robust automated brain extraction. *Hum. Brain Mapp.* 17, 143–155. doi: 10.1002/hbm.10062

Smith, S. M., Jenkinson, M., Johansen-Berg, H., Rueckert, D., Nichols, T. E., Mackay, C. E., et al. (2006). Tract-based spatial statistics: voxelwise analysis of multi-subject diffusion data. *Neuroimage* 31, 1487–1505. doi: 10.1016/j.neuroimage.2006.02.024

Swick, D., Ashley, V., and Turken, U. (2008). Left inferior frontal gyrus is critical for response inhibition. *BMC Neurosci.* 9:102. doi: 10.1186/1471-2202-9-102

Takeuchi, H., Taki, Y., Sassa, Y., Hashizume, H., Sekiguchi, A., Fukushima, A., et al. (2010). Regional gray matter volume of dopaminergic system associate with creativity: evidence from voxel-based morphometry. *Neuroimage* 51, 578–585. doi: 10.1016/j.neuroimage.2010.02.078

Tesche, C. D., and Karhu, J. J. (2000). Anticipatory cerebellar responses during somatosensory omission in man. *Hum. Brain Mapp.* 9, 119–142. doi: 10.1002/(SICI)1097-0193(200003)9:3<119::AID-HBM2>3.0.CO;2-R

Tomer, R., Fisher, T., Giladi, N., and Aharon-Perez, J. (2002). Dissociation between spontaneous and reactive flexibility in early Parkinson's disease. *Neuropsychiatry Neuropsychol. Behav. Neurol.* 15, 106–112.

Vandervert, L. R., Schimpf, P. H., and Liu, H. (2007). How working memory and the cerebellum collaborate to produce creativity and innovation. *Creat. Res. J.* 19, 1–18. doi: 10.1080/10400410709336873

Venditti, S., Verdone, L., Pesce, C., Tocci, N., Caserta, M., and Ben-Soussan, T. D. (2014). Creating well-being: increased creativity and proNGF decrease following Quadrato Motor Training. *Biomed. Res. Int.* 2015:275062.

Vestergaard-Poulsen, P., van Beek, M., Skewes, J., Bjarkam, C. R., Stubberup, M., Bertelsen, J., et al. (2009). Long-term meditation is associated with increased gray matter density in the brain stem. *Neuroreport* 20, 170–174. doi: 10.1097/WNR.0b013e328320012a

Wayne, P. M., Gow, B. J., Costa, M. D., Peng, C. K., Lipsitz, L. A., Hausdorff, J. M., et al. (2014). Complexity-based measures inform effects of Tai Chi training on standing postural control: cross-sectional and randomized trial studies. *PLoS ONE* 9:e114731. doi: 10.1371/journal.pone.0114731

Wei, G. X., Xu, T., Fan, F. M., Dong, H. M., Jiang, L. L., Li, H. J., et al. (2013). Can tai chi reshape the brain? A brain morphometry study. *PLoS ONE* 8:e61038. doi: 10.1371/journal.pone.0061038

Zabelina, D. L., Robinson, M. D., Ostafin, B. D., and Council, J. R. (2011). Manipulating mindfulness benefits creative elaboration at high levels of neuroticism. *Empir. Stud. Arts* 29, 243–255. doi: 10.2190/EM.29.2.g

18

Attenuated Sensitivity to the Emotions of others by Insular Lesion

Yuri Terasawa[1], Yoshiko Kurosaki[2], Yukio Ibata[3], Yoshiya Moriguchi[4] and Satoshi Umeda[1]*

[1] Department of Psychology, Keio University, Tokyo, Japan, [2] Department of Communication Disorders, Health Sciences University of Hokkaido, Hokkaido, Japan, [3] Department of Neurosurgery, Nasu Red Cross Hospital, Tochigi, Japan, [4] Department of Psychophysiology, National Institute of Mental Health, National Center of Neurology and Psychiatry, Tokyo, Japan

***Correspondence:**
Yuri Terasawa,
Department of Psychology,
Keio University, 4-1-1 Hiyoshi,
Kohoku-ku, Yokohama,
Kanagawa 223-8521, Tokyo, Japan
terasawa@flet.keio.ac.jp

The insular cortex has been considered to be the neural base of visceral sensation for many years. Previous studies in psychology and cognitive neuroscience have accumulated evidence indicating that interoception is an essential factor in the subjective feeling of emotion. Recent neuroimaging studies have demonstrated that anterior insular cortex activation is associated with accessing interoceptive information and underpinning the subjective experience of emotional state. Only a small number of studies have focused on the influence of insular damage on emotion processing and interoceptive awareness. Moreover, disparate hypotheses have been proposed for the alteration of emotion processing by insular lesions. Some studies show that insular lesions yield an inability for understanding and representing disgust exclusively, but other studies suggest that such lesions modulate arousal and valence judgments for both positive and negative emotions. In this study, we examined the alteration in emotion recognition in three right insular and adjacent area damaged cases with well-preserved higher cognitive function. Participants performed an experimental task using morphed photos that ranged between neutral and emotional facial expressions (i.e., anger, sadness, disgust, and happiness). Recognition rates of particular emotions were calculated to measure emotional sensitivity. In addition, they performed heartbeat perception task for measuring interoceptive accuracy. The cases identified emotions that have high arousal level (e.g., anger) as less aroused emotions (e.g., sadness) and a case showed remarkably low interoceptive accuracy. The current results show that insular lesions lead to attenuated emotional sensitivity across emotions, rather than category-specific impairments such as to disgust. Despite the small number of cases, our findings suggest that the insular cortex modulates recognition of emotional saliency and mediates interoceptive and emotional awareness.

Keywords: insula, interoceptive accuracy, facial expression, arousal level, heartbeat perception task

Introduction

Our emotional experience is well associated with internal bodily change. We have a racing pulse or sweaty palms when we are excited. The peripheral theory of emotion (James, 1884; Lange, 1885/1992) proposed that "our feeling of the same changes as they occur is the emotion." Although the theory triggered long discussion, the findings of recent psychological and brain imaging studies have indicated that we refer to our internal bodily state when we are aware of our emotional state. Additionally, our internal state modulates our emotional experience (Bechara et al., 1996; Pollatos et al., 2005; Lane, 2008; Dunn et al., 2010; Terasawa et al., 2013a,b).

The perception of afferent information arising from anywhere and everywhere within the body has been termed "interoception" (Sherrington, 1906; Cameron, 2001). Some methods have been developed as measurements of an individual's sensitivity to perceive interoception, such as the heartbeat perception task (Schandry, 1981) and water load test (Herbert et al., 2012). Some questionnaires such as the Autonomic Perception Questionnaire (Mandler et al., 1958), the Body Perception Questionnaire (Porges, 1993), and The Multidimensional Assessment of Interoceptive Awareness (Mehling et al., 2012) are known as indices of interoceptive sensibility (Garfinkel et al., 2015). Some studies have suggested that sensitivity to interoception represents the disposition of a person's emotional experience, such as its intensity (Critchley et al., 2004; Pollatos et al., 2005; Werner et al., 2009) or the tendency to focus on arousal (Barrett et al., 2004). Furthermore, a previous study of ours showed that interoceptive sensitivity predicts sensitivity to the emotions of others (Terasawa et al., 2014). These findings indicate that the way in which interoceptive information is processed can affect emotional experience in daily life.

Neuroimaging studies also support the notion that interoceptive sensitivity is closely connected with emotional experience. Activation of anterior insular cortex is commonly observed in studies investigating emotional experience or empathy (Lamm and Singer, 2010; Lee and Siegle, 2012; Gu et al., 2013). Anterior insular activation is also found when participants are aware of their internal bodily state (Evans et al., 2002; Critchley et al., 2004; Hamaguchi et al., 2004). Our previous study confirmed that both interoceptive and emotional awareness recruited right anterior insular activation, and the activation mediated interoception and disposition of social anxiety (Terasawa et al., 2013b). These findings support the hypothesis that the integration of interoception and the interpretation of environmental information yields the subjective feeling of emotion and that the insular cortex is one of the critical regions housing this mechanism.

If interoception plays a role in the foundation of subjective emotion, what changes can be observed when the insular cortex is damaged? Case NK had a selective left middle insula lesion by cerebral infarction and showed difficulties in recognizing and experiencing disgust, even though memory and intelligence were well preserved (Calder et al., 2000). Several studies support this finding (Adolphs et al., 2003; Borg et al., 2013). The origin of disgust is considered to be a biological signal for preventing the intake of rotten foods (Susskind et al., 2008). Furthermore, Penfield found that people felt sensation in their gastrointestinal organs when electrical stimulation was applied to the insular cortex during a neurosurgical operation (Penfield and Faulk, 1955). Reported abdominal sensation, pain, and nausea by stimulation of the insula suggest that the insula plays a key role in representing interoception. These facts indicate that the insular cortex is important not only in interoceptive processing, but also in generating emotional experience that relates to bodily sensation.

However, the accumulation of neuropsychological studies on emotional processing revealed that not all areas of the insular cortex simply subserve the feeling of disgust. A recent neuropsychological study of 15 insular cortex lesion cases showed that the ability to recognize fear, happiness, and surprise was affected, although there was no change in disgust (Boucher et al., 2015). Adolphs et al. (2000) and Dal Monte et al. (2013) examined the recognition of six basic emotions in over 100 patients with lesions and identified several regions that when damaged led to the decline of emotion recognition performance; these included the medial prefrontal cortex, cingulate cortex, supramarginal gyrus, and insular cortex. Patients with insular lesions showed lower performance on discrimination between negative emotions, rather than a specific deficit for feeling disgust.

Recent studies presented findings those were contrary to expectations from previous studies. For example, Damasio et al. (2012) reported the bilateral insular lesion cases showed normal ability to feel emotion and self-awareness. Furthermore, Couto et al. (2013) found a minimal impact of insular cortical lesions when compared subcortical lesion cases. These studies suggest that the insular cortex is not solely involved in the neural substrates of emotional awareness, but rather that subcortical neural networks between the insula and other regions, such as the thalamus, should be considered. Furthermore, these studies highlight the importance of accurate documentation of the effects of insular lesions on emotional recognition ability beyond the performance of basic emotion category recognition.

A previous study asked patients with insular lesions to evaluate the arousal level of emotion-inducing pictures and observed that the patients reported a lower arousal level for the pictures regardless of valence (Berntson et al., 2011). Because arousal level and the activity of the sympathetic nervous system are closely connected, deviation from the homeostatic state of sympathetic activity and perceiving the deviation can lead to recognition of the arousal level. The circumplex model of emotion posits two dimensions for defining emotions, arousal, and valence (Russell, 1980; Kuppens et al., 2013). Emotions are plotted on the plane defined by arousal and valence. Sadness, anger, and disgust are plotted closely to each other on the valence dimension; however, arousal level is the difference between these emotions. The arousal level for anger is the highest, but for sadness it is the lowest and disgust is in the middle. The insular cortex is considered to be one neural substrate for representing arousal level (Lewis et al., 2007), and decreased blood flow in this region yields blunting of interoception (Khalsa et al., 2009). These findings may support the hypotheses that insular cortical and/or subcortical lesions impair interoception, which underlies the recognition of arousal level, and leads to lower performance for discrimination between negative emotions.

On the basis of this hypothesis, we examined the effects on emotional experience following brain lesions including those in the right anterior insula cortex. We measured sensitivity to other's emotion using morphed continua photos that ranged between a neutral and an emotional facial expression, and revealed that the sensitivity was predicted by the interoceptive accuracy of individuals (Terasawa et al., 2014). These results may suggest that emotional experience is affected by the perception of internal bodily changes triggered by emotional expression. In the present study, we examined changes in emotional experience in patients with right anterior insular lesions by the experimental task, and discuss its influence on their daily life.

Materials and Methods

The study was performed with the approval of the Keio University Research Ethics Committee (No. 09006). Before participating in the study, all cases were informed (1) of the purpose and procedure of the study and (2) that they were able to cease their participation in the study at any time. All cases signed a written informed consent form.

Cases

Three patients with lesions participated the study. They were outpatients of the neurosurgery department of the Nasu Red Cross Hospital.

Case A

Case A, a right-handed 61-year-old man, experienced convulsive seizure by cerebral infarction when he was 59 years old. Right dorsolateral prefrontal cortex (DLPFC) and anterior insula cortex were resected by a hematoma evacuation operation (**Figure 1A**). The lesion area included the cortical and subcortical anterior to middle insula and ranged from ventral to dorsal parts. Three-dimensional magnetic resonance angiography showed no occlusions of the major cerebral arteries. Case A has a past history of myocardial infarct, diabetes mellitus, hypertension, paroxysmal atrial fibrillation, and symptomatic epilepsy. Motor function, perception, and language abilities are well preserved. Orientation, memory, working memory, and intelligence are also preserved, but Perseveration error on the Wisconsin card-sorting task (WCST) was observed [Category achieved (CA) : 0, Perseverative errors of the Nelson type (PEN) : 25, Difficulties of maintaining set (DMS) : 0].

Case A reported no change in memory and motor function, but he described himself as having more impatience and being duller compared with his premorbid level. He is still working at a supermarket 5 days a week. According to his family, Case A became easy to anger, disinhibited, and apathetic. Case A showed a poor understanding of others' intentions or facial expressions. His thermal sensations became dull and he never felt hot. These reports indicated that even though his memory ability and intelligence were preserved, Case A had difficulty in controlling his own emotions.

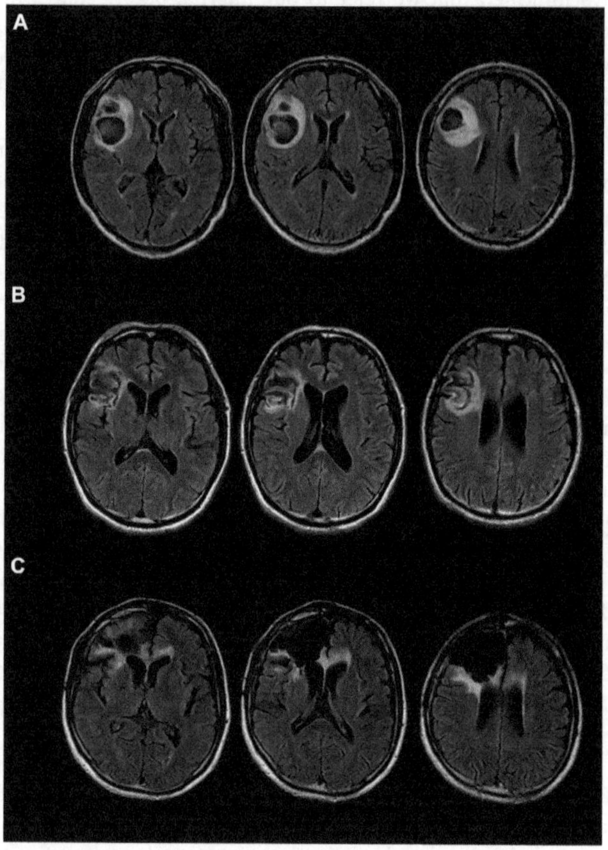

FIGURE 1 | Transverse magnetic resonance imaging with FLAIR acquisition of the brains of the cases. Left sides of images correspond to the right side of the brain. **(A)** Case A, **(B)** Case B, and **(C)** Case C.

Case B

Case B, a right-handed 45-year-old man, experienced aneurysmal subarachnoid hemorrhage of the middle cerebral artery when he was 44 years old. He underwent surgical clipping. Magnetic resonance imaging showed hemorrhagic contusion of the temporal pole and decreased cerebral perfusion was observed at the right dorsolateral frontal cortex and the anterior insular cortex (**Figure 1B**). The lesion area included the cortical and subcortical anterior to middle insula and ranged from the ventral to dorsal parts. Motor function, perception, and language abilities are well preserved. Orientation, memory, working memory, and intelligence are also preserved, but Perseveration error on the WCST was observed (CA: 2, PEN: 25, DMS: 1). Lower performances of verbal fluency test and delayed memory recall were observed.

Case B reported that he has mellowed and is easily tired. He never feels anxiety about anything, and cannot understand people who feel anxiety at all. According to his family and doctor, Case B shows a poor understanding of others' facial expressions. Additionally, he has an attitude of indifference toward his own clinical condition. The indifference may coincide with the lower anxiety trait.

Case C

Case C, a right-handed 66-year-old man, experienced convulsive seizure and left hemiparesis when he was 60 years old. Brain edema of the right frontal white matter was observed by computed tomography scan. Pathological changes were observed on the surface of the right frontal and premotor areas, and the right frontal lobe including the right anterior insula was resected (**Figure 1C**). The lesion area included the cortical and subcortical anterior insula and ranged from the ventral to dorsal parts. Then, Case C was diagnosed with oligodendroglioma. Orientation, memory, working memory and response inhibition, and sustained attention are preserved, but lower performance on the WCST (CA: 0, PEN: 21, DMS: 0) and verbal fluency task were identified. Motor function, perception, language abilities, and communication abilities are well preserved. He went back to his work.

Case C reported that he frequently experienced action slips and sometimes lost his intention for action. For example, he required much more effort for concentrating on reading books compared with symptom onset. However, he and his family thought this impairment did not significantly impact on his daily life. He had myocardial infarction four times for 4 years since the neurosurgical procedure. However, he traveled through Japan by bicycle on his own and also trained for a marathon, regardless of the heart attack risk. Surprisingly, he reported that he hardly felt anxious or at risk when performing such an exercise.

Age Matched Healthy Control

Fourteen males (mean age 56.7 ± 3.63 years) participated and completed the heartbeat perception task and the emotional sensitivity task. No participants had any psychiatric disorders. Mini Mental State Examination (MMSE) was conducted and confirmed that their score exceeded 27 points (mean 28.9 ± 0.95), indicating no subjects with cognitive impairment.

Materials and Procedures

Heartbeat Perception Task

Only Cases A and C participated the Heartbeat Perception Task. The task was based on a task developed by Schandry (1981) and Ehlers and Breuer (1992) for examining interoceptive accuracy. The task has been used in many previous studies. A detailed description of the task can be found in our previous study (Terasawa et al., 2014).

Heartbeats were measured using a pulse oximeter (Polymate AP1542, TEAC, Tokyo, Japan) on the fingertips during specific periods of time. Participants were asked to count the number of times that they felt their own heartbeat during the measurement period. Discrepancies between the number of reported and actual heartbeats during the measurement period were calculated based on the formula used to define heartbeat perception error rates by Ehlers and Breuer (1992): (actual heartbeats – reported heartbeats/actual heartbeats) × 100. Each case completed six trials at 35 s × 2, 25 s × 2, and 45 s × 2, with the order randomized between subjects.

The Time Estimation Task

The time estimation task was developed by Dunn et al. (2010) to overcome the influence of time estimation on interoceptive accuracy. In the task, participants are asked to count the number of seconds during a given period, and then the reported length was compared with the actual duration. Time estimation error rates were calculated in a manner similar to that of the heartbeat perception error rate and checked whether the error rates for both task were dissociated. Each case completed six trials at 23 s × 2, 40 s × 2, and 56 s × 2, with the order randomized between subjects.

Emotional Sensitivity Task

To estimate the emotional sensitivity of the cases, we prepared an experimental task that was identical to the task used in Terasawa et al. (2014). A detailed description of the task can be found in Terasawa et al. (2014). In this task, the cases were presented with a facial expression photo and were asked whether or not they felt emotion from the photos, and were asked to report the name of the emotion if they did feel emotion.

We selected five photos each of a male individual and a female individual, with the following facial expressions: angry, sad, disgust, happiness, and neutral, from the Advanced Telecommunications Research Institute International Facial Expression Database (DB99). In addition to the original photos, morphed photos were prepared for the task. These photos were made in nine variations, with each variation having different percentages of the neutral and emotion expressions (angry, sad, disgust, and happiness), ranging from 10% neutral to 100% of each emotion. For example, Neutral (N) 90% – Happiness (H):10%, N:80% – H:20%, N:70% – H:30%, N:60% – H:40%, N:50% – H:50%, N:40% – H:60%, N:30% – H:70%, N:20% – H:80%, N:10% – H:90%, and N:0% – H:100% (**Figure 2A**). We did not use the original neutral picture (N:100% – H:0%) in this task, as every subject answered that they did not feel emotion from the picture.

In each trial, (1) a stimulus was presented for 2 s. (2) Participants were asked whether the stimulus made them feel an emotion or not. (3) If participants responded "Yes," then four options were presented on the screen: anger, sadness, disgust, and happiness, and they were asked to choose the appropriate one for the emotion expressed by the stimulus. Because it was important to determine if the participants felt emotion, we asked them to respond "Yes" if they actually felt any emotion from the stimulus. Participants were clearly instructed to categorize emotion that the model expressed, not the one that they were experiencing. If they responded "No," the options did not appear and the task moved on to the next trial after the presentation of a fixation point for 5 s (**Figure 2B**).

Each stimulus was presented five times in random order, thus there were 200 trials in total (4 emotion × 10 steps × 5 times). Original stimuli that fully expressed a certain emotion (e.g., 100% anger) were labeled as having an "emotion value of 10," and neutral stimuli (e.g., 100% neutral) were labeled as having an "emotion value of 1." Each step between emotion values 10 and 1, e.g., from 9 to 2, was labeled depending on the percentage of the emotional value present in the photo (**Figure 2A**).

The number of times that participants reported feeling emotions as a result of viewing each stimulus were calculated. We classified those stimuli that made participants feel an appropriate emotion at least three times out of five (i.e., at least 60% of the trials) as having sufficient emotional value to produce the emotional response.

The threshold of emotional value was posited to be located near the midpoint between stimuli that produced an emotional

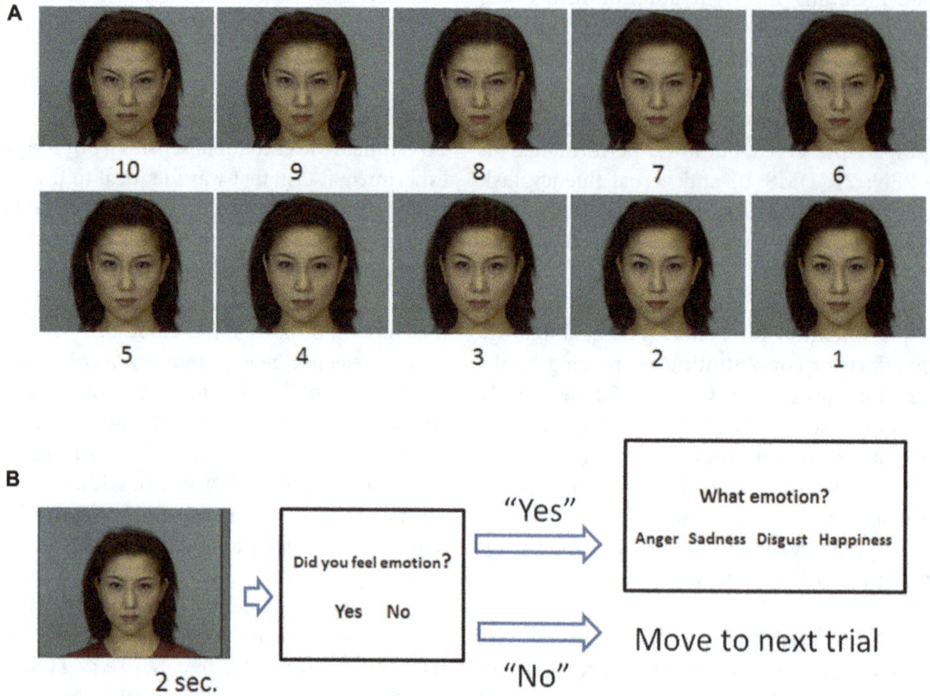

FIGURE 2 | The stimuli used in the task (A) and outline of a trial (B). (A) Morphed continua photos ranging between neutral (stimulus 1) and 100% of each facial expression (stimulus 10), i.e., angry. **(B)** A stimulus was followed by a judgment about whether the stimulus elicited an emotion. If participants responded "yes," options were presented and they were asked to choose the most appropriate emotion category.

response less than three times and stimuli that produced an emotional response three times and more. When participants reported that they felt a certain emotion three times when viewing a stimulus with an emotion value of six and two times when viewing the stimulus with an emotion value of 5, we considered their threshold for emotional response to be 5.5.

Results

Emotional Sensitivity Task

Case A's responses toward each stimuli and thresholds of emotional value are shown in **Figure 3A**. The red arrows show the place where the thresholds appear. Yellow arrows show averaged thresholds of the age matched control group. The thresholds are also shown in **Table 1**. The responses of Cases B and C are shown in **Figures 3B,C**, respectively. As the three figures and the table show, all cases needed a higher emotional value to recognize anger than healthy participants. The same was true for sadness and disgust. Case C made too many error responses, such as false identification of disgust as sadness or sadness as neutral, which prevented the definition of thresholds for disgust and sadness.

Although the numbers of case and control participants was small, we compared the thresholds of cases and the control group statistically. Crawford and Garthwaite (2002) and Crawford et al. (2010) reported statistical methods to compare scores of a single case and control group, and provided a program called Singlims_ES, which can be downloaded from http://homepages.abdn.ac.uk/j.crawford/pages/dept/Single_Case_Effect_Sizes.htm. This allowed us to test whether an individuals' score was significantly different from a control sample. As a result, the threshold for anger in Cases B and C showed a trend for being higher than in the control group ($t = 1.46$, $p < 0.10$; **Table 1**). Case A showed a significantly higher threshold for sadness ($t = 1.99$, $p < 0.05$), while Case B showed a higher threshold for disgust ($t = 2.28$, $p < 0.05$). Not all thresholds of the cases were statistically higher than those of the control group, although the overall trends were the same, except for happiness.

The results support the idea that that insula lesions do not lead to impaired recognition of a specific category of emotion such as disgust, but rather that sensitivity to emotional value is blunted. However, the recognition of happy emotions was not affected by insula lesions because the cases showed similar sensitivity to happy faces as healthy participants.

The Heartbeat Perception Task

Cases A and C completed the heartbeat perception task and the time estimation task, but B did not participate in the tasks. The error rates on the heartbeat perception task were 0.67 (Case A) and 0.99 (Case C), thus both cases showed very high error rates on this task. However, remarkably low error rates were observed on the time estimation task: 0.01 for Case A and 0.06 for Case C. High error rates on the heartbeat perception task indicate that lower interoceptive accuracy underlies the blunted sensitivity to others' emotion measured by the emotional sensitivity task.

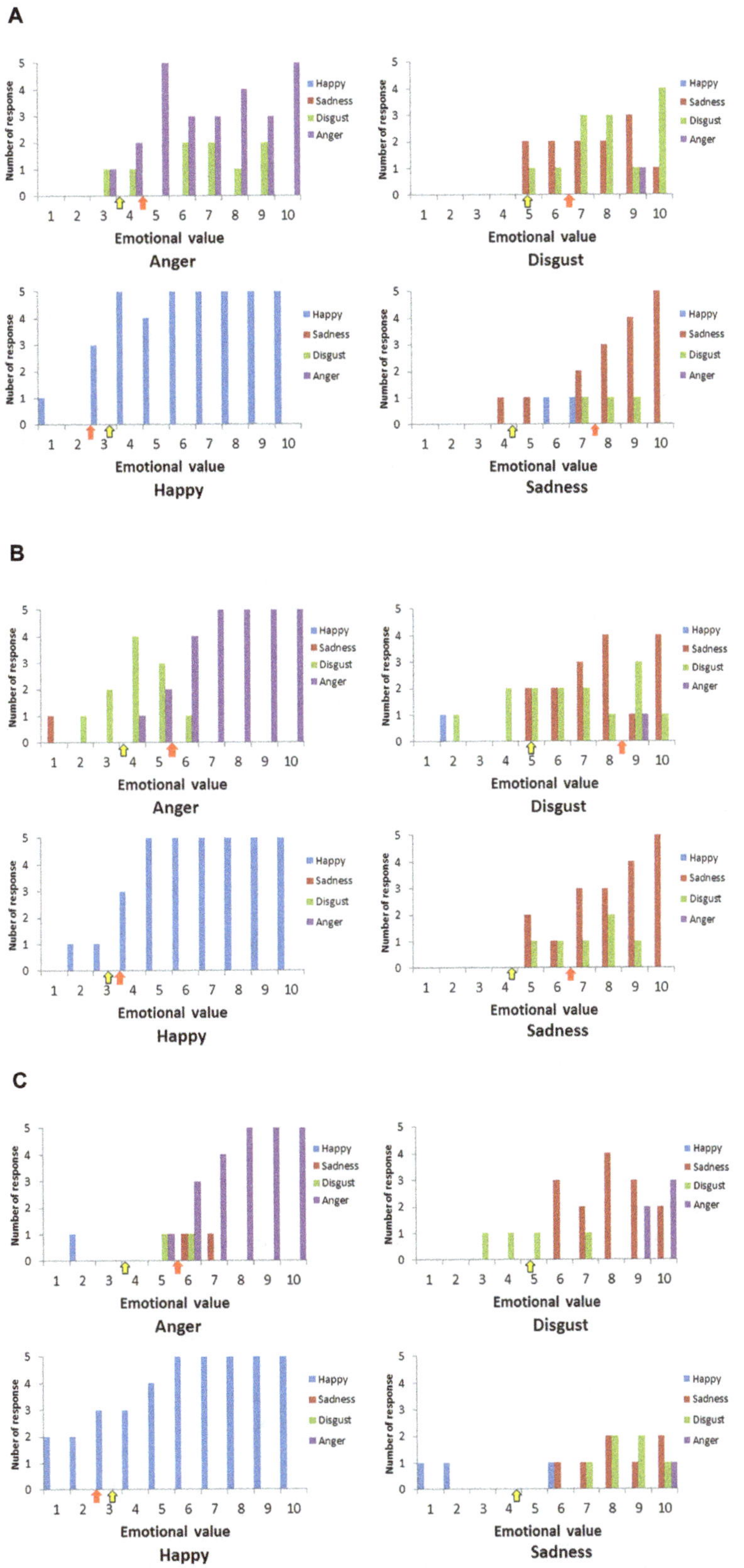

FIGURE 3 | Numbers of responses for happiness (blue), sadness (red), disgust (green), and anger (purple) for each stimulus. (A) Case A, **(B)** Case B, and **(C)** Case C. Red arrows show the place where the thresholds of cases appeared, and yellow arrows show the averaged thresholds of healthy participants.

TABLE 1 | Thresholds of emotional value of the cases and healthy participants.

		Threshold	*t*-value	*p*	Zcc	Control
Case A	Anger	4.5	0.68	0.26	0.70	3.64 (1.23)
	Disgust	6.5	1.00	0.17	1.03	4.94 (1.51)
	Happiness	2.5	−0.55	0.30	−0.57	3.18 (1.20)
	Sadness	7.5*	1.99	0.03	2.06	4.43 (1.49)
Case B	Anger	5.5+	1.46	0.08	1.51	
	Disgust	8.5*	2.28	0.02	2.36	
	Happiness	3.5	0.26	0.40	0.27	
	Sadness	6.5	1.34	0.10	1.39	
Case C	Anger	5.5+	1.46	0.08	1.51	
	Disgust	–	–	–	–	
	Happiness	2.5	−0.55	0.30	−0.57	
	Sadness	–	–	–	–	

SD in healthy participants are shown in parentheses. Zcc, effect size for simple t-test. $^{*}p < 0.05$, $^{+}p < 0.10$.

The control group also completed the heartbeat perception task, with a mean error rate of 0.70 ± 0.23. There were no differences between Case A compared with control group, while Case C showed significantly lower performance.

Discussion

In this study, we examined the effects on emotional experience following brain lesions including the right anterior/middle insula cortex. Findings from the emotional sensitivity task and the heartbeat perception task indicated that the cases had blunted sensitivity to emotional expression and lower interoceptive accuracy. Some previous studies have suggested that insular lesions lead to selective impairment of disgust recognition (Calder et al., 2000; Adolphs et al., 2003; Borg et al., 2013); however, other studies have suggested that such lesions yield lower performance for discrimination between negative emotions (Adolphs et al., 2000; Dal Monte et al., 2013). Furthermore, other studies have proposed reduced arousal to emotional stimuli regardless of emotional valence (Berntson et al., 2011).

At the very least, our findings do not support the hypothesis positing that insula lesions cause the selective impairment of disgust recognition. Though Cases A and B needed higher emotional value to recognize disgust compared with healthy participants, they could properly recognize disgust when the stimuli had higher values. Case C could not recognize disgust, but also sadness, thus the impairment cannot be considered selective. However, all cases showed some degree of difficulty in recognizing emotions from stimuli.

Unique patterns of error response can be seen when focused on **Figures 3A–C**; these are (i) misidentification of anger as disgust or sadness and (ii) misidentification of disgust as sadness. Based on a circumplex theory of emotion and insular function for recognizing arousal level, these patterns are very compelling. According to the circumplex theory of emotion (Russell, 1980), emotions can be plotted on a plane defined by two dimensions: arousal and valence. Anger, disgust, and sadness have similar emotional valence, but differences in arousal level discriminate these three emotions. Attenuation of recognized arousal level may lead to the error patterns that can be seen in the present study, because the arousal level of anger is the highest, disgust is middle, and sadness is the lowest. As the insular cortex is related to the recognition of arousal level, the error patterns may suggest that the cases could understand the emotional valence of stimuli, but that they had difficulty in sensing arousal level from the stimuli. Thus, they identified emotions (e.g., anger) that have high arousal level as less aroused emotions (e.g., sadness). Arousal level is closely connected to bodily responses such as cardiovascular activity, respiration, and body temperature. If interoceptive awareness for these responses underlies recognition of arousal level, it is reasonable to assume that insular lesions lead to reduced arousal level, and then to the altered emotional recognition performance observed in the present study.

Few previous studies have examined whether insular lesions lead to a decline in interoceptive accuracy. Khalsa et al. (2009) revealed that heartbeat sensation was not significantly affected by bilateral insular lesion, until an anesthetic was otherwise applied to the skin. However, Ronchi et al. (2015) reported that resection of selective right neoplastic insular lesions led to an increment of error rates on the heartbeat perception task, a result identical to that of our study. In particular, the resection boosted the error rates from 0.18 to 0.39. The patient of Ronchi et al. (2015) had higher interoceptive accuracy before resection, but that was not the case after resection. We did not measure interoceptive accuracy of cases prior to the lesion in the present study, and the heartbeat perception task showed rather large individual differences. In fact, although Case C showed remarkably low performance in the heartbeat perception task, the performance of Case A was identical to the control group. Thus, the high error rates of Case C could not be easily attributed to the right anterior insular lesion. As suggested by Damasio et al. (2012) and Couto et al. (2013), we considered the function of the insular subcortical network for the performance. The lesion area of Case A was rather selective, suggesting that the subcortical network should be largely preserved. The effect of coexisting somatosensory routes of interoception should also be considered (Khalsa et al., 2009; Couto et al., 2014). Findings from a series of previous studies and our results remain insufficient to fully elucidate the role of the insula for interoception and emotional awareness. Further testing of insular lesion cases is required to understand whether insular lesions lead to a decline in interoceptive accuracy and emotional awareness.

The reported episodes of the cases' daily life are also interesting. None of the three cases had a major problem with their memory and intelligence, but all reported episodes that indicate altered emotional experience. An absence of feelings of anxiety, observed in Cases B and C, is remarkable, because anxiety is considered to be closely related to interoception and the insular cortex is one of the neural correlates for this relationship (Paulus and Stein, 2006; Simmons et al., 2011; Terasawa et al., 2013a). These studies propose that amplified sensitivity to bodily information enables us to detect subtle internal bodily changes and evokes anxiety, because anxiety states are closely related to sympathetic activities. If the exaggerated attention to interoceptive information is assumed to yield anxious feelings, then conversely the unavailability of interoceptive information may reduce anxiety.

However, there were no significant differences between the cases and healthy participants for the thresholds of emotional value for happiness. If attenuated interoceptive accuracy lowers arousal level, this trend should also be observed for happiness recognition. We posit three possible reasons for this finding. The first hypothesis is that the right insula specifically underlies negative emotion processing, not that for positive emotion. Craig (2005) suggested this asymmetrical function of the insula, and proposed associations of the right anterior insula with energy expenditure (arousal) feelings, such as risk and depression, and the left insula with energy enrichment (affiliation) feelings, such as social bonding. His suggestion may support the notion that positive emotion is related to left insular activity rather than the right insula. However, some studies have reported that highly aroused emotions recruit anterior insula activity regardless of valence (Iaria et al., 2008; Moriguchi et al., 2014). These studies imply that the laterality of insular function is still controversial, and this idea seems insufficient for explaining our findings about happiness. The second hypothesis is that people have a tendency to process happy faces preferentially more than other emotions. Calvo and Nummenmaa (2008) observed that happy faces need less time to be detected precisely compared with fearful, angry, sad, surprised, and disgusted faces. The findings suggest that sensitivity to happy faces predominates over the blunting of emotion recognition based on the insula lesion, thus recognition of happy emotions was maintained. The third hypothesis concerns a technical issue, that there is only one option for positive valence emotion. If cases could understand valence properly, there is no possibility to misidentify happy as other emotions, but the possibilities were high for negative emotions, because there are three options for negative emotions. Our results are not detailed enough to decide which is the most plausible hypothesis, but the findings suggest that the cases could understand emotional valence properly and that the effect of interoception would be rather small on recognition of the valence.

All three cases showed lower performance in the WCST, and the number of perseveration error responses was quite high in particular. Neuroimaging studies designate the involvement of the insula cortex in various cognitive functions such as attention or memory, thus it is possible to posit that insula lesions affected performance through basic cognitive functions (Kurth et al., 2010). However, we need to note that the cases' lesions extended to the right DLPFC, which is well known for neural correlates for executing the WCST (Buchsbaum et al., 2005; Lie et al., 2006). The effects of the cases' DLPFC lesions would be to induce lower performance on the WCST. Since DLPFC is known as an important region for appraisal or regulation processes of emotion, we should consider possibility that the DLPFC lesion may affect the performance of the emotional sensitivity task. Recent neuropsychological study (Jenkins et al., 2014) which used morphed expression photos showed the effect of DLPFC lesion on recognition of emotional expression was restricted. Though the effect of DLPFC lesion on the emotional sensitivity task would be limited, further careful investigations should be conducted to clarify the possible sole effect of DLPFC lesion.

Thus, extended lesion areas in the cases prevent us from concluding that the findings originated from the insular lesion selectively, which is a limitation of the present study. Future studies with more selective insula lesion cases are required. Recent reports suggest that the insular cortex can be divided into anterior/middle/posterior and ventral/dorsal regions, with differing cognitive functions in these regions. For example, Kurth et al. (2010) suggested that the ventral region was important for social-emotional function, the anterior dorsal region for cognitive function, and the medial to posterior regions for sensorimotor function. However, it was also reported that interoceptive processing is reliant on the posterior region, while exteroceptive processing relies on the anterior region (Farb et al., 2013). These inconsistent findings suggest that further neuropsychological studies focusing on the lesion area in the insula and its impact on cognitive function are required. Unfortunately, the lesion areas of the three cases in the present study were not restricted to a particular region.

Despite the small number of cases, our findings suggest that perceiving interoceptive information may modulate the contents of emotional experience, and the notion is consistent with some previous findings. In the future, studies with more lesion cases will be beneficial for understanding the causality between interoception and subjective emotion through neural substrates.

Acknowledgments

This work was supported by the Japan Society for the Promotion of Science and the KAKENHI Grant No. 26350990 by the Ministry of Education, Culture, Sports, Science and Technology of Japan. We are grateful to the three patients who participated in this study.

References

Adolphs, R., Damasio, H., Tranel, D., Cooper, G., and Damasio, A. R. (2000). A role for somatosensory cortices in the visual recognition of emotion as revealed by three-dimensional lesion mapping. *J. Neurosci.* 20, 2683–2690.

Adolphs, R., Tranel, D., and Damasio, A. R. (2003). Dissociable neural systems for recognizing emotions. *Brain Cogn.* 52, 61–69.

Barrett, L. F., Quigley, K. S., Bliss-Moreau, E., and Aronson, K. R. (2004). Interoceptive sensitivity and self-reports of emotional experience. *J. Pers. Soc. Psychol.* 87, 684–697. doi: 10.1037/0022-3514.87.5.684

Bechara, A., Tranel, D., Damasio, H., and Damasio, A. R. (1996). Failure to respond autonomically to anticipated future outcomes following damage to prefrontal cortex. *Cereb. Cortex* 6, 215–225. doi: 10.1093/cercor/6.2.215

Berntson, G. G., Norman, G. J., Bechara, A., Bruss, J., Tranel, D., and Cacioppo, J. T. (2011). The insula and evaluative processes. *Psychol. Sci.* 22, 80–86. doi: 10.1177/0956797610391097

Borg, C., Bedoin, N., Peyron, R., Bogey, S., Laurent, B., and Thomas-Anterion, C. (2013). Impaired emotional processing in a patient with a left posterior insula-SII lesion. *Neurocase* 19, 592–603. doi: 10.1080/13554794.2012.713491

Boucher, O., Rouleau, I., Lassonde, M., Lepore, F., Bouthillier, A., and Nguyen, D. K. (2015). Social information processing following resection of the insular cortex. *Neuropsychologia* 71, 1–10. doi: 10.1016/j.neuropsychologia.2015.03.008

Buchsbaum, B. R., Greer, S., Chang, W. L., and Berman, K. F. (2005). Meta-analysis of neuroimaging studies of the Wisconsin card-sorting task and component processes. *Hum. Brain Mapp.* 25, 35–45. doi: 10.1002/hbm.20128

Calder, A. J., Keane, J., Manes, F., Antoun, N., and Young, A. W. (2000). Impaired recognition and experience of disgust following brain injury. *Nat. Neurosci.* 3, 1077–1078. doi: 10.1038/80586

Calvo, M. G., and Nummenmaa, L. (2008). Detection of emotional faces: salient physical features guide effective visual search. *J. Exp. Psychol. Gen.* 137, 471–494. doi: 10.1037/a0012771

Cameron, O. G. (2001). Interoception: the inside story–a model for psychosomatic processes. *Psychosom. Med.* 63, 697–710. doi: 10.1097/00006842-200109000-00001

Couto, B., Sedeno, L., Sposato, L. A., Sigman, M., Riccio, P. M., Salles, A., et al. (2013). Insular networks for emotional processing and social cognition: comparison of two case reports with either cortical or subcortical involvement. *Cortex* 49, 1420–1434. doi: 10.1016/j.cortex.2012.08.006

Couto, B., Salles, A., Sedeno, L., Peradejordi, M., Barttfeld, P., Canales-Johnson, A., et al. (2014). The man who feels two hearts: the different pathways of interoception. *Soc. Cogn. Affect. Neurosci.* 9, 1253–1260. doi: 10.1093/scan/nst108

Craig, A. D. (2005). Forebrain emotional asymmetry: a neuroanatomical basis? *Trends Cogn. Sci.* 9, 566–571. doi: 10.1016/j.tics.2005.10.005

Crawford, J. R., and Garthwaite, P. H. (2002). Investigation of the single case in neuropsychology: confidence limits on the abnormality of test scores and test score differences. *Neuropsychologia* 40, 1196–1208. doi: 10.1016/S0028-3932(01)00224-X

Crawford, J. R., Garthwaite, P. H., and Porter, S. (2010). Point and interval estimates of effect sizes for the case-controls design in neuropsychology: rationale, methods, implementations, and proposed reporting standards. *Cogn. Neuropsychol.* 27, 245–260. doi: 10.1080/02643294.2010.513967

Critchley, H. D., Wiens, S., Rotshtein, P., Ohman, A., and Dolan, R. J. (2004). Neural systems supporting interoceptive awareness. *Nat. Neurosci.* 7, 189–195. doi: 10.1038/nn1176

Dal Monte, O., Krueger, F., Solomon, J. M., Schintu, S., Knutson, K. M., Strenziok, M., et al. (2013). A voxel-based lesion study on facial emotion recognition after penetrating brain injury. *Soc. Cogn. Affect. Neurosci.* 8, 632–639. doi: 10.1093/scan/nss041

Damasio, A., Damasio, H., and Tranel, D. (2012). Persistence of feelings and sentience after bilateral damage of the insula. *Cereb. Cortex* 23, 833–846. doi: 10.1093/cercor/bhs077

Dunn, B. D., Galton, H. C., Morgan, R., Evans, D., Oliver, C., Meyer, M., et al. (2010). Listening to your heart: how interoception shapes emotion experience and intuitive decision making. *Psychol. Sci.* 21, 1835–1844. doi: 10.1177/0956797610389191

Ehlers, A., and Breuer, P. (1992). Increased cardiac awareness in panic disorder. *J. Abnorm. Psychol.* 101, 371–382. doi: 10.1037/0021-843X.101.3.371

Evans, K. C., Banzett, R. B., Adams, L., McKay, L., Frackowiak, R. S. J., and Corfield, D. R. (2002). BOLD fMRI identifies limbic, paralimbic, and cerebellar activation during air hunger. *J. Neurophysiol.* 88, 1500–1511.

Farb, N. A., Segal, Z. V., and Anderson, A. K. (2013). Attentional modulation of primary interoceptive and exteroceptive cortices. *Cereb. Cortex* 23, 114–126. doi: 10.1093/cercor/bhr385

Garfinkel, S. N., Seth, A. K., Barrett, A. B., Suzuki, K., and Critchley, H. D. (2015). Knowing your own heart: distinguishing interoceptive accuracy from interoceptive awareness. *Biol. Psychol.* 104, 65–74. doi: 10.1016/j.biopsycho.2014.11.004

Gu, X., Hof, P. R., Friston, K. J., and Fan, J. (2013). Anterior insular cortex and emotional awareness. *J. Comp. Neurol.* 521, 3371–3388. doi: 10.1002/cne.23368

Hamaguchi, T., Kano, M., Rikimaru, H., Kanazawa, M., Itoh, M., Yanai, K., et al. (2004). Brain activity during distention of the descending colon in humans. *Neurogastroenterol. Motil.* 16, 299–309. doi: 10.1111/j.1365-2982.2004.00498.x

Herbert, B. M., Muth, E. R., Pollatos, O., and Herbert, C. (2012). Interoception across modalities: on the relationship between cardiac awareness and the sensitivity for gastric functions. *PLoS ONE* 7:e36646. doi: 10.1371/journal.pone.0036646

Iaria, G., Committeri, G., Pastorelli, C., Pizzamiglio, L., Watkins, K. E., and Carota, A. (2008). Neural activity of the anterior insula in emotional processing depends on the individuals' emotional susceptibility. *Hum. Brain Mapp.* 29, 363–373. doi: 10.1002/hbm.20393

James, W. (1884). What is an emotion? *Mind* 19, 188–205. doi: 10.1093/mind/os-IX.34.188

Jenkins, L. M., Andrewes, D. G., Nicholas, C. L., Drummond, K. J., Moffat, B. A., Phal, P., et al. (2014). Social cognition in patients following surgery to the prefrontal cortex. *Psychiatry Res.* 224, 192–203. doi: 10.1016/j.pscychresns.2014.08.007

Khalsa, S. S., Rudrauf, D., Feinstein, J. S., and Tranel, D. (2009). The pathways of interoceptive awareness. *Nat. Neurosci.* 12, 1494–1496. doi: 10.1038/nn.2411

Kuppens, P., Tuerlinckx, F., Russell, J. A., and Barrett, L. F. (2013). The relation between valence and arousal in subjective experience. *Psychol. Bull.* 139, 917–940. doi: 10.1037/a0030811

Kurth, F., Zilles, K., Fox, P. T., Laird, A. R., and Eickhoff, S. B. (2010). A link between the systems: functional differentiation and integration within the human insula revealed by meta-analysis. *Brain Struct. Funct.* 214, 519–534. doi: 10.1007/s00429-010-0255-z

Lamm, C., and Singer, T. (2010). The role of anterior insular cortex in social emotions. *Brain Struct. Funct.* 214, 579–591. doi: 10.1007/s00429-010-0251-3

Lane, R. D. (2008). Neural substrates of implicit and explicit emotional processes: a unifying framework for psychosomatic medicine. *Psychosom. Med.* 70, 214–231. doi: 10.1097/PSY.0b013e3181647e44

Lange, C. G. (1885/1992). *The Emotions: A Psychophysiological Study, The emotions.* Baltimore: Williams and Wilkins.

Lee, K. H., and Siegle, G. J. (2012). Common and distinct brain networks underlying explicit emotional evaluation: a meta-analytic study. *Soc. Cogn. Affect. Neurosci.* 7, 521–534. doi: 10.1093/scan/nsp001

Lewis, P. A., Critchley, H. D., Rotshtein, P., and Dolan, R. J. (2007). Neural correlates of processing valence and arousal in affective words. *Cereb. Cortex* 17, 742–748. doi: 10.1093/cercor/bhk024

Lie, C. H., Specht, K., Marshall, J. C., and Fink, G. R. (2006). Using fMRI to decompose the neural processes underlying the Wisconsin Card Sorting Test. *Neuroimage* 30, 1038–1049. doi: 10.1016/j.neuroimage.2005.10.031

Mandler, G., Mandler, J. M., and Uviller, E. T. (1958). Autonomic feedback: the perception of autonomic antivity. *J. Abnorm. Psychol.* 56, 367–374. doi: 10.1037/h0048083

Mehling, W. E., Price, C., Daubenmier, J. J., Acree, M., Bartmess, E., and Stewart, A. (2012). The multidimensional assessment of interoceptive awareness (MAIA). *PLoS ONE* 7:e48230. doi: 10.1371/journal.pone.0048230

Moriguchi, Y., Touroutoglou, A., Dickerson, B. C., and Barrett, L. F. (2014). Sex differences in the neural correlates of affective experience. *Soc. Cogn. Affect. Neurosci.* 9, 591–600. doi: 10.1093/scan/nst030

Paulus, M. P., and Stein, M. B. (2006). An insular view of anxiety. *Biol. Psychiatry* 60 , 383–387. doi: 10.1016/j.biopsych.2006.03.042

Penfield, W., and Faulk, M. E. Jr. (1955). The insula; further observations on its function. *Brain* 78, 445–470. doi: 10.1093/brain/78.4.445

Pollatos, O., Kirsch, W., and Schandry, R. (2005). On the relationship between interoceptive awareness, emotional experience, and brain processes. *Brain Res. Cogn. Brain Res.* 25, 948–962. doi: 10.1016/j.cogbrainres.2005.09.019

Porges, S. (1993). *Body Perception Questionnaire.* College Park: Laboratory of Developmental Assessment, University of Maryland.

Ronchi, R., Bello-Ruiz, J., Lukowska, M., Herbelin, B., Cabrilo, I., Schaller, K., et al. (2015). Right insular damage decreases heartbeat awareness and alters cardio-visual effects on bodily self-consciousness. *Neuropsychologia* 70C, 11–20. doi: 10.1016/j.neuropsychologia.2015.02.010

Russell, J. A. (1980). A circumplex model of affect. *J. Pers. Soc. Psychol.* 39, 1161–1178. doi: 10.1037/h0077714

Schandry, R. (1981). Heart beat perception and emotional experience. *Psychophysiology* 18, 483–488. doi: 10.1111/j.1469-8986.1981.tb02486.x

Sherrington, C. S. (1906). *The Integrative Action of the Nervous System*. New Haven: Yale University Press.

Simmons, A. N., Stein, M. B., Strigo, I. A., Arce, E., Hitchcock, C., and Paulus, M. P. (2011). Anxiety positive subjects show altered processing in the anterior insula during anticipation of negative stimuli. *Hum. Brain Mapp.* 32, 1836–1846. doi: 10.1002/hbm.21154

Susskind, J. M., Lee, D. H., Cusi, A., Feiman, R., Grabski, W., and Anderson, A. K. (2008). Expressing fear enhances sensory acquisition. *Nat. Neurosci.* 11, 843–850. doi: 10.1038/nn.2138

Terasawa, Y., Fukushima, H., and Umeda, S. (2013a). How does interoceptive awareness interact with the subjective experience of emotion? An fMRI study. *Hum. Brain Mapp.* 34, 598–612. doi: 10.1002/hbm.21458

Terasawa, Y., Shibata, M., Moriguchi, Y., and Umeda, S. (2013b). Anterior insular cortex mediates bodily sensibility and social anxiety. *Soc. Cogn. Affect. Neurosci.* 8 , 259–266. doi: 10.1093/scan/nss108

Terasawa, Y., Moriguchi, Y., Tochizawa, S., and Umeda, S. (2014). Interoceptive sensitivity predicts sensitivity to the emotions of others. *Cogn. Emot.* 28, 1435–1448. doi: 10.1080/02699931.2014.888988

Werner, N. S., Jung, K., Duschek, S., and Schandry, R. (2009). Enhanced cardiac perception is associated with benefits in decision-making. *Psychophysiology* 46, 1123–1129. doi: 10.1111/j.1469-8986.2009.00855.x

Permissions

All chapters in this book were first published by Frontiers; hereby published with permission under the Creative Commons Attribution License or equivalent. Every chapter published in this book has been scrutinized by our experts. Their significance has been extensively debated. The topics covered herein carry significant findings which will fuel the growth of the discipline. They may even be implemented as practical applications or may be referred to as a beginning point for another development.

The contributors of this book come from diverse backgrounds, making this book a truly international effort. This book will bring forth new frontiers with its revolutionizing research information and detailed analysis of the nascent developments around the world.

We would like to thank all the contributing authors for lending their expertise to make the book truly unique. They have played a crucial role in the development of this book. Without their invaluable contributions this book wouldn't have been possible. They have made vital efforts to compile up to date information on the varied aspects of this subject to make this book a valuable addition to the collection of many professionals and students.

This book was conceptualized with the vision of imparting up-to-date information and advanced data in this field. To ensure the same, a matchless editorial board was set up. Every individual on the board went through rigorous rounds of assessment to prove their worth. After which they invested a large part of their time researching and compiling the most relevant data for our readers.

The editorial board has been involved in producing this book since its inception. They have spent rigorous hours researching and exploring the diverse topics which have resulted in the successful publishing of this book. They have passed on their knowledge of decades through this book. To expedite this challenging task, the publisher supported the team at every step. A small team of assistant editors was also appointed to further simplify the editing procedure and attain best results for the readers.

Apart from the editorial board, the designing team has also invested a significant amount of their time in understanding the subject and creating the most relevant covers. They scrutinized every image to scout for the most suitable representation of the subject and create an appropriate cover for the book.

The publishing team has been an ardent support to the editorial, designing and production team. Their endless efforts to recruit the best for this project, has resulted in the accomplishment of this book. They are a veteran in the field of academics and their pool of knowledge is as vast as their experience in printing. Their expertise and guidance has proved useful at every step. Their uncompromising quality standards have made this book an exceptional effort. Their encouragement from time to time has been an inspiration for everyone.

The publisher and the editorial board hope that this book will prove to be a valuable piece of knowledge for researchers, students, practitioners and scholars across the globe.

List of Contributors

Vivien Ainley, Lara Maister and Manos Tsakiris
Lab of Action and Body, Department of Psychology, Royal Holloway University of London, Egham, UK

Christine Ma-Kellams
Department of Psychology, University of La Verne, La Verne, CA, USA

Norman Farb
University of Toronto Mississauga, Mississauga, ON, Canada

Jennifer Daubenmier
University of California San Francisco, San Francisco, CA, US

Cynthia J. Price
University of Washington, Seattle, WA, USA

Catherine Kerr
Brown University, Providence, RI, USA

Barnaby D. Dunn
University of Exeter, Exeter, UK

Anne Carolyn Klein
Rice University, Houston, TX, USA

Martin P. Paulus
University of California San Diego, La Jolla, CA, USA

Giuseppe Calì
Laboratory of Neuropsychology and Cognitive Neuroscience, Department of Neuroscience, Imaging and Clinical Sciences, University G. d'Annunzio, Chieti, Italy

Ettore Ambrosini and Giorgia Committeri
Laboratory of Neuropsychology and Cognitive Neuroscience, Department of Neuroscience, Imaging and Clinical Sciences, University G. d'Annunzio, Chieti, Italy
Institute for Advanced Biomedical Technologies, University G. d'Annunzio, Chieti, Italy

Laura Picconi
Department of Psychological, Health and Territory Sciences, University G. d'Annunzio, Chieti, Italy

Wolf E. Mehling
Department of Family and Community Medicine, Osher Center for Integrative Medicine, University of California San Francisco, San Francisco, CA, USA

Marasha de Jong
Depression Clinical and Research Program, Massachusetts General Hospital, Harvard Medical School, Boston, MA, USA
Department of Psychiatry and Psychology, School of Mental Health and Neuroscience, Maastricht University Medical Center, Maastricht, Netherlands
Mondriaan, Maastricht, Netherlands
Department of Psychiatry, Massachusetts General Hospital, Harvard Medical School, Boston, MA, USA

Kiran Hug
Department of Psychosomatic Medicine, Medical Faculty, Medical Center-University of Freiburg, Freiburg, Germany
Division of Clinical Psychology and Psychotherapy, Faculty of Psychology, University of Basel, Basel, Switzerland

Britta K. Hölzel
Department of Psychiatry, Massachusetts General Hospital, Harvard Medical School, Boston, MA, USA
Department of Neuroradiology, Klinikum rechts der Isar, Technical University of Munich, Munich, Germany

Alexander T. Sack
Department of Cognitive Neuroscience, Faculty of Psychology and Neuroscience, Maastricht University, Maastricht, Netherlands

Frenk Peeters
Department of Psychiatry and Psychology, School of Mental Health and Neuroscience, Maastricht University Medical Center, Maastricht, Netherlands

Heidi Ashih and David Mischoulon
Depression Clinical and Research Program, Massachusetts General Hospital, Harvard Medical School, Boston, MA, USA

Tim Gard
Department of Psychiatry, Massachusetts General Hospital, Harvard Medical School, Boston, MA, USA
Institute for Complementary and Integrative Medicine, University Hospital Zurich, University of Zurich, Zurich, Switzerland
Translational Neuromodeling Unit, Institute for Biomedical Engineering, University of Zurich and Swiss Federal Institute of Technology (ETH), Zurich, Switzerland
Maastricht University, Maastricht, Netherlands

Sahib S. Khalsa
Department of Neurology, University of Iowa, Iowa City, IA, USA
Laureate Institute for Brain Research, Tulsa, OK, USA
Faculty of Community Medicine, University of Tulsa, Tulsa, OK, USA

David Rudrauf
Department of Neurology, University of Iowa, Iowa City, IA, USA
Laboratory of Functional Imaging, Institut National de la Santé et de la Recherche Médicale U678s/Université Pierre et Marie Curie, Paris, France

Richard J. Davidson
Department of Psychology, University of Wisconsin-Madison, Madison, WI, USA

Daniel Tranel
Department of Neurology, University of Iowa, Iowa City, IA, USA
Department of Psychology, University of Iowa, Iowa City, IA, USA

Pei-Ying S. Chan and Chia-Hsiung Cheng
Department of Occupational Therapy, College of Medicine, Chang Gung University, Taoyuan, Taiwan
Healthy Ageing Research Center, Chang Gung University, Taoyuan, Taiwan

Shih-Chieh Hsu
Department of Psychiatry, Chang Gung Memorial Hospital, Taoyuan, Taiwan
Department of Medicine, Chang Gung University, Taoyuan, Taiwan

Chia-Yih Liu
Department of Psychiatry, Chang Gung Memorial Hospital, Taoyuan, Taiwan
Department of Traditional Chinese Medicine, Chang Gung University, Taoyuan, Taiwan

Paul W. Davenport
Department of Physiological Sciences, College of Veterinary Medicine, University of Florida, Gainesville, FL, USA

Christine Wiebking
Cluster of Excellence in Cognitive Sciences, Department of Sociology of Physical Activity and Health, University of Potsdam, Potsdam, Germany

Stefan Schmidt
Department of Psychosomatic Medicine, University Medical Center Freiburg, Freiburg, Germany
Institute for Transcultural Health Studies, European University Viadrina, Frankfurt (Oder), Germany

Georg Northoff
Mind, Brain Imaging and Neuroethics, Institute of Mental Health Research, University of Ottawa, Ottawa, ON, Canada
Graduate Institute of Humanities in Medicine, Taipei Medical University, Taipei, Taiwan
Taipei Medical University-Shuang Ho Hospital, Brain and Consciousness Research Center, New Taipei City, Taiwan
Department of Psychology, National Chengchi University, Taipei, Taiwan
Center for Cognition and Brain Disorders, Normal University Hangzhou, Hangzhou, China

Simone Otten, Eva Schötz and Karin Meissner
Meissner Lab, Institute of Medical Psychology, Ludwig-Maximilians-University, Munich, Germany

Marc Wittmann
Institute for Frontier Areas of Psychology and Mental Health, Freiburg, Germany

Niko Kohls
Division Integrative Health Promotion, University of Applied Sciences and Arts, Coburg, Germany

Peter Payne and Mardi A. Crane-Godreau
Microbiology/Immunology, Geisel Medical School at Dartmouth, Hanover, NH, USA

Peter A. Levine
Somatic Experiencing Trauma Institute, Boulder, CO, USA

Mariachiara Longarzo, Francesca D'Olimpio, Angela Chiavazzo, Luigi Trojano and Dario Grossi
Laboratory of Neuropsychology, Department of Psychology, Second University of Naples, Caserta, Italy

Gabriella Santangelo
Laboratory of Neuropsychology, Department of Psychology, Second University of Naples, Caserta, Italy
Hermitage Capodimonte, Napoli, Italy

Sibylle Petersen
Research Group on Health Psychology, KU Leuven, Leuven, Belgium
Institute for Health and Behaviour, University of Luxembourg, Luxembourg, Luxembourg

Ken Van Staeyen
Research Group on Health Psychology, KU Leuven, Leuven, Belgium

Meike Pappens and Evelien Vandenbossche
Health Psychology, University of Leuven, Leuven, Belgium

Omer Van den Bergh
Research Group on Health Psychology, KU Leuven, Leuven, Belgium
Health Psychology, University of Leuven, Leuven, Belgium

Olga Pollatos and Ellen Matthias
Health Psychology, Institute of Psychology and Education, University of Ulm, Ulm, Germany

Johannes Keller
Social Psychology, Institute of Psychology and Education, University of Ulm, Ulm, Germany

Nick Medford
Department of Psychiatry, Brighton and Sussex Medical School, Brighton, UK
Sackler Centre for Consciousness Science, University of Sussex, Brighton, UK

Mauricio Sierra, Argyris Stringaris, Vincent Giampietro, Michael J. Brammer and Anthony S. David
Institute of Psychiatry, Psychology, and Neuroscience, London, UK

Eleana Georgiou
Health Psychology, Institute of Psychology and Education, Ulm University, Ulm, Germany

Susanne Kobel, Sarah Kettner and Jürgen M. Steinacker
Research Group "Join the Healthy Boat – Primary School", Sports and Rehabilitation Medicine, Ulm University, Ulm, Germany

Jens Dreyhaupt
Institute of Epidemiology and Medical Biometry, Ulm University, Ulm, Germany

André Schulz and Claus Vögele
Institute for Health and Behaviour, Integrative Research Unit on Social and Individual Development, University of Luxembourg, Walferdange, Luxembourg

M. Cornelia Stoeckel, Roland W. Esser, Matthias Gamer and Christian Büchel
Department of Systems Neuroscience, University Medical Center Hamburg-Eppendorf, Hamburg, Germany

Tal D. Ben-Soussan
The Leslie and Susan Gonda (Goldschmied) Multidisciplinary Brain Research Center, Bar-Ilan University, Ramat Gan, Israel
Research Institute for Neuroscience, Education and Didactics, Patrizio Paoletti Foundation, Assisi, Italy

Andreas von Leupoldt
Department of Systems Neuroscience, University Medical Center Hamburg-Eppendorf, Hamburg, Germany
Research Group Health Psychology, Faculty of Psychology and Educational Sciences, University of Leuven, Leuven, Belgium

Aviva Berkovich-Ohana
Department of Neurobiology, Weizmann Institute of Science, Rehovot, Israel
Department of Physiology and Pharmacology, Sapienza University of Rome, Rome, Italy

Claudia Piervincenzi
Department of Physiology and Pharmacology, Sapienza University of Rome, Rome, Italy
Institute for Advanced Biomedical Technologies, Università degli Studi Gabriele D'Annunzio, Chieti, Italy

Joseph Glicksohn
The Leslie and Susan Gonda (Goldschmied) Multidisciplinary Brain Research Center, Bar-Ilan University, Ramat Gan, Israel
Department of Criminology, Bar-Ilan University, Ramat Gan, Israel

Filippo Carducci
Department of Physiology and Pharmacology, Sapienza University of Rome, Rome, Italy

Yuri Terasawa and Satoshi Umeda
Department of Psychology, Keio University, Tokyo, Japan

Yoshiya Moriguchi
Department of Psychophysiology, National Institute of Mental Health, National Center of Neurology and Psychiatry, Tokyo, Japan

Yukio Ibata
Department of Neurosurgery, Nasu Red Cross Hospital, Tochigi, Japan

Index